London Mathematical Society Lecture Note Series: 470

Surveys in Combinatorics 2021

Edited by

KONRAD K. DABROWSKI
Durham University

MAXIMILIEN GADOULEAU
Durham University

NICHOLAS GEORGIOU
Durham University

MATTHEW JOHNSON
Durham University

GEORGE B. MERTZIOS
Durham University

DANIËL PAULUSMA
Durham University

CAMBRIDGE
UNIVERSITY PRESS

University Printing House, Cambridge CB2 8BS, United Kingdom

One Liberty Plaza, 20th Floor, New York, NY 10006, USA

477 Williamstown Road, Port Melbourne, VIC 3207, Australia

314–321, 3rd Floor, Plot 3, Splendor Forum, Jasola District Centre,
New Delhi – 110025, India

103 Penang Road, #05–06/07, Visoncrest Commercial, Singapore 238467

Cambridge University Press is part of the University of Cambridge.

It furthers the University's mission by disseminating knowledge in the pursuit of education, learning, and research at the highest international levels of excellence.

www.cambridge.org
Information on this title: www.cambridge.org/9781009018883
DOI: 10.1017/9781009036214

© Cambridge University Press 2021

This publication is in copyright. Subject to statutory exception and to the provisions of relevant collective licensing agreements, no reproduction of any part may take place without the written permission of Cambridge University Press.

First published 2021

Printed in the United Kingdom by TJ Books Limited, Padstow Cornwall

A catalogue record for this publication is available from the British Library.

ISBN 978-1-009-01888-3 Paperback

Cambridge University Press has no responsibility for the persistence or accuracy of URLs for external or third-party internet websites referred to in this publication and does not guarantee that any content on such websites is, or will remain, accurate or appropriate.

LONDON MATHEMATICAL SOCIETY LECTURE NOTE SERIES

Managing Editor: Professor Endre Süli, Mathematical Institute, University of Oxford,
Woodstock Road, Oxford OX2 6GG, United Kingdom

The titles below are available from booksellers, or from Cambridge University Press at
www.cambridge.org/mathematics

360 Zariski geometries, B. ZILBER
361 Words: Notes on verbal width in groups, D. SEGAL
362 Differential tensor algebras and their module categories, R. BAUTISTA, L. SALMERÓN & R. ZUAZUA
363 Foundations of computational mathematics, Hong Kong 2008, F. CUCKER, A. PINKUS & M.J. TODD (eds)
364 Partial differential equations and fluid mechanics, J.C. ROBINSON & J.L. RODRIGO (eds)
365 Surveys in combinatorics 2009, S. HUCZYNSKA, J.D. MITCHELL & C.M. RONEY-DOUGAL (eds)
366 Highly oscillatory problems, B. ENGQUIST, A. FOKAS, E. HAIRER & A. ISERLES (eds)
367 Random matrices: High dimensional phenomena, G. BLOWER
368 Geometry of Riemann surfaces, F.P. GARDINER, G. GONZÁLEZ-DIEZ & C. KOUROUNIOTIS (eds)
369 Epidemics and rumours in complex networks, M. DRAIEF & L. MASSOULIÉ
370 Theory of p-adic distributions, S. ALBEVERIO, A.YU. KHRENNIKOV & V.M. SHELKOVICH
371 Conformal fractals, F. PRZYTYCKI & M. URBAŃSKI
372 Moonshine: The first quarter century and beyond, J. LEPOWSKY, J. MCKAY & M.P. TUITE (eds)
373 Smoothness, regularity and complete intersection, J. MAJADAS & A. G. RODICIO
374 Geometric analysis of hyperbolic differential equations: An introduction, S. ALINHAC
375 Triangulated categories, T. HOLM, P. JØRGENSEN & R. ROUQUIER (eds)
376 Permutation patterns, S. LINTON, N. RUŠKUC & V. VATTER (eds)
377 An introduction to Galois cohomology and its applications, G. BERHUY
378 Probability and mathematical genetics, N. H. BINGHAM & C. M. GOLDIE (eds)
379 Finite and algorithmic model theory, J. ESPARZA, C. MICHAUX & C. STEINHORN (eds)
380 Real and complex singularities, M. MANOEL, M.C. ROMERO FUSTER & C.T.C WALL (eds)
381 Symmetries and integrability of difference equations, D. LEVI, P. OLVER, Z. THOMOVA & P. WINTERNITZ (eds)
382 Forcing with random variables and proof complexity, J. KRAJÍČEK
383 Motivic integration and its interactions with model theory and non-Archimedean geometry I, R. CLUCKERS, J. NICAISE & J. SEBAG (eds)
384 Motivic integration and its interactions with model theory and non-Archimedean geometry II, R. CLUCKERS, J. NICAISE & J. SEBAG (eds)
385 Entropy of hidden Markov processes and connections to dynamical systems, B. MARCUS, K. PETERSEN & T. WEISSMAN (eds)
386 Independence-friendly logic, A.L. MANN, G. SANDU & M. SEVENSTER
387 Groups St Andrews 2009 in Bath I, C.M. CAMPBELL et al (eds)
388 Groups St Andrews 2009 in Bath II, C.M. CAMPBELL et al (eds)
389 Random fields on the sphere, D. MARINUCCI & G. PECCATI
390 Localization in periodic potentials, D.E. PELINOVSKY
391 Fusion systems in algebra and topology, M. ASCHBACHER, R. KESSAR & B. OLIVER
392 Surveys in combinatorics 2011, R. CHAPMAN (ed)
393 Non-abelian fundamental groups and Iwasawa theory, J. COATES et al (eds)
394 Variational problems in differential geometry, R. BIELAWSKI, K. HOUSTON & M. SPEIGHT (eds)
395 How groups grow, A. MANN
396 Arithmetic differential operators over the p-adic integers, C.C. RALPH & S.R. SIMANCA
397 Hyperbolic geometry and applications in quantum chaos and cosmology, J. BOLTE & F. STEINER (eds)
398 Mathematical models in contact mechanics, M. SOFONEA & A. MATEI
399 Circuit double cover of graphs, C.-Q. ZHANG
400 Dense sphere packings: a blueprint for formal proofs, T. HALES
401 A double Hall algebra approach to affine quantum Schur–Weyl theory, B. DENG, J. DU & Q. FU
402 Mathematical aspects of fluid mechanics, J.C. ROBINSON, J.L. RODRIGO & W. SADOWSKI (eds)
403 Foundations of computational mathematics, Budapest 2011, F. CUCKER, T. KRICK, A. PINKUS & A. SZANTO (eds)
404 Operator methods for boundary value problems, S. HASSI, H.S.V. DE SNOO & F.H. SZAFRANIEC (eds)
405 Torsors, étale homotopy and applications to rational points, A.N. SKOROBOGATOV (ed)
406 Appalachian set theory, J. CUMMINGS & E. SCHIMMERLING (eds)
407 The maximal subgroups of the low-dimensional finite classical groups, J.N. BRAY, D.F. HOLT & C.M. RONEY-DOUGAL
408 Complexity science: the Warwick master's course, R. BALL, V. KOLOKOLTSOV & R.S. MACKAY (eds)
409 Surveys in combinatorics 2013, S.R. BLACKBURN, S. GERKE & M. WILDON (eds)
410 Representation theory and harmonic analysis of wreath products of finite groups, T. CECCHERINI-SILBERSTEIN, F. SCARABOTTI & F. TOLLI
411 Moduli spaces, L. BRAMBILA-PAZ, O. GARCÍA-PRADA, P. NEWSTEAD & R.P. THOMAS (eds)
412 Automorphisms and equivalence relations in topological dynamics, D.B. ELLIS & R. ELLIS

413	Optimal transportation, Y. OLLIVIER, H. PAJOT & C. VILLANI (eds)
414	Automorphic forms and Galois representations I, F. DIAMOND, P.L. KASSAEI & M. KIM (eds)
415	Automorphic forms and Galois representations II, F. DIAMOND, P.L. KASSAEI & M. KIM (eds)
416	Reversibility in dynamics and group theory, A.G. O'FARRELL & I. SHORT
417	Recent advances in algebraic geometry, C.D. HACON, M. MUSTAŢĂ & M. POPA (eds)
418	The Bloch–Kato conjecture for the Riemann zeta function, J. COATES, A. RAGHURAM, A. SAIKIA & R. SUJATHA (eds)
419	The Cauchy problem for non-Lipschitz semi-linear parabolic partial differential equations, J.C. MEYER & D.J. NEEDHAM
420	Arithmetic and geometry, L. DIEULEFAIT *et al* (eds)
421	O-minimality and Diophantine geometry, G.O. JONES & A.J. WILKIE (eds)
422	Groups St Andrews 2013, C.M. CAMPBELL *et al* (eds)
423	Inequalities for graph eigenvalues, Z. STANIĆ
424	Surveys in combinatorics 2015, A. CZUMAJ *et al* (eds)
425	Geometry, topology and dynamics in negative curvature, C.S. ARAVINDA, F.T. FARRELL & J.-F. LAFONT (eds)
426	Lectures on the theory of water waves, T. BRIDGES, M. GROVES & D. NICHOLLS (eds)
427	Recent advances in Hodge theory, M. KERR & G. PEARLSTEIN (eds)
428	Geometry in a Fréchet context, C.T.J. DODSON, G. GALANIS & E. VASSILIOU
429	Sheaves and functions modulo p, L. TAELMAN
430	Recent progress in the theory of the Euler and Navier–Stokes equations, J.C. ROBINSON, J.L. RODRIGO, W. SADOWSKI & A. VIDAL-LÓPEZ (eds)
431	Harmonic and subharmonic function theory on the real hyperbolic ball, M. STOLL
432	Topics in graph automorphisms and reconstruction (2nd Edition), J. LAURI & R. SCAPELLATO
433	Regular and irregular holonomic D-modules, M. KASHIWARA & P. SCHAPIRA
434	Analytic semigroups and semilinear initial boundary value problems (2nd Edition), K. TAIRA
435	Graded rings and graded Grothendieck groups, R. HAZRAT
436	Groups, graphs and random walks, T. CECCHERINI-SILBERSTEIN, M. SALVATORI & E. SAVA-HUSS (eds)
437	Dynamics and analytic number theory, D. BADZIAHIN, A. GORODNIK & N. PEYERIMHOFF (eds)
438	Random walks and heat kernels on graphs, M.T. BARLOW
439	Evolution equations, K. AMMARI & S. GERBI (eds)
440	Surveys in combinatorics 2017, A. CLAESSON *et al* (eds)
441	Polynomials and the mod 2 Steenrod algebra I, G. WALKER & R.M.W. WOOD
442	Polynomials and the mod 2 Steenrod algebra II, G. WALKER & R.M.W. WOOD
443	Asymptotic analysis in general relativity, T. DAUDÉ, D. HÄFNER & J.-P. NICOLAS (eds)
444	Geometric and cohomological group theory, P.H. KROPHOLLER, I.J. LEARY, C. MARTÍNEZ-PÉREZ & B.E.A. NUCINKIS (eds)
445	Introduction to hidden semi-Markov models, J. VAN DER HOEK & R.J. ELLIOTT
446	Advances in two-dimensional homotopy and combinatorial group theory, W. METZLER & S. ROSEBROCK (eds)
447	New directions in locally compact groups, P.-E. CAPRACE & N. MONOD (eds)
448	Synthetic differential topology, M.C. BUNGE, F. GAGO & A.M. SAN LUIS
449	Permutation groups and cartesian decompositions, C.E. PRAEGER & C. SCHNEIDER
450	Partial differential equations arising from physics and geometry, M. BEN AYED *et al* (eds)
451	Topological methods in group theory, N. BROADDUS, M. DAVIS, J.-F. LAFONT & I. ORTIZ (eds)
452	Partial differential equations in fluid mechanics, C.L. FEFFERMAN, J.C. ROBINSON & J.L. RODRIGO (eds)
453	Stochastic stability of differential equations in abstract spaces, K. LIU
454	Beyond hyperbolicity, M. HAGEN, R. WEBB & H. WILTON (eds)
455	Groups St Andrews 2017 in Birmingham, C.M. CAMPBELL *et al* (eds)
456	Surveys in combinatorics 2019, A. LO, R. MYCROFT, G. PERARNAU & A. TREGLOWN (eds)
457	Shimura varieties, T. HAINES & M. HARRIS (eds)
458	Integrable systems and algebraic geometry I, R. DONAGI & T. SHASKA (eds)
459	Integrable systems and algebraic geometry II, R. DONAGI & T. SHASKA (eds)
460	Wigner-type theorems for Hilbert Grassmannians, M. PANKOV
461	Analysis and geometry on graphs and manifolds, M. KELLER, D. LENZ & R.K. WOJCIECHOWSKI
462	Zeta and L-functions of varieties and motives, B. KAHN
463	Differential geometry in the large, O. DEARRICOTT *et al* (eds)
464	Lectures on orthogonal polynomials and special functions, H.S. COHL & M.E.H. ISMAIL (eds)
465	Constrained Willmore surfaces, Á.C. QUINTINO
466	Invariance of modules under automorphisms of their envelopes and covers, A.K. SRIVASTAVA, A. TUGANBAEV & P.A. GUIL ASENSIO
467	The genesis of the Langlands program, J. MUELLER & F. SHAHIDI
468	(Co)end calculus, F. LOREGIAN
469	Computational cryptography, J.W. BOS & M. STAM

Contents

Preface *page* vii
Konrad K. Dabrowski, Maximilien Gadouleau,
Nicholas Georgiou, Matthew Johnson, George B. Mertzios,
Daniël Paulusma

1 **The partition complex: an invitation to combinatorial commutative algebra** 1
Karim Adiprasito and Geva Yashfe

2 **Hasse-Weil type theorems and relevant classes of polynomial functions** 43
Daniele Bartoli

3 **Decomposing the edges of a graph into simpler structures** 103
Marthe Bonamy

4 **Generating graphs randomly** 133
Catherine Greenhill

5 **Recent advances on the graph isomorphism problem** 187
Martin Grohe and Daniel Neuen

6 **Extremal aspects of graph and hypergraph decomposition problems** 235
Stefan Glock, Daniela Kühn and Deryk Osthus

7 **Borel combinatorics of locally finite graphs** 267
Oleg Pikhurko

8 **Codes and designs in Johnson graphs with high symmetry** 321
Cheryl E. Praeger

9 **Maximal subgroups of finite simple groups: classifications and applications** 343
Colva M. Roney-Dougal

Preface

The Twenty-Eighth British Combinatorial Conference is to be delivered online from Durham University from 5th July to 9th July 2021. The British Combinatorial Committee has invited nine distinguished combinatorialists to give survey lectures in areas of their expertise, and this volume contains survey articles on which these lectures are based.

In compiling this volume, we are indebted to the authors for preparing their articles so accurately and professionally, and to the referees for their rapid responses and constructive comments. We would also like to thank Tom Harris and Anna Scriven at Cambridge University Press. Finally, without the previous efforts of editors of earlier Surveys and the guidance of the British Combinatorial Committee, the preparation of this volume would have been somewhat daunting.

This conference is organised in partnership with the Clay Mathematics Institute, the Institute of Combinatorics and its Applications and Durham University; we thank each of these organisations for their generous involvement and support.

Konrad K. Dabrowski
Maximilien Gadouleau
Nicholas Georgiou
Matthew Johnson
George B. Mertzios
Daniël Paulusma
Durham University

January 2021

The partition complex: an invitation to combinatorial commutative algebra

Karim Adiprasito and Geva Yashfe

Abstract

We provide a new foundation for combinatorial commutative algebra and Stanley-Reisner theory using the partition complex introduced in [1]. One of the main advantages is that it is entirely self-contained, using only a minimal knowledge of algebra and topology. On the other hand, we also develop new techniques and results using this approach. In particular, we provide

1. A novel, self-contained method of establishing Reisner's theorem and Schenzel's formula for Buchsbaum complexes.
2. A simple new way to establish Poincaré duality for face rings of manifolds, in much greater generality and precision than previous treatments.
3. A "master-theorem" to generalize several previous results concerning the Lefschetz theorem on subdivisions.
4. Proof for a conjecture of Kühnel concerning triangulated manifolds with boundary.

1 Introduction

Starting with the work of Hochster, Reisner and Stanley, powerful methods from commutative algebra developed by algebraic geometers could be used to provide a new and powerful way to study face numbers of simplicial and polyhedral complexes [9, 26].

However, using these powerful tools came with a drawback. First, they made the theory harder to access without background in commutative algebra. Second, even many of those applying them often used them as a black box, and the tools themselves became a distraction, leading to missed results and open questions that would otherwise have been simple.

And so, as a tourist might use an expensive lens to capture a vista, doing so suboptimally because he does not grasp its pros and cons, the physics of its makeup, we are left with pictures that feel somewhat lacking, blurry or hiding the important, leaving us dissatisfied.

So our goal here is twofold: To show how basic household means can take a much simpler, more gratifying picture, without sacrificing any of the generality. We then go a step further, and use the new methods to generalize the results with ease, using only the ingredients that can be found within the first algebra books you can find in your kitchen, and just a smidge of algebraic topology you find in every spice rack. As for combinatorics, we shall assume nothing beyond the most basic familiarity with simplicial complexes.

Hence, this is not so much a survey, as it is an attempt to build better and more powerful foundations, as well as offer newcomers a road towards research in the area, that is at the heart of new developments between combinatorics and Hodge Theory [18, 12, 2, 1]. Additionally, we offer also researchers in combinatorial commutative algebra a more consistent and stronger set of tools. We are therefore a little curt on direct combinatorial applications, for which we refer to the initial sections of [1], and instead offer an focused introduction to the techniques.

1.1 Overview

Before we begin discussing the details, let us provide a little motivation. We want to understand various combinatorial invariants of simplicial complexes. Most basic among these is the face vector, counting the number of vertices, edges, and so on. We may wish to restrict the class of complexes under investigation: for example, to look only at planar graphs, or at simplicial complexes that triangulate a surface. The restrictions we place are usually homological in nature.

The issue is then how the combinatorics and topology come together. The trick is to use rings which contain information from both worlds.

Indeed, one of the key observations of combinatorial commutative algebra was the realization that the homological properties of a simplicial complex are encoded in its so called *face ring* in a variety of ways, often first glimpsed and disseminated as unpublished ideas and results of Hochster[1]. The first key result here is Reisner's theorem (discussed in Section 4), that connects the vanishing of homology over a fixed field to the Cohen-Macaulay property of the associated face ring. Here, not only the global homology of the simplicial complex comes into play, but also the homology of principal filters in the face lattice.

The essentially only proof available for this theorem goes via the local cohomology as introduced by Grothendieck in the 1960s, and most of the following research has similarly employed the same tool. We instead use the partition complex here, a significantly more down-to-earth tool that has several direct benefits, most of all that one can see what happens in a surgical way.

We also obtain the generalization to manifolds, due to Schenzel [22], which is relatively transparent at least to experts, but has the drawback that it is, in parts, only available in his German thesis. We provide this in Section 6.

Our next stop in the way is a new way to address and understand a fundamental property of intersection rings that arises in the context of combinatorial Hodge theory: Poincaré duality. Again we offer a new transparent proof of Poincaré duality for the face rings of spheres, and then proceed to provide generalizations to arbitrary manifolds, discussed in Section 7.

Finally, we discuss some new applications to face number problems for manifolds. In Section 9, we discuss the connection to subdivisions and Lefschetz properties, and provide a far-reaching subdivision theorem, providing a common generalization of previous works in one swoop. We also discuss related conjectures of Kühnel, concerning small triangulations of manifolds.

Acknowledgements We would like to thank Zuzka Patáková and Hailun Zheng for an attentive reading of our paper, helping us correct many typos and provide useful remarks to improve understanding. We also thank the anonymous referee for useful remarks. Karim Adiprasito is supported by the European Research Council under the European Unions Seventh Framework Programme ERC Grant agreement

---

[1]This seems to justify the old adage that discoveries are never named after their discoverer, for the other name of face rings, Stanley-Reisner rings, makes no mention of Hochster

ERC StG 716424 - CASe, a DFF grant 0135-00258B and the Israel Science Foundation under ISF Grant 1050/16. Geva Yashfe is supported by the European Research Council under the European Unions Seventh Framework Programme ERC Grant agreement ERC StG 716424 - CASe and the Israel Science Foundation under ISF Grant 1050/16.

2 Preliminaries

In this section we set up some basic notation and definitions. Experienced readers can skip most of the text, but may still wish to look at the notation and at definitions for relative simplicial complexes, as well as the corresponding modules over face rings.

2.1 Simplicial complexes and face rings

2.1.1 Simplicial complexes
We begin by recalling some common definitions.

Definition A simplicial complex Δ is a downwards-closed family of subsets of a finite set called the ground set. The ground set is usually left implicit or taken to be $[n] = \{1, \ldots, n\}$ for some n. Being downwards-closed means that if $\tau \in \Delta$ and $\rho \subset \tau$ then $\rho \in \Delta$.

In particular, if a simplicial complex is nonempty, it contains $\emptyset$ as a face. Thus the complex $\{\emptyset\}$ contains no nonempty faces, but is different than the void complex $\emptyset$.

A subcomplex of a simplicial complex is a subset which is itself a simplicial complex.

An element $\tau \in \Delta$ is called a simplex or a face. Its dimension is $\dim(\tau) = |\tau| - 1$, and the dimension of Δ is $\max_{\tau \in \Delta} \dim(\tau)$. A face of Δ is called a facet if its dimension equals $\dim(\Delta)$. Faces of dimension zero and one are called vertices and edges respectively.

Definition Let Δ be a simplicial complex. The *star* of a simplex τ is the simplicial complex $\text{st}_\tau(\Delta) = \{\rho \in \Delta \mid \tau \cup \rho \in \Delta\}$. The *link* of τ is $\text{lk}_\tau(\Delta) = \{\rho \in \Delta \mid \tau \cup \rho \in \Delta, \tau \cap \rho = \emptyset\}$.

The k-faces of Δ are denoted by $\Delta^{(k)} = \{\tau \in \Delta \mid \dim(\tau) = k\}$, and the k-skeleton $\Delta^{(\leq k)}$ is the subcomplex consisting of faces of dimension at most k.

In one or two places we use the join and subtraction operations. For simplicial complexes Δ_1, Δ_2 on disjoint ground sets, the join is $\Delta_1 * \Delta_2 = \{\tau \cup \rho \mid \tau \in \Delta_1, \rho \in \Delta_2\}$, a simplicial complex on the union of the ground sets of Δ_1 and Δ_2.

If Δ is a simplicial complex and τ is a face, $\Delta - \tau$ is the maximal subcomplex which does not contain τ. Its faces are $\{\sigma \in \Delta \mid \sigma \cap \tau = \emptyset\}$.

It is worth noting that simplicial complexes are not equivalent to semi-simplicial sets (sometimes called Δ-complexes by Hatcher).

2.1.2 Relative simplicial complexes
We work with relative simplicial complexes analogously to how one often works with pairs of topological spaces. The theory generalizes smoothly to this setting, which is sometimes cleaner. See also [3, 1].

Definition A relative simplicial complex $\Psi = (\Delta, \Gamma)$ is a pair consisting of a simplicial complex Δ and a subcomplex Γ. Its faces are $\Delta \setminus \Gamma$, i.e. the non-faces of Γ. In particular, $\dim(\Psi) = \max_{\tau \in \Psi} \dim(\tau)$ can be smaller than $\dim(\Delta)$, and it is possible for $\emptyset$ not to be a face. Any simplicial complex Δ can be treated in this language as the relative complex $(\Delta, \emptyset)$.

The star of a simplex τ within Ψ is $\mathrm{st}_\tau \Psi = (\mathrm{st}_\tau \Delta, \mathrm{st}_\tau \Gamma)$. Similarly, the link is $\mathrm{lk}_\tau \Psi = (\mathrm{lk}_\tau \Delta, \mathrm{lk}_\tau \Gamma)$.

A relative complex Ψ is pure if all its maximal faces have the same dimension.

Many basic lemmas about simplicial complexes work for relative complexes as well, and we will often extend definitions from absolute to relative without further mention. For instance, if $\Psi = (\Delta, \Gamma)$ and $\tau \in \Delta$ then $\mathrm{st}_\tau \Psi = \tau^* \mathrm{lk}_\tau \Psi = (\tau^* \mathrm{lk}_\tau \Delta, \tau^* \mathrm{lk}_\tau \Gamma)$. Note that the join of any complex with the void complex is void.

The open star of a face τ in a simplicial complex is usually defined to be the set of faces containing τ. This is not a subcomplex in the usual sense, but we can define a relative complex to fill the same role.

Definition Let Δ be a simplicial complex. The open star of a face τ is $\mathrm{st}_\tau^\circ \Delta = (\mathrm{st}_\tau \Delta, \mathrm{st}_\tau \Delta - \tau)$.

2.1.3 Homology of complexes

The cohomology $H^*(\Psi; \Bbbk)$ of a relative complex $\Psi = (\Delta, \Gamma)$ is the simplicial cohomology of the pair with coefficients in $\Bbbk$. For a complex Δ, we consider $\emptyset \in \Delta$ as a face (of dimension -1) for this purpose. Thus our $H^*(\Delta)$ is what is often denoted $\tilde{H}^*(\Delta)$. In particular the void complex $\emptyset$ has vanishing cohomology in all dimensions, but $\Delta = \{\emptyset\}$ has

$$H^i(\Delta; \Bbbk) = \begin{cases} \Bbbk & i = -1 \\ 0 & \text{otherwise.} \end{cases}$$

2.1.4 Face rings

Face rings, or Stanley-Reisner rings, are main object of the paper. Our treatment is standard except for the relative case, in which we follow [3] and [1].

Fix a field $\Bbbk$. Except in Section 3, $\Bbbk$ is assumed to be infinite. This is a harmless assumption, as field extensions change no property that interests us in this context.

Definition Let Δ be a simplicial complex. Define the polynomial ring $\Bbbk[x_v \mid v \in \Delta^{(0)}]$, with variables indexed by vertices of Δ. The Stanley-Reisner ideal (or non-face ideal) I_Δ of Δ is the ideal generated by all elements of the form $x_{v_1} \cdot x_{v_2} \cdot \ldots \cdot x_{v_j}$ where $\{v_1, \ldots, v_j\}$ is not a face of Δ.

The Stanley-Reisner ring (or face ring) of Δ is

$$\Bbbk[\Delta] := \Bbbk[x_v \mid v \in \Delta^{(0)}]/I_\Delta.$$

If $\Psi = (\Delta, \Gamma)$ is a relative complex, the *relative face module* of Ψ is defined by I_Γ / I_Δ. This is an ideal of $\Bbbk[\Delta]$.

Two main types of maps between face rings and modules are used in this paper. If $\Psi = (\Delta, \Gamma)$ and $\Psi' = (\Delta, \Gamma')$ are relative complexes such that $\Gamma' \subset \Gamma$, there is an inclusion map
$$\Bbbk[\Psi] \hookrightarrow \Bbbk[\Psi'].$$
Similarly, if $\Psi = (\Delta, \Gamma)$ and $\Psi' = (\Delta', \Gamma)$ such that $\Delta' \subset \Delta$ is a subcomplex, there is a restriction map
$$\Bbbk[\Psi] \twoheadrightarrow \Bbbk[\Psi'].$$
In general, maps do not exist in the opposite direction. Two particularly relevant examples are the inclusion of an open star into a complex and the restriction to the star of a face. Explicitly, for $\Psi = (\Delta, \Gamma)$ and any $\tau \in \Delta$, these are maps
$$\Bbbk[\operatorname{st}_\tau^\circ \Psi] \simeq \Bbbk[\Delta, \operatorname{st}_\tau \Gamma \cup (\Delta - \tau)] \to \Bbbk[\Psi]$$
and
$$\Bbbk[\Psi] \to \Bbbk[\operatorname{st}_\tau \Psi]$$
respectively.

2.1.5 Gradings of face rings Face rings can be *graded* by monomial degree. That is, if Δ is a complex, we can write
$$\Bbbk[\Delta] = \bigoplus_{n \geq 0} \Bbbk[\Delta]_n,$$
where the direct sum is a sum of vector spaces over $\Bbbk$, and $\Bbbk[\Delta]_n$ is the subspace spanned by monomials of degree n. This is called the coarse grading. An element of $\Bbbk[\Delta]$ is homogeneous if it is in a single graded piece, or in other words, if it is a linear combination of monomials having the same degree.

There is also a finer grading, by the exponent vectors of monomials. If Δ has vertices $\{v_1, \ldots, v_k\}$ and $x^\alpha = x_{v_1}^{\alpha_1} \cdot \ldots \cdot x_{v_k}^{\alpha_k}$ is a monomial, its exponent vector is $(\alpha_1, \ldots, \alpha_k) \in \mathbb{Z}_{\geq 0}^{\Delta^{(0)}}$. The piece of $\Bbbk[\Delta]$ in degree α is the span of this monomial. This is the fine grading.

Note that in both cases, the degree of a product of homogeneous elements is the sum of their degrees.

Given a graded module or algebra, one can encode the dimensions of the graded pieces in a generating function. This is called the *Hilbert series*, and we shall focus mostly on the Hilbert series of a face ring with respect to the coarse grading
$$H(\Bbbk[\Delta])(t) = \sum_{i=0}^\infty \dim_\Bbbk((\Bbbk_i[\Delta])) \cdot t^i.$$

Motivation This interests us because the Hilbert series of a simplicial complex is also combinatorial:
$$H(\Bbbk[\Delta])(t) = \frac{1}{(1-t)^n} \sum_{i=0}^d f_{i-1} t^i (1-t)^{n-i}.$$

The same discussion applies verbatim to relative face modules. The relevance of this is that maps between modules often preserve the degree. In this case, we can often understand a complex of maps most easily by examining each graded piece separately.

2.2 Chain complexes

We discuss some definitions and basic lemmas for chain and double complexes, and provide a basic introduction. If you have not seen chain complexes before, we recommend Hatcher for a basic introduction [8].

Definition [Chain complexes and tensor products] All our complexes are cohomologically graded. That is, our chain complexes are denoted C^*, with differential $C^* \to C^{*+1}$. It is convenient to call $H^i(C^*)$ the i-th homology, rather than cohomology, of C^*.

To shift the index by p, we write C^{*+p} (and $(C^{*+p})^i = C^{i+p}$).

If (B^*, d) and (C^*, d') are chain complexes, their tensor product is the double complex $T^{*,*}$ defined by

$$T^{i,j} = B^i \otimes C^j$$

together with maps $d^h = d \otimes \mathrm{id} : T^{i,j} \to T^{i+1,j}$ and $d^v = \mathrm{id} \otimes d' : T^{i,j} \to T^{i,j+1}$. If B^*, C^* are complexes of modules over some ring R, the tensor product is of R modules, i.e. it is $B^i \otimes_R C^j$. Note the convention here is that the squares of the complex commute.

A small piece of $T^{*,*}$ can be pictured as follows.

$$\begin{array}{ccccc}
\vdots & & \vdots & & \\
\uparrow {\scriptstyle d^v} & & \uparrow {\scriptstyle d^v} & & \\
\cdots \xrightarrow{d^h} B^i \otimes C^{j+1} \xrightarrow{d^h} & B^{i+1} \otimes C^{j+1} \xrightarrow{d^h} & \cdots \\
\uparrow {\scriptstyle d^v} & & \uparrow {\scriptstyle d^v} & & \\
\cdots \xrightarrow{d^h} B^i \otimes C^j \xrightarrow{d^h} & B^{i+1} \otimes C^j \xrightarrow{d^h} & \cdots \\
\uparrow {\scriptstyle d^v} & & \uparrow {\scriptstyle d^v} & & \\
\vdots & & \vdots & &
\end{array}$$

2.3 Double complexes

The main proofs of the paper are established using the homology of double complexes, the homological way to perform what combinatorialists know well as double counting. To do this in a manner as accessible as possible, without leaving too much for the reader, we use mapping cones very extensively.

Everything we need is introduced below.

We begin with some notation.

Definition Let $C^{*,*}$ be a double complex with commuting differentials d^h and d^v (our double complexes always have commuting differentials). Each row $C^{*,j}$ and each column $C^{i,*}$ is a chain complex with differential induced from d^h or d^v respectively.

The *total complex* of C is a chain complex given by $\mathrm{Tot}(C)^k = \bigoplus_{i+j=k} C^{i,j}$ and differential $d^k_{\mathrm{Tot}} : \mathrm{Tot}(C)^k \to \mathrm{Tot}(C)^{k+1}$ defined by either $d^h + (-1)^k d^v$ or $d^v + (-1)^k d^h$. These give equivalent homology, and it is convenient to have both (an alternative is to transpose the complex, but both are used for the same double complex here).

We denote elements $\alpha \in \mathrm{Tot}(C)^k$ by sums $\alpha = \sum_{i+j=k} \alpha^{i,j}$, where it is understood that $\alpha^{i,j} \in C^{i,j}$.

The *truncation* $C^{*\geq i_0,*}$ is a double complex defined by

$$(C^{*\geq i_0,*})^{i,j} = C^{i\geq i_0,j} = \begin{cases} C^{i,j} & i \geq i_0 \\ 0 & \text{otherwise,} \end{cases}$$

with the same differentials as $C^{*,*}$, and 0 for $i < i_0$.

The truncation $C^{*,*\leq j_1}$ is defined analogously.

Our goal for the rest of this section is to produce exact sequences tying together the rows, columns, and total complex of a double complex. We do this using mapping cones. The idea is introduced after a little preparation.

Lemma 2.1 *Let $C^{*,*}$ be a bounded double complex. For $H^k(\mathrm{Tot}(C))$ to vanish, it suffices that the homology in the vertical direction of $C^{i,k-i}$ is zero for each i, i.e. that $H^{k-i}(C^{i,*}) = 0$ for all i. Similarly, it suffices that $H^{k-i}(C^{*,i}) = 0$ for all i.*

Proof We show this for the vertical case, the horizontal one being analogous. Let $\sum_{i+j=k} \alpha^{i,j} \in \mathrm{Tot}(C)^k$ be a cycle, and let i_0 be the minimial index such that $\alpha^{i_0,k-i_0} \neq 0$. Then $d^v(\alpha^{i_0,k-i_0}) = 0$, so by assumption there is some $\beta = \beta^{i_0,k-i_0-1}$ mapping to α^{i_0,k_0} under d^v.

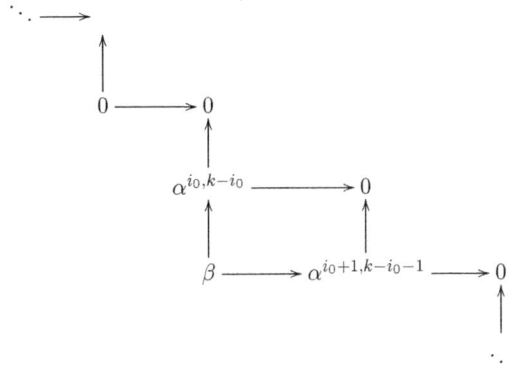

Thus $\alpha' = \alpha - ((-1)^{k-1}d^h + d^v)(\beta)$ differs from α by a boundary. Replacing α by α' increases the minimal nonvanishing index i_0, and after finitely many steps the process terminates because $C^{*,*}$ is bounded. □

Corollary 2.2 *If all rows or all columns of a double complex are exact then the total complex is acyclic.*

We introduce maps $\mathfrak{R}, \mathfrak{U}$ (for "right" and "up") between columns (respectively rows) of a double complex and the total complexes of certain truncations.

Definition Let $C^{*,*}$ be a double complex. There is a chain map
$$\mathfrak{R}^i : C^{i,*} \to \mathrm{Tot}(C^{*\geq i+1,*})^{*+i+1},$$
from the i-th column to the total complex of a truncation of $C^{*,*}$, which is given by
$$C^{i,j} \to \mathrm{Tot}(C^{*\geq i+1,*})^{i+j+1}$$
$$\alpha \mapsto d^h(\alpha) \in C^{i+1,j} \subset \bigoplus_{\substack{r+s=i+j+1,\\ r\geq i+1}} C^{r,s}.$$
For the signs make $\mathfrak{R}$ commute with the differentials, the differential of the total complex is taken to be $d^v + (-1)^k d^h$.

This is illustrated below, with summands of $\mathrm{Tot}(C^{*\geq i+1,*})^{i+j+1}$ underlined.

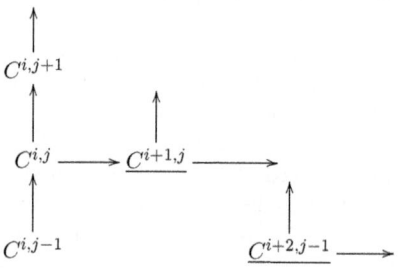

There is a similar chain map
$$\mathfrak{L}^j : \mathrm{Tot}(C^{*,*\leq j})^* \to C^{*-j,j+1},$$
from the total complex of a truncation of $C^{*,*}$ to the $j+1$-th row. On an element
$$\alpha = \sum_{r+s=k} \alpha^{r,s} \in \mathrm{Tot}(C^{*,*\leq j})^k$$
we define it by
$$\mathfrak{L}^j(\alpha) = d^v(\alpha^{k-j,j}).$$
Here the differential of the total complex should be taken to be $d^h + (-1)^k d^v$.

Definition [Mapping cones] Let $f : (C^*, \partial) \to (C'^*, \partial')$ be a map of chain complexes. The mapping cone of f is the chain complex $(M(f)^*, d)$, where $M(f)^i = C^i \oplus C'^{i-1}$ and $d^i(\alpha, \beta) = (\partial\alpha, \partial'\beta + (-1)^i f\alpha)$.

Given f, we can construct a map of chain complexes $\iota : C'^{*-1} \to M(f)^*$ by $\beta \mapsto (0, \beta)$. This fits into a short exact sequence
$$0 \to C'^{*-1} \xrightarrow{\iota} M(f)^* \to C^* \to 0,$$
which gives rise to a long exact sequence
$$\ldots \to H^i(C) \to H^i(C') \to H^{i+1}(M(f)) \to H^{i+1}(C) \to \ldots$$
in which the connecting homomorphism is induced by f.

The next lemma is the essential point.

Lemma 2.3 *Let $C^{*,*}$ be a double complex with commuting vertical and horizontal maps d^h, d^v. There are isomorphisms*

$$M(\mathfrak{R}^i) \simeq \operatorname{Tot}(C^{*\geq i,*})^{*+i}$$
$$M(\mathfrak{U}^j) \simeq \operatorname{Tot}(C^{*,*\leq j+1})^*.$$

Proof First consider $f = \mathfrak{R}^i : C^{i,*} \to \operatorname{Tot}(C^{*\geq i+1,*})^{*+i+1}$. By definition,

$$M(f)^j = C^{i,j} \oplus \operatorname{Tot}(C^{*\geq i+1,*})^{i+j+1-1} = \bigoplus_{\substack{r+s=i+j \\ r \geq i}} C^{r,s} = \operatorname{Tot}(C^{*\geq i,*})^{i+j},$$

and the differential of $M(f)^*$ is essentially the same as that of $\operatorname{Tot}(C^{*\geq i,*})$.
Now consider $f = \mathfrak{U}^j : \operatorname{Tot}(C^{*,*\leq j})^* \to C^{*-j,j+1}$. This time

$$M(f)^i = \operatorname{Tot}(C^{*,*\leq j})^i \oplus C^{i-1-j,j+1} = \bigoplus_{\substack{r+s=i \\ s \leq j+1}} C^{r,s} = \operatorname{Tot}(C^{*,*\leq j+1})^i,$$

and the differential of $M(f)^*$ is the same as that of the total complex on the right hand side above if j is even. If j is odd, it is harmless to modify the differential of the mapping cone to be

$$d^i(\alpha, \beta) = (\partial \alpha, \partial' \beta + (-1)^{i-1} f\alpha)$$

instead of the expression above: the two expressions give isomorphic complexes $M(f)^*$. □

Remark Mapping cones are a construction in homological algebra, motivated by a similar construction in algebraic topology. They are found in most textbooks on homological algebra, sometimes with slightly different indexing or sign conventions. The topological construction from which they originate is described, for instance, in chapter 0 of Hatcher's text [8].

3 Cohen-Macaulay Complexes and why we care

Let us now turn to the little bit of commutative algebra necessary for our purposes. We refer to [5] for a general account, and [6] for something a little more specialized to our situation.

3.1 The Basic Idea

Consider a simplicial complex Δ and its face ring $\Bbbk[\Delta]$: if Δ has at least one vertex v, this is a graded ring with $\Bbbk[\Delta]_i \neq 0$ for each i. Indeed $x_v^i \in \Bbbk[\Delta]_i$. Thus, as a vector space over $\Bbbk$, each graded piece has finite dimension, but the entire ring is always infinite dimensional. It is useful to work with a finite-dimensional $\Bbbk$-algebra instead, provided it preserves enough information about $\Bbbk[\Delta]$. The idea then is to "peel" $\Bbbk[\Delta]$ by quotienting out an ideal which is as large as reasonably possible.

That $\Bbbk[\Delta]$ is Cohen-Macaulay means this peeling can be performed especially nicely, as we shall soon see.

The importance of all this is due to the fact that $\Bbbk[\Delta]$ is always Cohen-Macaulay if Δ is the link or star of a face in any triangulation of a manifold with boundary (and in particular if Δ is a disk or a sphere). This is a shadow of the fact that each point of a manifold has a neighborhood with trivial topology.

We put rings and modules on an equal footing, so these tools are later available for relative face modules.

Definition Let R be a ring and M an R-module. A *regular sequence* on M, or *M-sequence*, is a sequence of elements $(\theta_1, \ldots, \theta_n)$ in R such that:

1. Each θ_i is a nonzerodivisor on $M/\langle \theta_1, \ldots, \theta_{i-1} \rangle$, and

2. $M/\langle \theta_1, \ldots, \theta_n \rangle \neq 0$.

If R is a graded ring, a sequence as above is called homogeneous if each θ_i is.

We care mainly about homogeneous regular sequences, and among them mainly about those in which all elements have degree 1. We will see that if the field $\Bbbk$ is infinite, A and M are graded with A generated in degree 1 (in particular $A_0 = \Bbbk$), and there exists an M-sequence of length n, then there also exists an M-sequence consisting of degree-1 elements.

Quotienting a graded $\Bbbk$-algebra A by a regular sequence of degree-1 elements is an operation which is well-behaved with respect to the *Hilbert series* of A, which we previously encountered in the case $A = \Bbbk[\Delta]$. For a graded $\Bbbk$-vector space V, the Hilbert series is

$$H_V(t) = \sum_{i=0}^{\infty} \dim_{\Bbbk}(V_i) \cdot t^i.$$

Consider the ideal $\langle \theta_1 \rangle$ generated by a nonzerodivisor of A having degree 1. Since the multiplication map by θ_1 is an injection of vector spaces $A \to A$ which increases the degree by 1, we have

$$\dim_{\Bbbk}(\langle \theta_1 \rangle_i) = \dim_{\Bbbk}(A_{i-1}),$$

so

$$H_{\langle \theta_1 \rangle}(t) = \sum_{i=0}^{\infty} \dim_{\Bbbk}(A_{i-1}) \cdot t^i = t \cdot H_{A(t)}.$$

In particular, we find that

$$H_{A/\langle \theta_1 \rangle}(t) = H_A(t) - H_{\langle \theta_1 \rangle}(t) = (1-t) \cdot H_A(t).$$

Modding out by the ideal $\langle \Theta \rangle$ generated by a regular sequence $\Theta = (\theta_1, \ldots, \theta_n)$ consisting of degree-1 elements therefore gives

$$H_{A/\langle \Theta \rangle}(t) = (1-t)^n \cdot H_A(t).$$

All this works in just the same way if we instead work with a regular sequence of degree-1 elements on an A-module M.

Definition Let A be a finitely-generated graded $\Bbbk$-algebra and let M be a finitely-generated A-module. A *homogeneous system of parameters* (h.s.o.p.) for M is a sequence of homogeneous elements $\Theta = (\theta_1, \ldots, \theta_n)$ of A of minimal length among those sequences satisfying that the quotient $M/\langle\Theta\rangle$ is finite dimensional over $\Bbbk$.

The length of a h.s.o.p. as above is the Krull dimension of the support of M (while this fact gives important context, we will have no further use for it). For a face ring $\Bbbk[\Delta]$, it is always $\dim(\Delta) + 1$, or equivalently the maximum cardinality of a face. For a relative face module $\Bbbk[\Delta, \Gamma]$, it is the maximum cardinality of a face of Δ which is not contained in Γ.

Definition Let A be a finitely-generated graded $\Bbbk$-algebra, M a finitely-generated graded A-module. Then M is *Cohen-Macaulay* if it has a homogeneous system of parameters which is an M-sequence.

It is not difficult to show that if M is Cohen-Macaulay and has h.s.o.p. of length n, then any M-sequence of length n is also an h.s.o.p., and no longer M-sequence can exist. Therefore, again assuming k is infinite and A is generated in degree 1, this sequence may be chosen to consist of degree 1 elements of A.

Hence, under all these assumptions, a Cohen-Macaulay M can be "peeled" as nicely as can be hoped: there is an M-sequence $\Theta = (\theta_1, \ldots, \theta_n)$ such that

$$(1-t)^n \cdot H_M(t) = H_{M/\langle\Theta\rangle}(t)$$

is a polynomial. That the left-hand side is a polynomial is true even if the Cohen-Macaulay assumption is omitted. It is the equality to the Hilbert series on the right that is exceptional. A numerical consequence is that the coefficients of this polynomial are positive. More important for us is that the dimensions of graded pieces of $M/\langle\Theta\rangle$ are related to those of M by an explicit formula depending on n alone, and that $M/\langle\Theta\rangle$ is finite dimensional.

3.1.1 An explicit calculation
Let us see what these dimensions are. First, the Hilbert series $H_{\Bbbk[\Delta]}$ is controlled by the f-vector of Δ in the following way. Each non-vanishing monomial in $\Bbbk[\Delta]$ is supported on a unique face of Δ, and the set of monomials with given support $\{v_{i_1}, \ldots, v_{i_m}\}$ is

$$\left\{ \prod_{j=1}^{m} x_{i_j}^{s_j} \;\middle|\; 1 \leq s_1, \ldots, s_m \right\}.$$

The span of this set is a graded $\Bbbk$-vector space with Hilbert series equal to the rational function

$$\frac{x^m}{(1-x)^m} = \frac{x^m(1-x)^{d+1-m}}{(1-x)^{d+1}}$$

where $d = \dim(\Delta)$. Summing over faces (and keeping in mind the difference between face dimension and cardinality) we obtain

$$H_{\Bbbk[\Delta]} = \frac{\sum_{i=-1}^{d} f_i \cdot x^{i+1}(1-x)^{d-i}}{(1-x)^{d+1}}.$$

Thus if $\Bbbk[\Delta]$ is Cohen-Macaulay, its quotient by a regular sequence of length $d+1$ has Hilbert series (now polynomial) $\sum_{i=-1}^{d} f_i \cdot x^{i+1}(1-x)^{d-i}$, and one can verify it equals $\sum_{i=0}^{d+1} h_i x^i$ where $(h_0, \ldots, h_{d+1})$ is the h-vector of Δ. The entries of the h-vector have tremendous meaning, and tell us more intimately what a simplicial complex is about than the face numbers or the Betti numbers do. For instance, they display some fundamental symmetries of the face vector. When we prove the Poincaré duality theorem for face rings (Theorem 7.3), this will imply the famous Dehn-Sommerville relations [24]: for spheres Δ of dimension $d-1$, we have the fundamental symmetry

$$h_i = h_{d-i}.$$

Think of this as a generalization of the fact that the alternating sum of the face numbers equals the Euler characteristic. This fact can be seen as a special case: it is equivalent to the relation $h_0 = h_d$.

3.1.2 Associated primes and prime avoidance

We need the following notions from commutative algebra.

Let A be a finitely generated $\Bbbk$-algebra and let M be an A-module.

Definition A prime ideal $\mathfrak{p} \subset A$ is *associated* to M if there exists an $m \in M$ such that

$$\{a \in A \mid am = 0\} = \mathfrak{p},$$

i.e. if $\mathfrak{p}$ is the annihilator of m.

Lemma 3.1 *The set of zerodivisors on M equals the union of the primes associated to M. That is,*

$$\{a \in A \mid am = 0 \text{ for some } 0 \neq m \in M\} = \bigcup_{\substack{\mathfrak{p} \subset A \text{ prime} \\ \mathfrak{p} \text{ is associated to } M}} \mathfrak{p}.$$

Lemma 3.2 *Let $I \subset A$ be an ideal and let $\mathfrak{p}_1, \ldots \mathfrak{p}_n \subset A$ be prime ideals. If*

$$I \subset \bigcup_i \mathfrak{p}_i$$

then $I \subset \mathfrak{p}_j$ for some $1 \leq j \leq n$.

The last lemma is an analogue of the fact that if V is a vector space over an infinite field $\Bbbk$ and $W_1, \ldots, W_n$ are proper subspaces, then $\bigcup_i W_i \subsetneq V$ and a generic element of V is not in $\bigcup_i W_i$.

That some property holds for a generic element of a vector space means it holds for a member of a dense open subset of the vector space with respect to an appropriate topology. One such topology here is the Zariski topology obtained by identifying V with an affine space over $\Bbbk$.

Remark [some technicalities] The Cohen-Macaulay property has an important role in commutative algebra and algebraic geometry. The discussion here is specialized to the context of face rings: our definition is not the common one outside combinatorics. That it is a specialization of other definitions is a theorem we will not need or prove.

The partition complex

Some amount of dimension theory for commutative rings could not be avoided in the discussion on systems of parameters. Short computational proofs of the necessary facts can be given, and the general case can be found in most textbooks on commutative algebra. We recommend the unfamiliar reader simply accept them for now.

3.2 The Koszul Complex

The Koszul complex is a homological tool. It can be used computationally, for instance to find the length of the longest regular sequence contained in an ideal. Conversely, given a regular sequence in A generating an ideal I, the Koszul complex gives a free resolution of A/I. This is a chain complex of free modules such that A/I is its last (and only nonvanishing) cohomology group. In a sense, it spreads the quotienting operation into simpler layers.

The point of this will become apparent when the partition complex is introduced, and we begin dealing with all stars of faces of a simplicial complex as a cohesive unit.

Let A be a $\Bbbk$-algebra and let $\Theta = (\theta_1, \ldots, \theta_n)$ be a sequence of elements in A. The Koszul complex $K^* = K^*(\Theta)$ is the chain complex

$$0 \to K^0 \xrightarrow{\partial_0} K^1 \xrightarrow{\partial_1} \ldots \to K^n \xrightarrow{\partial_n} 0$$

where $K^0 = A$ and in general $K^i = \bigwedge^i A^n = \bigwedge^i \left(\bigoplus_{j=1}^n A \cdot e_j \right)$. The maps $\partial^i : K^i \to K^{i+1}$ are defined by

$$z \mapsto \left(\sum_{i=1}^n \theta_i e_i \right) \wedge z.$$

This is largely consistent with the notation used by Eisenbud in [7].

We collect some basic facts and observations. These culminate in Theorem 3.3.

3.2.1 Exterior powers It is helpful to recall that $\bigwedge^i \left(\bigoplus_{j=1}^n A \cdot e_j \right)$ is a free A-module with basis all wedges of the form $e_{j_1} \wedge \ldots \wedge e_{j_i}$ where $1 \leq j_1 < \ldots < j_i \leq n$. Thus as A-modules,

$$K^i \otimes M = \left(\bigwedge^i A^n \right) \otimes M \simeq \bigoplus_{1 \leq j_1 < \ldots < j_i \leq n} M \cdot e_{j_1} \wedge \ldots \wedge e_{j_i}.$$

The maps of the complex $K^* \otimes M$, being induced from K^*, are defined by the same expression for each ∂^i.

3.2.2 The top cohomology Note that $K^n = A \cdot e_1 \wedge \ldots \wedge e_n$ is a free module of rank 1, and thus $K^n \otimes M \simeq M$. Further, the image of the map $K^{n-1} \otimes M \xrightarrow{\partial_{n-1}} K^n \otimes M$ is $\langle \Theta \rangle \cdot M$: on generators of K^{n-1} we have

$$\partial_{n-1} \left(\bigwedge_{j \neq i} e_j \right) = (-1)^{i+1} \theta_i \cdot e_1 \wedge \ldots \wedge e_n.$$

This implies $H^n(K^* \otimes M) = M/\langle \Theta \rangle M$.

3.2.3 The action of $z \in \langle \Theta \rangle$ on homology
Suppose $z \in \langle \Theta \rangle$. Then z induces the zero map on $H^i(K^* \otimes M)$ for each i. To see this, write

$$z = \sum_{i=1}^{n} a_i \theta_i,$$

and define the following maps $f^t : K^t \to K^{t-1}$:

$$e_{j_1} \wedge \ldots \wedge e_{j_t} \mapsto \sum_{i=1}^{t} (-1)^{i-1} a_{j_i} \cdot e_{j_1} \wedge \ldots \wedge \hat{e}_{j_i} \wedge \ldots \wedge e_{j_t},$$

where $\hat{e}_{j_i}$ denotes omission. Then the map $\partial^{t-1} \circ f^t + f^{t+1} \circ \partial^t$ acts on K^t as the multiplication map by z, and is thus a chain homotopy between $z\cdot$ and the zero map. This remains true on tensoring K^* with M.

Let us compute $\partial^{i-1} \circ f^i + f^{i+1} \circ \partial^i$ to verify this claim. Without loss of generality, consider the basis element $b = e_1 \wedge \ldots \wedge e_i$. Then

$$\partial^{i-1} \circ f^i(b) = \partial^{i-1} \left(\sum_{j=1}^{i} (-1)^{j-1} a_j \cdot e_1 \wedge \ldots \wedge \hat{e}_j \wedge \ldots \wedge e_i \right)$$

$$= \sum_{j=1}^{i} \left[a_j \theta_j \cdot e_1 \wedge \ldots \wedge e_i \right.$$

$$\left. + \sum_{s=i+1}^{n} (-1)^{i+j-2} a_j \theta_s \cdot e_1 \wedge \ldots \wedge \hat{e}_j \wedge \ldots \wedge e_i \wedge e_s \right].$$

Similarly,

$$f^{i+1} \circ \partial^i(b) = f^{i+1} \left(\sum_{s=i+1}^{n} (-1)^i \theta_s \cdot e_1 \wedge \ldots \wedge e_i \wedge e_s \right)$$

$$= \sum_{s=i+1}^{n} \left[\sum_{j=1}^{i} (-1)^{i+j-1} a_j \theta_s \cdot e_1 \wedge \ldots \wedge \hat{e}_j \wedge \ldots \wedge e_i \wedge e_s \right.$$

$$\left. + a_s \theta_s \cdot e_1 \wedge \ldots \wedge e_i \right].$$

We see the coefficients cancel in the sum to give $\sum_{j=1}^{n} a_j \theta_j \cdot e_1 \wedge \ldots \wedge e_i = z \cdot e_1 \wedge \ldots \wedge e_i$ as desired.

3.2.4 Koszul homology and regular sequences
We are ready to prove that the homology of the Koszul complex $K^* \otimes M$ tells us the maximal length of an M-sequence in $\langle \Theta \rangle$.

Recall that the irrelevant ideal of a graded ring is the ideal generated by homogeneous elements of positive degree. For a face ring $\Bbbk[\Delta]$, this is simply the ideal $\langle x_v \mid v \in \Delta^{(0)} \rangle$.

Theorem 3.3 *Let A be a graded $\Bbbk$-algebra. If M is a finitely-generated graded A-module and $\Theta = (\theta_1, \ldots, \theta_n)$ is a sequence of homogeneous elements contained in the irrelevant ideal of A, the minimal i for which $H^i(K^*(\Theta) \otimes M) \neq 0$ is the maximal length of an M-sequence contained in $\langle \Theta \rangle$. This length is at most n.*

Further, if $\Bbbk$ is infinite, generic linear combinations of $\theta_1, \ldots, \theta_n$ give regular sequences.

Remark We state this theorem in the graded setting, but it is true for Noetherian rings and Noetherian modules in general. The condition that Θ consists of homogeneous elements contained in the irrelevant ideal should be replaced by the condition $\langle \Theta \rangle \cdot M \neq M$. Generalizing the proof to this situation is an exercise.

Proof By induction on i. For $i = 0$, $H^0(K^*(\Theta) \otimes M) \neq 0$ iff the map $M \to \bigoplus_{i=1}^n M \cdot e_i$ given by $m \mapsto \sum_{i=1}^n \theta_i m \cdot e_i$ has a nontrivial kernel, and this occurs iff some $m \in M$ is in the kernel of each multiplication map $\theta_i \cdot : M \to M$ (equivalently, some m is in the kernel of each $z \in \langle \Theta \rangle$).

Suppose the minimal i for which $H^i(K^*(\Theta) \otimes M) \neq 0$ is positive, and assume the claim is true up to $i - 1$. Then in particular, $H^0(K^*(\Theta) \otimes M) = 0$, and there is no $m \in M$ in the kernel of all $\langle \Theta \rangle$. Thus $\langle \Theta \rangle$ is not contained in any associated prime (or it would annihilate an element of M by Lemma 3.1). There is then some $z_1 \in \langle \Theta \rangle$ which is a nonzerodivisor on M, and if $\Bbbk$ is infinite we may take z_1 to be a linear combination of $\theta_1, \ldots \theta_n$ (a generic one works). Note $M/z_1 M \neq 0$, or equivalently $M \neq z_1 M$, since z_1 has positive degree, and for the minimal degree d such that $M_d \neq 0$ we have $M_d \cap z_1 M = 0$.

Since each K^j is free over A, its tensor product with the short exact sequence

$$0 \to M \xrightarrow{z_1 \cdot} M \to M/z_1 M \to 0$$

is exact. This gives a short exact sequence of complexes:

$$0 \to K^*(\Theta) \otimes M \xrightarrow{z_1 \cdot} K^*(\Theta) \otimes M \to K^*(\Theta) \otimes M/z_1 M \to 0,$$

resulting in a long exact sequence in homology:

$$\ldots \to H^j(K^* \otimes M) \to H^j(K^* \otimes (M/z_1 M)) \to H^{j+1}(K^* \otimes M) \to \ldots$$

where $H^{j+1}(K^* \otimes M) = 0$ for $j + 1 < i$, proving that $H^j(K^* \otimes (M/z_1 M)) = 0$ for $j < i - 1$. In particular, by the inductive assumption applied to $M/z_1 M$, there is an $M/z_1 M$-sequence of length $i - 1$ in $\langle \Theta \rangle$. Denoting it $z_2, \ldots, z_i$, we have that $z_1, z_2, \ldots, z_i$ is an M-sequence.

We will be finished if we prove $H^{i-1}(K^* \otimes (M/z_1 M)) \neq 0$, since applying the inductive claim to $M/z_1 M$ then shows that there is no longer M-sequence starting with z_1 in $\langle \Theta \rangle$, where z_1 is an arbitrary nonzerodivisor on M.

A piece of the long exact sequence from before reads:

$$\ldots \to 0 \to H^{i-1}(K^* \otimes (M/z_1 M)) \to H^i(K^* \otimes M) \xrightarrow{z_1 \cdot} H^i(K^* \otimes M) \to \ldots,$$

where the first 0 is $H^{i-1}(K^* \otimes M)$. Thus if $H^{i-1}(K^* \otimes (M/z_1 M)) = 0$ then z_1 is a nonzerodivisor on $H^i(K^* \otimes M) \neq 0$. This is a contradiction, since each element of $\langle \Theta \rangle$ acts as the 0 map on $H^i(K^* \otimes M)$. $\square$

Corollary 3.4 *Under the conditions of the theorem, all maximal M-sequences in $\langle \Theta \rangle$ have the same length.*

Proof This follows from the proof above, which picks an arbitrary nonzerodivisor at each step of the induction. There is no possibility of getting stuck in the construction of one sequence earlier than in another. □

Remark [further notes] The explicit chain homotopy used earlier in this section can be replaced by a longer but more illuminating method, which we indicate. The Koszul complexes $K_n^* = K^*(\theta_1, \ldots, \theta_n)$ and $K_{n+1}^* = K^*(\theta_1, \ldots, \theta_{n+1})$ can be related in the following way: the mapping cone of the map $\theta_{n+1} \cdot : K_n^* \to K_n^*$ can be naturally identified with K_{n+1}^*. Hence there is a short exact sequence of chain complexes

$$0 \to K_n^{*-1} \to K_{n+1}^* \to K_n^* \to 0.$$

If $\theta_{n+1} - \theta'_{n+1} \in \langle \theta_1, \ldots, \theta_n \rangle$, the associated short exact sequences are isomorphic, essentially by a change of basis in each $K_{n+1}^i = \bigwedge^i(A^{n+1})$. In particular, for $\theta_{n+1} \in \langle \theta_1, \ldots, \theta_n \rangle$, one can take $\theta'_{n+1} = 0$. The associated long exact sequences are then isomorphic by the naturality of the snake lemma. For θ_{n+1}, the long exact sequence is:

$$\ldots \to H^{i-1}(K_n^*) \to H^i(K_{n+1}^*) \to H^i(K_n^*) \xrightarrow{\theta_{n+1} \cdot} H^i(K_n^*) \to \ldots,$$

and the existence of an isomorphic sequence in which θ_{n+1} is replaced by 0 proves the claim.

4 The Partition Complex & Reisner's Theorem

In this section we introduce our central tool, the partition complex. As a first application we prove Reisner's theorem, which provides a topological characterization of Cohen-Macaulay complexes.

4.1 The Partition Complex

In many branches of geometry homological algebra provides tools for piecing together local data and understanding global behavior. In conjunction with the Koszul complex, the partition complex is a tool suitable for understanding face rings and their quotients by systems of parameters.

Definition Let $\Psi = (\Delta, \Gamma)$ be a relative complex. We define the (unreduced) partition complex $P^* = P^*(\Psi)$ by

$$P^* = 0 \to \Bbbk[\Psi] \xrightarrow{d^{-1}} \bigoplus_{v \in \Delta^{(0)}} \Bbbk[\operatorname{st}_v \Psi] \xrightarrow{d^0} \ldots \to \bigoplus_{\sigma \in \Delta^{(d)}} \Bbbk[\operatorname{st}_\sigma \Psi] \to 0,$$

with indexing such that $P^{-1} = \Bbbk[\Psi]$ and $P^i = \bigoplus_{\tau \in \Delta^{(i)}} \Bbbk[\operatorname{st}_\tau \Psi]$ for $i \geq 0$.

The maps are given by maps of the Čech complex corresponding to the cover of (Δ, Γ) by the open stars of vertices.

The partition complex

More explicitly, choose some ordering of the vertices. Then if $\rho \subset \tau = \rho \cup \{v_{i_j}\}$ is a pair of faces, and $\tau = \{v_{i_1}, \ldots, v_{i_k}\}$ with $i_1 < \ldots < i_k$, then denoting by $\alpha \cdot e_\rho$ the element α in the summand $\Bbbk[\mathrm{st}_\rho \Psi]$ (and similarly with e_τ) we define

$$d(\alpha \cdot e_\rho) = (-1)^j \alpha \cdot e_\tau.$$

In particular, the map $P^{-1} \to P^0$ is $\alpha \mapsto \sum_{v \in \Delta^{(0)}} \alpha \cdot e_v$.

Note that if x^α is a monomial in $\Bbbk[\Psi]$, it is a nonvanishing element of $\Bbbk[\mathrm{st}_\tau \Psi]$ precisely when the support of α is a face of $\mathrm{st}_\tau(\Delta)$ (it is automatic that it is not a face of Γ, hence it is not in $\mathrm{st}_\tau(\Gamma)$). The next proposition uses this observation to compute the cohomology of P^*.

Proposition 4.1 *For each i, $H^i(P^*) = H^i(\Psi; \Bbbk)$.*

Proof Each P^i inherits the fine grading of $\Bbbk[\Psi]$, and the differential d^i has (fine) degree 0. Thus, as a complex of vector spaces over $\Bbbk$, P^* splits into the direct sum of its fine-graded pieces P^*_α, and this induces a corresponding decomposition of the homology.

Consider a monomial $x^\alpha \in \Bbbk[\Psi]$: for any $\tau \in \Delta$, x^α is in $\Bbbk[\mathrm{st}_\tau \Psi]$ if and only if $\tau \cup \mathrm{supp}(\alpha)$ is a face of Δ but not of Γ, or equivalently if τ is in the first but not the second member of the pair $\mathrm{st}_{\mathrm{supp}(\alpha)} \Psi$.

Using this, we see P^*_α is a Čech complex of $\mathrm{st}_{\mathrm{supp}(\alpha)} \Psi$, which is acyclic if $\mathrm{supp}(\alpha) \neq \emptyset$ because stars of faces are contractible. For $x^\alpha = 1$, the complex computes the cohomology of the pair (Δ, Γ). □

The general method is to use this together with the Koszul complex: we form the double complex $P^*(\Psi) \otimes K^*(\Theta)$ for Θ some generic sequence of linear forms in $\Bbbk[\Delta]$, and perform homological calculations on the resulting double complex.

The same idea works with other complexes of $\Bbbk[\Delta]$-modules constructed to compute the cohomology. An example can be seen in Theorem 9.1. Some of this paper can be generalized in such a direction, but we refrain from introducing more formalism at this point.

4.2 Reisner's Theorem

Reisner's theorem [20] gives a link between the algebra of face rings and the topology of simplicial complexes: it tells us a complex has a Cohen-Macaulay face ring exactly when it is sufficiently well-connected, both locally and globally.

Theorem 4.2 (Reisner) *Let $\Psi = (\Delta, \Gamma)$ be a relative complex. Then $\Bbbk[\Psi]$ is Cohen-Macaulay if and only if for all faces τ of Δ (including $\tau = \emptyset$):*

$$H^i(\mathrm{lk}_\tau \Psi; \Bbbk) = 0 \quad \forall -1 \leq i < \dim(\Psi) - \dim(\tau).$$

We shall give a new and elementary proof of this theorem here, though for simplicity we restrict to the "if" direction critical for us.

Remark In the non-relative setting, the right-hand side of the last inequality is sometimes replaced by $\dim(\mathrm{lk}_\tau \Delta)$. This is never larger than $\dim(\Delta) - \dim(\tau)$, so our condition is stronger. They are equivalent in the non-relative setting, modulo the topological fact that a (non-relative) complex satisfying the weaker condition is pure. However, this is not true in the relative case. An example is given by

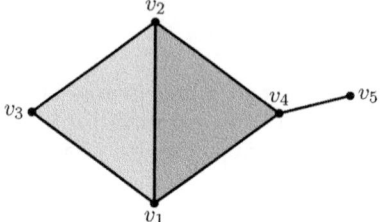

where the complex is relative to the gray (right) triangle.

The stronger assumption ensures purity, because if $\tau \in \Psi$ has lower than maximal dimension then $\mathrm{lk}_\tau \Psi$ cannot have nontrivial cohomology in dimension -1, so $\mathrm{lk}_\tau \Psi \neq (\{\emptyset\}, \emptyset)$ (note it is automatic that $\mathrm{lk}_\tau(\Gamma) = \emptyset$ since $\tau \in (\Delta, \Gamma)$). Thus Ψ has a face properly containing τ.

Often, one writes "Δ is Cohen-Macaulay" and means whichever of the two equivalent conditions is more convenient for the situation at hand. In this section, to indicate the condition on $\Bbbk[\Delta]$ we write that Δ is algebraically Cohen-Macaulay, and to indicate the homological condition we write that Δ is topologically Cohen-Macaulay.

Here is a simple application of Reisner's theorem first mentioned in Section 3.1, the importance of which is difficult to overstate: if M is a triangulated manifold, the star of each face is Cohen-Macaulay.

Remark A second, more combinatorial application is of course that of face numbers of simplicial complexes. Notice that the face numbers of a simplicial complex are a non-negative combination of its h-numbers, and that these are nonnegative for Cohen-Macaulay complexes by Reisner's theorem (as they correspond to dimensions of vectorspaces).

On the other hand, the h-numbers also cannot be too large, as a polynomial ring cannot grow too quickly. For instance, a polynomial ring on n variables cannot generate, in degree k, a space of dimension larger than $\binom{n+k-1}{j}$. Macaulay [15] used this to provide a complete characterization of face numbers of Cohen-Macaulay complexes.

We begin with the following observations.

Lemma 4.3 *If Ψ is topologically Cohen-Macaulay and $\tau \in \Psi$ then $\mathrm{lk}_\tau \Psi$ is topologically Cohen-Macaulay.*

Proof Let σ be a face of $\mathrm{lk}_\tau \Psi$. We can verify directly that $\mathrm{lk}_\sigma(\mathrm{lk}_\tau \Psi)$ is the link of $\sigma \cup \tau$ in $\Psi = (\Delta, \Gamma)$: it suffices to verify this in Δ and in Γ separately. So by assumption its homology vanishes beneath the top dimension. □

Lemma 4.4 *A relative simplicial complex* $\Psi = (\Delta, \Gamma)$ *is algebraically Cohen-Macaulay if and only if the cone* $v * \Psi = (v * \Delta, v * \Gamma)$ *is algebraically Cohen-Macaulay.*

Proof The result follows from the fact that x_v is a nonzerodivisor on $M = \Bbbk[(v * \Delta, v * \Gamma)]$ and $M/\langle x_v \rangle M = \Bbbk[\Psi] =: N$.

In one direction, if N is algebraically Cohen-Macaulay with $\Theta = (\theta_1, \ldots, \theta_n)$ a regular sequence of parameters, then so is M with the sequence $(x_v, \theta_1, \ldots, \theta_n)$. On the other hand, if M is Cohen-Macaulay, then by Corollary 3.4 it also has a regular sequence of parameters $(x_v, \theta_1, \ldots, \theta_n)$, and so N is Cohen-Macaulay. $\square$

This is all we need in order to prove the first half of Reisner's theorem.

Theorem 4.5 *If* $\Psi = (\Delta, \Gamma)$ *is topologically Cohen-Macaulay then it is algebraically Cohen-Macaulay.*

Proof By induction on dimension. Consider first a non-relative complex Δ of dimension $d = 0$. Then $\Bbbk[\Delta] \simeq \Bbbk[x_1, \ldots, x_n]/\langle x_i x_j \mid i \neq j \rangle$ for some n, and $\Bbbk[\Delta]/\langle \sum_i x_i \rangle$ is finite-dimensional over $\Bbbk$: each monomial of degree 2 vanishes in the quotient, since $x_j^2 = x_j \cdot \sum_j x_j$. Further, $\sum_i x_i$ is a nonzerodivisor: each non-vanishing element of $\Bbbk[\Delta]$ is of the form $c_0 + \sum_i c_i x_i^{t_i}$ with $c_0, \ldots, c_n \in \Bbbk$, and its product by $\sum_i x_i$ does not vanish.

In the relative case, suppose $d = \dim(\Psi) = 0$. Then unless $\Gamma = \emptyset$ (the non-relative situation) we have $\Bbbk[\Psi] = \langle x_v \mid v \in \Psi \rangle$, and again quotienting by the ideal generated by $\sum_{v \in \Psi} x_v$ (a nonzerodivisor) gives a finite dimensional vector space.

Suppose the theorem holds for relative complexes of dimension up to $d-1$, and suppose Ψ is topologically Cohen-Macaulay of dimension d. Then the link of each face τ is algebraically Cohen-Macaulay of dimension $d - \dim(\tau)$, and by the coning lemma $\text{st}_\tau \Psi$ is algebraically Cohen-Macaulay as well.

Let Θ be a generic sequence of $d + 1$ linear forms in $\Bbbk[\Delta]$, and consider $C^{*,*} = P^*(\Psi) \otimes K^*(\Theta)$. By 2.3 the mapping cone of $\mathfrak{R}^{-1}$ gives us a short exact sequence

$$0 \to \text{Tot}(C^{* \geq 0,*})^{*-1} \to \text{Tot}(C^{* \geq -1,*})^{*-1} = \text{Tot}(C^{*,*})^{*-1} \to C^{-1,*} \to 0.$$

The corresponding long exact sequence is

$$\ldots \to H^i(C^{-1,*}) \to H^i(\text{Tot}(C^{* \geq 0,*})^*) \to H^i(\text{Tot}(C^{*,*})^*) \to H^{i+1}(C^{-1,*}) \to$$
$$\ldots \to H^{d-1}(\text{Tot}(C^{*,*})^*) \to H^d(C^{-1,*}) \to H^d(\text{Tot}(C^{* \geq 0,*})^*) \to \ldots$$

Notice that if $k < d$ then $H^{k-i}(C^{*,i}) = 0$: for $i < 0$ this is automatic because $C^{*,i}$ is the 0 complex, and otherwise it occurs because the row $C^{*,i} = P^* \otimes K^i(\Theta)$ is exact until the d-th place, since P^* is (and K^i is a free module). By Lemma 2.1, this implies $H^k(\text{Tot}(C^{*,*})^*) = 0$ for $k < d$.

This implies the maps $C^{-1,*} \to \text{Tot}(C^{* \geq 0,*})^*$ induce isomorphisms on the k-th homology for all $k < d$, and an injection on H^d.

Since the star of each face is algebraically Cohen-Macaulay, each column of $C^{* \geq 0,*}$ has $H^i(C^{j,*}) = 0$ for all $i < d+1$, so another application of Lemma 2.1 proves $H^i(\text{Tot}(C^{* \geq 0,*})^*) = 0$ for $i < d+1$ as well. This implies that $H^i(C^{-1,*}) = H^i(\Bbbk[\Psi] \otimes K^*(\Theta)) = 0$ for all such i, and the theorem follows. $\square$

5 Partition of Unity

Let $\Psi = (\Delta, \Gamma)$ be a triangulated d-manifold with boundary, and let $\Gamma = \emptyset$ or the boundary subcomplex (that the boundary always is in fact a subcomplex is a simple result in topology). Then for each $\tau \in \Delta$, $\mathrm{st}_\tau \Psi$ is Cohen-Macaulay of dimension d. This is the situation in which the partition complex is most useful: computations of interest can be reduced to the homology of $\mathrm{Tot}(P^* \otimes K^*)$. To motivate the computation of this homology we begin with an application, the partition of unity theorem.

For a generic sequence of linear elements $\Theta = (\theta_1, \ldots, \theta_{d+1})$ in $\Bbbk[\Delta]$, we wish to understand $P^*(\Psi)/\langle \Theta \rangle$. The partition of unity theorem computes the dimension of the kernel of the first differential of this complex, i.e. of

$$\Bbbk[\Psi]/\langle \Theta \rangle \to \bigoplus_{v \in \Delta^{(0)}} \Bbbk[\mathrm{st}_v \Psi]/\langle \Theta \rangle,$$

in terms of $H^*(\Delta, \Gamma)$.

Here is a special case: if $(\Delta, \emptyset)$ is a sphere then this map is injective except in the top degree $d+1$, where the kernel has dimension one (and in fact equals the entire $(d+1)$-th graded piece).

We state the theorem in slightly greater generality, recovering a key result of [1].

Theorem 5.1 (partition of unity) *Let* $\Psi = (\Delta, \Gamma)$ *be a relative complex, pure of dimension d, such that* $\mathrm{st}_\tau \Psi$ *is Cohen-Macaulay for each nonempty* $\tau \in \Delta$. *Let* Θ *be a generic sequence of $d+1$ elements in* $\Bbbk[\Delta]_1$. *Then*

$$\dim_\Bbbk H^i(P^*/\langle \Theta \rangle)_j = \binom{d+1}{j} \cdot \dim_\Bbbk H^{i+j}(\Psi; \Bbbk).$$

Complexes satisfying the conditions of the theorem are called Buchsbaum complexes. Note the subscript j indicates the j-th graded piece is being taken, where the grading is induced from $\Bbbk[\Delta]$.

Proof Recall $H^{d+1}(\Theta) = \Bbbk[\Psi]/\langle \Theta \rangle$: this is shown in Section 3.2.2. Define $\tilde{K}^*(\Theta)$ to be the *augmented Koszul complex*. This is the complex

$$K^0(\Theta) \to \ldots \to K^{d+1}(\Theta) \to \tilde{K}^{d+2}(\Theta) = H^{d+1}(\Theta) = \Bbbk[\Delta]/\langle \Theta \rangle \to 0,$$

where the map $\tilde{K}^{d+1}(\Theta) \to H^{d+1}(\Theta)$ is the natural quotient map $K^{d+1} \to K^{d+1}/\mathrm{im}(K^d)$.

We will work with the complex $C^{*,*} = P^* \otimes \tilde{K}^*(\Theta)$.

Since the last nonzero map of $\tilde{K}^*$ is the cokernel of the map before it, $H^{d+1}(\tilde{K}^* \otimes M) = H^{d+2}(\tilde{K}^* \otimes M) = 0$ for any $\Bbbk[\Delta]$-module M. Therefore the Cohen-Macaulayness of the stars of Ψ implies $\tilde{K}^*(\Theta) \otimes \Bbbk[\mathrm{st}_\tau \Psi]$ is exact for each $\tau \in \Delta$ (homologies up to the d-th position are equal for the augmented and un-augmented complex). In particular, the columns $C^{i,*}$ are exact for each $i \geq 0$. The column $C^{-1,*} = \Bbbk[\Psi] \otimes \tilde{K}^*(\Theta)$ is only exact if Ψ is Cohen-Macaulay.

Consider the map $\mathfrak{U}^{d+1} : \mathrm{Tot}(C^{*,*\leq d+1})^* \to C^{*-d-1,d+2}$. The exact sequence associated to its mapping cone is

$$0 \to C^{*-d-2,d+2} \to \mathrm{Tot}(C^{*,*\leq d+2})^* = \mathrm{Tot}(C^{*,*})^* \to \mathrm{Tot}(C^{*,*\leq d+1})^* \to 0. \quad (*)$$

Note that $H^{d-i}(C^{*,i}) = H^{d+1-i}(C^{i,*}) = 0$ for all i. For $i \geq 0$ this is automatic, since the entire i-th column is exact. For $i = -1$, it follows from the fact that $H^{d+1}(C^{-1,*}) = H^{d+2}(C^{-1,*}) = 0$, as both are equal to the respective homologies of $\tilde{K}^*$. Therefore Lemma 2.1 implies

$$H^d(\mathrm{Tot}(C^{*,*})^*) = H^{d+1}(\mathrm{Tot}(C^{*,*})^*) = 0.$$

Similarly, $H^t(\mathrm{Tot}(C^{*,*})^*)$ for all $t > d+1$, since the (-1)-th column is not involved. Therefore for all $t \geq d$ the short sequence $(*)$ above yields the exact sequences in homology

$$\ldots 0 \to H^t(\mathrm{Tot}(C^{*,*\leq d+1})^*) \to H^{t+1}(C^{*-d-2,d+2}) \to 0,$$

where the zeros on the left and right end are the t and $(t+1)$-th homologies of $\mathrm{Tot}(C^{*,*})^*$ respectively.

Set $s = i+d+2$. Since the row $C^{*,d+2}$ is $P^*/\langle \Theta \rangle$, $H^s(C^{*-d-2,d+2}) = H^i(P^*/\langle \Theta \rangle)$. We now know it is isomorphic to

$$H^{s-1}(\mathrm{Tot}(C^{*,*\leq d+1})^*) = H^{i+d+1}(\mathrm{Tot}(C^{*,*\leq d+1})^*),$$

where $C^{*,*\leq d+1} \simeq P^* \otimes K^*(\Theta)$. The claim is now reduced to Theorem 5.2. From its statement we conclude the identity

$$\dim_\Bbbk H^i(P^*/\langle \Theta \rangle)_j = \dim_\Bbbk H^{i+d+1}(\mathrm{Tot}(P^* \otimes K^*)^*)_j$$
$$= \binom{d+1}{j} \cdot \dim_\Bbbk H^{i+j}(\Psi; \Bbbk).$$

□

5.1 The homology of $\mathrm{Tot}(P^* \otimes K^*)$

As earlier, $\Psi = (\Delta, \Gamma)$ is relative complex, pure of dimension d, in which the star of each $\tau \in \Delta$ is Cohen-Macaulay. Let $\Theta = (\theta_1, \ldots, \theta_{d+1})$ a generic sequence in $\Bbbk[\Delta]_1$ and denote $C^{*,*} = P^*(\Psi) \otimes K^*(\Theta)$.

5.1.1 Grading the double complex

We wish to understand the homology of $\mathrm{Tot}(C^{*,*})$ as a graded $\Bbbk[\Delta]$-module. It is convenient to grade each $C^{i,j}$ such that the differentials of the double complex are maps of graded modules, or in other words take homogeneous elements to homogeneous elements of the same degree.

We grade K^* as follows. Each K^i has a basis of the form

$$\{e_{j_1} \wedge \ldots e_{j_i} \mid j_1 < j_2 < \ldots < j_i\}.$$

Set each such basis element to be of degree $d+1-i$. In particular, $K^{d+1} \simeq \Bbbk[\Delta]$ is a free module with basis $e_1 \wedge e_2 \wedge \ldots \wedge e_{d+1}$, and this basis element has degree 0. Similarly, $K^0 = \Bbbk[\Delta]$ is a free module with basis $\{1\}$, re-graded so that 1 has degree $d+1$ (the common notation for this is $K^0(\Theta) = \Bbbk[\Delta](-d-1)$).

Thus if $\alpha \in \Bbbk[\Delta]_t$ is homogeneous of degree t, $\alpha \cdot e_{j_1} \wedge \ldots \wedge e_{j_i}$ has degree $t+d+1-i$. The differential, $z \mapsto (\sum \theta_i e_i) \wedge z$, is then a map of graded modules. This induces the required grading on $P^* \otimes K^*$. More generally it induces a grading on $K^* \otimes M$ for M any graded $\Bbbk[\Delta]$-module, in which $m \cdot e_{j_1} \wedge \ldots \wedge e_{j_i}$ has degree $t + d + 1 - i$ for any $m \in M_t$.

Since the kernel, image, and cokernel of a map of graded modules are graded modules, there is an induced grading on the homology of the rows, columns, and total complex of $C^{*,*}$. In fact, if we only want to compute dimensions, these can be computed separately for each graded piece of $C^{*,*}$ (no longer a $\Bbbk[\Delta]$-module but a $\Bbbk$-vector space).

5.1.2 The grading and $P^*/\langle \theta \rangle$

Our choice of grading for K^* is motivated by the following consideration. The augmented complex $\tilde{K}^*$ has $\tilde{K}^{d+1} \simeq \Bbbk[\Delta]$ (the isomorphism now being of graded modules), and the image of $\tilde{K}^d$ in $\tilde{K}^{d+1}$ is precisely $\langle \Theta \rangle$. Thus the cokernel $\tilde{K}^{d+2} = H^{d+1}(K^*)$ of $\tilde{K}^d \to \tilde{K}^{d+1}$ is precisely $\Bbbk[\Delta]/\langle \Theta \rangle$, and has the same grading.

These considerations apply just as well when $\tilde{K}^*$ is replaced with $\tilde{K}^* \otimes M$ for M a graded $\Bbbk[\Delta]$-module. Hence all computations on $P^*(\Psi)/\langle \Theta \rangle$ performed using these methods respect its grading automatically.

5.1.3 Computation of the homology

With the above grading in hand, we can compute $H^i(\mathrm{Tot}(C^{*,*}))_j$ for any i, j.

Theorem 5.2 *For any i and j, we have*

$$\dim_{\Bbbk} H^i(\mathrm{Tot}(C^{*,*}))_j = \binom{d+1}{j} \cdot \dim_{\Bbbk} H^{i+j-d-1}(\Psi; \Bbbk).$$

Proof As often happens, the proof writes itself once the correct short exact sequence of complexes is found. In the interest of making this process as transparent as possible, let us begin by explicitly unraveling the degrees and indices involved.

Observe that for $j \in \mathbb{Z}$ any module in the row $C_j^{*,s}$ is spanned by elements of the form $\alpha \cdot e_{j_1} \wedge \ldots \wedge e_{j_s}$, where $\alpha \in \Bbbk[\mathrm{st}_\tau \Psi]$ and

$$\deg(\alpha) + d + 1 - s = j,$$

or $\deg(\alpha) = j + s - d - 1$.

We now recall Proposition 4.1. Its proof shows P_t^* is exact for all $t > 0$. This means the row $C_j^{*,s}$ is exact unless $d + 1 - j - s = 0$, so the interesting row is $s = d + 1 - j$. In any lower row $s' < s$, the complex $C_j^{*,*}$ vanishes entirely: elements there are sums of expressions $\alpha \cdot e_{j_1} \wedge \ldots \wedge e_{j_{s'}}$, where $\alpha \in \Bbbk[\mathrm{st}_\tau \Psi]$ has negative degree. That the lowest row is also the only one carrying nontrivial homology suggests relating $H^*(C^{*,s})_j$ with $H^*(\mathrm{Tot}(C^{*,*\geq s})^*)_j$ and with $H^*(\mathrm{Tot}(C^{*,*\geq s+1})^*)_j = 0$.

Taking the transpose $(\mathfrak{R}^i)^\dagger$ of $\mathfrak{R}^i$ (exchanging the two upper indices, along with the vertical and horizontal differentials) provides a map

$$(\mathfrak{R}^\dagger)^s : C_j^{*,s} \to \mathrm{Tot}(C^{*,*\geq s+1})_j^{*+s+1},$$

with mapping cone $\mathrm{Tot}(C^{*,*\geq s})_j^{*+s}$. Thus there is a long exact sequence in homology:

$$\ldots \to H^i(\mathrm{Tot}(C^{*,*\geq s+1})_j^{*+s}) \to H^i(\mathrm{Tot}(C^{*,*\geq s})_j^{*+s}) \to$$

$$H^i(C_j^{*,s}) \to H^{i+1}(\text{Tot}(C^{*,*\geq s+1})_j^{*+s}) \to \cdots$$

where $\text{Tot}(C^{*,*\geq s+1})_j^{*+s}$ is exact. We obtain isomorphisms

$$H^{i+s}(\text{Tot}(C^{*,*})_j^*) = H^i(\text{Tot}(C^{*,*\geq s})_j^{*+s}) \simeq H^i(C_j^{*,s}),$$

or

$$H^i(\text{Tot}(C^{*,*})_j^*) = H^{i-s}(C_j^{*,s}).$$

Since $s = d+1-j$, the right hand side is

$$H^{i+j-d-1}(P^* \otimes K^{d+1-j}) = H^{i+j-d-1}(P^*) \otimes K^{d+1-j},$$

which we know by Proposition 4.1 to have dimension

$$\binom{d+1}{j} \cdot \dim_{\Bbbk} H^{i+j-d-1}(\Psi; \Bbbk)$$

as claimed. □

6 Schenzel's Formula

We now know enough to compute $\dim_{\Bbbk}(\Bbbk[\Psi]/\langle\Theta\rangle)_j$ for Ψ a d-dimensional triangulated manifold with boundary. This is a rather powerful fact, and we recover a formula of Schenzel [22, 23], which allows one to generalize the characterization of face numbers to manifolds and Buchsbaum complexes.

As in the previous section, Γ may be the boundary subcomplex or $\emptyset$, independently of whether Δ has a nonempty boundary. The double complex $C^{*,*}$ is $P^*(\Psi) \otimes K^*(\Theta)$.

Recall that if $\Psi = (\Delta, \Gamma)$ is Cohen-Macaulay, $\dim_{\Bbbk}(\Bbbk[\Psi]/\langle\Theta\rangle)_j = h_j(\Psi)$ is an entry of the h-vector, and hence determined by the f-vector. For general Ψ this is not the case. To perform the computation we use the map $\mathfrak{R}^{-1} : C^{-1,*} \to \text{Tot}(C^{*\geq 0,*})^*$ as in the proof of Reisner's theorem, obtaining once again the long exact sequence

$$\cdots \to H^i(C^{-1,*}) \to H^i(\text{Tot}(C^{*\geq 0,*})^*) \to H^i(\text{Tot}(C^{*,*})^*) \to H^{i+1}(C^{-1,*}) \to$$
$$\cdots \to H^{d-1}(\text{Tot}(C^{*,*})^*) \to H^d(C^{-1,*}) \to H^d(\text{Tot}(C^{*\geq 0,*})^*) \to \cdots$$

We know $H^i(\text{Tot}(C^{*\geq 0,*})^*) = 0$ for each $i \leq d$, since columns with nonnegative indices are exact until the $(d+1)$-th place. This gives isomorphisms

$$H^{i-1}(\text{Tot}(C^{*,*})^*)_j \simeq H^i(C^{-1,*})_j$$

for each $i \leq d$, so

$$\dim_{\Bbbk} H^i(C^{-1,*})_j = \binom{d+1}{j} \cdot \dim_{\Bbbk} H^{i+j-d-2}(\Psi; \Bbbk)$$

for such i. Now, it is a general fact that

$$\sum_{i=0}^{d+1}(-1)^{d+1-i} \dim_{\Bbbk} H^i(C^{-1,*})_j = \sum_{i=0}^{d+1}(-1)^{d+1-i} \dim_{\Bbbk}(C^{-1,i})_j, \qquad (*)$$

these being two expressions for the Euler characteristic of the complex, multiplied by $(-1)^{d+1}$. The left hand side of $(*)$ is

$$\dim_\Bbbk H^{d+1}(C^{-1,*})_j + \sum_{i=0}^{d}(-1)^{d+1-i}\dim_\Bbbk H^i(C^{-1,*})_j$$

$$= \dim_\Bbbk(\Bbbk[\Delta,\Gamma]/\langle\Theta\rangle)_j + \binom{d+1}{j}\sum_{i=0}^{d}(-1)^{d+1-i}\dim_\Bbbk H^{i+j-d-2}(\Delta,\Gamma;\Bbbk)$$

$$= \dim_\Bbbk(\Bbbk[\Delta,\Gamma]/\langle\Theta\rangle)_j + \binom{d+1}{j}\sum_{i=0}^{j-2}(-1)^{i+j+1}\dim_\Bbbk H^i(\Delta,\Gamma;\Bbbk)$$

by our homological computations above. Note that in the Cohen-Macaulay case this is just $\dim_\Bbbk(\Bbbk[\Psi]/\langle\Theta\rangle)_j$.

To compute the right hand side of $(*)$, note each $C^{-1,i} = \Bbbk[\Psi] \otimes K^i(\Theta)$ is fully known. In fact the alternating sum yields $h_j(\Psi)$: this is easiest to compute using the fact that the Hilbert series of $C^{-1,i}$ is

$$\binom{d+1}{i}x^{d+1-i}H_{\Bbbk[\Psi]}(x).$$

Substituting, we find

$$\dim_\Bbbk(\Bbbk[\Psi]/\langle\Theta\rangle)_j = h_j(\Psi) + \binom{d+1}{j}\sum_{i=0}^{j-2}(-1)^{i+j}\cdot\dim_\Bbbk H^i(\Psi;\Bbbk),$$

a formula originally due to Schenzel.

Example 6.1 The unique minimal triangulation Δ of the torus has 7 vertices, 21 edges, and 14 triangles. Thus the Hilbert series of its face ring is

$$\frac{(1-x)^3 + 7x(1-x)^2 + 21x^2(1-x) + 14x^3}{(1-x)^3} = \frac{1+4x+10x^2-x^3}{(1-x)^3},$$

and the h-vector is $(1,4,10,-1)$. Recall the dimensions of the cohomology groups are $0,2,1$ in dimensions $0,1,2$ respectively. Schenzel's formula tells us that for Θ a generic linear system of parameters, the dimensions of the graded pieces of $\Bbbk[\Delta]/\langle\Theta\rangle$ are 1, 4, 10 and 1.

7 Poincaré duality

Let Δ be a closed, connected, and orientable triangulated manifold, Θ a linear system of parameters. We shall prove a certain ring associated with Δ is a Poincaré duality algebra: in the special case of Δ a sphere, this algebra is just $\Bbbk[\Delta]/\langle\Theta\rangle$.

The theorem requires a little preparation.

The partition complex

7.1 Poincaré duality algebras in general

Definition A finitely generated graded $\Bbbk$-algebra A is a Poincaré duality algebra of degree n if:

1. $A_i = 0$ unless $0 \leq i \leq n$,

2. $A_n \simeq \Bbbk$,

3. For any $0 \leq i \leq n$ the multiplication map induces a non-degenerate bilinear pairing
$$A_i \times A_{n-i} \to A_n.$$

The last statement means that for any nonzero $x \in A_i$ there exists a $y \in A_{n-i}$ such that $xy \in A_n$ is nonzero. This implies that $A_i \simeq A_{n-i}$ as vector spaces over $\Bbbk$.

The case in which A is generated in degree 1 is particularly nice.

Lemma 7.1 *Let A be a finitely generated graded $\Bbbk$-algebra generated in degree 1. Then the following conditions are equivalent.*

1. *A is a Poincaré duality algebra of degree n.*

2. *A vanishes above degree n, and $\{a \in A \mid ax = 0 \text{ for all } x \in A_1\} = A_n$.*

Proof If A is a Poincaré duality algebra of degree n and $i < n$, let $a \in A_i$. There is a $y \in A_{n-i}$ such that $ay \neq 0$. Since A is generated in degree 1, y is an expression in degree-1 elements, and one of these has a nonzero product with a. Thus for any $x \in A$, if $xA_1 = 0$ then the degree-i part of x is zero for all $i < n$, or in other words $x \in A_n$.

If A vanishes above degree n and $\{a \in A \mid ax = 0 \text{ for all } x \in A_1\} = A_n$, let us show each $x \in A_j$ has some $y \in A_{n-j}$ such that $xy \neq 0$ by descending induction on j. For $j = n-1$, this is just the assumption: if $x \in A_{n-1}$ has $xy = 0$ for all $y \in A_1$ then $x \in A_n$, but then $x \in A_{n-1} \cap A_n = 0$.

Suppose the statement is known for some $j \geq 1$ and let $x \in A_{j-1}$. Then again there is some $y \in A_1$ such that $xy \neq 0$ as before, and $xy \in A_j$. Thus there is a $z \in A_{n-j}$ such that $(xy)z \neq 0$. The product $x(yz)$ is nonzero and in A_n. $\square$

7.2 Poincaré duality for face rings of manifolds

Let Δ be a triangulation of a closed, connected, orientable manifold of dimension d, and let $\Theta = (\theta_1, \ldots, \theta_{d+1})$ be a generic sequence in $\Bbbk[\Delta]_1$.

Observe that $\Bbbk[\Delta]/\langle\Theta\rangle$ vanishes above degree $d+1$, and the degree $d+1$ part has dimension 1, by Schenzel's formula and the fact that $h_{d+1}(\Delta) = (-1)^d \chi(\Delta)$. However, $\Bbbk[\Delta]/\langle\Theta\rangle$ is generally not a Poincaré duality algebra. For instance, above we used Schenzel's formula to show $\dim \Bbbk[\Delta]_1 \neq \dim \Bbbk[\Delta]_2$ for Δ the minimal triangulation of a torus.

For convenience, we denote
$$A := A(\Delta) := \Bbbk[\Delta]/\langle\Theta\rangle,$$

$$A(\mathrm{st}_v\Delta) := \Bbbk[\mathrm{st}_v\Delta]/\langle\Theta\rangle,$$

and similarly $A(\mathrm{st}_v^\circ\Delta) = \Bbbk[\mathrm{st}_v\Delta, \mathrm{lk}_v\Delta]/\langle\Theta\rangle$.

Let us cut straight to the point of the idea, so as not to lose the forest for the trees. Details are provided further in this section. The goal is to introduce a nice quotient $B = B(\Delta)$ of A in which Poincaré duality holds. Its defining properties are the following:

1. B is a graded quotient of A, and $B_{d+1} \simeq A_{d+1} \simeq \Bbbk$.

2. For each $v \in \Delta^{(0)}$ the restriction $A(\Delta) \to A(\mathrm{st}_v\Delta)$ factors through B. In particular restriction maps $B \to A(\mathrm{st}_v\Delta)$ are defined.

3. For each $\alpha \in B$ of degree less than d, there is some $v \in \Delta^{(0)}$ such that the restriction of α to $A(\mathrm{st}_v\Delta)$ is nonzero.

4. For each $v \in \Delta^{(0)}$ the composition $A(\mathrm{st}_v^\circ\Delta) \to A \to B$ is injective.

Once B is constructed and the properties above are shown to hold, the theorem is rather short. We need one last, basic lemma.

Lemma 7.2 (the cone lemma) *For each $v \in \Delta^{(0)}$, there is an isomorphism*

$$A(\mathrm{st}_v\Delta) \to A(\mathrm{st}_v^\circ\Delta)$$

$$\alpha \mapsto x_v\alpha.$$

Proof First consider $\mathrm{lk}_v\Delta$ as a subcomplex of $\mathrm{st}_v\Delta$. Each nonface of the link contains the vertex v, so the non-face ideal of the link is just $\langle x_v \rangle$. Thus $\Bbbk[\mathrm{st}_v\Delta, \mathrm{lk}_v\Delta] = \langle x_v \rangle / I_{\mathrm{st}_v\Delta}$, and the map

$$\Bbbk[\mathrm{st}_v\Delta] \to \Bbbk[\mathrm{st}_v\Delta, \mathrm{lk}_v\Delta]$$

$$\alpha \mapsto x_v \cdot \alpha$$

is an isomorphism, since it is surjective and injective, x_v being a nonzerodivisor in $\Bbbk[\mathrm{st}_v\Delta]$. The lemma then follows since $A(\mathrm{st}_v\Delta)$ and $A(\mathrm{st}_v^\circ\Delta)$ are quotients of these two isomorphic modules by their product with the same ideal $\langle\Theta\rangle$. □

The following goes back to unpublished early work of Hochster [26], with partial results in a series of papers of Novik and Swartz [19].

Theorem 7.3 (Poincaré duality) *The ring $B(\Delta)$ is a Poincaré duality algebra.*

Proof Let $\alpha \in B_i$ for $i < d+1$. Then by Proposition 3 of B there is a $v \in \Delta^{(0)}$ such that the restriction of α to $A(\mathrm{st}_v\Delta)$ is nonzero. Consider the maps

$$B \to A(\mathrm{st}_v\Delta) \to A(\mathrm{st}_v^\circ\Delta) \to B,$$

and denote their composition by f. Observe $f(\alpha) \neq 0$ is nonzero: the restriction of α to $A(\mathrm{st}_v\Delta)$ is nonzero by assumption. The next map, to $A(\mathrm{st}_v^\circ\Delta)$, has no kernel because it is an isomorphism. The last map is injective by Proposition 4.

Since f is a map of $\Bbbk[\Delta]$-modules, it is determined by the image of the generator 1 of B. This image is x_v by definition. Hence f is just the multiplication map by x_v, and $x_v\alpha \neq 0$. Since B is generated in degree 1 (it is a quotient of A, hence of $\Bbbk[\Delta]$), the theorem follows from Lemma 7.1. □

Remark Our notation A, B for the rings in this section follows [1]. However, that paper uses upper indices for the grading (lower indices take a different role there, denoting a dual object). We denote the grading with lower indices, as in the rest of this paper.

7.2.1 The algebra $B(\Delta)$ and its properties In this section we contruct $B(\Delta)$ for Δ a triangulated manifold as above and prove the necessary properties hold. The reader may wish to follow the proof for the case of Δ a sphere at first. In this case, $B(\Delta) = A(\Delta)$ and the only missing piece is Proposition 4, proved below in Proposition 7.5.

Definition The ideal $J = J(\Delta) \subset A(\Delta)$ is 0 in degree $d+1$, and in degrees $i < d+1$ it is

$$H^{-1}(P^*/\langle\Theta\rangle)_i = \ker\left[A(\Delta) \to \bigoplus_{v \in \Delta^{(0)}} A(\mathrm{st}_v \Delta)\right]_i.$$

We define $B(\Delta) = A(\Delta)/J$.

To show this is well defined, we need to prove J is in fact an ideal. Clearly it is a vector subspace in each degree. The product of an element in the kernel of the restriction map to all vertex stars and another element of $A(\Delta)$ is clearly again in the kernel, so all that remains to show is that no such product has a nonzero component in degree $d + 1$. We show something stronger.

Lemma 7.4 *Let $\alpha \in A$ such that the restriction of α to $A(\mathrm{st}_v \Delta)$ vanishes for each $v \in \Delta^{(0)}$, and let $\beta \in A$ be homogeneous of degree at least 1. Then $\alpha\beta = 0$.*

Proof It suffices to prove the lemma for β of degree 1, since any homogeneous element of positive degree is a polynomial expression in such elements. Any β of degree 1 is a linear combination of variables x_v, so it suffices to prove the claim for $\beta = x_v$, where $v \in \Delta$ is an arbitrary vertex. Since we have already established multiplication by x_v is equivalent to the composition:

$$A \xrightarrow{\text{restriction}} A(\mathrm{st}_v \Delta) \xrightarrow{x_v} A(\mathrm{st}_v^\circ \Delta) \to A,$$

the claim follows from the assumption that α vanishes in $A(\mathrm{st}_v \Delta)$. □

We prove Proposition 1 holds for B. At the beginning of 7.2 we sketched a computation that $A_{d+1} \simeq \Bbbk$. Let us carry it out in more detail. Schenzel's formula states

$$\dim_\Bbbk (\Bbbk[\Delta]/\langle\Theta\rangle)_{d+1} = h_{d+1}(\Delta) + \sum_{i=0}^{d-1}(-1)^{i+d+1} \cdot \dim_\Bbbk H^i(\Delta; \Bbbk).$$

The sum on the right is $(-1)^{d+1}(\chi(\Delta) - (-1)^d \dim_\Bbbk H^d(\Delta; \Bbbk))$, where $\dim_\Bbbk H^d(\Delta; \Bbbk) = 1$ since Δ is a closed orientable manifold of dimension d. Together with the fact $h_{d+1}(\Delta) = (-1)^d \chi(\Delta)$ (as follows directly from the definition $\chi(\Delta) = \sum_i (-1)^i f_i(\Delta)$ and our calculation at the end of Section 3.1), this implies the claim. The same holds for B, since $J_{d+1} = 0$ by definition.

Proposition 2 is a consequene of the Noether isomorphism theorems: let $v \in \Delta^{(0)}$. Since B is a quotient of A by an ideal contained in the kernel of the restriction map, the map factors through B. Similarly, Proposition 3 follows directly from the definition: each $\alpha \in B$ of degree at most d has a nonzero image in $\bigoplus_{v \in \Delta^{(0)}} A(\mathrm{st}_v \Delta)$, and in particular in $A(\mathrm{st}_v \Delta)$ for some vertex v, else $\alpha = 0$ since its preimage in A is an element of J.

Finally, we establish Proposition 4 in two steps: first we show $A(\mathrm{st}_v^\circ \Delta)$ injects into A, then we show it also injects into B.

Proposition 7.5 *For Δ a manifold and $v \in \Delta^{(0)}$, the map $A(\mathrm{st}_v^\circ \Delta) \to A = A(\Delta)$ is injective.*

Proof For any simplicial complex Σ and vertex $v \in \Sigma$ there is a short exact sequence
$$0 \to \Bbbk[\mathrm{st}_v^\circ \Sigma] \to \Bbbk[\Sigma] \to \Bbbk[\Sigma - v] \to 0,$$
where $\Sigma - v = \{\tau \in \Delta \mid v \notin \tau\}$ is the anti-star of v. Applying this to $\Sigma = \mathrm{st}_\tau \Delta$ for each $\tau \in \Delta$ (including $\emptyset$) gives the short exact sequence of double complexes
$$0 \to P^*(\mathrm{st}_v^\circ \Delta) \otimes K^*(\Theta) \to P^*(\Delta) \otimes K^*(\Theta) \to P^*(\Delta - v) \otimes K^*(\Theta) \to 0.$$

What this means is that these maps of double complexes commute with the differentials, and restrict to exact sequences
$$0 \to P^i(\mathrm{st}_v^\circ \Delta) \otimes K^j \to P^i(\Delta) \otimes K^j \to P^i(\Delta - v) \otimes K^j \to 0$$
for each i, j. Thus the short exact sequence of double complexes restricts to a short exact sequence of rows, for each row. It also restricts to a short exact sequence of total complexes, and to a short exact sequence of the (-1)-th column. We shall use each of these in turn.

Rows of the double complex sequence For each i we have a short exact sequence of the i-th rows:
$$0 \to P^*(\mathrm{st}_v^\circ \Delta) \otimes K^i(\Theta) \to P^*(\Delta) \otimes K^i(\Theta) \to P^*(\Delta - v) \otimes K^i(\Theta) \to 0,$$
which yields a long exact sequence in cohomology. Let us understand what this sequence really is: it suffices to examine the case $i = 0$, because the i-th row is isomorphic, as a vector space, to the $\binom{d+1}{i}$-th power of the 0-th. Up to a change in grading, this is just a short exact sequence of the partition complexes:
$$0 \to P^*(\mathrm{st}_v^\circ \Delta) \to P^*(\Delta) \to P^*(\Delta - v) \to 0,$$
which is exact in positive degrees. Thus to understand the homology it suffices to restrict it to degree 0.

Now, $P^*(\Delta)_0$ is the Čech complex of Δ covered by its stars of vertices, which is isomorphic to the complex of simplicial cochains. The same holds for $P^*(\Delta - v)_0$, and the map
$$P^*(\Delta)_0 \to P^*(\Delta - v)_0,$$

induced from the restriction map $\Bbbk[\Delta] \to \Bbbk[\Delta - v]$, is precisely the map of Čech complexes induced from the simplicial map corresponding to the inclusion $\Delta - v \to \Delta$. This in particular means the map

$$H^i(P^*(\Delta)_0) \to H^i(P^*(\Delta - v)_0)$$

is the map $H^i(\Delta; \Bbbk) \to H^i(\Delta - v; \Bbbk)$ in simplicial cohomology induced by the inclusion $\Delta - v \to \Delta$. It is then a topological result that this is a surjection for every i: this is not difficult to prove using the Mayer-Vietoris sequence for cohomology, using the decomposition $\Delta = (\Delta - v) \cup \mathrm{st}_v \Delta$ and the fact $H^d(\Delta - v) = 0$. Similarly, it is an injection except at $i = d$.

Denote $C^{i,j}(\Sigma) = P^i(\Sigma) \otimes K^j(\Theta)$ for each complex Σ among $\{\mathrm{st}_v^\circ \Delta, \Delta, \Delta - v\}$. For each row j we obtain an exact sequence:

$$\ldots \xrightarrow{0} H^i(C^{*,j}(\mathrm{st}_v^\circ \Delta)) \to H^i(C^{*,j}(\Delta)) \twoheadrightarrow H^i(C^{*,j}(\Delta - v)) \xrightarrow{0} H^{i+1}(C^{*,j}(\mathrm{st}_v^\circ \Delta)) \to \ldots$$

Total complexes and columns of the double complex sequence With our understanding of the rows in hand, we examine the total complexes. Recall that in the proof of Theorem 5.2 we saw that if $C^{*,*}(\Sigma) = P^*(\Sigma) \otimes K^*(\Theta)$, Σ being any of the three complexes under consideration, then there are isomorphisms

$$H^i(\mathrm{Tot}(C^{*,*})_j^*) \simeq H^{i-s}(C_j^{*,s}).$$

for each i, j and for $s = d + 1 - j$. These isomorphisms commute with the maps $C^{*,*}(\mathrm{st}_v^\circ \Delta) \to C^{*,*}(\Delta)$ and $C^{*,*}(\Delta) \to C^{*,*}(\Delta - v)$ by the naturality of the snake lemma. Thus there is a long exact sequence

$$\ldots \xrightarrow{0} H^i(\mathrm{Tot}(C^{*,*}(\mathrm{st}_v^\circ \Delta))_j^*) \to H^i(\mathrm{Tot}(C^{*,*}(\Delta))_j^*) \twoheadrightarrow H^i(\mathrm{Tot}(C^{*,*}(\Delta - v))_j^*) \xrightarrow{0} \ldots$$

corresponding to the long exact sequence on the homologies of the rows. This implies the maps

$$H^i(\mathrm{Tot}(C^{*,*}(\Delta))_j^*) \to H^i(\mathrm{Tot}(C^{*,*}(\Delta - v))_j^*)$$

are surjective for all i and injective except when $i + j = 2d + 1$, because the same occurs for the corresponding maps on the homologies of the rows.

Finally, in the proof of Schenzel's theorem we had isomorphisms

$$H^{i-1}(\mathrm{Tot}(C^{*,*})^*)_j \simeq H^i(C^{-1,*})_j,$$

again commuting with $C^{*,*}(\mathrm{st}_v^\circ \Delta) \to C^{*,*}(\Delta)$ and with $C^{*,*}(\Delta) \to C^{*,*}(\Delta - v)$. Note this means

$$H^i(C^{-1,*}(\Delta))_j \to H^i(C^{-1,*}(\Delta - v))_j$$

is surjective for all i, j and injective unless $i + j = 2d + 2$. Recalling that

$$H^{d+1}(C^{-1,*}(\Sigma))_j \simeq (\Bbbk[\Sigma]/\langle \Theta \rangle)_j = A(\Sigma)_j,$$

we obtain the exact sequence

$$H^d(C^{-1,*}(\Delta))_j \xrightarrow{f_1} H^d(C^{-1,*}(\Delta - v))_j \xrightarrow{f_2} A(\mathrm{st}_v^\circ \Delta)_j \xrightarrow{f_3} A(\Delta)_j \to A(\Delta - v)_j \to 0,$$

where f_1 is surjective for each j, so by exactness $f_2 = 0$ and f_3 is an injection.
□

The last piece of the proof of Proposition 4 is simply another step similar to those above.

Proposition 7.6 *For Δ a manifold and $v \in \Delta^{(0)}$, the map $A(\mathrm{st}_v^\circ \Delta) \to B = B(\Delta)$ is injective.*

Proof Consider the following commutative diagram, in which the ideals J are those from the definition of B.

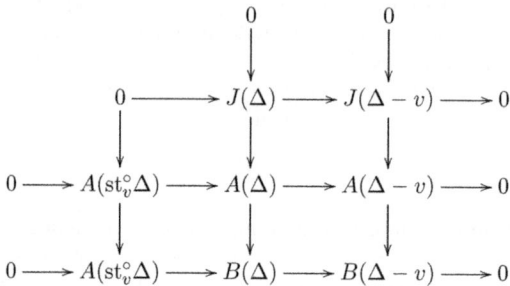

The columns of the diagram are exact by definition, and the middle row
$$0 \to A(\mathrm{st}_v^\circ \Delta)_j \to A(\Delta)_j \to A(\Delta - v)_j \to 0$$
was shown to be exact at the end of the previous proof. Thus if we show the map $J(\Delta) \to J(\Delta - v)$ is an isomorphism, the snake lemma will prove the last row is exact as well, establishing the proposition. During our proof of the partition of unity theorem we proved
$$H^{-1}(P^*/\langle \Theta \rangle) \simeq H^d(\mathrm{Tot}(P^* \otimes K^*)^*)$$
for Buchsbaum complexes. In the previous proof, we established
$$H^d(\mathrm{Tot}(P^*(\Delta) \otimes K^*)^*) \to H^d(\mathrm{Tot}(P^*(\Delta - v) \otimes K^*)^*)$$
is surjective in all degrees and injective except in degree $d+1$. Since J is defined to be precisely $H^{-1}(P^*/\langle \Theta \rangle)$ in all but degree $d+1$ (in which it is 0), $J(\Delta) \simeq J(\Delta - v)$ by the map induced from the restriction, as required. □

7.3 Further remarks on Poincaré duality

First, let us notice that Theorem 7.3 describes a unique way to get a Poincaré duality algebra. Indeed, to any element γ of a polynomial ring A generated in degree one, one can associate a unique quotient algebra with fundamental class $[\gamma]$. So, the face ring of a closed orientable pseudomanifold of dimension $d-1$ has a unique quotient that is a Poincaré algebra with respect to the unique fundamental class in degree d.

However, the phenomenon extends further. As proven in [1], Theorem 7.3 and its proof extend immediately to prove that

Theorem 7.7 *For a triangulated orientable manifold M of dimension $d-1$, we have a perfect pairing*
$$B_k(M) \times B_{d-k}(M, \partial M) \to B_d(M, \partial M)$$

The partition complex   31

8 Applications: Triangulations and a conjecture of Kühnel

We now will spend two sections providing basic, and mostly inductive, combinatorial applications of partition of unity, although more sophisticated applications are left out here, and relegated to [1]. As the ideas employed are adaptations of some of those in earlier parts of the paper, we allow ourselves to use the methods freely at this point. We hope the proof demonstrates the simplicity and versatility of the tools involved. An important property in this context is the Lefschetz property.

Definition A Cohen-Macaulay complex Δ of dimension d has the generic strong Lefschetz property if for a generic sequence $\Theta = (\theta_1, \ldots, \theta_{d+1})$ in $\Bbbk[\Delta]_1$ and a generic $\ell \in (\Bbbk[\Delta]/\langle\Theta\rangle)_1$, the map

$$\ell^{d+1-2j} \cdot : (\Bbbk[\Delta]/\langle\Theta\rangle)_j \to (\Bbbk[\Delta]/\langle\Theta\rangle)_{d+1-j}$$

is injective for each $j \leq \frac{d}{2}$.

The generic "almost Lefschetz property" is the weaker demand that ℓ^{d-2j}· be an injection for each j, under the same conditions.

The Hard Lefschetz theorem is the statement that the strong Lefschetz property holds for some Θ.

For Δ the face ring of a sphere, the Hard Lefschetz theorem was proved in [1], with the caveat that the linear system has to be chosen sufficiently generic. Special cases were known earlier, a main one being the result for face rings of simplicial polytopes: in this case geometric tools are available. See [25, 16].

8.1 The inductive principle and partition of unity

The most basic application of partition of unity is a tool for induction. Indeed, assume that we know the strong Lefschetz property for spheres of dimension $d-1$. Then we can conclude the almost Lefschetz property for closed (not necessarily orientable) manifolds of dimension d. Indeed, this is simple: We have

$$\begin{array}{ccc} B_j(M) & \xrightarrow{\cdot \ell^{d-2j}} & B_{d-j}(M) \\ \downarrow & & \downarrow \\ \bigoplus_{v \in M^{(0)}} A_j(\mathrm{st}_v M) & \xrightarrow{\cdot \ell^{d-2j}} & \bigoplus_{v \in M^{(0)}} A_{d-j}(\mathrm{st}_v M) \end{array}$$

8.2 Kühnel's efficient triangulations of manifolds

Poincaré duality combines the importance of two maps: The map

$$A(\Delta) \twoheadrightarrow \bigoplus_{v \in \Delta^{(0)}} A(\mathrm{st}_v \Delta)$$

elevation of the trivial surjection $A(\Delta) \to A(\mathrm{st}_v \Delta)$, which is described by the partition theorem, and the map $A(\mathrm{st}_v^\circ \Delta) \to A(\Delta)$ whose properties are described by our analysis of Schenzel's work. We can also elevate this map to the sum

$$\bigoplus_{v \in \Delta^{(0)}} A(\mathrm{st}_v^\circ \Delta) \to A(\Delta)$$

and once again, we trivially obtain a surjection in every positive degree.

This is quite powerful: Kühnel [14] famously asked how small a triangulation of a manifold can be chosen. We provide such a bound here, on the number of vertices. For closed, orientable manifolds, this is a result of Murai [17], though our proof is simpler.

Consider now any element η of $A_{d-2j+2}(M)$, for instance, the $(d-2j+2)$-th power of a degree one element. Consider the diagram

$$\begin{array}{ccc} A_j(M) & \xrightarrow{\cdot \eta} & A_{d-j+2}(M) \\ \uparrow & & \uparrow \\ \bigoplus_{v \in M^{(0)}} A_{j-1}(\mathrm{st}_v M) & \xrightarrow{\cdot \eta} & \bigoplus_{v \in M^{(0)}} A_{d-j+1}(\mathrm{st}_v M) \end{array}$$

where the vertical maps are the cone lemmas, given by the composition

$$A_{j-1}(\mathrm{st}_v M) \simeq A_j(\mathrm{st}_v^\circ M) \longrightarrow A_j(M)$$

where the last map is the inclusion of ideals. Now, following the Lefschetz property, the bottom map is a surjection. Hence, the top horizontal map is a surjection.

By the partition of unity theorem, the kernel of the top map is of dimension at least $\binom{d+1}{j} \dim_{\Bbbk} H^{j-1}(M; \Bbbk)$, and the image is of dimension at least

$$\binom{d+1}{j-1} \dim_{\Bbbk} H^{d-j+1}(M; \Bbbk).$$

It follows that $A_j(M)$ is of dimension at least

$$\binom{d+1}{j} \dim_{\Bbbk} H^{j-1}(M; \Bbbk) + \binom{d+1}{j-1} \dim_{\Bbbk} H^{d-j+1}(M; \Bbbk).$$

In particular, following 4.2, we have

$$\binom{d+1}{j} b_{j-1}(M) + \binom{d+1}{j-1} b_{d-j+1}(M) \leq \binom{n-d+j}{j} \quad \text{for } 1 \leq j \leq \frac{d+1}{2}.$$

9 Applications: Subdivisions and the almost-Lefschetz property

In this section we prove a general subdivision theorem, showing that under relatively mild conditions a subdivision of a Cohen-Macaulay complex has the "almost-Lefschetz" property.

Theorem 9.1 *Let Σ be a Cohen-Macaulay (finite) ball complex of dimension d (not necessarily simplicial), and let Δ be any subdivision of Σ satisfying the following condition: for each face $\sigma \in \Sigma$ of dimension at least $\frac{d}{2}$, the subdivision induced on $\partial \sigma$ by Δ is an induced subcomplex of Δ. Then Δ has the almost-strong Lefschetz property: for a generic linear system of parameters $\langle \Theta \rangle$ in $\Bbbk[\Delta]$, there is a linear form ω such that*

$$\omega^{d-2i-1} : (\Bbbk[\Delta]/\langle \Theta \rangle)_i \to (\Bbbk[\Delta]/\langle \Theta \rangle)_{d-i-1}$$

is injective.

The partition complex

This generalizes several results, and applies to a wide range of subdivisions, for instance barycentric subdivisions [13] (who proved this only for a very special case of shellable complexes), antiprism subdivisions introduced by Izmestiev and Joswig [10], and interval subdivisions of [4]; and even for these subdivisions, there was no Lefschetz property known in this generality.

Among the more sophisticated connections, it generalizes the main result of [11] (by seeing balanced subdivisions of the sphere as derived subdivisions of a cell complex) and, for ball subdivisions of spheres, it is in turn a a special case of the biased pairing theorem [1, Section 5.4].

The use of the Hard Lefschetz theorem is again inductive, on the links of vertices in the interiors of (subdivided) faces $\sigma \in \Sigma$.

The next several subsections carry out necessary preparations. These are, in order:

1. Definitions of complexes and subdivisions,

2. A partition of unity theorem for disks with induced boundary,

3. Discussion of the Koszul complex and the Lefschetz property for spheres, and

4. Introduction of a modified partition complex.

The proof is easily put together once these pieces are ready. The idea is that for a facet K of Σ, the simplicial subdivision $K' \subset \Delta$ has the almost Lefschetz property: this is a general theorem for simplicial disks with induced boundary. Once it is known, we use the Cohen-Macaulay property to deduce the global statement for Σ.

In other words, we can formulate the key auxiliary result as follows, from which the result follows via inductive principle:

Theorem 9.2 *With notation as in Theorem 9.1, we have*

$$\Bbbk[\Delta]/\langle \Theta \rangle \longrightarrow \bigoplus_v \Bbbk[\mathrm{st}_v \Delta]/\langle \Theta \rangle$$

is injective for all degrees at most $\frac{d}{2}$, and where v ranges over vertices in the interior of facets of Σ.

9.1 Complexes and subdivisions

We use the following definition of a ball complex, which is common in PL topology (we refer to [21] for a basic introduction). By polyhedron we mean a topological space homeomorphic to the realization of a finite simplicial complex.

Definition A ball complex is a finite set Σ of closed disks covering a polyhedron such that if $\sigma_1, \sigma_2 \in \Sigma$ then:

1. $\sigma_1^\circ \cap \sigma_2^\circ = \emptyset$, and

2. Each of $\partial \sigma_1$ and $\sigma_1 \cap \sigma_2$ is a union of disks in Σ.

More general definitions are possible. For instance, one can relax finiteness to local finiteness.

Definition A simplicial subdivision Δ of a complex Σ is a simplicial complex triangulating the polyhedron $\bigcup_{\sigma \in \Sigma} \sigma$ in such a way that each face of Σ is the image of a closed subcomplex.

The subdivision *has induced boundaries* if for each $\sigma \in \Sigma$ the subdivision Δ induces on $\partial \sigma$ is an induced subcomplex of Δ.

9.2 Partition of unity for disks with induced boundary

Let Δ be a triangulated disk of dimension d, and suppose $\partial \Delta$ is an induced subcomplex. Then a strengthened partition of unity theorem holds: fewer direct summands are required in the partition complex. In particular, it suffices to replace P^0 with a sum over stars of interior vertices only. This is useful for us because the links of such vertices have the Lefschetz property, being homology spheres.

We denote $\Delta^\circ = (\Delta, \partial \Delta)$. A *strongly interior face* $\tau \in \Delta$ is one for which $\tau^{(0)} \subset \Delta^\circ$. That is, each vertex of τ is an interior vertex (this is stronger than τ being an interior face, which merely means it is not entirely contained in $\partial \Delta$). The interior partition complex $P^*_{\text{int}} = P^*_{\text{int}}(\Delta)$ is

$$0 \to \Bbbk[\Delta] \to \bigoplus_{\substack{v \in \Delta^{(0)}, \\ v \in \Delta^\circ}} \Bbbk[\mathrm{st}_v \Delta] \to \bigoplus_{\substack{e \in \Delta^{(1)}, \\ e^{(0)} \subset \Delta^\circ}} \Bbbk[\mathrm{st}_e \Delta] \to \ldots \to \bigoplus_{\substack{\tau \in \Delta^{(d)}, \\ \tau^{(0)} \subset \Delta^\circ}} \Bbbk[\mathrm{st}_\tau \Delta] \to 0,$$

with maps induced by those of the corresponding Čech complex.

Theorem 9.3 *Let Δ a triangulated disk of dimension d with $\partial \Delta$ induced and Θ a generic linear system of parameters in $\Bbbk[\Delta]_1$. Then P^*_{int} is exact, and the map*

$$\Bbbk[\Delta]/\langle \Theta \rangle \to \bigoplus_{v \in \Delta^{(0)}, v \in \Delta^\circ} \Bbbk[\mathrm{st}_v \Delta]/\langle \Theta \rangle$$

is injective beneath degree $d + 1$.

Proof Once we prove exactness, the theorem is clear using our previous results: for instance, the proof of partition of unity applies verbatim (and can be simplified: all total complexes are exact because all rows are).

Consider a monomial $x^\alpha \in \Bbbk[\Delta]$ of degree j and denote $\rho = \mathrm{supp}(\alpha)$, noting $\dim \rho = j - 1$. Then the restriction of x^α to $\Bbbk[\mathrm{st}_\tau \Delta]$ does not vanish precisely when $\rho \cup \tau \in \Delta$. Let us consider the set of all such strongly interior faces τ: it is clearly downwards closed, so it forms a subcomplex Σ of Δ. Since it contains the trivial face it is never the void complex. We split the proof into cases.

- If ρ is an interior face, it has at least one interior vertex v: if all vertices are in $\partial \Delta$ then $\rho \subset \partial \Delta$, since $\partial \Delta$ is induced. In this case $\Sigma = \mathrm{st}_v \Sigma$: each $\sigma \in \Sigma$ satisfies $\sigma \cup \rho \in \Delta$, and in particular $\sigma \cup \{v\} \in \Delta$. Thus Σ deformation retracts onto v and is acyclic.

- If ρ is a boundary face, its link contains at least one interior vertex (otherwise $\mathrm{st}_\rho \Delta = \rho * \mathrm{lk}_\rho \Delta$ is entirely contained in the boundary, which has dimension $d-1$, but a triangulated disk is pure). The union of the open stars of interior vertices of $\mathrm{lk}_\rho \Delta$ within $\mathrm{st}_\rho \Delta$ is then the entire $|\Delta|^\circ \cap |\mathrm{st}_\rho \Delta|$. Indeed, if if $x \in |\Delta|^\circ$ is an interior point, it is in the interior of some face $\tau \in \Delta$ which must have an interior vertex (else $|\tau|$ is contained in the boundary by the inducedness assumption). It follows that

$$\{\mathrm{st}_\tau^\circ(\mathrm{lk}_\rho \Delta) \mid \tau = \{v_{i_1}, \ldots, v_{i_k}\} \text{ is a set of interior vertices of } \mathrm{lk}_\rho \Delta\}$$
$$= \{\mathrm{st}_\tau^\circ(\mathrm{lk}_\rho \Delta) \mid \tau \in \Sigma\}$$

covers $|\Delta|^\circ \cap |\mathrm{lk}_\rho \Delta|$. Further, this collection of open stars forms a good cover, since the intersection of open stars of strongly interior faces of $\mathrm{lk}_\rho \Delta$ is again the open star of a strongly interior face of $\mathrm{lk}_\rho \Delta$, and open stars are contractible. Since $|\mathrm{lk}_\rho \Delta|$ is homotopy equivalent to its interior $|\Delta|^\circ \cap |\mathrm{lk}_\rho \Delta|$, the complex Σ is contractible.

□

Remark To see the homotopy equivalence between $|\Delta|^\circ \cap |\mathrm{lk}_\rho \Delta|$ and $|\mathrm{lk}_\rho \Delta|$, notice $|\mathrm{st}_\rho \Delta| - |\rho| \stackrel{\text{homeo.}}{\cong} |\mathrm{lk}_\rho \Delta| \times |\rho| \times (0,1]$. This is in turn homotopy equivalent to both $|\mathrm{lk}_\rho \Delta|$ and $|\mathrm{lk}_\rho \Delta| \times |\rho| \times (0,1)$. The latter can be identified with the open subset $|\mathrm{st}_\rho^\circ \Delta| - |\rho|$ of $|\Delta|$, and is thus a manifold with boundary, so the collaring theorem applies. Its interior is $(|\Delta|^\circ \cap |\mathrm{lk}_\rho \Delta|) \times |\rho| \times (0,1)$, which is homotopy equivalent to $|\Delta|^\circ \cap |\mathrm{lk}_\rho \Delta|$.

9.3 The Koszul complex and the Lefschetz property

Our goal is to establish the following lemma, a strengthening of Theorem 3.3.

Lemma 9.4 *Suppose Δ is a Cohen-Macaulay complex of dimension k and Θ is a generic sequence of $d+1 \geq k+1$ elements in $\Bbbk[\Delta]_1$. Consider $K^*(\Theta)$ with the grading in which $\alpha \cdot e_{i_1} \wedge \ldots \wedge e_{i_t}$ has degree $\deg \alpha$ for any $\alpha \in \Bbbk[\Delta]$. If $\Bbbk[\Delta]$ has the generic Lefschetz property, then $H^i(K^*(\Theta)) = 0$ for all $i \leq k$ and $H^i(K^*)_j = 0$ for all $i > k$ and $j \leq \frac{k}{2}$.*

We will soon be dealing with double complexes in which the columns are not exact up to the last position, but we still want a partition of unity theorem to hold, at least in low degrees. The lemma gives the necessary exactness property. It is simplest to prove with the "naive" grading described in the statement - the corollary contains the statement we will use later.

Proof We return to a remark at the end of Section 3.2. There is a simple relation between $K^*(\theta_1, \ldots, \theta_j)$ and $K^*(\theta_1, \ldots, \theta_{j+1})$: the latter is isomorphic to the mapping cone of the chain map

$$K^*(\theta_1, \ldots, \theta_j) \to K^*(\theta_1, \ldots, \theta_j)$$

induced from multiplication by θ_{j+1}. More explicitly, there is a short exact sequence of chain complexes of the form

$$0 \to K^{*-1}(\theta_1,\ldots,\theta_j) \xrightarrow{\iota} K^*(\theta_1,\ldots,\theta_{j+1}) \xrightarrow{\pi} K^*(\theta_1,\ldots,\theta_j) \to 0$$

where the maps are given by

$$\iota(\alpha \cdot e_{i_1} \wedge \ldots \wedge e_{i_{t-1}}) = \alpha \cdot e_{i_1} \wedge \ldots \wedge e_{i_{t-1}} \wedge e_{j+1}$$

and, on basis elements,

$$\pi(e_{i_1} \wedge \ldots \wedge e_{i_t}) = \begin{cases} 0 & j+1 \text{ is among } i_1,\ldots,i_t \\ e_{i_1} \wedge \ldots \wedge e_{i_t} & \text{otherwise.} \end{cases}$$

It can be verified the connecting homomorphism is θ_{j+1}· up to sign. Denoting $K_j^* = K^*(\theta_1,\ldots,\theta_j)$ and similarly K_{j+1}^*, this means there is a long exact sequence:

$$\ldots \to H^i(K_j^*) \xrightarrow{\pm\theta_{j+1}\cdot} H^i(K_j^*) \to H^{i+1}(K_{j+1}^*) \to H^{i+1}(K_j^*) \xrightarrow{\mp\theta_{j+1}\cdot} H^{i+1}(K_j^*) \to \ldots$$

In particular, for $j = k+1$, $H^i(K_{k+1}^*) = 0$ unless $i = k+1$, in which case the homology is $\Bbbk[\Delta]/\langle\theta_1,\ldots,\theta_{k+1}\rangle$. We see $H^i(K_{k+2}^*) \neq 0$ only if $i = k+1$ or $i = k+2$, in which case it fits into a short exact sequence in which the other nonzero members are either 0, or the kernel and cokernel of multiplication by θ_{k+1}. These are, more explicitly,

$$0 \to H^{k+1}(K_{k+2}^*) \to \ker\left[\Bbbk[\Delta]/\langle\theta_1,\ldots,\theta_{k+1}\rangle \xrightarrow{\theta_{k+2}\cdot} \Bbbk[\Delta]/\langle\theta_1,\ldots,\theta_{k+1}\rangle\right] \to 0$$

and

$$0 \to \mathrm{coker}\left[\Bbbk[\Delta]/\langle\theta_1,\ldots,\theta_{k+1}\rangle \xrightarrow{\theta_{k+2}\cdot} \Bbbk[\Delta]/\langle\theta_1,\ldots,\theta_{k+1}\rangle\right] \to H^{k+1}(K_{k+2}^*) \to 0.$$

For $j > k+1$, by induction, each $H^i(K_{j+1}^*)$ fits into a short exact sequence of $\Bbbk[\Delta]$-modules which vanish in each degree $\leq \frac{k}{2}$, and is 0 unless $i \geq k+1$. □

Corollary 9.5 *Under the conditions of the previous lemma, consider $K^*(\Theta)$ with the grading in which $\deg(\alpha \cdot e_{i_1} \wedge \ldots \wedge e_{i_t}) = \deg(\alpha) + d + 1 - t$. Then $H^i(K^*)_j = 0$ for all $i \leq k+1$ and $j \leq \frac{d}{2}$.*

Proof Consider a cycle z in $H^i(K^*)_j$: it is a sum of elements of the form $\alpha e_{t_1} \wedge \ldots \wedge e_{t_i}$, with $\deg(\alpha) = i + j - d - 1$. We have $\deg(\alpha) > \frac{k}{2}$ by the previous theorem, so

$$i + j > d + 1 + \frac{k}{2}.$$

If $i < k+1$ we already know $H^i(K^*) = 0$. If $i = k+1$, this implies $j > d - \frac{k}{2} \geq \frac{d}{2}$. □

9.4 A modified partition complex

Consider complex Σ as in the theorem. It has a cellular chain complex, which computes its homology. We will work with a similar chain complex, with "augmentation" at the wrong end. For $\sigma \in \Sigma$, let us denote by $\Bbbk[\sigma]$ the face ring of the subdivision induced on σ by Δ. The complex is:

$$\tilde{P}^*(\Sigma) = 0 \to \Bbbk[\Delta] \to \bigoplus_{\tau \in \Sigma^{(d)}} \Bbbk[\tau] \to \bigoplus_{\rho \in \Sigma^{(d-1)}} \Bbbk[\rho] \to \ldots \to \bigoplus_{v \in \Sigma^{(0)}} \Bbbk \to 0.$$

Note that without the $\Bbbk[\Delta]$ at the beginning, the 0-th graded piece computes $H_*(\Sigma; \Bbbk)$. Nevertheless we keep $\tilde{P}^*$ cohomologically graded in order to make preceding arguments easier to use in this setting. The main result on this complex is the following.

Theorem 9.6 *Let Θ be a generic sequence of $d+1$ elements in $\Bbbk[\Delta]_1$. The kernel $H^{-1}(\tilde{P}^*/\langle\Theta\rangle)$ of $\Bbbk[\Delta]/\langle\Theta\rangle \to \bigoplus_{\tau \in \Sigma^{(d)}} \Bbbk[\tau]/\langle\Theta\rangle$ vanishes in each degree $i \leq \frac{d}{2}$.*

The idea is similar to the proof of the partition of unity theorem. We begin by discussing the needed modifications.

9.4.1 The unreduced partition complex
Consider a single monomial x^α, and the summands of $\tilde{P}^*$ in which x^α is nonzero: these are the faces containing $\rho = \mathrm{supp}(\alpha)$, together with $\Bbbk[\Delta] = \tilde{P}^{-1}$. As a poset with the inclusion relation, the collection of faces corresponding to these summands (other than to $\Bbbk[\Delta]$ itself) is homotopy equivalent to $\mathrm{lk}_\rho \Delta$, and has no homology beneath the top dimension $s = d - |\rho|$. Observe that a face of dimension s in $\mathrm{lk}_\rho \Delta$ corresponds to the summand of some $\sigma \in \Sigma^{(s+|\rho|)}$, and that if $\rho \neq \emptyset$, it is the reduced homology of $\mathrm{lk}_\rho \Delta$ which is computed by these summands since ρ itself is contained in some unique minimal face of Σ. Thus $H^i(\tilde{P}^*) = 0$ unless $i = 0$ or $i = d$.

9.4.2 Homology of the total complex
Set $C^{*,*} = \tilde{P}^* \otimes K^*(\Theta)$. We follow the strategy of the partition of unity theorem, and ideas from that proof are used freely here. The grading of $C^{*,*}$ is also from that part of the paper; it is the same grading recalled in Corollary 9.5.

The main differences between what follows and the proof of partition of unity are that we work only with the part of $C^{*,*}$ in degree $\leq \frac{d}{2}$, and that it is possible that $H^0(\tilde{P}^*) \neq 0$.

Two observations are necessary at this point. The first is clear: in low degrees, all columns of the augmented double complex (with an additional top row $\tilde{P}^*/\langle\Theta\rangle$) are exact. The second is the next lemma.

Lemma 9.7 *For each $i \leq \frac{d}{2}$, $H^d(\mathrm{Tot}(C^{*,*})^*)_i = 0$.*

Proof Let $i \leq \frac{d}{2}$. Consider the short exact sequence associated with the mapping cone of $\mathfrak{U}^d : \mathrm{Tot}(C^{*,*\leq d})^* \to C^{*-d,d+1}$: it is

$$\ldots \to H^j(C^{*-d-1,d+1}) \to H^j(\mathrm{Tot}(C^{*,*})^*) \cdots$$
$$\cdots \to H^j(\mathrm{Tot}(C^{*,*\leq d})^*) \xrightarrow{\partial} H^{j+1}(C^{*-d,d+1}) \to \ldots,$$

where the connecting homomorphism ∂ is induced by the map

$$\alpha = \sum_{\substack{r+s=j, \\ s \leq d}} \alpha^{r,s} \mapsto d^v(\alpha^{j-d,d}).$$

Suppose $\alpha \in H^j(\mathrm{Tot}(C^{*,*\leq d})^*)_i$ has $\partial \alpha = 0$. Then α is in the image of

$$H^j(\mathrm{Tot}(C^{*,*})^*)_i$$

in an obvious way: since $d^v(\alpha^{j-d,d}) = 0$, the sum

$$\sum_{\substack{r+s=j, \\ s \leq d}} \alpha^{r,s}$$

is already a cycle of $\mathrm{Tot}(C^{*,*})^*_i$, so it represents an element of $H^j(\mathrm{Tot}(C^{*,*})^*)_i$. However, this α has no summand in the top row, and all columns of $C^{*,*}$ are exact beneath the top row in degree i. From Lemma 2.1 it follows α is zero in $H^j(\mathrm{Tot}(C^{*,*})^*)_i$, hence its image in $H^j(\mathrm{Tot}(C^{*,*\leq d})^*)_i$ is zero also.

We can conclude ∂ is injective in degrees $i \leq \frac{d}{2}$. Take a piece around

$$H^d(\mathrm{Tot}(C^{*,*})^*)_i$$

of the exact sequence from above:

$$\ldots \to H^d(C^{*-d-1,d+1})_i \xrightarrow{f_1} H^d(\mathrm{Tot}(C^{*,*})^*)_i \cdots$$
$$\cdots \xrightarrow{f_2} H^d(\mathrm{Tot}(C^{*,*\leq d})^*)_i \xrightarrow{\partial} H^{d+1}(C^{*-d,d+1})_i \to \ldots$$

Since ∂ is injective, $f_2 = 0$ and f_1 is surjective. Thus $H^d(C^{*-d-1,d+1})_i = H^{-1}(C^{*,d+1})_i = 0$ surjects onto $H^d(\mathrm{Tot}(C^{*,*})^*)_i$. □

Together with these observations, directly following the proof of the partition of unity theorem gives the result.

9.5 Proof of the subdivision theorem

Lemma 9.8 *For each $\sigma \in \Sigma^{(d)}$, $\Bbbk[\sigma]$ has the generic almost-Lefschetz property.*

Proof Pick a generic system of parameters $\Theta = (\theta_1, \ldots, \theta_{d+1})$ for σ. The map

$$\Bbbk[\sigma]/\langle \Theta \rangle \to \bigoplus_{v \in \sigma^{(0)}, v \in \sigma^\circ} \Bbbk[\mathrm{st}_v \sigma]/\langle \Theta \rangle$$

is injective in each degree $\leq d$ by Theorem 9.3 (injectivity up to degree $\frac{d}{2}$ suffices). For each interior vertex v, the cone lemma gives

$$\Bbbk[\mathrm{st}_v \sigma]/\langle \Theta \rangle \simeq \Bbbk[\mathrm{lk}_v \sigma]/\langle \Theta' \rangle,$$

a sphere of dimension $d - 1$ with the generic Lefschetz property, (Θ' is a system of d parameters depending on Θ: see the cone lemma).

Choosing a generic $\ell \in \Bbbk[\sigma]/\langle\Theta\rangle$, the map

$$\ell^{d-2j} : (\Bbbk[\operatorname{st}_v\sigma]/\langle\Theta\rangle)_j \to (\Bbbk[\operatorname{st}_v\sigma]/\langle\Theta\rangle)_{d-j}$$

is therefore injective for each $j \leq \frac{d}{2}$. For such j, the injectivity of the bottom horizontal map in the commutative diagram

$$\begin{array}{ccc} (\Bbbk[\sigma]/\langle\Theta\rangle)_j & \xrightarrow{\ell^{d-2j}\cdot} & (\Bbbk[\sigma]/\langle\Theta\rangle)_{d-2j} \\ \downarrow & & \downarrow \\ \bigoplus_{v\in\sigma^{(0)}, v\in\sigma^{\circ}} (\Bbbk[\operatorname{st}_v\sigma]/\langle\Theta\rangle)_j & \xrightarrow{\ell^{d-2j}\cdot} & \bigoplus_{v\in\sigma^{(0)}, v\in\sigma^{\circ}} (\Bbbk[\operatorname{st}_v\sigma]/\langle\Theta\rangle)_{d-2j} \end{array}$$

implies multiplication ℓ^{d-2j} is injective on $(\Bbbk[\sigma]/\langle\Theta\rangle)_j$, as required. □

Proof This is completely analogous to the previous lemma. The map $\Bbbk[\Delta]/\langle\Theta\rangle \to \bigoplus_{\tau \in \Sigma^{(d)}} \Bbbk[\tau]/\langle\Theta\rangle$ is injective up to degree $\frac{d}{2}$. Each summand $\Bbbk[\tau]$ has the generic almost Lefschetz property, and for generic $\ell \in \Bbbk[\sigma]/\langle\Theta\rangle$ the bottom horizontal map in the commutative diagram

$$\begin{array}{ccc} (\Bbbk[\Delta]/\langle\Theta\rangle)_j & \xrightarrow{\ell^{d-2j}\cdot} & (\Bbbk[\Delta]/\langle\Theta\rangle)_{d-2j} \\ \downarrow & & \downarrow \\ \bigoplus_{\tau\in\Sigma^{(d)}} (\Bbbk[\tau]/\langle\Theta\rangle)_j & \xrightarrow{\ell^{d-2j}\cdot} & \bigoplus_{\tau\in\Sigma^{(d)}} (\Bbbk[\tau]/\langle\Theta\rangle)_{d-2j} \end{array}$$

yields the result. □

References

[1] Karim Adiprasito, *Combinatorial Lefschetz theorems beyond positivity*, 2018, preprint, arXiv:1812.10454. 1, 2, 3, 4, 20, 27, 30, 31, 33

[2] Karim Adiprasito, June Huh, and Eric Katz, *Hodge theory for combinatorial geometries.*, Ann. Math. (2) **188** (2018), no. 2, 381–452. 1

[3] Karim Adiprasito and Raman Sanyal, *Relative Stanley-Reisner theory and upper bound theorems for Minkowski sums.*, Publ. Math., Inst. Hautes Étud. Sci. **124** (2016), 99–163. 3, 4

[4] Imran Anwar and Shaheen Nazir, *The f- and h-vectors of interval subdivisions.*, J. Comb. Theory, Ser. A **169** (2020), 22 (English), Id/No 105124. 33

[5] Michael F. Atiyah and I. G. Macdonald, *Introduction to commutative algebra.*, student economy edition ed., Boulder: Westview Press, 2016 (English). 9

[6] W. Bruns and J. Herzog, *Cohen-Macaulay rings*, Cambridge Studies in Advanced Mathematics, vol. 39, Cambridge University Press, Cambridge, 1993. 9

[7] David Eisenbud, *Commutative algebra with a view toward algebraic geometry*, Graduate Texts in Mathematics, vol. 150, Springer-Verlag, New York, 1995. 13

[8] A. Hatcher, *Algebraic topology*, Cambridge University Press, Cambridge, 2002. 6, 9

[9] Melvin Hochster, *Cohen-Macaulay rings, combinatorics, and simplicial complexes.*, Ring Theory II, Proc. 2nd Okla. Conf. 1975, 171-223 (1977)., 1977. 1

[10] Ivan Izmestiev and Michael Joswig, *Branched coverings, triangulations, and 3-manifolds.*, Adv. Geom. **3** (2003), no. 2, 191–225. 33

[11] Martina Juhnke-Kubitzke and Satoshi Murai, *Balanced generalized lower bound inequality for simplicial polytopes*, preprint, arXiv:1503.06430 (2015). 33

[12] Kalle Karu, *Hard Lefschetz theorem for nonrational polytopes*, Invent. Math. **157** (2004), no. 2, 419–447. 1

[13] Martina Kubitzke and Eran Nevo, *The lefschetz property for barycentric subdivisions of shellable complexes*, Transactions of the American Mathematical Society **361** (2009), no. 11, 6151–6163. 33

[14] Wolfgang Kühnel, *Tight Polyhedral Submanifolds and Tight Triangulations*, Lecture Notes in Mathematics, vol. 1612, Springer-Verlag, Berlin, 1995. 32

[15] F. S. Macaulay, *Some properties of enumeration in the theory of modular systems.*, Proc. Lond. Math. Soc. (2) **26** (1927), 531–555. 18

[16] Peter McMullen, *On simple polytopes.*, Invent. Math. **113** (1993), no. 2, 419–444. 31

[17] Satoshi Murai, *Tight combinatorial manifolds and graded Betti numbers.*, Collect. Math. **66** (2015), no. 3, 367–386. 32

[18] Satoshi Murai and Eran Nevo, *On the generalized lower bound conjecture for polytopes and spheres.*, Acta Math. **210** (2013), no. 1, 185–202. 1

[19] Isabella Novik and Ed Swartz, *Gorenstein rings through face rings of manifolds.*, Compos. Math. **145** (2009), no. 4, 993–1000. 26

[20] G. A. Reisner, *Cohen-Macaulay quotients of polynomial rings*, Advances in Math. **21** (1976), no. 1, 30–49. 17

[21] Colin Patrick Rourke and Brian Joseph Sanderson, *Introduction to piecewise-linear topology*, Springer Study Edition, Springer-Verlag, Berlin, 1982, Reprint. 33

[22] P. Schenzel, *On the number of faces of simplicial complexes and the purity of Frobenius*, Math. Z. **178** (1981), no. 1, 125–142. 2, 23

[23] Peter Schenzel, *Dualisierende Komplexe in der lokalen Algebra und Buchsbaum-Ringe*, Lecture Notes in Mathematics, vol. 907, Springer-Verlag, Berlin, 1982, With an English summary. 23

[24] D. M. Y. Sommerville, *The relations connecting the angle-sums and volume of a polytope in space of n dimensions.*, Proc. R. Soc. Lond., Ser. A **115** (1927), 103–119. 12

[25] Richard Stanley, *The number of faces of a simplicial convex polytope*, Adv. in Math. **35** (1980), no. 3, 236–238. 31

[26] _____, *Combinatorics and commutative algebra*, second ed., Progress in Mathematics, vol. 41, Birkhäuser Boston Inc., Boston, MA, 1996. 1, 26

Einstein Institute of Mathematics
Hebrew University of Jerusalem
91904 Jerusalem, Israel
adiprasito@math.huji.ac.il, geva.yashfe@mail.huji.ac.il

Hasse-Weil type theorems and relevant classes of polynomial functions

Daniele Bartoli

Abstract

Several types of functions over finite fields have relevant applications in applied areas of mathematics, such as cryptography and coding theory. Among them, planar functions, APN permutations, permutation polynomials, and scattered polynomials have been widely studied in the last few years.

In order to provide both non-existence results and explicit constructions of infinite families, sometimes algebraic varieties over finite fields turn out to be a useful tool. In a typical argument involving algebraic varieties, the key step is estimating the number of their rational points over some finite field. For this reason, Hasse-Weil type theorems (such as Lang-Weil's and Serre's) play a fundamental role.

1 Introduction

"The fundamental problem in number theory is surely how to solve equations in integers. Since this question is still largely inaccessible, we shall content ourselves with the problem of solving polynomial congruences modulo p."

N. Katz [137]

Counting the number of solutions of equations over finite fields has been an intriguing problem in modern mathematics. Gauss himself wanted to estimate the number of solutions (x,y) of the equation $ax^3 - by^3 \equiv_p 1$. In our modern language, he was interested in counting the number of affine $\mathbb{F}_p$-rational points of the curve $aX^3 - bY^3 - 1 = 0$. This concept was later extended to non-prime fields and to any (finite) system of equations over finite fields.

In recent years a number of results concerning different types of polynomials have been obtained using connections with specific curves or varieties over finite fields. Usually, such polynomials are not only relevant for theoretical reasons, but they are also related with interesting questions in applied areas of mathematics, such as coding theory, combinatorics, graph theory, and cryptography.

The first theoretical result connecting algebraic curves over finite fields and relevant combinatorial objects dates back to 1955 when Segre [201, 203] proved his famous result on ovals in projective Galois planes in odd characteristic and showed that an algebraic curve (the algebraic envelope) can be associated to any arc. In subsequent decades, most of the constructions of complete arcs in projective planes (and generalizations of these objects) were based on the following idea of Segre and Lombardo-Radice [202, 158]: "the points of the k-arc should be chosen, with some exceptions, among the points of a conic, or generally from the points of a high-order algebraic curve". The fundamental idea in this approach is to translate the collinearity condition between an external point to the arc and a pair of distinct points of it into the investigation of a specific algebraic curve $\mathcal{C}$ attached to the arc itself. In particular, the key point is to prove the existence of a suitable $\mathbb{F}_q$-rational point of $\mathcal{C}$ by means of the Hasse-Weil theorem. We refer to [214, 118, 116] and the references therein for a more comprehensive introduction to constructions of arcs, their generalizations and their connections with algebraic varieties over finite fields.

In the last few decades a number of problems in coding theory, combinatorics, graph theory, or cryptography have been addressed using a similar approach: the existence (or the non-existence) of a specific relevant object is investigated through an algebraic curve (or, in general, a variety) over a finite field. Usually, one wants to establish whether such a variety possesses a suitable $\mathbb{F}_q$-rational point.

The aim of this survey is to collect results on relevant classes of polynomials which have been obtained via their connection with algebraic varieties over finite fields. In particular, we will focus on permutation polynomials and rational functions, minimal value sets polynomials and rational functions, o-polynomials, scattered polynomials, APN functions, planar functions, and Kloosterman polynomials.

The main application of permutation polynomials is cryptography [147, 148, 153, 154, 155, 178], although they also have useful applications in sonar and radar communications via circular Costas arrays [103], in check digit systems [207, 233], in uniformly distributed sequences connected with Monte-Carlo methods and cryptography [62], and in mutually orthogonal Latin squares and related cryptosystems [145].

Scattered polynomials (see [208]) are useful to construct rank-metric codes, a particular type of error-correcting code where the codewords are taken from the set of matrices over a finite field and which have applications to network coding.

Almost perfect nonlinear functions, planar functions and their generalization are related with differential cryptanalysis [34, 183, 184, 48], relative difference sets [95], error-correcting codes [47, 48]. Also, planar functions are connected with commutative semifields (see e.g. [244]), whereas APN functions are related with dimensional dual hyperovals [85].

In this survey, we also want to provide an overview of the different techniques that can be used to study varieties associated with the families of functions described above. Particular emphasis is given to three different approaches (function fields, branch investigation, covers of curves) in order to describe the wide spectrum of possible tools one can use when dealing with these kinds of problems.

The survey is organized as follows.

Sections 2 and 3 contain preliminaries on Hasse-Weil type theorems and algebraic varieties. Links between algebraic curves and corresponding function fields are also provided.

Section 4 lists all the known (up to our knowledge) techniques and approaches one can use to prove the absolute irreducibility of a variety or even the existence of an absolutely irreducible $\mathbb{F}_q$-rational component. This is usually the most problematic step in the investigation of a variety associated with the functions that are the object of this manuscript.

Section 5 provides a survey about the results on permutation polynomials, exceptional polynomials, permutation rational functions, and complete permutation polynomials obtained using this approach.

Section 6 deals with minimal value set polynomials and minimal value set rational functions, where connections with Galois theory are also discussed.

Kloostermann polynomials and o-polynomials are considered in Sections 7 and 8.

Scattered polynomials can also be investigated through varieties of dimension larger than one; see Section 9.

Finally, most of the results described in Sections 10, 11, and 12, concerning planar functions in odd and even characteristic and APN functions, are mainly obtained thanks to their connections with specific surfaces in three-dimensional spaces.

2 Hasse-Weil type theorems

Quite surprisingly, the problem of counting the number of $\mathbb{F}_q$-rational points of a variety over a finite field is connected with the Riemann hypothesis, one of the most challenging problems in mathematics. The Riemann zeta function $\zeta(s) = \sum_{n=1}^{\infty} n^{-s}$ in the form of an Euler product can be defined into a more general setting, as algebras of finite type over $\mathbb{Z}$; see [204, 205].

Artin [5], in analogy with the Dedekind zeta function for number fields, proposed a definition of the zeta function for an algebraic curve over a finite field as well as a Riemann-type hypothesis for this kind of zeta function. Later, the definition of zeta function was generalized to any non-singular variety over finite fields.

As a notation, $\mathbb{P}^r(\mathbb{F}_q)$ and $\mathbb{A}^r(\mathbb{F}_q)$ denote the projective and the affine space of dimension $r \in \mathbb{N}$ over the finite field $\mathbb{F}_q$, q a prime power. A variety and more specifically a curve, i.e. a variety of dimension 1, are described by a certain set of equations with coefficients in a finite field $\mathbb{F}_q$. We say that a variety $\mathcal{V}$ is *absolutely irreducible* if there are no varieties $\mathcal{V}'$ and $\mathcal{V}''$ defined over the algebraic closure of $\mathbb{F}_q$ and different from $\mathcal{V}$ such that $\mathcal{V} = \mathcal{V}' \cup \mathcal{V}''$. If a variety $\mathcal{V} \subset \mathbb{P}^r(\mathbb{F}_q)$ is defined by $F_i(X_0, \ldots, X_r) = 0$, for $i = 1, \ldots, s$, an $\mathbb{F}_q$-rational point of $\mathcal{V}$ is a point $(x_0 : \ldots : x_r) \in \mathbb{P}^r(\mathbb{F}_q)$ such that $F_i(x_0, \ldots, x_r) = 0$, for $i = 1, \ldots s$. A point is affine if $x_0 \neq 0$. The set of the $\mathbb{F}_q$-rational points of $\mathcal{V}$ is usually denoted by $\mathcal{V}(\mathbb{F}_q)$. We usually denote by the same symbol homogenized polynomials and their dehomogenizations, if the context is clear. For a more comprehensive introduction to algebraic varieties and curves we refer to [117, 108].

Definition 2.1 Let $\mathcal{V} \subset \mathbb{P}^n(\mathbb{F}_q)$ be a non-singular projective variety. Let $\mathcal{V}(\mathbb{F}_{q^m})$ be the set of $\mathbb{F}_{q^m}$-rational points of $\mathcal{V}$. The zeta function $Z_{\mathcal{V}}(T)$ of $\mathcal{V}$ is

$$Z_{\mathcal{V}}(T) = \exp\left(\sum_{m=1}^{\infty} \frac{\#\mathcal{V}(\mathbb{F}_{q^m})}{m} T^m \right).$$

In the case $\mathcal{V}$ is a non-singular curve, properties of such a zeta function were initially investigated by Schmidt and Hasse: the rationality was proved by Schmidt [199], whereas Hasse [109, 110] proved the Riemann hypothesis in the case when $\mathcal{V}$ is an elliptic curve over a finite field.

Theorem 2.2 *[109] For an elliptic curve $\mathcal{E}$ over $\mathbb{F}_q$,*

$$Z_{\mathcal{E}}(T) = \frac{(1 - \alpha T)(1 - \beta T)}{(1 - T)(1 - qT)},$$

where $\alpha, \beta \in \mathbb{C}$ satisfy $|\alpha| = |\beta| = \sqrt{q}$.

Since
$$\frac{TZ'_\mathcal{E}(T)}{Z_\mathcal{E}(T)} = \sum_{m>0} \#\mathcal{V}(\mathbb{F}_{q^m})T^m,$$
$\#\mathcal{E}(\mathbb{F}_{q^m}) = 1 - \alpha^m - \beta^m + q^m$ and therefore
$$\#\mathcal{E}(\mathbb{F}_q) \in [1 + q - 2\sqrt{q}, 1 + q + 2\sqrt{q}].$$

This result was generalized a few years later by Weil [228] to curves of any genus.

Theorem 2.3 *[228] Let $\mathcal{C}$ be a projective absolutely irreducible non-singular curve of (geometric) genus g defined over $\mathbb{F}_q$. Then*
$$Z_\mathcal{C}(T) = \frac{\prod_{i=1}^{2g}(1 - \alpha_i T)}{(1-T)(1-qT)},$$
where $\alpha_i \in \mathbb{C}$ satisfy $|\alpha_i| = \sqrt{q}$.

As a corollary, one gets the famous Hasse-Weil bound.

Theorem 2.4 *[Hasse-Weil bound for curves] Let $\mathcal{C} \subset \mathbb{P}^n(\mathbb{F}_q)$ be a projective absolutely irreducible non-singular curve of genus g defined over $\mathbb{F}_q$. Then*
$$q + 1 - 2g\sqrt{q} \leq \#\mathcal{C}(\mathbb{F}_q) \leq q + 1 + 2g\sqrt{q}. \tag{2.1}$$

If $\mathcal{C}$ is a non-singular *plane* curve, then $g = (d-1)(d-2)/2$, where d is the degree of the curve $\mathcal{C}$, and (2.1) reads
$$q + 1 - (d-1)(d-2)\sqrt{q} \leq \#\mathcal{C}(\mathbb{F}_q) \leq q + 1 + (d-1)(d-2)\sqrt{q}. \tag{2.2}$$

Using some algebraic number theory Serre [206] improved this to the following.

Theorem 2.5 *[Serre's improvement of Hasse-Weil bound] Let $\mathcal{C} \subset \mathbb{P}^n(\mathbb{F}_q)$ be a projective absolutely irreducible non-singular curve of genus g defined over $\mathbb{F}_q$. Then*
$$q + 1 - g\lfloor 2\sqrt{q} \rfloor \leq \#\mathcal{C}(\mathbb{F}_q) \leq q + 1 + g\lfloor 2\sqrt{q} \rfloor. \tag{2.3}$$

If the curve $\mathcal{C}$ is singular, there is some ambiguity in defining what an $\mathbb{F}_q$-rational point of $\mathcal{C}$ actually is. For this reason often the function field version is also used; see Section 3.2 and Theorem 3.3. Clearly, if $\mathcal{C}$ is non-singular, then there is a bijection between $\mathbb{F}_q$-rational places (or branches) of the function field associated with $\mathcal{C}$ and $\mathbb{F}_q$-rational points of $\mathcal{C}$. In the singular case, this is no more true. Consider a singular (absolutely irreducible) curve $\mathcal{C} \subset \mathbb{P}^2(\mathbb{F}_q)$ defined by $F(X,Y,T) = 0$. The difference between the number of $\mathbb{F}_q$-rational points of a non-singular model $\mathcal{C}' \subset \mathbb{P}^r(\mathbb{F}_q)$, for some integer r, of $\mathcal{C}$ and the number of "true" $\mathbb{F}_q$-rational points $(x_0 : y_0 : t_0) \in \mathbb{P}^2(\mathbb{F}_q)$ of $\mathcal{C}$ is at most $(d-1)(d-2)/2 - g$; see [117, Lemma 9.55]. We refer the interested readers to [117, Section 9.6] where other relations are investigated. We point out that actually the bound (2.2) holds even for singular (absolutely irreducible) curves.

Theorem 2.6 (Aubry-Perret bound) *[8, Corollary 2.5] Let $\mathcal{C} \subset \mathbb{P}^n(\mathbb{F}_q)$ be an absolutely irreducible curve which is a complete intersection of $n-1$ hypersurfaces of degrees $d_1, \ldots, d_{n-1}$ and set $d = \prod_{i=1}^{n-1} d_i$. Then*
$$q + 1 - (d-1)(d-2)\sqrt{q} \leq \#\mathcal{C}(\mathbb{F}_q) \leq q + 1 + (d-1)(d-2)\sqrt{q}. \qquad (2.4)$$

Some relevant problems connected with functions over finite fields are related not just to algebraic curves but, more generally, to varieties of higher dimensions, often surfaces. Again, the key point in non-existence or classification results is to prove that such varieties possess suitable $\mathbb{F}_q$-rational points. To this end, generalizations of Hasse-Weil type bounds for algebraic curves are needed.

Weil himself [229] started the investigation of the Riemann zeta function of algebraic varieties over finite fields. His conjectures on the rationality of $Z_V(T)$, on its functional equation, and on the Riemann hypothesis for it yielded a series of fundamental results by Dwork, Artin, Deligne [76, 77, 78, 83, 6].

The first estimate on the number of $\mathbb{F}_q$-rational points of an algebraic variety was given by Lang and Weil [142] in 1954.

Theorem 2.7 *[Lang-Weil Theorem] Let $\mathcal{V} \subset \mathbb{P}^N(\mathbb{F}_q)$ be an absolutely irreducible variety of dimension n and degree d. Then there exists a constant C depending only on N, n, and d such that*
$$\left| \#\mathcal{V}(\mathbb{F}_q) - \sum_{i=0}^{n} q^i \right| \leq (d-1)(d-2)q^{n-1/2} + Cq^{n-1}. \qquad (2.5)$$

Although the constant C was not computed in [142], explicit estimates have been provided for instance in [44, 101, 100, 156, 200, 39] and they have the general shape $C = f(d)$ provided that $q > g(n,d)$, where f and g are polynomials of (usually) small degree. We refer to [44] for a survey on these bounds. Excellent surveys on Hasse-Weil and Lang-Weil type theorems are [99, 100]. See also [185] for more details on zeta functions for varieties over finite fields.

In all the results presented above (Hasse-Weil bound, Lang-Weil bound, Aubry-Perret bound) the variety must necessarily be (at least) absolutely irreducible. In most applications described in this survey, only the existence of a suitable absolutely irreducible $\mathbb{F}_q$-rational component is actually required; see Section 4. It is worth pointing out that results obtained via such theorems are always "asymptotic", i.e. they hold when the size of the field is large with respect to the degree of the associated variety. Thus, in some of the applications, different varieties of lower degree must be investigated in order to apply Hasse-Weil type theorems; see for instance Section 5.3.

3 Preliminaries on algebraic varieties and function fields

As already mentioned in the previous section, one of the main issues when using an approach based on Hasse-Weil type theorems is to prove the absolutely irreducibility of the variety under investigation or at least the existence of a suitable absolutely irreducible $\mathbb{F}_q$-rational component in it. Here we provide an overview on plane curves, function fields, and their connection.

3.1 Plane curves

Let $F(X,Y) \in \mathbb{K}[X,Y]$, $\mathbb{K}$ a field, be a polynomial defining an affine plane curve $\mathcal{C}$, let $P = (u,v) \in \mathbb{A}^2(\mathbb{K})$ be a point in the plane, and write

$$F(X+u, Y+v) = F_0(X,Y) + F_1(X,Y) + F_2(X,Y) + \cdots,$$

where F_i is either zero or homogeneous of degree i. The *multiplicity* of $P \in \mathcal{C}$, written as $m_P(\mathcal{C})$, is the smallest integer m such that $F_m \neq 0$ and $F_i = 0$ for $i < m$; the polynomial F_m is the *tangent cone* of $\mathcal{C}$ at P. A linear divisor of the tangent cone is called a *tangent* of $\mathcal{C}$ at P. The point P is on the curve $\mathcal{C}$ if and only if $m_P(\mathcal{C}) \geq 1$. If P is on $\mathcal{C}$, then P is a *simple* point of $\mathcal{C}$ if $m_P(\mathcal{C}) = 1$, otherwise P is a *singular* point of $\mathcal{C}$. It is possible to define in a similar way the multiplicity of an ideal point of $\mathcal{C}$, that is a point of the curve lying on the line at infinity.

Given two plane curves $\mathcal{A}$ and $\mathcal{B}$ and a point P on the plane, the *intersection number* $I(P, \mathcal{A} \cap \mathcal{B})$ of $\mathcal{A}$ and $\mathcal{B}$ at the point P can be defined by seven axioms. We do not include its precise and long definition here. For more details, we refer to [92] and [117] where the intersection number is defined equivalently in terms of local rings and in terms of resultants, respectively.

Concerning the intersection number, the following two classical results can be found in most of the textbooks on algebraic curves.

Lemma 3.1 *Let $\mathcal{A}$ and $\mathcal{B}$ be two plane curves. For any affine point P, the intersection number satisfies the inequality*

$$I(P, \mathcal{A} \cap \mathcal{B}) \geq m_P(\mathcal{A}) m_P(\mathcal{B}),$$

with equality if and only if the tangents at P to $\mathcal{A}$ are all distinct from the tangents at P to $\mathcal{B}$.

Theorem 3.2 (Bézout's Theorem) *Let $\mathcal{A}$ and $\mathcal{B}$ be two projective plane curves over an algebraically closed field $\mathbb{K}$, having no component in common. Let A and B be the polynomials associated with $\mathcal{A}$ and $\mathcal{B}$ respectively. Then*

$$\sum_P I(P, \mathcal{A} \cap \mathcal{B}) = \deg A \cdot \deg B,$$

where the sum runs over all points in the projective plane $\mathbb{P}^2(\mathbb{K})$.

3.2 Link with function field theory

Consider a projective, absolutely irreducible, algebraic curve $\mathcal{C}$ over the algebraically closed field $\mathbb{K} = \overline{\mathbb{F}_q}$, embedded in a projective space $\mathbb{P}^r(\mathbb{K})$ with homogeneous coordinates $(X_0 : \ldots : X_r)$ and not contained in the hyperplane at infinity $H_\infty : X_0 = 0$. Let $\mathcal{I}(\mathcal{C})$ be the ideal of $\mathcal{C}$. Denote by $\mathbb{K}(\mathcal{C})$ the function field of $\mathcal{C}$, i.e. the field of $\mathbb{K}$-rational functions on $\mathcal{C}$. Clearly, $\mathbb{K}(\mathcal{C})$ is generated over $\mathbb{K}$ by the coordinate functions $x_1, \ldots, x_r$ with $x_i = \frac{X_i + \mathcal{I}(\mathcal{C})}{X_0 + \mathcal{I}(\mathcal{C})}$, and $\mathbb{K}(\mathcal{C}) : \mathbb{K}$ is a field extension of transcendence degree 1.

The set of places of $\mathcal{C}$ is denoted by $\mathbb{P}(\mathcal{C})$ and consists of the set of places of its function field $\mathbb{K}(\mathcal{C})$. For every $P \in \mathbb{P}(\mathcal{C})$ and every non-zero $z \in \mathbb{K}(\mathcal{C})$, we denote by

$v_P(z) \in \mathbb{Z}$ the valuation of z at P; see [212, Chapter 1] for its definition (it can be computed via branch representations, as explained below). The place P is a *zero* or a *pole* of z as $v_P(z) > 0$ or $v_P(z) < 0$.

Suppose that $\mathcal{C}$ is defined over $\mathbb{F}_q$, i.e. $\mathcal{I}(\mathcal{C})$ is generated by polynomials over $\mathbb{F}_q$. Then $\mathbb{F}_q(\mathcal{C})$ denotes the $\mathbb{F}_q$-rational function field of $\mathcal{C}$, i.e. the field of $\mathbb{F}_q$-rational functions on $\mathcal{C}$. The $\mathbb{F}_q$-rational places of $\mathcal{C}$ are those places $P \in \mathbb{P}(\mathcal{C})$ which are defined over $\mathbb{F}_q$; i.e., $\mathbb{F}_q$-rational places of $\mathcal{C}$ are the places of degree 1 in $\mathbb{F}_q(\mathcal{C})$, which are exactly the restriction to $\mathbb{F}_q(\mathcal{C})$ of the places of $\mathbb{K}(\mathcal{C})$ in the constant field extension $\mathbb{K}(\mathcal{C}) : \mathbb{F}_q(\mathcal{C})$. The center of an $\mathbb{F}_q$-rational place is an $\mathbb{F}_q$-rational point of $\mathcal{C}$; conversely, if P is a simple $\mathbb{F}_q$-rational point of $\mathcal{C}$, then the only place centered at P is $\mathbb{F}_q$-rational, and may be identified with P. On the other hand, if P is an m-fold singular point of $\mathcal{C}$ then there exist $n \in \{1, \ldots, m\}$ places $P_1, \ldots, P_n \in \mathbb{P}(\mathcal{C})$ centered at P and in principle all of them may not be $\mathbb{F}_q$-rational even if P is $\mathbb{F}_q$-rational.

A useful tool when investigating algebraic plane curves is provided by branches. Consider a curve $\mathcal{C} \subset \mathbb{P}^2(\mathbb{F}_q)$ defined by $F(X, Y) = 0$ and let $F^*(X, Y, T) = 0$ be its homogenization. Let $\overline{\mathbb{F}_q}((t))$, denote the field of rational functions of the formal power series in the indeterminate t. For a place in the function field $\overline{\mathbb{F}_q}(\mathcal{C})$, we can always associate a primitive branch representation $(x(t), y(t), z(t))$ for some $x(t), y(t), z(t) \in \overline{\mathbb{F}_q}((t))$ such that $F^*(x(t), y(t), z(t)) = 0$; see [117, Chapter 4]. The *center* of the place equals $(x(0), y(0), z(0)) \in \mathbb{P}^2(\overline{\mathbb{F}_q})$. A branch is said $\mathbb{F}_q$-rational if $x(t), y(t), z(t) \in \mathbb{F}_q((t))$. A branch is merely the geometric counterpart of a place which is a field-theoretic concept. For a given place P and a non-zero element $u \in \overline{\mathbb{F}_q}(\mathcal{C})$, plugging the corresponding branch representation of P into u, one gets $u(t) = \sum a_i t^i \in \overline{\mathbb{F}_q}((t))$. Then the valuation $v_P(u) = \min\{i : a_i \neq 0\}$. We refer to [117] for more detailed links between branches and places, and their properties.

The genus of a function field $\mathbb{F}_q(\mathcal{C})$ associated with a curve $\mathcal{C}$ is defined to be the same as the genus of $\mathcal{C}$. We refer to [117, 212] for further details on the link between algebraic curves and function fields.

In the context of function fields, Theorem 2.4 can be restated as follows.

Theorem 3.3 *[Hasse-Weil bound for function fields] Let $\mathcal{C}$ be an absolutely irreducible projective curve of genus g defined over $\mathbb{F}_q$ and denote by $\mathcal{F}$ the function field $\mathbb{K}(\mathcal{C})$. Then its number of $\mathbb{F}_q$-rational places $N_1(\mathcal{F})$ satisfies*

$$q + 1 - 2g\sqrt{q} \leq N_1(\mathcal{F}) \leq q + 1 + 2g\sqrt{q}. \tag{3.1}$$

Remark There is a close relationship between the number of $\mathbb{F}_q$-rational places of $\mathcal{F}$ and the number $R_q(\mathcal{C})$ of points of $\mathcal{C}$ which lie in $\mathbb{P}^2(\mathbb{F}_q)$, when the curve $\mathcal{C}$ is singular. More details can be found in [117, Section 9.6]. For practical applications, these two numbers basically almost coincide, since most of the results obtained via the investigation of a suitable algebraic variety are valid "for q large enough".

3.3 Equivalence of curves

Given two affine plane curves $\mathcal{C}$ and $\mathcal{C}'$, a *rational map* $\phi : \mathcal{C} \to \mathcal{C}'$ is defined by

$$(x, y) \mapsto \left(\frac{f(x, y)}{h(x, y)}, \frac{g(x, y)}{h(x, y)} \right),$$

for some $f, g, h \in \mathbb{F}[X, Y]$, $\mathbb{F}$ any field, satisfying that h does not vanish on $\mathcal{C}$. We say that $\mathcal{C}$ and $\mathcal{C}'$ are *birationally equivalent* if there are rational maps $\phi_1 : \mathcal{C} \to \mathcal{C}'$ and $\phi_2 : \mathcal{C}' \to \mathcal{C}$ such that $\phi_1 \circ \phi_2$ and $\phi_2 \circ \phi_1$ are the identity maps on $\mathcal{C}'$ and $\mathcal{C}$; note that $\mathcal{C}$ and $\mathcal{C}'$ are birationally equivalent if and only if their function fields are $\mathbb{F}$-isomorphic.

If a rational map $\phi : \mathcal{C} \to \mathcal{C}'$ is defined at all points $(x, y) \in \mathcal{C}$, which means at least one of $f(x, y)$, $g(x, y)$ and $h(x, y)$ is not zero, then it is called a *morphism*. If there is another morphism $\psi : \mathcal{C}' \to \mathcal{C}$ such that $\phi \circ \psi$ and $\psi \circ \phi$ are the identity maps on $\mathcal{C}'$ and $\mathcal{C}$, then ϕ is an *isomorphism*, and $\mathcal{C}$ and $\mathcal{C}'$ are said to be isomorphic. If two curves are isomorphic, then they are birationally equivalent, but the converse in not true in general; see [117, Chapter 5].

Isomorphisms preserve the genus and the number of absolutely irreducible components of a curve, but not the degree or the degrees of components. The concepts of birational equivalence and isomorphism can be given for non-plane curves in a similar way. Note that if the two curves $\mathcal{C}$ and $\mathcal{C}'$ are defined over $\mathbb{F}_q$ and the isomorphism $\phi : \mathcal{C} \to \mathcal{C}'$ is defined over $\mathbb{F}_q$, i.e. all the rational functions which define the components of ϕ have coefficients in $\mathbb{F}_q$, then ϕ sends absolutely irreducible $\mathbb{F}_q$-rational components of $\mathcal{C}$ into absolutely irreducible $\mathbb{F}_q$-rational components of $\mathcal{C}'$. In some applications, this can be enough to show that the starting curve $\mathcal{C}$ possesses suitable $\mathbb{F}_q$-rational points.

3.4 Kummer or Artin-Schreier covers of curves

Consider two curves $\mathcal{C}$ and $\mathcal{C}'$ defined over $\mathbb{F}_q$ and a non-constant $\mathbb{F}_q$-rational map $\psi : \mathcal{C}' \to \mathcal{C}$ of degree $d = \deg(\psi) = [\overline{\mathbb{F}_q}(\mathcal{C}') : \overline{\mathbb{F}_q}(\mathcal{C})]$. Such a map is called a *covering* and it induces a map $\mathbb{P}(\mathcal{C}') \to \mathbb{P}(\mathcal{C})$ which sends $\mathbb{F}_q$-rational places of $\mathcal{C}'$ to $\mathbb{F}_q$-rational places of $\mathcal{C}$. For a place $P \in \overline{\mathbb{F}_q}(\mathcal{C})$ denote by $v_P(u)$ the valuation at P of a function $u \in \overline{\mathbb{F}_q}(\mathcal{C})$.

Theorem 3.4 *Let $\mathcal{C} : F(X, Y) = 0$ be an absolutely irreducible plane curve defined over a finite field $\mathbb{F}_q$ of characteristic p.*

1. *[212, Corollary 3.7.4] Let m be a positive integer with $\gcd(m, p) = 1$ and $f(X, Y) \in \mathbb{F}_q[X, Y]$ be such that there exists a place $Q \in \overline{\mathbb{F}_q}(\mathcal{C})$ such that $\gcd(v_Q(f(x, y)), m) = 1$. Let*

$$\mathcal{C}' \subset \mathbb{P}^3(\mathbb{F}_q) : \begin{cases} F(X, Y) = 0 \\ Z^m = f(X, Y) \end{cases}.$$

   Then $\mathcal{C}'$ is an absolutely irreducible curve defined over $\mathbb{F}_q$ and it is called a Kummer cover of $\mathcal{C}$.

2. *[212, Theorem 3.7.10] Let $L(x) \in \mathbb{F}_q[T]$ be a separable $\mathbb{F}_p$-linearized polynomial with all its roots in $\mathbb{F}_q$. Consider a rational function $u(X, Y) \in \mathbb{F}_q(X, Y)$ and suppose that for each place $P \in \overline{\mathbb{F}_q}(\mathcal{C})$ there exists $\omega_P \in \overline{\mathbb{F}_q}(\mathcal{C})$ such that*

$$v_P(u - L(\omega_P)) \geq 0 \quad \text{or} \quad v_P(u - L(\omega_P)) = -m < 0, \ \gcd(m, p) = 1.$$

   Let

$$\mathcal{C}' \subset \mathbb{P}^3(\mathbb{F}_q) : \begin{cases} F(X, Y) = 0 \\ L(Z) = u(X, Y) \end{cases}.$$

Suppose that there exists $Q \in \overline{\mathbb{F}_q}(\mathcal{C})$ such that $v_Q(u - L(\omega_Q)) = -m < 0$, $\gcd(m,p) = 1$, for some $\omega_Q \in \overline{\mathbb{F}_q}(\mathcal{C})$. Then $\mathcal{C}'$ is an absolutely irreducible curve defined over $\mathbb{F}_q$ and it is called an Artin-Schreier cover of $\mathcal{C}$.

Remark The above theorem contains little information about the curves $\mathcal{C}'$. In [212, Corollary 3.7.4] and [212, Theorem 3.7.10] the genus of $\mathcal{C}'$ and the ramification index of places of $\mathcal{C}'$ are computed. Note that when $\mathcal{C} : X = 0$, the curve $\mathcal{C}'$ is a plane curve and Theorem 3.4 above applies as well.

4 Strategies to investigate the algebraic variety

In this section we summarize all the methods and criteria used so far in the literature to prove the absolutely irreducibility of a variety or the existence of a suitable absolutely irreducible $\mathbb{F}_q$-rational component in it. We start with criteria on algebraic curves.

Theorem 4.1 *[11, Lemma 7][4, Lemma 2.1] Let P be a simple $\mathbb{F}_q$-rational point of the curve $\mathcal{C}$. Then there exists an $\mathbb{F}_q$-rational absolutely irreducible component of $\mathcal{C}$ passing through P.*

The criterion above can be generalized to hypersurfaces too; see [225, Proposition 3].

If the $\mathbb{F}_q$-rational point P of $\mathcal{C}$ is singular, we need extra conditions.

Theorem 4.2 *[202], [11, Lemma 8] Let $\mathcal{C}$ be a curve of degree d. Suppose that there exists a point $P \in \mathcal{C}$ such that*

- *there is no linear component of $\mathcal{C}$ through P;*

- *there exists a tangent line t_P at P to $\mathcal{C}$ intersecting $\mathcal{C}$ in P with multiplicity exactly d;*

- *the tangent line t_P is non-repeated among all the tangent lines at P to $\mathcal{C}$.*

Then $\mathcal{C}$ is absolutely irreducible.

If P is a singular $\mathbb{F}_q$-rational point, a possible strategy consists in resolving the singularity by a birational transformation; see for instance [225]. In this direction, a key fact is the following. Suppose that there is a birational transformation ϕ defined over $\mathbb{F}_q$ between two plane curves $\mathcal{C}$ and $\mathcal{D}$ both defined over $\mathbb{F}_q$ such that the plane curve $\mathcal{D}$ has a nonsingular $\mathbb{F}_q$-rational point Q. Then, by Theorem 4.1, there exists an absolutely irreducible $\mathbb{F}_q$-rational component of $\mathcal{D}$ containing Q. Since $\mathcal{C}$ and $\mathcal{D}$ are birationally equivalent, it follows that $\mathcal{C}$ has an absolutely irreducible $\mathbb{F}_q$-rational component through $\phi^{-1}(Q)$.

When dealing with resolution of singularities, branch analysis is a useful tool. Note that local quadratic transformations (see [117, Section 4] and [28, Section 2]) are in particular $\mathbb{F}_q$-birational transformations.

Consider a curve $\mathcal{C}$ defined by

$$F(X,Y) = F_r(X,Y) + F_{r+1}(X,Y) + \cdots = 0,$$

where each $F_i(X,Y)$ is zero or homogeneous in X and Y and of degree i. First, we can suppose that the singular point under examination is the origin $O = (0,0)$ and that $X = 0$ is not a tangent line at O. Let r be its multiplicity. The *geometric transform* of a curve $\mathcal{C}$ is the curve $\mathcal{C}'$ given by $F'(X,Y) = F(X, XY)/X^r$. Note that if $Y = 0$ is not a tangent line at O then we can also consider $\mathcal{C}'$ defined by $F'(X,Y) = F(XY, Y)/Y^r$. By [117, Theorem 4.44], there exists a bijection between the branches of $\mathcal{C}$ centered at the origin and the branches of $\mathcal{C}'$ centered at an affine point on $X = 0$. A finite number of local quadratic transformations can be be performed to determine the total number of branches centered at a point. In particular, if r is coprime with the characteristic and the tangent cone $F_r(X,Y)$ at O splits into non-repeated linear factors (over the algebraic closure) distinct from X then there are precisely r distinct branches centered at O. In fact, distinct linear factors of $F_r(X,Y)$ correspond to distinct affine points of $\mathcal{C}'$ on $X = 0$.

We refer to [117] for a more comprehensive introduction to local quadratic transformations.

Theorem 4.3 *Let P be a singular $\mathbb{F}_q$-rational point of the curve $\mathcal{C}$. Suppose that there exists a tangent line t_P through P which is non-repeated and defined over $\mathbb{F}_q$. Then there exists a component of $\mathcal{C}$ which is absolutely irreducible and defined over $\mathbb{F}_q$.*

Proof Without loss of generality we can suppose that P is the origin and that t_p is the line $Y = 0$. Let r be the multiplicity of P, this means that the polynomial defining the curve $\mathcal{C}$ has the homogeneous part of smallest degree given by $Yg(X,Y)$, with $Y \nmid g(X,Y)$, and $g(X,Y)$ homogeneous and of degree $r-1$. The local quadratic transformation ψ sending $F(X,Y)$ to $F(X, XY)/X^r$ maps $\mathcal{C}$ into an algebraic curve $\mathcal{C}'$ for which the origin is a simple ($\mathbb{F}_q$-rational) point. By Theorem 4.1, $\mathcal{C}'$ has an absolutely irreducible $\mathbb{F}_q$-rational component $\mathcal{D}'$. The corresponding component $\mathcal{D} = \psi^{-1}(\mathcal{D}')$ of $\mathcal{C}$ is absolutely irreducible and defined over $\mathbb{F}_q$. □

The above criterion can be generalized to the following.

Theorem 4.4 *(See also [11, Lemma 7]) Suppose that there exists an $\mathbb{F}_q$-rational branch centered at an $\mathbb{F}_q$-rational point of a curve $\mathcal{C}$. Then there exists a component of $\mathcal{C}$ which is absolutely irreducible and defined over $\mathbb{F}_q$. In particular, if the branch is unique, then it is $\mathbb{F}_q$-rational.*

In some cases, it is possible to deduce the uniqueness of the branch centered at a singular point.

Proposition 4.5 *[224, Lemma 6] Let $P = (\eta, \xi)$ be a point of the curve $\mathcal{C}$ defined by $F(X,Y) = 0$. Suppose that*

$$F(X+\eta, Y+\xi) = aX^u + bY^v + \sum_{i+j > \max\{u,v\}} c_{i,j} X^i Y^j,$$

where $\gcd(u,v) = 1$. Then there exists a unique branch centered at P.

The above criterion has been used in [224] to prove the Carlitz Conjecture in the case $n = 2p$, p prime.

The knowledge of the number and the type of the branches centered at singular points can be useful when investigating specific curves.

Proposition 4.6 *[21, Proposition 2.3] Let $\mathcal{C}$ be a curve of the affine equation*

$$Y^q + \alpha X^q + X^{q^r - q^{r-1} + q - 1} Y + L(X, Y),$$

where all the monomials in $L(X,Y)$ have degree at least $q^{r+1} + q - 1$. Then there is a unique branch centered at the origin.

Proposition 4.7 *[21, Proposition 2.2] Let $\mathcal{C}$ be the curve defined by $F(X,Y) = 0$, where*

$$F(X,Y) = AX^m + BY^n + \sum a_{ij} X^i Y^j, \tag{4.1}$$

with $n < m$, $a_{m0} a_{0n} \neq 0$, and

$$a_{ij} = 0 \quad \text{if} \quad \begin{cases} 0 < i < m; \text{ or} \\ i = 0, j \leq n. \end{cases} \tag{4.2}$$

If $p \nmid \gcd(n,m)$ then $\mathcal{C}$ has $\gcd(n,m)$ branches centered at the origin.

Eisenstein's Irreducibility Criterion for univariate polynomial over $\mathbb{Z}$ can be generalized to bivariate polynomials over any field.

Theorem 4.8 *[212, Proposition 3.1.15] Let $\mathcal{C}$ be a curve defined by*

$$\phi(X,Y) := a_n(X) Y^n + a_{n-1}(X) Y^{n-1} + \cdots + a_1(X) Y + a_0(X) = 0,$$

with a_i in a field $\mathbb{F}$. Assume that there exists $\alpha \in \overline{\mathbb{F}} \cup \{\infty\}$ such that one of the following conditions holds.

1. *$a_n(\alpha) \neq 0$, $a_i(\alpha) = 0$ for $i = 0, \ldots, n-1$, $(a_i/a_0)(\alpha) = 0$, and α is a zero of $a_0(X)$ with multiplicity coprime with n.*

2. *$a_n(\alpha) \neq 0$, $a_i(\alpha) \in \mathbb{F}$ for $i = 1, \ldots, n-1$, and α is a zero of $1/a_0(X)$ with multiplicity coprime with n.*

Then $\mathcal{C}$ is absolutely irreducible.

Remark The above theorem actually holds in the more general context of function fields; see [212, Proposition 3.1.15] for more details.

Curves defined by a polynomial $F(X,Y)$ which is a trinomial in X or in Y have been investigated in [198].

Theorem 4.9 *Let $n \geq 2m$ be two positive integers. Consider $n_1 = n/\gcd(n,m)$, $m_1 = m/\gcd(n,m)$ and suppose $p \nmid nm(n-m)$. Consider $a(X), b(X) \in \mathbb{F}_q(X)^*$ such that $a^{-n} b^{n-m} \notin \mathbb{F}_q$. Then*

$$Y^n + a(X) Y^m + b(X)$$

is reducible over $\overline{\mathbb{F}_q}(X)$ if and only if either

1. $Y^{n_1} + a(X)Y^{m_1} + b(X)$ has a proper linear or quadratic factor over $\overline{\mathbb{F}_q}(X)$; or

2. there exists an integer ℓ such that

$$(n/\ell, m/\ell) \in \bigcup_{p \text{ prime}} \{(2p,p)\} \cup \{(6,1), (6,2), (7,1), (8,2), (8,4),$$
$$(9,3), (10,2), (10,4), (12,2), (12,3), (12,4), (15,5)\},$$

$a = u^{(n-m)/\ell}a'(v)$, $b = u^{n/\ell}b'(v)$ with $u, v \in \overline{\mathbb{F}_q}(X)$, and a', b' are described in [198, Table 1].

Irreducibility criteria can be also deduced by the so-called Newton polytopes associated with curves. For any $S \subset \mathbb{R}^n$ the convex hull of S is defined as

$$conv(S) = \left\{ \sum_{i=1}^k \lambda_i x_i \ : \ x_i \in S, \lambda_i \geq 0, \sum_{i=1}^k \lambda_i = 1 \right\}.$$

The convex hull of finitely many points in $\mathbb{R}^n$ is called a *polytope*. The Minkowski sum of two sets $A, B \in \mathbb{R}^n$ is defined as $A + B = \{a + b : a \in A, b \in B\}$. A point in $\mathbb{R}^n$ is called *integral* if its coordinates are integers. A polytope in $\mathbb{R}^n$ is called integral if all of its vertices are integral. An integral polytope C is called *integrally decomposable* if there exist integral polytopes A and B such that $C = A + B$ where both A and B have at least two points. In this case A and B are called summands of C. Otherwise, C is called *integrally indecomposable*.

Let $\mathbb{F}$ be any field and consider any polynomial

$$f(X_1, X_2, \ldots, X_n) = \sum c_{e_1 e_2 \ldots e_n} X_1^{e_1} X_2^{e_2} \cdots X_n^{e_n} \in \mathbb{F}[X_1, X_2, \ldots, X_n].$$

The *Newton polytope* P_f of f is defined as the convex hull in $\mathbb{R}^n$ of the points

$$\{(e_1, e_2, \ldots, e_n) \ : \ c_{e_1 e_2 \ldots e_n} \neq 0\}.$$

A fundamental result linking Newton polytopes and irreducibility of hypersurfaces is the following.

Lemma 4.10 *[93, Lemma 2.1] Let $f, g, h \in \mathbb{F}[X_1, \ldots, X_n]$ with $f = gh$. Then $P_f = P_g + P_h$.*

As a corollary, any non-zero polynomial $f \in \mathbb{F}[X_1, \ldots, X_n]$ not divisible by any X_i and such that its Newton polytope P_f is integrally indecomposable is absolutely irreducible over $\mathbb{F}$.

Lemma 4.11 *[157, Lemma 2.11], [94, Lemma 13] Let P be an integral polygon having the edge sequence $\{c_i e_i\}_{1 \leq i \leq n}$, where $e_i \in \mathbb{Z}^2$ are primitive vectors. Then, an integral polytope Q is a summand of P if and only if it has the edge sequence of the form $\{d_i e_i\}_{1 \leq i \leq n}$, $0 \leq d_i \leq c_i$, where $\sum_{i=1}^n d_i e_i = (0,0)$.*

4.1 A method based on intersection multiplicities

In this subsection we describe a machinery based on the investigation of the intersection multiplicities at singular points of plane curves. Such a method has been successfully used in [133, 135, 113, 114, 146, 247, 51, 52, 53, 193, 197, 28, 21] to prove that a certain curve contains a suitable absolutely irreducible component defined over the same field.

Consider an algebraic curve $\mathcal{C}$ defined over $\mathbb{F}_q$. Suppose, by way of contradiction, that $\mathcal{C}$ has no absolutely irreducible components over $\mathbb{F}_q$.

1. Find all the singular points of $\mathcal{C}$.

2. Assume that $\mathcal{C}$ splits into two components $\mathcal{A}$ and $\mathcal{B}$ sharing no common irreducible component. Note that $\mathcal{A}$ and $\mathcal{B}$ intersect at singular points of $\mathcal{C}$. An upper bound on the total intersection number of $\mathcal{A}$ and $\mathcal{B}$ is then obtained. At this step, many tools can be used, as for instance branch investigation via quadratic transformations.

3. Under the assumption that $\mathcal{A}$ and $\mathcal{B}$ are not defined over $\mathbb{F}_q$ a lower bound on $\deg \mathcal{A} \cdot \deg \mathcal{B}$ is obtained.

4. Finally, by using Bézout's Theorem (see Theorem 3.2), a contradiction between the two bounds arises.

This method can be summarized in the following lemma.

Theorem 4.12 *[146, Lemma 2] Let $\mathcal{C}$ be a curve of degree n and let $\mathcal{S}$ be the set of its singular points. Also, let $i(P)$ denote the maximum possible intersection multiplicity of two putative components of $\mathcal{C}$ at $P \in \mathcal{C}$. If*

$$\sum_{P \in \mathcal{S}} i(P) < \frac{2n^2}{9},$$

then $\mathcal{C}$ possesses at least one absolutely irreducible component defined over $\mathbb{F}_q$.

Estimates on the intersection numbers can be easily provided in some cases. In particular, if there is a unique branch centered at point P of a curve $\mathcal{C}$ (see for instance Propositions 4.4, 4.5, 4.6, 4.7), then the intersection multiplicity of two putative components of $\mathcal{C}$ at P is 0. Other results in this direction are listed below.

Lemma 4.13 *[133, Proposition 2] Let $F(X,Y) \in \mathbb{F}_q[X,Y]$ and suppose that*

$$F(X,Y) = A(X,Y)B(X,Y).$$

Let $P = (u,v)$ be a point in the affine plane $\mathbb{A}^2(\mathbb{F}_q)$ and write

$$F(X+u, Y+v) = F_m(X,Y) + F_{m+1}(X,Y) + \cdots,$$

where F_i is zero or homogeneous of degree i and $F_m \neq 0$. Let L be a linear polynomial and suppose that $F_m = L^m$ and $L \nmid F_{m+1}$. Then $I(P, \mathcal{A} \cap \mathcal{B}) = 0$, where $\mathcal{A}$ and $\mathcal{B}$ are the curves defined by $A(X,Y)$ and $B(X,Y)$.

The next result was proved in [197, Lemma 4.3] for q even. Actually it still holds when q is odd and its proof is almost the same.

Lemma 4.14 *[197, Lemma 4.3] [28, Lemma 2.5] Let $F(X,Y) \in \mathbb{F}_q[X,Y]$ and suppose that $F(X,Y) = A(X,Y)B(X,Y)$. Let $P = (u,v)$ be a point in the affine plane $\mathbb{A}^2(\mathbb{F}_q)$ and write*

$$F(X+u, Y+v) = F_m(X,Y) + F_{m+1}(X,Y) + \cdots,$$

where F_i is zero or homogeneous of degree i and $F_m \neq 0$. Let L be a linear polynomial and suppose that $F_m = L^m$, $L \mid F_{m+1}$, $L^2 \nmid F_{m+1}$. Then $I(P, \mathcal{A} \cap \mathcal{B}) = 0$ or m, where $\mathcal{A}$ and $\mathcal{B}$ are the curves defined by $A(X,Y)$ and $B(X,Y)$ respectively.

4.2 Connection with algebraic hypersurfaces

We conclude this section with the following result which links varieties of different dimensions.

Lemma 4.15 *[7, Lemma 2.1] Let $\mathcal{H}$ be a projective hypersurface and $\mathcal{X}$ a projective variety of dimension $n-1$ in $\mathbb{P}^n(\mathbb{F}_q)$. If $\mathcal{X} \cap \mathcal{H}$ has a non-repeated absolutely irreducible component defined over $\mathbb{F}_q$ then $\mathcal{X}$ has a non-repeated absolutely irreducible component defined over $\mathbb{F}_q$.*

Usually, in order to prove that a hypersurface $\mathcal{H}$ has a suitable $\mathbb{F}_q$-rational point one investigates the variety $\mathcal{X} \cap \mathcal{H}$ obtained intersecting $\mathcal{H}$ with specific hyperplanes $\mathcal{X}$. Note that the existence of absolutely irreducible $\mathbb{F}_q$-rational components in $\mathcal{X} \cap \mathcal{H}$ does not ensure a priori the existence of suitable $\mathbb{F}_q$-rational points, since points lying on the hyperplane $\mathcal{X}$ often cannot be considered.

5 Permutation polynomials and permutation rational functions

5.1 Basic definitions and connections with algebraic curves

Let $q = p^h$ be a prime power. A polynomial $f(x) \in \mathbb{F}_q[x]$ is a *permutation polynomial* (PP) if it is a bijection of the finite field $\mathbb{F}_q$ into itself. Permutation polynomials were first studied by Hermite and Dickson; see [80, 112].

Permutation polynomials $f(x) \in \mathbb{F}_q[x]$ which are also permutations over infinitely many extensions $\mathbb{F}_{q^m}$ of $\mathbb{F}_q$ are called *exceptional polynomials*.

In principle, to construct a permutation polynomial is straightforward. First note that the set of all PPs of $\mathbb{F}_q$ form a group which is isomorphic to the symmetric group $\mathcal{S}_q$ of order $q!$. One can start from any bijection ψ of the field $\mathbb{F}_q$, that is an element of $\mathcal{S}_q$, and then construct the unique polynomial f of degree at most $q-1$ such that $\psi(x) = f(x)$ for any $x \in \mathbb{F}_q$. Such a polynomial f can be found by the Lagrange Interpolation Formula, that is

$$f(x) = \sum_{a \in \mathbb{F}_q} \psi(a)(1 - (x-a)^{q-1});$$

see [176, Remark 8.1.3]. On the other hand, particular, simple structures or additional extraordinary properties are usually required by applications of PPs in other areas of mathematics and engineering, such as cryptography, coding theory, or combinatorial designs. In this case, permutation polynomials meeting these criteria are usually difficult to find. For a deeper introduction to the connections of PPs with other fields of mathematics we refer to [120, 176, 35] and the references therein.

There are several families of well known PPs, such as monomials, Dickson polynomials, linearized polynomials which satisfy particular constraints. Since the literature on PPs increases dramatically year by year, recently a new way to classify PPs has been proposed by A. Akbary, D. Ghioca, Q. Wang; see [227, 2].

Every polynomial $f(x) \in \mathbb{F}_q[x]$ has the form $ax^r h(x^s) + b$ for some positive integers r, s such that $s \mid (q - 1)$. The authors in [2] pointed out that any non-constant polynomial $f(x) \in \mathbb{F}_q[x]$ of degree at most $q - 1$ can be written uniquely as
$$ax^r h(x^{(q-1)/\ell}) + b,$$
where the integer ℓ is the smallest possible and it is called *index*. Namely, if
$$f(x) = a_1 x^{d=n_1} + a_2 x^{n_2} + \cdots + a_s x^{r=n_s} + b, \qquad (5.1)$$
with $n_1 > n_2 > \cdots > n_s$, $a_i \neq 0$, then ℓ is defined as
$$(q-1)/\gcd(q-1, n_1 - r, n_2 - r, \ldots, n_{s-1} - r). \qquad (5.2)$$

Given a polynomial $f(x) \in \mathbb{F}_q[x]$, let us consider the curve $\mathcal{C}_f$ with affine equation
$$\mathcal{C}_f : \frac{f(X) - f(Y)}{X - Y} = 0. \qquad (5.3)$$
A standard approach to the problem of deciding whether $f(x)$ is a PP is the investigation of the set of $\mathbb{F}_q$-rational points of $\mathcal{C}_f$. In fact, if $a \neq b$ are two distinct elements of $\mathbb{F}_q$ such that $f(a) = f(b)$, then the $\mathbb{F}_q$-rational point (a, b) belongs to $\mathcal{C}_f$ and does not lie on the line $X - Y = 0$. On the other hand, if (a, b) is an $\mathbb{F}_q$-rational point of $\mathcal{C}_f$ not lying on $X - Y = 0$, then $f(a) = f(b)$ and so $f(x)$ is not a PP. Therefore we have proved the following.

Proposition 5.1 *Let $f(x) \in \mathbb{F}_q[x]$ and $\mathcal{C}_f$ be the curve (5.3). Then $f(x)$ is a PP if and only if there are no (affine) $\mathbb{F}_q$-rational points of $\mathcal{C}_f$ off the line $X - Y = 0$.*

Example 5.2 Let us consider the polynomial $f(x) = x^3 + x \in \mathbb{F}_q[x]$, $char(\mathbb{F}_q) \neq 3$. In this case, the affine equation of the curve $\mathcal{C}_f$ reads $X^2 + XY + Y^2 + 1 = 0$, which defines an absolutely irreducible conic. It is well known that $\mathcal{C}_f$ possesses exactly $q + 1$ $\mathbb{F}_q$-rational points in $\mathbb{P}^2(\mathbb{F}_q)$, 4 of which at most lying on the line at infinity or on $X - Y = 0$. This means that there are at least $q - 3$ affine $\mathbb{F}_q$-rational points off the line $X - Y = 0$. If $q > 3$ then such a number is positive and by Proposition 5.1 the polynomial $f(x) = x^3 + x$ is not a PP.

First, the degree of the polynomial $f(x)$ and thus of the curve $\mathcal{C}_f$ must be small enough (roughly smaller than the fourth root of the size of the field) in order to have a non-empty lower bound in Theorem 2.4 on the number of the $\mathbb{F}_q$-rational points of $\mathcal{C}_f$.

Secondly, the existence of an absolutely irreducible component of $\mathcal{C}_f$ defined over $\mathbb{F}_q$ is often sufficient to prove that a certain polynomial $f(x)$ is *not* a PP, as the following proposition shows.

Proposition 5.3 Let $f(x) \in \mathbb{F}_q[x]$, $q \geq 7$, be a PP. Then $\mathcal{C}_f$ does not contain any absolutely irreducible component defined over $\mathbb{F}_q$ of degree d smaller than $\sqrt[4]{q}+2$ and different from $X - Y = 0$.

Proof Suppose that $\mathcal{D}$ is an absolutely irreducible component of $\mathcal{C}_f$ defined over $\mathbb{F}_q$ of degree $d \leq \sqrt[4]{q}+1$. By the Aubry-Perret bound 2.6 the number of its $\mathbb{F}_q$-rational points in $\mathbb{P}^2(\mathbb{F}_q)$ is at least

$$\mathcal{D}(\mathbb{F}_q) \geq q + 1 - \sqrt[4]{q}(\sqrt[4]{q} - 1)\sqrt{q} = q^{3/4} + 1.$$

There are at most $2d \leq 2\sqrt[4]{q}+2$ $\mathbb{F}_q$-rational points of $\mathcal{D}$ lying on the line at infinity or on $X - Y = 0$. Since

$$q^{3/4} + 1 - 2\sqrt[4]{q} + 2 = q^{3/4} - 2\sqrt[4]{q} - 1 > 0$$

whenever $q \geq 7$, the curve $\mathcal{D}$ (and so $\mathcal{C}_f$) possesses an affine $\mathbb{F}_q$-rational point off the line $X - Y = 0$. By Proposition 5.1, $f(x)$ is not a PP, a contradiction. □

Let $h(x) = f(x)/g(x) \in \mathbb{F}_q(x)$, $f(x)$ and $g(x)$ coprime, be a rational function and suppose that $g(x)$ has no roots in $\mathbb{F}_q$. We can associate to $h(x)$, similarly to what we have done for a polynomial $f(x) \in \mathbb{F}_q[x]$, a curve $\mathcal{C}_h$ defined by

$$\mathcal{C}_h : \frac{f(X)g(Y) - f(Y)g(X)}{X - Y} = 0. \tag{5.4}$$

Note that without loss of generality, we can suppose that $\deg(f(x)) < \deg(g(x))$ and in this case the degree of $\mathcal{C}_h$ is $\deg(f(x)) + \deg(g(x)) - 1$.

Proposition 5.4 Let $h(x) = f(x)/g(x) \in \mathbb{F}_q(x)$, with $g(x) \neq 0$ for any $x \in \mathbb{F}_q$, and $\mathcal{C}_h$ be the curve in (5.4). Then $f(x)/g(x)$ is a permutation of $\mathbb{F}_q$ if and only if there are no (affine) $\mathbb{F}_q$-rational points of $\mathcal{C}_h$ off the line $X - Y = 0$.

Similar considerations for permutation polynomials $f(x) \in \mathbb{F}_q[x]$ (Proposition 5.3) can be done for rational functions $f(x)/g(x) \in \mathbb{F}_q(x)$, with $g(x) \neq 0$ for any $x \in \mathbb{F}_q$.

5.2 Exceptional polynomials and Carlitz Conjecture

There exists a clear connection between exceptional polynomials $f(x)$ and curves $\mathcal{C}_f$. In the literature, some authors call *exceptional* those polynomials $f(x)$ for which $\mathcal{C}_f$ possesses no absolutely irreducible component defined over $\mathbb{F}_q$; see for instance [70, Section 8] for the case $q = p$ or [57]. Such a definition was successively generalized by Wan in [224] where he defined "generalized exceptional" polynomials as polynomials $f(x) = g\left(x^{p^\ell}\right)$ for some $g(x) \in \mathbb{F}_q[x]$, $g'(x) \neq 0$, g exceptional over $\mathbb{F}_q$ or of degree 1. This was done to overcome the problem of the multiplicity of the component $X - Y$ in $\mathcal{C}_f$, since in this case

$$\mathcal{C}_f : \frac{f(X) - f(Y)}{X - Y} = \frac{g^{p^\ell}(X) - g^{p^\ell}(Y)}{X - Y} = (X - Y)^{p^\ell - 1} \left(\frac{g(X) - g(Y)}{X - Y}\right)^{p^\ell}.$$

All these definitions can be unified by the following theorem. Actually, such a result was established by a more arithmetical approach based on the investigation of the Galois group attached to the polynomial $f(x) - t$.

Theorem 5.5 *[57] A polynomial $f \in \mathbb{F}_q[x]$ is exceptional over $\mathbb{F}_q$ if and only if every absolutely irreducible component of C_f defined over $\mathbb{F}_q$ is $X - Y = 0$ (up to a scalar factor).*

Theorem 5.6   1. *[107], [224, Theorem 2.2], [70], [40],[216] There exists a sequence $c_1, c_2, \ldots$ of positive integers such that for any finite field $\mathbb{F}_q$ of order $q > c_n$ with $(n, q) = 1$, the following statement holds: if $f(x) \in \mathbb{F}_q[x]$ is a permutation polynomial of $\mathbb{F}_q$ with degree n then f is exceptional over $\mathbb{F}_q$.*

2. *[224, Theorem 2.4] There exists a sequence $c_1, c_2, \ldots$ of positive integers such that for any finite field $\mathbb{F}_q$ of order $q > c_n$ the following statement holds: $f(x) \in \mathbb{F}_q[x]$ is a permutation polynomial of $\mathbb{F}_q$ with degree greater than 2 if and only if $f(x)$ is a generalized exceptional polynomial over $\mathbb{F}_q$.*

This fact together with Proposition 5.3 yield that each permutation polynomial of degree d small with respect to q (roughly $d \leq \sqrt[4]{q}$) is indeed an exceptional polynomial; see also [168, 232, 106, 57].

More precisely the following holds.

Theorem 5.7 *Let $f(x) \in \mathbb{F}_q[x]$ be a permutation polynomial of degree n.*

1. *[231] There exists $q_0 = q_0(n)$ such that if $q > q_0$, q is large compared with n, and $|f(\mathbb{F}_q)| = q + O(1)$ then $f(x)$ is exceptional.*

2. *[98, Theorem 1] If $q \geq n^4$ then $f(x)$ is exceptional.*

3. *[54, Theorem 3.1 and Remark 3.4] If f is separable and*

$$q \geq B_n = \left(\frac{(n-2)(n-3) + \sqrt{(n-2)^2(n-3)^2 + 2(n^2-1)}}{2} \right)^2$$

then f is exceptional.

All the above results are obtained carefully considering the estimates provided by the Hasse-Weil theorem or the Aubry-Perret bound.

More on exceptional polynomials can be found in [176, Chapter 8.4]; see also the references therein.

When in 1897 Dickson [80] classified PPs of degrees less than 7, he found that, except for a few cases, permutation polynomials of a given degree fall into a finite number of well-defined categories. Also, his results showed that if $p \neq 2$, apart from a few exceptions, there are no permutation polynomials of degree 2, 4, and 6. This yielded the following conjecture by Carlitz.

Carlitz Conjecture. Given a positive integer $2n$, there is a constant C_{2n}, such that if $q > C_{2n}$, then there are no permutation polynomials of degree $2n$ over $\mathbb{F}_q$.

Dickson's results show that Carlitz's conjecture is true for $2n = 2, 4, 6$. In 1967 Hayes [107] introduced some geometric ideas in order to address Carlitz's conjecture and he was able to verify it for $2n = 8, 10$ and when the characteristic of the field does not divide $2n$.

In this latter case, the proof relies on a nice observation made in [107, Theorem 3.1], where he proves the existence of a simple $\mathbb{F}_q$-rational point P at infinity under the assumption that $(q-1, 2n) > 1$ and $p \nmid d$. This directly shows that by Lemma 4.1 there exists an absolutely irreducible $\mathbb{F}_q$-rational component through P. As a corollary, if the characteristic is an odd prime p not dividing $2n$, then $(q-1, 2n) \geq 2$ and therefore by the Hasse-Weil theorem or the Aubry-Perret bound, if q is large enough with respect to $2n$ then $f(x)$ is not a permutation nor an exceptional polynomial.

For the cases $2n = 8, 10$, the arguments rely on the investigation of the possible factorizations of $\mathcal{C}_f$ and to this end, an interesting connection between the curve $\mathcal{C}_f$ and its (putative) linear components is given in [107, Theorem 5.2], which was also generalized by Wan in [224, Lemma 3.1].

Theorem 5.8 *[107, Theorem 5.2] Let $f(x) \in \mathbb{F}_q[x]$ of degree d not divisible by the characteristic p. Then $\mathcal{C}_f$ has a line as a component if and only if $f(x) = g((x + a/d)^s)$, where $g(x)$ is a polynomial of degree $d_1 < d$ with $sd_1 = d$.*

Wan [224] proved Carlitz's conjecture for degrees $2n = 12$ and $2n = 14$; see [224, Theorem 3.5].

A few years later, Carlitz's conjecture was proved to be true by Wan [225] for the case $2n = 2r$, where r is an odd prime. Note that if $p \neq r$ then the result was already proved by Hayes [107] and therefore only the case $2n = 2p$ was new in [225]. Up to our knowledge, it was the first time that arguments based on singular points were used to address problems concerning permutation polynomials. In more detail, when dealing with the case $r = p$ the point $(1 : -1 : 0)$ is a singular $\mathbb{F}_q$-rational point of $\mathcal{C}_f$. Using implicitly local quadratic transformations, via Proposition 4.5 Wan successfully established Carlitz's conjecture for the case $2n = 2p$. It is worth mentioning that the same result was obtained independently by Cohen [59] using a different method based on the investigation of the Galois group of indecomposable exceptional polynomials.

Finally, using the classification of finite simple groups and Galois covers of the projective line over $\overline{\mathbb{F}_q}$, Fried, Guralnick and Saxl [91] showed in 1993 that there is no exceptional polynomial of even degree over $\mathbb{F}_q$ when q is odd, solving Carlitz's conjecture.

Later on Carlitz's conjecture was generalized to the so-called Carlitz-Wan conjecture.

Conjecture 5.9 *[226] There are no exceptional polynomials of degree d over $\mathbb{F}_q$ if $\gcd(d, q - 1) > 1$.*

Such a conjecture was proved to be true by Lenstra and by Cohen and Fried [61] using arguments involving Galois groups.

5.3 A way to decrease the degree of the curve $\mathcal{C}_f$

In general it is difficult to prove the existence of suitable absolutely irreducible components in $\mathcal{C}_f$. A collection of possible strategies can be found in Section 4. Anyway, these methods do not provide any information about the degree of a putative absolutely irreducible $\mathbb{F}_q$-rational component of $\mathcal{C}_f$. Thus, Proposition 5.3 can be applied with $d = \deg(\mathcal{C}_f) = \deg(f(x)) - 1$. In many concrete applications such a value d exceeds $\sqrt[4]{q} + 2$ and no direct information on $\mathcal{C}_f$ can be obtained.

On the other hand, a useful criterion due to different authors can be applied; see [2, 186, 246]. As already mentioned previously, any polynomial $f(x)$ of degree smaller than $q-1$, with $f(x) = 0$, can be uniquely written as $x^r h(x^{(q-1)/\ell})$ for some divisor ℓ of $q-1$ and some polynomial $h(x) \in \mathbb{F}_q[x]$ according to (5.1) and (5.2).

Theorem 5.10 *The polynomial $x^r h(x^{(q-1)/\ell})$ is a PP of $\mathbb{F}_q$ if and only if*

1. $\gcd(r, (q-1)/\ell) = 1$; and

2. $x^r h(x)^{(q-1)/\ell}$ permutes the set μ_ℓ of the ℓ-th roots of unity in $\mathbb{F}_q$.

Such a result can be seen as a special case of the so-called "AGW criterion"; see [2].

In some cases the degree of the polynomial $x^r h(x)^{(q-1)/\ell}$ seen as polynomial in μ_ℓ can be dramatically smaller that $\deg(f(x))$. Also, the structure of μ_ℓ sometimes allows useful consideration.

5.4 PPs of index $\ell = q+1$ in $\mathbb{F}_{q^2}$ and generalized Niho exponents

Consider a polynomial $f(x) \in \mathbb{F}_{q^2}[x]$ of index $q+1$, q odd, that is

$$f(x) = x^r \sum_{i=0}^{s} a_i x^{i(q-1)}, \qquad a_s \neq 0. \tag{5.5}$$

Sometimes, polynomials in this form are also called "polynomials from generalized Niho exponents"; see for instance [152]. As a definition, a positive integer d is called a *Niho exponent* with respect to the finite field $\mathbb{F}_{2^{2k}}$ if $d \equiv 2j \pmod{2k-1}$ for some non-negative integer j. In particular, when $j=0$, the integer d is then called a *normalized Niho exponent*. Niho exponents were originally introduced by Niho [182] who investigated the cross-correlation between an m-sequence and its d-decimation.

If $f(x)$ as in (5.5) is a PP then by Theorem 5.10, r must be coprime with $(q^2-1)/\ell = q-1$. Also, $g(x) = x^r \left(\sum_{i=0}^s a_i x^i\right)^{q-1}$ permutes μ_{q+1}, the sets of the $(q+1)$-roots of unity in $\mathbb{F}_{q^2}$. Note that $\sum_{i=0}^s a_i x^s$ cannot vanish in μ_{q+1}; otherwise $g(x)$ cannot be a permutation. Now, for any $x \in \mu_{q+1}$,

$$g(x) = x^r \left(\sum_{i=0}^s a_i x^i\right)^{q-1} = x^r \frac{\sum_{i=0}^s a_i^q x^{iq}}{\sum_{i=0}^s a_i x^i} = x^{r-s} \frac{\sum_{i=0}^s a_{s-i}^q x^i}{\sum_{i=0}^s a_i x^i},$$

since $x^q = 1/x$ for any $x \in \mu_{q+1}$. Since we are assuming that $f(x)$ is a PP in $\mathbb{F}_{q^2}$, the curve (defined over $\mathbb{F}_{q^2}$) $\mathcal{C}_g$ associated with the rational function $g(x)$ does not have affine $\mathbb{F}_{q^2}$-rational points off $X - Y = 0$ whose coordinates belong to μ_{q+1}. So, one could directly investigate $\mathcal{C}_g$. Unfortunately, the existence of an absolutely irreducible $\mathbb{F}_{q^2}$-rational component in $\mathcal{C}_g$ does not ensure (a priori) the existence of points $(\overline{x}, \overline{y}) \in \mathcal{C}_g$, with $\overline{x} \neq \overline{y} \in \mu_{q+1}$.

A way to bypass this problem is to consider the following function $\phi(X)$ from $\mathbb{F}_q$ into $\mu_{q+1} \setminus \{1\}$ defined by

$$\phi(X) = \frac{X+e}{X-e},$$

where e is a fixed element of $\mathbb{F}_{q^2} \setminus \mathbb{F}_q$ such that $e^q = -e$. It is readily seen that ϕ is invertible and
$$\phi^{-1}(X) = e\frac{X+1}{X-1}.$$
Also, $\phi(X)$ and $\phi^{-1}(X)$ can be easily extended to functions $\overline{\phi}(X)$ and $\overline{\phi}^{-1}(X)$ between $\mathbb{F}_q \cup \{\infty\}$ and μ_{q+1} by considering $\phi(\infty) = 1$.

Now, for any $x \in \mathbb{F}_q$,
$$\overline{g}(x) = g(\phi(x)) = \frac{(x+e)^{r-s}\sum_{i=0}^s a_{s-i}^q (x+e)^i (x-e)^{s-i}}{(x-e)^{r-s}\sum_{i=0}^s a_i (x+e)^i (x-e)^{s-i}}.$$

Consider the curve $\mathcal{C}_{\overline{g}}$ of the affine equation $F(X,Y)/(X-Y) = 0$, where $F(X,Y)$ reads as follows:
$$(X+e)^{r-s}(Y-e)^{r-s}\sum_{i=0}^s a_{s-i}^q (X+e)^i(X-e)^{s-i} \sum_{i=0}^s a_i (Y+e)^i (Y-e)^{s-i}$$
$$-(Y+e)^{r-s}(X-e)^{r-s}\sum_{i=0}^s a_{s-i}^q(Y+e)^i(Y-e)^{s-i}\sum_{i=0}^s a_i (X+e)^i(X-e)^{s-i}.$$

Such a curve is clearly defined over $\mathbb{F}_q$. Also, there is a bijection between absolutely irreducible components of $\mathcal{C}_{\overline{g}}$ and $\mathcal{C}_g$, since the two curves are $\mathbb{F}_{q^2}$-birationally equivalent.

Proposition 5.11 *Let $f(x)$, $g(x)$, $\overline{g}(x)$ be defined as above. Let $2r-1 \leq \sqrt[4]{q}+1$ and $q \geq 7$ odd prime power. If $f(x)$ is a PP of $\mathbb{F}_{q^2}$ then $\mathcal{C}_{\overline{g}}$ does not have absolutely irreducible components defined over $\mathbb{F}_q$ other than $X-Y=0$.*

Proof Let $\mathcal{D}$ be a degree-d component of $\mathcal{C}_{\overline{g}}$ absolutely irreducible and defined over $\mathbb{F}_q$, distinct from $X-Y=0$. Then $d \leq 2r-1 \leq \sqrt[4]{q}+1$. By the Aubry-Perret bound 2.6 there are at least
$$|\mathcal{D}(\mathbb{F}_q)| - 2d \geq q + 1 - (d-1)(d-2)\sqrt{q} - 2d$$
affine $\mathbb{F}_q$-rational points of $\mathcal{D}$ off the line $X-Y=0$. Since $q \geq 7$, $|\mathcal{D}(\mathbb{F}_q)| - 2d$ is positive. Let (x_0, y_0) be one of such points. Then $\overline{g}(x_0) = \overline{g}(y_0)$ and $g\left(\frac{x_0+e}{x_0-e}\right) = g\left(\frac{y_0+e}{y_0-e}\right)$. Since $\frac{x_0+e}{x_0-e} \neq \frac{y_0+e}{y_0-e} \in \mu_{q+1}$, $g(x)$ is not a bijection of μ_{q+1} and $f(x)$ is not a permutation of $\mathbb{F}_{q^2}$, a contradiction. □

A similar argument can be used in the case q even.

Proposition 5.12 *Let $f(x) = x^r \sum_{i=0}^s a_i x^{i(q-1)} \in \mathbb{F}_{q^2}[x]$, $a_s \neq 0$ of index $q+1$, $2r-1 \leq \sqrt[4]{q}+1$, $q \geq 8$ even. For any $x \in \mathbb{F}_q$, consider $\psi(x) = \frac{x+e}{x+e+1}$, where $e \in \mathbb{F}_{q^2} \setminus \mathbb{F}_q$ is such that $e^q = e+1$. Let*
$$g(x) = x^{r-s}\frac{\sum_{i=0}^s a_{s-i}^q x^i}{\sum_{i=0}^s a_i x^i},$$
$$\overline{g}(x) = g(\psi(x)) = \frac{(x+e)^{r-s}\sum_{i=0}^s a_{s-i}^q (x+e)^i(x+e+1)^{s-i}}{(x+e+1)^{r-s}\sum_{i=0}^s a_i(x+e)^i(x+e+1)^{s-i}}.$$

If $f(x)$ is a PP of $\mathbb{F}_{q^2}$ then $\mathcal{C}_{\overline{g}}$ does not have absolutely irreducible components defined over $\mathbb{F}_q$ other than $X-Y=0$.

It is worth noting that the degree of the curve $\mathcal{C}_{\bar{g}}$ is in general much smaller than the degree of the curve $\mathcal{C}_f$ associated with $f(x)$ as in (5.5).

Example 5.13 Let $q = 3^k$, k even, and consider $f(x) = x^{\ell q+\ell+5} + x^{(\ell+5)q+\ell} - x^{(\ell-1)q+\ell+6}$, where $\gcd(5+2\ell, q-1) = 1$. It has been conjectured that $f(x)$ is a permutation polynomial over $\mathbb{F}_{q^2}$; see [151, Conjecture 5.1]. Clearly, the curve $\mathcal{C}_f$ has degree roughly equal to ℓq which is too high with respect to q^2 and therefore, even if we can prove the existence of an absolutely irreducible $\mathbb{F}_{q^2}$-rational component of $\mathcal{C}_f$, its degree would be too large to deduce, by the Hasse-Weil theorem or the Aubry-Perret bound, the existence of an $\mathbb{F}_{q^2}$-rational point $(x_0, y_0) \in \mathcal{C}_f$ with $x_0 \neq y_0$. By Theorem 5.10, it is enough to check permutational properties of the rational function $g(x) = \frac{-x^7+x^6+x}{x^6+x-1}$ over μ_{q+1}. Since the degree of $\mathcal{C}_g$ is small and the coefficients do not depend on the choice of the field $\mathbb{F}_q$, it can be factorized into two components defined over $\mathbb{F}_9$. It can be proved that such components do not have points $(x_0, y_0) \in \mu_{q+1}^2$ with $x_0 \neq y_0$; see [17, Theorem 2.3]. On the other hand, one can consider the curve $\mathcal{C}_{\bar{g}}$ which is now defined over $\mathbb{F}_q$ and prove that the two components of this curve are not defined over $\mathbb{F}_q$. In general this second approach is more handleable, since determining points over μ_{q+1} could be difficult in principle.

Results similar to the previous example can be found in [17]; the same results have been obtained in [150] using completely different techniques, determining some quadratic factors of a degree-5 polynomial and a degree-7 polynomial.

It is also worth noting that in this approach useful tools from packages for symbolic computation as MAGMA [43] could be helpful.

Examples of results obtained for polynomials over $\mathbb{F}_{q^2}$ of index $\ell = q+1$ using curves investigation are the following.

Theorem 5.14   1. *[122, Theorem 1.1] Assume that $r > 2$, $q \geq 2^8(r-1)^4$, $a \in \mathbb{F}_{q^2}^*$, and $a^{q+1} \neq 1$. Then $f(x) = x(a + x^{r(q-1)})$ is not a PP of $\mathbb{F}_{q^2}$.*

2. *[122, Theorem 1.2] Let q be odd and let a and e be integers such that $2 < a \leq e/4+1$. Then $f_{a,q}(x) = x^{q-2} + x^{q^2-2} + \cdots + x^{q^{a-1}-2}$ is not a PP of $\mathbb{F}_{q^e}$.*

In the above cases, irreducibility of the curve is obtained either using connections with function fields (Kummer extensions) or clever tricks relying on the particular shape of the polynomials under investigation.

In the last years much of the research on polynomials of index $q+1$ have been focusing on trinomials of the form

$$f_{(r,s_1,s_2),A,B}(x) = x^r(1 + Ax^{s_1(q-1)} + By^{s_2(q-1)}), \quad (5.6)$$

where r, s_1, s_2 are positive integers and $A, B \in \mathbb{F}_{q^2}$. Although such polynomials are the easiest case apart from monomials and binomials, it is very hard to classify for a given choice of r, s_1, s_2 the whole set of pairs (A, B) for which $f_{(r,s_1,s_2),A,B}(x)$ permutes $\mathbb{F}_{q^2}$. The integers s_1 and s_2 can be treated as elements modulo $q+1$ and therefore they can be considered as negative or fractional numbers. The following theorem summarizes what is known so far on these trinomials.

Theorem 5.15 *Consider the polynomial $f_{(1,q,2),A,B}(x) \in \mathbb{F}_{q^2}[x]$.*

1. If $p=2$ then $f_{(1,q,2),A,B}(x)$ is a PP if and only if

   - $B(1+A^{q+1}+B^{q+1})+A^{2q}=0$, $B^{q+1}\neq 1$, and $Tr_{q/2}\left(\frac{B^{q+1}}{A^{q+1}}\right)=0$; or
   - $Tr_{q/2}\left(1+\frac{1}{A^{q+1}}\right)=0$ and $B=A^{q-1}$.

   See [220] for the sufficient part and [12, 123] for the necessary part.

2. If $p=3$ then $f_{(1,q,2),A,B}(x)$ is a PP if and only if
   $$\begin{cases} A^q B^q = A(B^{q+1}-A^{q+1}), \\ 1-\left(\frac{B}{A}\right)^{q+1} \text{ is a square in } \mathbb{F}_q^*. \end{cases}$$

   See [218] for the sufficient part and [129] for the necessary part.

3. If $p>3$ then $f_{(1,q,2),A,B}(x)$ is a PP if and only if

   - $$\begin{cases} A^q B^q = A(B^{q+1}-A^{q+1}), \\ 1-4\left(\frac{B}{A}\right)^{q+1} \text{ is a square in } \mathbb{F}_q^*. \end{cases}$$;

   or

   - $$\begin{cases} A^{q-1}+3B=0, \\ -3\left(1-4\left(\frac{B}{A}\right)^{q+1}\right) \text{ is a square in } \mathbb{F}_q^*. \end{cases}$$

   See [218] for the sufficient part and [23] for the necessary part.

The main ingredient in the proofs of necessary parts [23, 12, 123, 129] is the Hasse-Weil theorem or the Aubry-Perret bound, although approaches in [123, 129] cannot be directly generalized to every characteristic. Note that in this case the curve $\mathcal{C}_{\bar{g}}$ has degree at most 5. By Proposition 5.11 or Proposition 5.12, if $f_{(1,q,2),A,B}(x)$ permutes $\mathbb{F}_{q^2}$ then necessarily $\mathcal{C}_{\bar{g}}$ does not have an absolutely irreducible $\mathbb{F}_q$-rational component apart from $X-Y=0$ and there are only a few possibilities for the decomposition of $\mathcal{C}_{\bar{g}}$.

Using Hermite's criterion, Hou [121] completely classified permutation polynomials $f_{(1,1,2),A,B}(x) \in \mathbb{F}_{q^2}[x]$ in both even and odd characteristic, whereas in [217] proofs for necessary and sufficient conditions on A,B for the case $(s_1,s_2)=(-1/2,1/2)$, q even, are based on Kloosterman sums. It is worth noting that the authors in [217] also provide sufficient conditions on the coefficients A,B for the case $(s_1,s_2)=(3/4,1/4)$, q even.

More recently, Hou [124] proved that conditions in [217] when $(s_1,s_2)=(3/4,1/4)$ are also necessary. The main ingredients here are results concerning equations of degree three and four over fields of even characteristic and interesting constraints on univariate polynomials $f(x)$ to be indecomposable over $\mathbb{F}_q$ based on the Hasse-Weil theorem; see [127, Theorem 1.1].

Note that, when $(s_1,s_2)=(3/4,1/4)$, up to linear equivalence, A can be chosen to be in $\mathbb{F}_q$. Using this simplification, necessary and sufficient conditions in this case read $A=B$; see [124, Theorem 1.1]. As pointed out in [124], the most difficult part is to show that $B \in \mathbb{F}_q^*$.

Recently, for specific values of the integers s_1 and s_2, necessary conditions for $f_{(s_1+s_2,s_1,s_2),A,B}(x)$ to permute $\mathbb{F}_{q^2}$ are provided in [26]. The main achievement is the proof that, under certain conditions, $B \in \mathbb{F}_q^*$. The approach is the one described in Subsection 4.1.

Theorem 5.16 *Let $(s_1, q+1) = 1$, $q = 2^{2s+1}$. Suppose that $f_{A,B,s_1,s_2}(X)$ is a PP of $\mathbb{F}_{q^2}$, $s_1 < s_2$, $A \in \mathbb{F}_q^*$, $B \in \mathbb{F}_{q^2}^*$. Let $s_1 \equiv j \pmod 8$ and $s_2 \equiv i \pmod 8$. Then one of the following conditions is satisfied.*

1. $2(s_1 + s_2) \geq \sqrt[4]{q}$.

2. $B \in \mathbb{F}_q$.

3. $A^2 + B^{q+1} \neq 0$, $B \notin \mathbb{F}_q$, and $s_2 < 5s_1$.

4. $A^2 + B^{q+1} = 0$, $B \notin \mathbb{F}_q$, $(i,j) \in \{(1,5), (3,7), (5,1), (7,3)\}$, and $s_2 < 38s_1$.

5. $A^2 + B^{q+1} = 0$, $B \notin \mathbb{F}_q$, $(i,j) \notin \{(1,5), (3,7), (5,1), (7,3)\}$.

5.5 Carlitz rank of a permutation polynomial

By a result of Carlitz [49] any permutation polynomial over $\mathbb{F}_q$, $q \geq 3$, is a composition of linear polynomials $ax + b$, $a, b \in \mathbb{F}_q$, $a \neq 0$, and x^{q-2}, i.e. any permutation f over $\mathbb{F}_q$ can be represented by a polynomial of the form

$$P_n(x) = \left(\cdots \left((a_0 x + a_1)^{q-2} + a_2 \right)^{q-2} \cdots + a_n \right)^{q-2} + a_{n+1}, \qquad (5.7)$$

for some $n \geq 0$, where $a_i \neq 0$ for $i = 0, 2, \ldots, n$. Also, $f(c) = P_n(c)$ for all $c \in \mathbb{F}_q$ although this representation is not unique and n is not necessary minimal. The smallest integer $n \geq 0$ for which $f(x)$ coincides with a $P_n(x)$ is called the Carlitz rank $Cr(f)$ of $f(x)$; see [3]. As pointed out in [4], (5.7) enables approximation of f by a degree-1 rational function $r_n(x) = (\alpha x + \beta)/(\alpha' x + \beta')$ in a way that $f(c) = r_n(c)$ for all $c \in \mathbb{F}_q \setminus \mathcal{O}_n$, where $\mathcal{O}_n \subset \mathbb{F}_q \cup \{\infty\}$ is called the set of poles of f and it has size at most n.

The knowledge of the Carlitz rank of a given polynomial f can help decrease the degree of the curve associated with f and this has been used in [4] to determine conditions on pairs of polynomials (f, g), $f, g \in \mathbb{F}_q[x]$, such that both f and $f+g$ are permutations. Such a problem can be seen as an extension of the investigation on the existence of CPPs; see Subsection 5.6.

Let f and g be two polynomials. If $Cr(f)$ is n then $f(c) + g(c) = r_n(c) + g(c)$ for any $c \in \mathbb{F}_q \setminus \mathcal{O}_n$. If n is small with respect to q the functions $f(x) + g(x)$ and $r_n(x) + g(x)$ coincide on a large subset of $\mathbb{F}_q$. This enables us to study the curve $\mathcal{C}_{f,g}$ of degree $\deg(g) + 1$ associated with $r_n(x) + g(x) = (\alpha x + \beta)/(\alpha' x + \beta') + g(x)$, that is defined by

$$\mathcal{C}_{f,g} : (\alpha' X + \beta')(\alpha' Y + \beta') \frac{g(X) - g(Y)}{X - Y} + \alpha \beta' - \alpha' \beta = 0.$$

Since $P_\infty = (1 : 0 : 0)$ is a simple $\mathbb{F}_q$-rational point of $\mathcal{C}_{f,g}$, by Theorem 4.2 together with the Hasse-Weil or the Aubry-Perret bounds the authors in [4] prove the following.

Theorem 5.17 *[4, Theorem 2.2] Let f and $f+g$ be permutation polynomials over $\mathbb{F}_q$, where $\deg(g) = k \in \{1, \ldots, q-2\}$. Then*

$$Cr(f) \geq \frac{1}{k+1}\left(q - \frac{k(k-1)}{2}\lfloor 2\sqrt{q}\rfloor - \gcd(k, q-1)\right).$$

5.6 Complete permutation polynomials and generalization

Let $f(x) \in \mathbb{F}_q[x]$ be a permutation polynomial of $\mathbb{F}_q$. The polynomial $f(x)$ is said to be a *complete permutation polynomial* of $\mathbb{F}_q$ if $f(x) + x$ is also a permutation polynomial of $\mathbb{F}_q$; see for instance [181]. CPPs are also related to bent and nega-bent functions which are studied for a number of applications in cryptography, combinatorial designs, and coding theory; see for instance [141, 175, 245, 211].

The connection between CPPs and algebraic curves is straightforward. Since both $f(x) \in \mathbb{F}_q[x]$ and $g(x) = f(x) + x \in \mathbb{F}_q[x]$ must be permutations of $\mathbb{F}_q$, the two curves $\mathcal{C}_f$ and $\mathcal{C}_g$ defined as in (5.3) must satisfy Proposition 5.1. Note that Example 5.2 shows that x^3 is never a CPP if q is larger that 3.

Based on the classification of indecomposable exceptional polynomials and on Theorem 5.5, a first result on the classification of CPPs is the following, which proves a conjecture by Chowla and Zassenhaus [56] in 1968. Clearly, such a result can be described as well as in a more geometrical way using the connection between exceptional polynomials and algebraic curves.

Theorem 5.18 *[58, Theorem 1] If $d \geq 2$ and $p > (d^2 - 3d + 4)^2$, then there is no CPP of degree d over $\mathbb{F}_p$.*

A first generalization of this result has been obtained by Cohen, Mullen and Shiue [60] in 1995.

Theorem 5.19 *[60, Theorem 2] Suppose f and $f+g$ are monic permutation polynomials over $\mathbb{F}_p$ of degree $d \geq 3$, $p > (d^2 - 3d + 4)^2$. Then either $\deg(g) = 0$ or $\deg(g) \geq 3d/5$.*

Connection with the Carlitz Rank (see Subsection 5.5) has been obtained in [131].

Theorem 5.20 *[131, Theorem 1] If $f(x)$ is a CPP of $\mathbb{F}_q$, then either $\mathcal{L}(f) := \max_{a,b \in \mathbb{F}_q} |\{c \in \mathbb{F}_q : f(c) = ac + b\}| \geq \lfloor (q+5)/2 \rfloor$ or $Cr(f) \geq \lfloor q/2 \rfloor$.*

The most studied class of CPPs is the monomial one. If there exists a complete permutation monomial of degree d over $\mathbb{F}_\ell$, then d is called a CPP exponent over $\mathbb{F}_\ell$. Complete permutation monomials have been investigated in a number of recent papers, especially for $\ell = q^n$ and $d = \frac{\ell-1}{q-1} + 1$. Note that for an element $\alpha \in \mathbb{F}_\ell^*$, the monomial αx^d is a CPP of $\mathbb{F}_\ell$ if and only if $\gcd(d, \ell-1) = 1$ and $\alpha x^d + x$ is a PP of $\mathbb{F}_\ell$. In [31, 32, 234, 235] PPs of type $f_b(x) = x^{\frac{q^n-1}{q-1}+1} + bx$ over $\mathbb{F}_{q^n}$ are thoroughly investigated for $n = 2$, $n = 3$, and $n = 4$. For $n = 6$, sufficient conditions for f_b to be a PP of $\mathbb{F}_{q^6}$ are provided in [234, 235] in the special cases of characteristic $p \in \{2, 3, 5\}$, whereas in [20] all a's for which $ax^{\frac{q^6-1}{q-1}+1}$ is a CPP over $\mathbb{F}_{q^6}$ are explicitly listed. The case $p = n + 1$ is dealt within [167]. Finally, in [19] complete permutation monomials of degree $\frac{q^n-1}{q-1} + 1$ over the finite field with q^n elements, for $n + 1$ a prime and $(n + 1)^4 < q$, have been considered.

One can immediately see that in the case $d = \frac{\ell-1}{q-1} + 1$, the curve associated with $f_b(x) = x^{\frac{q^n-1}{q-1}+1} + bx$ has degree larger than $\sqrt[4]{q^n}$ and therefore one cannot apply Proposition 5.3. On the other hand, using Theorem 5.10 one can reduce the degree of the curve, since $f_b(x) = x^{\frac{q^n-1}{q-1}+1} + bx = x\left(x^{\frac{q^n-1}{q-1}} + b\right)$ is a polynomial of index $\ell = q - 1$; see also Niederreiter-Robinson Criterion in [181]. So, f_b permutes $\mathbb{F}_{q^n}$ if and only $\widetilde{f}_b(x) = x(x + b)^{\frac{q^n-1}{q-1}}$ permutes $\mathbb{F}_q$. Now, for $x \in \mathbb{F}_q$,

$$g_b(x) := \widetilde{f}_b(x) = x(x+b)^{\frac{q^n-1}{q-1}} = x(x+b)^{q^{n-1}+q^{n-2}+\cdots+q+1}$$
$$= x \prod_{i=0}^{n-1} \left(x + b^{q^i}\right).$$

The curve $\mathcal{C}_{g_b}$ has now degree n and if n is small enough with respect to q, one can try to apply Proposition 5.3 to $\mathcal{C}_{g_b}$.

Whereas in [19] the investigation is performed directly through *good exceptional polynomials* (see [19, Definition 2.2]), in [20] the link with algebraic curves of degree 6 is explicitly used; see [20, Subsections 3.1 and 3.2].

5.7 Permutation rational functions

Much less is known for permutation rational functions, even of low degree; see for instance [89, 125, 126]. The "right" setting in this case is the projective line $\mathbb{P}^1(\mathbb{F}_q) = \mathbb{F}_q \cup \{\infty\}$. The authors use different techniques, as a function field based approach [89] for degree-3 rational functions and Hermite's criterion [125] for degree-3 rational functions and partial results for degree-4 rational functions.

The complete classification of degree-4 rational functions permuting $\mathbb{P}^1(\mathbb{F}_q)$ is provided in [126] where the author investigates the curve $\mathcal{C}_f$ associated with the rational function f, similarly to what is done in Proposition 5.4.

Let $f(x) = P(x)/Q(x)$, where $P, Q \in \mathbb{F}_q[x]$, $\deg P = 4$, $\deg Q = 3$, Q is irreducible over $\mathbb{F}_q$ and $Q \nmid P$. Up to equivalence, we may write

$$f(x) = x + (ax^2 + bx + x)/Q(x) \tag{5.8}$$

where $a, b, c \in \mathbb{F}_q$ are not all 0. If $char(\mathbb{F}_q) \neq 3$, we may assume that $Q(x) = x^3 + dx + e$, for some $d, e \in \mathbb{F}_q$; when $char(\mathbb{F}_q) = 3$, we may assume that $Q(x) = x^3 + dx + e$ or $Q(x) = x^3 + x^2 + e$.

Theorem 5.21 *Let $f(x)$ as in (5.8).*

1. *[126, Theorem 3.1] Suppose that $q > 13$. Then f is a permutation of $\mathbb{P}^1(\mathbb{F}_q)$ if and only if $a = -3d$, $b = -9e$ and $c = d$.*

2. *[126, Theorem 4.1] Suppose that $char(\mathbb{F}_q) = 3$, $q > 81$, and $Q(x) = x^3 + x^2 + e$. Then f is a permutation of $\mathbb{P}^1(\mathbb{F}_q)$ if and only if $a = 1$, $b = c = 0$.*

Other interesting results concern permutation rational functions of the type

$$f_b(x) = x + \frac{1}{(x^p - x + b)^s} \in \mathbb{F}_q[x], \qquad q = p^n, \tag{5.9}$$

where $s \in \mathbb{N}$, $Tr_{p^n/p}(b) \neq 0$; see [111, 149, 149, 219, 236, 237, 240, 241, 242, 120]. Since $f(\infty) = \infty$, permutational properties of such functions can be directly investigated in $\mathbb{F}_q$.

In particular, the case $s = 1$ has been addressed in [237]; see also [89, 125].

Theorem 5.22 *Let $f_b(x)$ be as in (5.9).*

1. *[237] If $p = 2, 3$ then $f_b(x)$ permutes $\mathbb{F}_{p^n}$ for all $n \geq 1$.*

2. *[128, Theorem 1.1] If $p > 3$ and $n \geq 5$ then $f_b(x)$ does not permute $\mathbb{F}_{p^n}$.*

3. *[14] If $p > 3$ and $n = 3, 4$ then $f_b(x)$ does not permute $\mathbb{F}_{p^n}$.*

Interestingly, the behavior of $f_b(x)$ strictly depends on the characteristic. To address the non-existence results in the theorem above, the authors in [128] studied the curve $\mathcal{C}_{f_b}$ attached to $f_b(x)$ by considering a chain of function field extensions (both of Kummer and Artin-Schreier type), whereas in [14] the problem is translated into the investigation of an $\mathbb{F}_p$-variety $\mathcal{V}_{f_b}$ in higher dimension. The main ingredient here is the proof of the existence of an absolutely irreducible $\mathbb{F}_p$-rational component in $\mathcal{V}_{f_b}$ which, together with Lang-Weil bound provides the desired result.

6 Minimal value set polynomials and minimal value set rational functions

Let $f(x) \in \mathbb{F}_q[x]$ of degree $d \geq 1$. Its value set is

$$V_f = \{f(\alpha) \mid \alpha \in \mathbb{F}_q\}.$$

By a simple counting argument $\#V_f \geq \lceil \frac{q}{d} \rceil$. Clearly, for a permutation polynomial $f(x)$ we have $V_f = \mathbb{F}_q$. Polynomials having the smallest possible value set $\lceil \frac{q}{d} \rceil$ are called *minimal value set polynomials*. Such polynomials have been investigated by many authors over the past decades [42, 50, 104, 105, 172]. The subject, important in its own right, is known to be relevant in other branches of mathematics.

There is a close relationship between minimal value set polynomials and algebraic curves. For instance, Frobenius non-classical curves [213] play an important role in [41].

The approach used in [105] is based on the investigation of the curve $\mathcal{C}_f : f(X) - f(Y) = 0$. If $\deg(f(x))$ is small enough with respect to q, roughly speaking the smaller the value set V_f, the larger the number of absolutely irreducible $\mathbb{F}_q$-rational components of $\mathcal{C}_f$; see also [231]. This is a direct application of the Aubry-Perret bound.

Theorem 6.1 *[105, Theorem 1, Lemma 4] Let $f(x) \in \mathbb{F}_q[x]$ be of degree $d < \sqrt[4]{q}$ with $\#V_f < 2q/d$. Then $\mathcal{C}_f$ has at least $d/2$ absolutely irreducible components. The possibilities are listed in [105, Table 1].*

Given a rational function $h(x) \in \mathbb{F}_q(x)$, similarly to the polynomial case, one can define its value set as
$$V_h = \{h(\alpha) \mid \alpha \in \mathbb{F}_q \cup \{\infty\}\}.$$
Clearly,
$$\left\lceil \frac{q+1}{d} \right\rceil \leq \#V_h \leq q+1.$$
When the upper bound is reached we say that $h(x)$ permutes $\mathbb{P}^1(\mathbb{F}_q)$; a rational function for which the lower bound above is achieved will be called *minimal value set rational function* (*m.v.s.r.f.*). Despite the analogy, it is worth noting that minimal value set polynomials $f(x) \in \mathbb{F}_q[x]$, considered as rational functions in $\mathbb{F}_q(x)$, may not give rise to *m.v.s.r.f.*. In fact, one can readily check that for such a polynomial f, the set $V_f = \{f(\alpha) \mid \alpha \in \mathbb{P}^1(\mathbb{F}_q)\}$ satisfies

$$\#V_f = \begin{cases} \left\lceil \frac{q+1}{d} \right\rceil & \text{if } d \text{ divides } q \\ \left\lceil \frac{q+1}{d} \right\rceil + 1 & \text{otherwise,} \end{cases}$$

where $d = \deg f$. An approach similar to that one in [105] is used in [15] to obtain a classification of rational function $h(x)$ of degree small with respect to q and having a "small" value set. In this case, a link with Galois group of the function field extension $\mathbb{F}_q(x)/\mathbb{F}_q(h(x))$ is explicitly provided. As for the polynomial case, the results are obtained investigating the structure of the curve $\mathcal{C}_f$.

Theorem 6.2 *Let $h(x) = f(x)/g(x) \in \mathbb{F}_q(x)$ be a rational function of degree d_1, where $f(x), g(x) \in \mathbb{F}_q[x]$ are coprime.*

1. *[15, Theorem 1.1] If $\mathbb{F}_q(x)/\mathbb{F}_q(h(x))$ is a Galois extension, then either $\#V_h = \left\lceil \frac{q+1}{d_1} \right\rceil$ or $\#V_h = \left\lceil \frac{q+1}{d_1} \right\rceil + 1$. Moreover, the latter case holds if and only if the action of $\mathrm{Gal}(\mathbb{F}_q(x)/\mathbb{F}_q(h(x)))$ on $h^{-1}(V_h)$ has at least two short orbits.*

2. *[15, Theorem 1.2] Suppose $d_2 = \min\{\deg f, \deg g\}$ is such that $d_1 + 2d_2 \leq \sqrt{q} + 1 + \frac{1}{\sqrt{q}}$. If*
$$\#V_h < \left\lceil \frac{q+1}{d_1} \right\rceil + \frac{q+1+d_1 - d_2(d_2+1)\sqrt{q}}{(d_1-1)d_1^2} - 1,$$
then $\mathbb{F}_q(x)/\mathbb{F}_q(h(x))$ is a Galois extension.

7 Kloosterman polynomials

Kloosterman polynomials have been introduced by Hollmann and Xiang [119] to prove several identities involving Kloosterman sums over $\mathbb{F}_{2^n}$.

Definition 7.1 [30, Definition 1.1] *Let n be a positive integer. Let $c_{n-1} \cdots c_0$ be the binary representation of $c \in \{0, 1, \ldots, 2^n - 1\}$ with digits $c_i \in \{0, 1\}$, and denote its weight by $\mathrm{wt}(c) = \sum_{i=0}^{n-1} c_i$. Given such numbers c and d, define the polynomials on $\mathbb{F}_{2^n}$:*

$$L_c(x) = \sum_{i=0}^{n-1} c_i x^{2^i} \quad \text{and} \quad L_{c,d}(x) = L_c(x) + L_d\left(x^{2^n - 2}\right).$$

The polynomial $L_{c,d}$ is called a *Kloosterman polynomial* on $\mathbb{F}_{2^n}$ if $\mathrm{wt}(d)$ is even and $L_{c,d}$ is injective on $T_1 := \{\gamma \in \mathbb{F}_{2^n} : Tr_{2^n}(\gamma) = 1\}$, where $Tr_{2^n}(\cdot)$ stands for the absolute trace map on $\mathbb{F}_{2^n}$.

In [119], the authors proved that for all n the polynomials $L_{1,3}(x), L_{1,6}(x), L_{1,10}(x)$ are Kloosterman polynomials over $\mathbb{F}_{2^n}$. Moreover, they conjectured that there are no more examples.

Conjecture 7.2 *[119, Conjecture 4.4] Let d be a positive number with $\mathrm{wt}(d)$ even. For all $n \geq 1$, we have that $L_{1,d}$ is a Kloosterman polynomial on $\mathbb{F}_{2^n}$ if and only if $d \in \{3, 6, 10\}$.*

This conjecture has been asymptotically settled in [30], first translating it into a problem concerning permutational properties of a specific rational function $\tilde{f}$ and then applying the usual machinery connected with the curve $\mathcal{C}_{\tilde{f}}$ as in (5.4). Finally, tools from function field theory (Artin-Schreier extensions) are applied to obtain the following result.

Theorem 7.3 *[30, Theorem 1.3] Let d be a positive integer with $\mathrm{wt}(d)$ even. If $n \geq 4\lceil \log_2 d \rceil$ and $L_{1,d}$ is a Kloosterman polynomial over $\mathbb{F}_{2^n}$, then $d \in \{3, 6, 10\}$.*

8 o-polynomials

Let q be a power of 2. A set $\mathcal{A} \subset \mathbb{P}^2(\mathbb{F}_q)$ is an arc if no three points of $\mathcal{A}$ are collinear. The arc is *complete* if each point in $\mathbb{P}^2(\mathbb{F}_q) \setminus \mathcal{A}$ is collinear with two distinct points of $\mathcal{A}$. For more details on arcs in projective planes and their connections with other areas of mathematics we refer to [118, 116] and the references therein. It is well known that $q + 2$, if q is even, is the largest possible size of a complete arc; see for instance [116, Chapter 8]. An arc of size $q + 2$ in $\mathbb{P}^2(\mathbb{F}_q)$, q even, is called a hyperoval. The classification of hyperovals is still open; see e.g. [55, 116, 223, 187].

Definition 8.1 A polynomial $f(x) \in \mathbb{F}_q[x]$ is called an o-polynomial if the set

$$\{(f(c) : c : 1) \ : \ c \in \mathbb{F}_q\} \cup \{(1:0:0), (0:1:0)\} \subset \mathbb{P}^2(\mathbb{F}_q)$$

is a hyperoval of $\mathbb{P}^2(\mathbb{F}_q)$. Exceptional o-polynomials are polynomials $f(x) \in \mathbb{F}_q[x]$ which are o-polynomials over infinitely many extensions $\mathbb{F}_{q^n}$ of $\mathbb{F}_q$.

Consider a polynomial $f(x) \in \mathbb{F}_q[x]$ and let $\mathcal{H}_f$ be the surface in $\mathbb{P}^3(\mathbb{F}_q)$ defined by
$$\frac{X(f(Y) + f(Z)) + Y(f(X) + f(Z)) + Z(f(X) + f(Y))}{(X+Y)(X+Z)(Y+Z)} = 0.$$

As observed in [52], $f(x)$ is an o-polynomial if and only if affine $\mathbb{F}_q$-rational points of $\mathcal{H}_f$ belong to $(X+Y)(X+Z)(Y+Z) = 0$; see also [114, Section 2] for the case $f(x)$ is a monomial. Since the planes $X + Y = 0$, $X + Z = 0$, $Y + Z = 0$ are not components of $\mathcal{H}_f$ (see [52, Proposition 2.4] and [114, Theorem 2]) the existence of an absolutely irreducible $\mathbb{F}_q$-rational component in $\mathcal{H}_f$ yields, by the Lang-Weil bound, that f is not an o-polynomial if its degree is small enough with respect to q and in particular $f(x)$ is not an exceptional o-polynomial.

Note that in the case $f(x) = x^k$, $\mathcal{H}_f$ is indeed a plane curve $\mathcal{C}_k$: investigation of its singular points and the approach described in Subsection 4.1 yields the main result of [114].

Theorem 8.2 *[114, Theorem 1] If $k \neq 6$ and $k \neq 2^i$ then $f(x)$ is not an exceptional o-polynomial.*

In the case $f(x)$ is a polynomial of degree k, one can study the intersection of $\mathcal{H}_f$ with the plane at infinity. Such a curve is precisely $\mathcal{C}_k$ and therefore, using Lemma 4.15 and [114, Theorem 1] the classification for k not 6 nor a power of two easily follows; see [52, Proposition 6]. The case k equals a power of two must be handled separately: a deeper analysis of the possible factorization of $\mathcal{H}_f$ is necessary; see [52, Proposition 2.8]. As a definition, two polynomials $f, g \in \mathbb{F}_q[x]$ with $f(0) = g(0) = 0$ and $f(1) = g(1) = 1$ are *equivalent* if there exists an $a \in \mathbb{F}_q$ such that $g(x) = (f(x+a) + f(a))/(f(1+a) + f(a))$. This equivalence defines an equivalence relation and it preserves the property of being an o-polynomial of $\mathbb{F}_q$.

Theorem 8.3 *[52, Theorem 1.2, Corollary 1.3] If $f(x)$ is an o-polynomial of $\mathbb{F}_q$ of degree less than $\sqrt[4]{q}/2$, then $f(x)$ is equivalent to either x^6 or x^{2^k} for a positive integer k. In particular, an exceptional o-polynomial is equivalent to either x^6 or x^{2^k} for a positive integer k.*

9 Scattered polynomials and maximum scattered linear sets

Another interesting class of polynomials which have deep connections with many other areas of mathematics are the so-called scattered polynomials; see [208].

Let us recall some basic definitions on linear sets first. Let q be a prime power and $r, n \in \mathbb{N}$. Let V be a vector space of dimension r over $\mathbb{F}_{q^n}$. For any k-dimensional $\mathbb{F}_q$-vector subspace U of V, the set $L(U)$ defined by the non-zero vectors of U is called an $\mathbb{F}_q$-*linear set* of $\Lambda = \mathrm{PG}(V, q^n)$ of *rank* k, i.e.

$$L(U) = \{\langle \mathbf{u} \rangle_{\mathbb{F}_{q^n}} : \mathbf{u} \in U \setminus \{\mathbf{0}\}\}.$$

It is notable that the same linear set can be defined by different vector subspaces. Consequently, we always consider a linear set and the vector subspace defining it simultaneously.

Let $\Omega = \mathrm{PG}(W, \mathbb{F}_{q^n})$ be a subspace of Λ and let $L(U)$ be an $\mathbb{F}_q$-linear set of Λ. We say that Ω has *weight i* in $L(U)$ if $\dim_{\mathbb{F}_q}(W \cap U) = i$. Thus a point of Λ belongs to $L(U)$ if and only if it has weight at least 1. Moreover, for any $\mathbb{F}_q$-linear set $L(U)$ of rank k,

$$|L(U)| \leq \frac{q^k - 1}{q - 1}.$$

When the equality holds, i.e. all the points of $L(U)$ have weight 1, we call $L(U)$ a *scattered* linear set. A scattered $\mathbb{F}_q$-linear set of highest possible rank is called a *maximum scattered* $\mathbb{F}_q$-*linear set*. See [36] for the possible ranks of maximum scattered linear sets.

Maximum scattered linear sets have various applications in Galois geometry, including blocking sets [10, 160, 163], two-intersection sets [36, 37], finite semifields

[46, 84, 162, 170], translation caps [18], translation hyperovals [82], etc. For more applications and related topics, see [143] and the references therein. For recent surveys on linear sets and particularly on the theory of scattered spaces, see [144, 188].

For every n-dimensional $\mathbb{F}_q$-subspace U of $\mathbb{F}_{q^n} \times \mathbb{F}_{q^n}$ there exist a suitable basis of $\mathbb{F}_{q^n} \times \mathbb{F}_{q^n}$ and an $\mathbb{F}_q$-linearized polynomial $f(x) = \sum A_i x^{q^i} \in \mathbb{F}_{q^n}[x]$ of degree less than q^n such that $U = \{(x, f(x)) : x \in \mathbb{F}_{q^n}\}$.

Following this description, maximum scattered linear sets in $\mathbb{P}^1(\mathbb{F}_{q^n})$ can be described via the so-called scattered polynomials; [208].

Also, the techniques of Section 4 apply to the intersection problem for linear sets in $\mathbb{P}^1(\mathbb{F}_{q^n})$; see [250].

Here we present a slightly more general definition; see [28].

Definition 9.1 Let $f(x) = \sum A_i x^{q^i} \in \mathbb{F}_{q^n}[x]$ be an $\mathbb{F}_q$-linearized polynomial over $\mathbb{F}_{q^n}$ and

$$U_t = \{(x^{q^t}, f(x)) : x \in \mathbb{F}_{q^n}\}. \tag{9.1}$$

If $L(U_t)$ is a maximum scattered linear set, then $f(x)$ is said to be a *scattered polynomial of index* t. A *scattered polynomial* is simply a scattered polynomial of index 0.

A necessary and sufficient condition for $L(U)$ to define a maximum scattered linear set in $\mathbb{P}^1(\mathbb{F}_{q^n})$ is

$$\frac{f(x)}{x} = \frac{f(y)}{y} \text{ if and only if } \frac{y}{x} \in \mathbb{F}_q, \quad \text{for } x, y \in \mathbb{F}_{q^n}^*. \tag{9.2}$$

We refer to [190, Section 2] (and references therein) for the list of all known scattered polynomials.

There is a very interesting link between scattered linear sets and the so-called maximum rank distance (MRD for short) codes; see [28, 66, 208, 36, 67, 18, 16, 27, 29, 68, 69, 169, 239, 249, 161, 179, 210].

To the best of our knowledge, up to $\Gamma L(2, q^n)$-equivalence of the associated $\mathbb{F}_q$-subspaces, most of the constructions of scattered polynomials for arbitrary n can be summarized as one family

$$f(x) = \delta x^{q^s} + x^{q^{n-s}}, \tag{9.3}$$

where s satisfies $\gcd(s, n) = 1$ and $\text{Norm}_{\mathbb{F}_{q^n}/\mathbb{F}_q}(\delta) = \delta^{(q^n-1)/(q-1)} \neq 1$.

See Section 9.2 for references on non-exceptional families.

The literature regarding scattered polynomials and MRD codes is quite large. Here, we do not intend to provide a comprehensive list of all the constructions of this kind of polynomials; we refer to [190, 209] for excellent surveys on this topic.

In recent years, an increasing number of results on scattered polynomials (and related maximum scattered linear sets) are based on investigation of curves or varieties associated with them; see for instance [16, 21, 28, 29, 189, 173, 169].

A first interesting connection between scattered polynomials and algebraic varieties is the following. For a given $\mathbb{F}_q$-linearized polynomial $f(x) = \sum_{i=0}^{n-1} a_i x^{q^i} \in$

$\mathbb{F}_{q^n}[x]$, its Dickson matrix is defined as

$$M(f) := \begin{pmatrix} a_0 & a_1 & a_2 & \cdots & a_{n-1} \\ a_{n-1}^q & a_0^q & a_1^q & \cdots & a_{n-2}^q \\ a_{n-2}^{q^2} & a_{n-1}^{q^2} & a_0^{q^2} & \cdots & a_{n-3}^{q^2} \\ \vdots & \vdots & \vdots & & \vdots \\ a_1^{q^{n-1}} & a_2^{q^{n-1}} & a_3^{q^{n-1}} & \cdots & a_0^{q^{n-1}} \end{pmatrix}. \qquad (9.4)$$

Then $f(x) \in \mathbb{F}_{q^n}[x]$ is scattered if and only if for every $m \in \mathbb{F}_{q^n}$ the polynomial $f(x) + mx$, seen as a linear transformation of $\mathbb{F}_{q^n}$, has kernel of dimension at most 1 over $\mathbb{F}_q$; see e.g. [67, Section 7] and [238, Section 1]. This is equivalent to require

$$\forall \, m \in \mathbb{F}_{q^n} \quad Rank(M(f(x) + mx)) \geq n - 1.$$

Denote $M_m := M(f(x) + mx)$ and consider M'_m the $(n-1) \times (n-1)$ submatrix of M_m obtained by considering the last $n-1$ columns and the first $n-1$ rows of M_m. By [65, Theorem 1.3], $Rank(M(f(x) + mx)) \geq n - 1$ if and only if either $\det(M_m) \neq 0$ or $\det(M_m) = 0$ and $\det(M'_m) \neq 0$.

Proposition 9.2 *The polynomial $f(x) \in \mathbb{F}_{q^n}[x]$ is not scattered if and only if there exists $m \in \mathbb{F}_{q^n}$ such that*

$$\det(M_m) = \det(M'_m) = 0. \qquad (9.5)$$

Using a normal basis $\{\xi, \xi^q, \ldots, \xi^{q^{n-1}}\}$ of $\mathbb{F}_{q^n}$ over $\mathbb{F}_q$, one can consider $a_i = \sum_{j=0}^{n-1} \alpha_i^{(j)} \xi^{q^j}$ and $m = \sum_{j=0}^{n-1} x_j \xi^{q^j}$ and therefore Equations (9.5) define an $\mathbb{F}_q$-variety $\mathcal{V}_f \subset \mathbb{P}^n(\mathbb{F}_q)$. Thus, Proposition 9.2 can be rephrased as follows.

Proposition 9.3 *The polynomial $f(x) \in \mathbb{F}_{q^n}[x]$ is not scattered if and only if $\mathcal{V}_f$ possesses an affine $\mathbb{F}_q$-rational point.*

Note that, in this approach, all the tools introduced for varieties or curves in the previous section can be adopted. Modifications of this general method can be found in Subsection 9.2.

Very recently, De Boeck and Van de Voorde [71] determine all possible weight distributions of $\mathbb{F}_q$-linear sets properly contained in $\mathbb{P}(\mathbb{F}_{q^5})$, investigating algebraic curves of degree three.

9.1 Exceptional scattered polynomials

As scattered polynomials appear to be very rare, it is natural to look for some classifications of them.

Definition 9.4 [28] Given an integer $0 \leq t \leq n-1$ and a q-polynomial f whose coefficients are in $\mathbb{F}_{q^n}$, if

$$U_{m,t} = \{(x^{q^t}, f(x)) : x \in \mathbb{F}_{q^{mn}}\} \qquad (9.6)$$

defines a maximum scattered linear set in $\mathbb{P}^1(\mathbb{F}_{q^{mn}})$ for infinitely many m, then we call f an *exceptional scattered polynomial of index t*.

We can describe the family (9.3) as an exceptional one. Taking $t = s$, from (9.3) we get
$$\{(x^{q^s}, x + \delta x^{q^{2s}}) : x \in \mathbb{F}_{q^{mn}}\}$$
which defines a maximum scattered linear set for all mn satisfying $\gcd(mn, s) = 1$. This means $X + \delta X^{q^{2s}}$ is an exceptional scattered polynomial of index s.

Definition 9.5 [28, 21] An exceptional scattered polynomial $f(x)$ is *normalized* if

1. the coefficient of x^{q^t} in $f(x)$ is always 0;

2. when $t > 0$, the coefficient of x in $f(x) = \sum A_i x^{q^i} \in \mathbb{F}_{q^n}[x]$ is non-zero;

3. $f(x)$ is monic.

So far, classification results on exceptional scattered polynomials are summarized in the following theorem.

Theorem 9.6 1. *[28, Corollary 3.4] For $q > 5$, x^{q^k} is the unique exceptional scattered monic polynomial of index 0.*

2. *[28, Corollary 3.7] The only exceptional scattered monic polynomials f of index 1 over $\mathbb{F}_{q^n}$ are x and $bx + x^{q^2}$ where $b \in \mathbb{F}_{q^n}$ satisfying $\mathrm{Norm}_{q^n/q}(b) \neq 1$. In particular, when $q = 2$, $f(x)$ must be x.*

3. *[21, Corollary 4.5] Let $f(x) \in \mathbb{F}_{q^r}[x]$ be a monic $\mathbb{F}_q$-linearized polynomial. Then $f(x)$ is exceptional scattered of index 2 if and only if r is odd and*
$$f(x) = x + \delta x^{q^4},$$
with $\mathrm{Norm}_{q^r/q}(\delta) \neq 1$.

4. *[90, Theorem 1.2] Let $t \geq 1$. Let $f(x)$ be a normalized exceptional scattered polynomial of index t of degree q^r, and let $d := \max\{r, t\}$. Suppose that d is an odd prime. Then $f(x) = x$.*

Whereas in [90] the argument is based on a clever investigation of the monodromy groups of scattered polynomials, exceptional scattered polynomials have been investigated in [28, 21] by means of connections with algebraic curves. The key point is the following straightforward link between scattered linear sets and algebraic curves.

Lemma 9.7 *[28, Lemma 2.1] The vector space $U = \{(x^{q^t}, f(x)) : x \in \mathbb{F}_{q^n}\}$ defines a maximum scattered linear set $L(U)$ in $\mathbb{P}^1(\mathbb{F}_{q^n})$ if and only if the curve $\mathcal{C}_f$ defined by*
$$\frac{f(X)Y^{q^t} - f(Y)X^{q^t}}{X^q Y - XY^q} \tag{9.7}$$
in $\mathbb{P}^2(\mathbb{F}_{q^n})$ contains no affine point (x, y) such that $\frac{y}{x} \notin \mathbb{F}_q$.

In both [21] and [28], the main idea relies on the approach described in Subsection 4.1 and careful investigation of branches centered at singular points of the curve $\mathcal{C}_f$ defined in (9.7). To this end, Propositions 4.6 and 4.7 are crucial in [21].

9.2 Families of non-exceptional scattered polynomials

Connection with algebraic varieties and algebraic curves in particular have been used to deal with families of non-exceptional scattered polynomials; see [16, 189, 173, 169, 27, 159].

In terms of scattered polynomials, in [189] the authors prove that the binomial $x^{q^s} + \delta x^{q^{n+s}}$ is not scattered if

$$n \geq \begin{cases} 4s+2 & \text{if } q=3 \text{ and } s>1, \text{or } q=2 \text{ and } s>2; \\ 4s+2 & \text{otherwise}; \end{cases}$$

see [189, Theorem 1.1]. The goal is achieved by considering space curves (different for q even and q odd). The corresponding function fields are then investigated, via Kummer or Artin-Schreier type extensions which turn out to be constant field extensions. Hasse-Weil bound for function fields provides the desired result.

The approach in [173] is slightly different. The authors prove that for q large enough, the polynomial $f(x) = x^q + bx^{q^2}$ is not scattered over $\mathbb{F}_{q^5}$ for any $b \in \mathbb{F}_{q^5}^*$; see [173, Proposition 3.3]. To this end, they investigate a hypersurface in $\mathcal{V} \subset \mathbb{A}^5(\mathbb{F}_q)$ of degree five via an $\mathbb{F}_{q^5}$-birationally equivalent hypersurface in $\mathbb{A}^5(\mathbb{F}_{q^5})$, showing that this last variety is irreducible. Thus they apply a Hasse-Weil type result (see [173, Lemma 3.2]) to deduce the existence of a suitable affine $\mathbb{F}_q$-rational point in $\mathcal{V}$ and therefore the non-scatteredness of $f(x)$.

Recently, an interesting link between the so-called h-scattered linear sets (see [210, 69, 161]) and Moore exponent sets has been established in [180].

Definition 9.8 Let $I = \{i_0, i_1, \ldots, i_{k-1}\} \subset \mathbb{Z}_{\geq 0}$. For every $A = (\alpha_0, \alpha_1, \ldots, \alpha_{k-1}) \in \mathbb{F}_{q^n}^k$, denote

$$M_{A,I} = \begin{pmatrix} \alpha_0^{q^{i_0}} & \alpha_0^{q^{i_1}} & \cdots & \alpha_0^{q^{i_{k-1}}} \\ \alpha_1^{q^{i_0}} & \alpha_1^{q^{i_1}} & \cdots & \alpha_0 1^{q^{i_{k-1}}} \\ \vdots & \vdots & & \vdots \\ \alpha_{k-1}^{q^{i_0}} & \alpha_{k-1}^{q^{i_1}} & \cdots & \alpha_{k-1}^{q^{i_{k-1}}} \end{pmatrix}.$$

The set I is a *Moore exponent set* for q and n if

$$\det(M_{A,I}) = 0 \text{ if and only if } \alpha_0, \alpha_1, \ldots, \alpha_{k-1} \text{ are } \mathbb{F}_q\text{-linearly dependent.}$$

Asymptotic behavior of Moore exponent sets is also related with MRD codes and then with $\mathbb{F}_q$-linearized polynomials; see [68, Questions 4.6 and 4.7].

The main achievement in this direction is the following.

Theorem 9.9 *[29, Theorem 1.1]. Assume that $q > 5$ and $I = \{0, i_1, i_2, \ldots, i_{k-1}\}$ with $0 < i_1 < \ldots < i_{k-1}$ is not an arithmetic progression. Then there exists an integer N depending only on I such that I is not a Moore exponent set for q and n provided that $n > N$.*

The above result has been obtained via the analysis of the two varieties $\mathcal{G}_k$:

$G_k(X_1, \ldots, X_k) = 0$ and $\mathcal{V}_I : \frac{F_I(X_1, \ldots, X_k)}{G_k(X_1, \ldots, X_k)} = 0$, where

$$F_I(X_1, \ldots, X_k) = \det \begin{pmatrix} X_1^{q^{i_0}} & X_1^{q^{i_1}} & \cdots & X_1^{q^{i_{k-1}}} \\ X_2^{q^{i_0}} & X_2^{q^{i_1}} & \cdots & X_2^{q^{i_{k-1}}} \\ \vdots & \vdots & \ddots & \vdots \\ X_k^{q^{i_0}} & X_k^{q^{i_1}} & \cdots & X_k^{q^{i_{k-1}}} \end{pmatrix}, \qquad (9.8)$$

and

$$G_k(X_1, \ldots, X_k) = \det \begin{pmatrix} X_1 & X_1^{q^1} & \cdots & X_1^{q^{k-1}} \\ X_2 & X_2^{q^1} & \cdots & X_2^{q^{k-1}} \\ \vdots & \vdots & \ddots & \vdots \\ X_k & X_k^{q^1} & \cdots & X_k^{q^{k-1}} \end{pmatrix}. \qquad (9.9)$$

The main result is proved by showing the existence for sufficiently large n of an affine $\mathbb{F}_{q^n}$-rational point in $\mathcal{V}_I$ which is not in $\mathcal{G}_k$. The problem is translated into a "curve problem" using inductively Lemma 4.15 and finally applying the machinery described in Subsection 4.1.

10 Planar functions in odd characteristic

Let q be an odd prime power.

Definition 10.1 A function $f : \mathbb{F}_q \to \mathbb{F}_q$ is *planar* or *perfect nonlinear* if, for each non-zero $\epsilon \in \mathbb{F}_q$, the function

$$x \mapsto f(x + \epsilon) - f(x) \qquad (10.1)$$

is a permutation on $\mathbb{F}_q$. If $f(x) \in \mathbb{F}_q[x]$ is planar over infinitely many extensions $\mathbb{F}_{q^m}$ of $\mathbb{F}_q$ then $f(x)$ is called an *exceptional planar polynomial*.

Note that we can always consider $f(x)$ normalized, since $f(x+\epsilon) - f(x)$ permutes $\mathbb{F}_q$ if and only if $(f(x + \epsilon) - f(x))/\alpha$, $\alpha \in \mathbb{F}_q^*$, does it. For similar reasons, if $f(x) = x^t$, t can be considered coprime with p.

Planar functions can be used to construct finite projective planes [79], relative difference sets [95], error-correcting codes [47], and S-boxes in block ciphers [183]. We refer to [191] for a more comprehensive introduction to planar functions.

To a planar function $f(x)$ in odd characteristic we can attach a surface in $\mathcal{S}_f \subset \mathbb{P}^3(\mathbb{F}_q)$ defined by

$$\mathcal{S}_f : \phi(X, Y, Z) = \frac{f(X + Z) - f(X) - f(Y + Z) + f(Y)}{(X - Y)Z} = 0. \qquad (10.2)$$

It is not difficult to see that $f(x)$ is planar on $\mathbb{F}_q$ if and only if all affine $\mathbb{F}_q$-rational points $\mathcal{S}_f$ satisfy $X = Y$ or $Z = 0$. Note that if $f(x)$ is a monomial of degree t then $\phi(X, Y, Z)$ is homogenous of degree $t - 1$ and therefore it describes a curve $\mathcal{C}_f : \frac{(X+1)^t - X^t - (Y+1)^t + Y^t}{X - Y} = 0$.

10.1 Planar monomials

Let us consider first the monomial case. In the case t odd, both 0 and -1 are roots of $(x+1)^t - x^t = 1$ and therefore $f(x) = x^t$ cannot be planar. Thus, only the case t even will be considered.

Proposition 10.2 *Let $f(x) = x^t \in \mathbb{F}_{p^n}[x]$, t a positive integer not divisible by p. Then $f(x)$ is planar in the following cases.*

1. $t = 2$.

2. $t = p^\ell + 1$, $\ell/\gcd(n,\ell)$ odd; see [63, 79].

3. $t = (3^\ell + 1)/2$, $p = 3$, ℓ odd, $\gcd(\ell, n) = 1$; [63].

Remark All the functions above are actually exceptional planar monomials.

A first step into the classification of exceptional planar monomials was given in [115, Theorem 1.1], where the authors provide a list of sufficient conditions for the curve $\mathcal{C}_f$ to have an $\mathbb{F}_p$-rational absolutely irreducible component. They are quite involved and they also depend, for the case $t \equiv 1 \pmod{p}$, on the non-negative integer ℓ such that $t = p^i \ell + r$, $r < p$. The proof of [115, Theorem 1.1] relies on deep investigations of pencil of projective curves via the positive characteristic version of Bertini's Theorem (see for instance [139]) together with an analysis of the singular points of $\mathcal{C}_f$.

Later, the case $t \not\equiv 1 \pmod{p}$ was completely solved in [247] using a different approach, based on the possible functional decomposition of the permutation polynomial $(x+1)^t - x^t$. In the case $t \leq \sqrt[4]{q}$ such a polynomial is exceptional and therefore, since $p \nmid t$, the classification of indecomposable exceptional polynomials can be used to obtain the desired result, improving [115, Theorem 1.1].

The case $t \equiv 1 \pmod{p}$ was finally solved in [146].

Theorem 10.3 *[146, Theorem 1] Let $t \equiv 1 \pmod{p}$ be such that $f(x) = x^t \in \mathbb{F}_{p^n}[x]$ is exceptional planar. Then $t = p^\ell + 1$ for some non-negative integer ℓ. In particular, Proposition 10.2 presents the complete list of exceptional planar monomials (up to composition with $g(x) = x^p$).*

This result has been obtained by applying to the curve $\mathcal{C}_f$ the machinery described in Subsection 4.1.

10.2 Planar polynomials

The case $f(x)$ not a monomial is more complicated. Mostly, planar functions which are not monomials are Dembowski-Ostrom (DO) polynomials. Actually, it has been even conjectured that all planar polynomials were of DO type; see [195]. The first counterexample of this conjecture was found by Coulter and Matthews [63] in characteristic 3; if $p > 3$ the conjecture is still open.

Lemma 10.4 *[230, Theorem 2.3] Let $f(x)$ be a DO polynomial over $\mathbb{F}_q$. Then $f(x)$ is planar if and only if $f(x)$ is 2-to-1.*

Connections between planar Dembowski-Ostrom polynomials and Dickson polynomials of first and second kind have been investigated in [64], whereas in [88], the authors investigate the curve $f(X+Y) - f(X) - f(Y) = 0$, where $f(x)$ is a DO polynomial which is also a reversed Dickson polynomial.

Let $f(x)$ be of degree d. Then the intersection of the surface $\mathcal{S}_f \subset \mathbb{P}^3(\mathbb{F}_q)$ (see (10.2)) with the plane at infinity is the curve $\mathcal{C}_f : \frac{(X+1)^t - X^t - (Y+1)^t + Y^t}{X-Y} = 0$. If such a curve possesses an absolutely irreducible $\mathbb{F}_q$-rational component which is not repeated, by Lemma 4.15 $\mathcal{S}_f$ contains an $\mathbb{F}_q$-rational absolutely irreducible component and by Lang-Weil Theorem (or similar results) $f(x)$ is not planar.

Theorem 10.5 [248] *Let p be an odd prime and let $f(x) \in \mathbb{F}_{p^n}[x]$ of degree d. If $f(x)$ is exceptional planar and $d \not\equiv 0, 1 \pmod{p}$, then up to equivalence the possibilities for $f(x)$ are listed in Proposition 10.2.*

For the case $d \equiv 1 \pmod{p}$, the authors in [53] obtained a partial classification. First they prove that the degree d of $f(x)$ must be $p^k + 1$ proving that otherwise the curve obtained intersecting $\mathcal{S}_f$ with the hyperplane at infinity contains a non-repeated $\mathbb{F}_q$-rational absolutely irreducible component. By Lemma 4.15 and the Lang-Weil bound this yields a contradiction. Analyzing the possible factorization of the polynomial $\phi(X, Y, Z)$ defining the surface $\mathcal{S}_f$ (10.2) they can also obtain some constraints on the second largest degree of monomials in $f(x)$.

Theorem 10.6 [53, Theorem 1] *Let $f(x) \in \mathbb{F}_{p^n}[x]$ be monic of degree $d \equiv 1 \pmod{p}$. If $f(x)$ is exceptional planar then $f(x) = x^{p^k+1} + h(x)$, for some integer k and some polynomial $h(x)$ of degree $e < p^k + 1$ with either $p \mid e$ or $p \mid e - 1$.*

A small improvement on the previous theorem can be obtained in the case $e \leq (p^k + 1)/2$, using the same technique together with an investigation of specific Kummer covers.

Theorem 10.7 [13] *Let $f(x) = x^{p^k+1} + h(x) = \sum_j a_j x^j \in \mathbb{F}_{p^n}[x]$ be exceptional planar with $\deg(h(x)) = e \leq (p^k+1)/2$. Then all the monomials in $f(x)$ have degree $j \equiv 0, 1 \pmod{p}$. Also, let m be the smallest degree of the monomials in $h(x)$. If $m \equiv 1 \pmod{p}$ then $m = p^\ell + 1$ for some non-negative integer ℓ. Let $J \subset \mathbb{N}$ be the set of the degrees of monomials in $f(x)$ and consider $g(x) = \sum_{j \in J \cap 2\mathbb{N}} a_j x^{j/2}$. Then $g(x)$ is an exceptional (permutation) polynomial.*

10.3 A generalization of planar functions

Recently, a generalization of Definition 10.1 has been given in [24]; see also [86], where it was motivated from practical differential cryptanalysis.

Definition 10.8 [24, Definition 1.1] *Let q be a prime power. Consider $\beta \in \mathbb{F}_q \setminus \{0, 1\}$. A function $f : \mathbb{F}_q \to \mathbb{F}_q$ is a β-planar function in $\mathbb{F}_q$ if*

$$\forall \gamma \in \mathbb{F}_q \quad f(x + \gamma) - \beta f(x) \text{ is a permutation of } \mathbb{F}_q.$$

If $f(x) = x^t$ is a β-planar monomial in $\mathbb{F}_q$, we call t a β-*exponent* in $\mathbb{F}_q$. If $f(x) \in \mathbb{F}_q[x]$ is β-planar over $\mathbb{F}_{q^r}$ for infinitely many r, then $f(x)$ is said an *exceptional β-planar function*.

Note that β-planar function can always be considered normalized.

In the case of β-exponents, the connection with algebraic curves is readily seen. For a monomial $f(x) = x^t$ and $\beta \in \mathbb{F}_q$, consider the curve $\mathcal{C}_{\beta,t}$ defined by

$$\mathcal{C}_{\beta,t} : \frac{(X+1)^t - (Y+1)^t - \beta(X^t - Y^t)}{X - Y} = 0. \tag{10.3}$$

It follows that $f(x) = x^t$ is β-planar if and only if $\mathcal{C}_{\beta,t}$ has no affine $\mathbb{F}_q$-rational points off $X - Y = 0$. Applying the machinery described in Subsection 4.1 to the curve $\mathcal{C}_{\beta,t}$ and using connections with Fermat curves [96, 97] a partial classification of β-exponents is obtained in [24, Theorem 4.5].

Theorem 10.9 *[24, Theorem 4.5] Let t be a positive integer and $\beta \in \mathbb{F}_{p^r} \setminus \{0, 1\}$, p a prime. Let q be the smallest power of p such that $q - 1$ is divisible by $t - 1$. Suppose that one of the following conditions holds.*

1. $\beta \in \mathbb{F}_{p^r} \setminus \{0,1\}$, $p \nmid t, t-1$, $p \nmid \prod_{m=1}^{7} \prod_{\ell=-7}^{7-m} m\frac{q-1}{t-1} + \ell$, $t \geq 470$;

2. $t = p^\alpha m + 1$, $(p, \alpha) \neq (3,1)$, $\alpha \geq 1$, $p \nmid m$, $m \neq p^r - 1$ $\forall\, r \mid \ell$, where ℓ is the smallest integer such that $m \mid p^\ell - 1$ and β is an m-th power in $\mathbb{F}_{p^\ell}$.

If $t \leq \sqrt[4]{p^r}$, then t is not a β-exponent in $\mathbb{F}_{p^r}$. In particular, $f(x) = x^t$ is exceptional β-planar.

11 Planar functions in even characteristic

Let q be even. In this case for any $a, \epsilon \in \mathbb{F}_q$ a function $f : \mathbb{F}_q \to \mathbb{F}_q$ cannot satisfy Definition 10.1 because $x = a$ and $x = a + \epsilon$ are mapped by (10.1) to the same image. This is the motivation to define a function $f : \mathbb{F}_q \to \mathbb{F}_q$ for even q to be *almost perfect nonlinear* (APN) if (10.1) is a 2-to-1 map; see Section 12. Such functions are highly relevant again for the construction of S-boxes in block ciphers [183]. However, there is no apparent link between APN functions and projective planes. Zhou [243] defined a natural analogue of planar functions on finite fields of characteristic two.

Definition 11.1 [243] *If q is even, a function $f : \mathbb{F}_q \to \mathbb{F}_q$ is planar if, for each non-zero $\epsilon \in \mathbb{F}_q$, the function*

$$x \mapsto \widehat{f}_\epsilon(x) := f(x + \epsilon) + f(x) + \epsilon x \tag{11.1}$$

is a permutation of $\mathbb{F}_q$.

As shown by Zhou [243] and Schmidt and Zhou [197], such planar functions have similar properties and applications as their counterparts in odd characteristic. Also in this case, we refer to [191] for an excellent survey of results for these functions.

As a definition, a polynomial $f : \mathbb{F}_q \to \mathbb{F}_q$ which is planar for infinitely many extensions $\mathbb{F}_{q^m}$ of $\mathbb{F}_q$ is called *exceptional planar*.

11.1 Planar monomials

Remark Without loss of generality we can suppose that the degree of each monomial in $f(x)$ is not a power of 2. In fact, for every $\mathbb{F}_2$-linearized polynomial $L : \mathbb{F}_q \to \mathbb{F}_q$

$$\widehat{(f+L)}_\epsilon(x) = f(x+\epsilon) + L(x+\epsilon) + f(x) + L(x) + \epsilon x$$
$$= f(x+\epsilon) + f(x) + \epsilon x + L(\epsilon) = \widehat{f}_\epsilon(x) + L(\epsilon),$$

and $\widehat{(f+L)}_\epsilon(x)$ permutes $\mathbb{F}_q$ if and only if so does $\widehat{f}_\epsilon(x)$.

We can attach to a planar function $f(x)$ in even characteristic a specific surface in $\mathcal{S}_f \subset \mathbb{P}^3(\mathbb{F}_q)$ as described in the following. Consider the polynomial

$$\phi(X, Y, Z) = \frac{f(X+Z) + f(X) + ZX + f(Y+Z) + f(Y) + ZY}{(X+Y)Z}.$$

By a direct consequence of the Definition 11.1, f is planar on $\mathbb{F}_q$ if and only if all affine $\mathbb{F}_q$-rational points $\mathcal{S}_f : \phi(X, Y, Z) = 0$ satisfy $X = Y$ or $Z = 0$. For practical reasons, it is often considered the projectively equivalent surface

$$\mathcal{S}'_f : \phi(X, Y, X+Z) = 1 + \frac{f(X) + f(Y) + f(Z) + f(X+Y+Z)}{(X+Y)(X+Z)} = 0.$$

Then $f(x)$ is planar on $\mathbb{F}_q$ if and only if all affine $\mathbb{F}_q$-rational points of $\mathcal{S}'_f$ satisfy $X = Y$ or $X = Z$.

It is clear that one can directly investigate the surfaces $\mathcal{S}_f$ or $\mathcal{S}'_f$ in order to prove that $f(x)$ is not planar on $\mathbb{F}_q$ using for instance Lang-Weil Theorem.

Proposition 11.2 *Let $f(x) \in \mathbb{F}_q[x]$. Suppose that $\mathcal{S}'_f$ possesses an absolutely irreducible $\mathbb{F}_q$-rational component not contained in $X = Y$, $X = Z$, or in the plane at infinity. If q is large enough then $f(x)$ is not planar on $\mathbb{F}_q$.*

A first result towards the classification of planar functions or at least exceptional planar functions is provided in [197].

Theorem 11.3 *[197, Theorem 1.3] Exceptional planar monomials have even degree.*

In their proof, the authors consider $f(x) = ax^t$, $a \neq 0$, and study the curve $\phi(X, Y, 1) = \frac{(X+1)^t + X^t + (Y+1)^t + Y^t + (X+Y)/a}{X+Y} = 0$ by applying the machinery described in Subsection 4.1 and proving that the curve $\phi(X, Y, 1) = 0$ possesses absolutely irreducible $\mathbb{F}_q$-rational components for any $a \in \mathbb{F}_q^*$. This is enough, together with the Aubry-Perret bound, to prove [197, Theorem 1.3].

This result has been extended in [177] to arbitrary exponents t, by studying an appropriate curve; see Result 11.7.

Theorem 11.4 *[177, Theorem 1.1] Let t be a positive integer such that $t^4 \leq q$, and let $a \in \mathbb{F}_q^*$. If $f(x) = ax^t$ is planar on $\mathbb{F}_q$ then t is a power of two. In particular, exceptional planar monomials are $\mathbb{F}_2$-linearized polynomials.*

11.2 Planar polynomials

A first result on exceptional planar polynomials was provided in [53, Theorem 2], by showing that the curve arising intersecting $\mathcal{S}'_f$ with the plane at infinity contains an $\mathbb{F}_q$-rational absolutely irreducible component. The desired conclusion is then obtained applying Lemma 4.15 and the Lang-Weil bound.

Theorem 11.5 *[53, Theorem 2] Let $f(x) \in \mathbb{F}_q[x]$ be of degree d. If $f(x)$ is exceptional planar, then either $d \in \{1, 2\}$ or $4 \mid d$.*

The strategy used in [22] to deal with the general polynomial case is slightly different. The authors consider the curve $\mathcal{C}'_f : \phi(X, Y, X+1) = 0$ obtained by intersecting $\mathcal{S}'_f$ with the plane $Z = X + Y$. Such a curve is projectively equivalent to $\mathcal{C}''_f : \overline{\phi}(X, 1, X+1, Y) = 0$, where $\overline{\phi}(X, Y, Z, T)$ is the homogenization of $\phi(X, Y, Z)$. If $f(x) = \sum_{i=0}^d A_i x^i$, then

$$\mathcal{C}''_f : Y^{d-2} + \sum_{i=3}^d A_i \frac{X^i + 1 + (X+1)^i}{X+1} Y^{d-i} = 0.$$

Also, if $f(x)$ is planar on $\mathbb{F}_q$, then all $\mathbb{F}_q$-rational points of $\mathcal{C}''_f$ satisfy $X = 1$ or $Y = 0$. Since $X = 1$, $Y = 0$, or the line at infinity are not components of $\mathcal{C}''_f$, by the Aubry-Perret bound if $\mathcal{C}''_f$ possesses an absolutely irreducible $\mathbb{F}_q$-rational component and the degree d is at most $\sqrt[4]{q}$ then $f(x)$ is not planar on $\mathbb{F}_q$; see [22, Proposition 2.1]. In the case in which $f(x)$ is not linearized, the existence of suitable component is obtained as a consequence of the existence of an $\mathbb{F}_q$-rational branch centered at the origin for $\mathcal{C}''_f$; see Theorem 4.4. The main ingredient is the use of local quadratic transformations. In particular, the authors prove, for any non-linearized polynomial $f(x)$, the existence of a chain of transformations $g(X, Y) \mapsto g(X, XY)/X^r$, $g(X, Y) \mapsto g(XY, Y)/Y^r$, $(X, Y) \mapsto (X, c_i X + Y)$, $c_i \in \mathbb{F}_q$, $r \in \mathbb{N}$, such that the origin is a simple point for the corresponding curve $\mathcal{C}'''_f$; see [22, Section 3].

Theorem 11.6 *[22, Theorem 1.1] Let $f(x) \in \mathbb{F}_q[x]$ be a polynomial of degree at most $q^{1/4}$. If $f(x)$ is planar on $\mathbb{F}_q$, then $f(x)$ is a $\mathbb{F}_2$-linearized polynomial.*

Result 11.7 *In the monomial case $f(x) = ax^t$, the curve $\mathcal{C}''_f$ reads*

$$Y^{t-2} + a \frac{X^d + 1 + (X+1)^d}{X+1} = 0$$

and this curve has been investigated in [177, Theorem 1.1] by showing that $\mathcal{C}''_f$ is a Kummer cover of $X = 0$; see Theorem 3.4. [177, Theorem 1.1] is implied by [22, Theorem 1.1].

It should be noted that there are examples of planar functions on $\mathbb{F}_q$ for even q that are not induced by $\mathbb{F}_2$-linearized polynomials, see [243, 197, 196, 130, 192]. All these examples have degree larger than $q^{1/4}$.

11.3 Non-exceptional planar polynomials

Among non-exceptional planar polynomials, quadratic ones have been mostly investigated.

Note that for a quadratic polynomial $f(x) = \sum_{j,k} a_{j,k} x^{q^j+q^k} \in \mathbb{F}_{q^n}[x]$ the function $\widehat{f}_\epsilon(x)$ in (11.1) reads $L_\epsilon(x) + \alpha_\epsilon$ for some $\mathbb{F}_q$-linearized polynomial $L_\epsilon(x)$ and $\alpha_\epsilon \in \mathbb{F}_{q^n}$. This means that $f(x)$ is planar if and only if $L_\epsilon(x)$ is a permutation of $\mathbb{F}_{q^n}$ for all $\epsilon \in \mathbb{F}_{q^n}^*$. This can be checked investigating the Dickson matrix (9.4) $M(L_\epsilon(x))$ of $L_\epsilon(x)$; see for instance [130, Proposition 3.2] and [25, Proposition 2.2]. In particular, for all $\epsilon \in \mathbb{F}_{q^n}^*$ we must have $\det(M(L_\epsilon(x))) \neq 0$. Now, let us define $X_i := \epsilon^{q^i}$, $i = 0, \ldots, n-1$. The condition $\det(M(L_\epsilon(x))) = 0$ can be written as

$$F_f(X_0, \ldots, X_{n-1}) = 0,$$

where the polynomial F_f depends on the coefficients $a_{j,k}$ of $f(x)$ and is homogenous of degree n. This means that it defines a hypersurface $\mathcal{X}_f$ of degree n in $\mathbb{P}^n(\mathbb{F}_{q^n})$. Consider the hypersurface $\mathcal{Y}_f$ obtained from $\mathcal{X}_f$ via the change of variables $X_i = \sum_i \xi^{q^i} Y_i$, where $\{\xi, \xi^q, \ldots, \xi^{q^{n-1}}\}$ is a normal basis of $\mathbb{F}_{q^n}$ over $\mathbb{F}_q$. It can be proved that $\mathcal{Y}_f \subset \mathbb{P}^n(\mathbb{F}_q)$. This immediately yields that if $\mathcal{X}_f$ is absolutely irreducible, so it is $\mathcal{Y}_f$ and therefore by Lang-Weil Theorem [142] it contains an $\mathbb{F}_q$-rational point $P \in \mathbb{P}^n(\mathbb{F}_q)$ which corresponds to a non-zero $\epsilon \in \mathbb{F}_{q^n}$ such that $\det(M(L_\epsilon(x))) = 0$ and therefore $f(x)$ is not planar. The same conclusion can be obtained if one can show the existence of an $\mathbb{F}_q$-rational absolutely irreducible component in $\mathcal{Y}_f$.

This approach is used for instance in [25] to classify planar binomials $f_{a,b}(x) = ax^{q^2+1} + bx^{q+1} \in \mathbb{F}_{q^3}[x]$.

Theorem 11.8 [25, Theorem 2.6] *Let m be a positive integer and denote 2^m by q. Then $f_{a,b}(x) = ax^{q^2+1} + bx^{q+1}$ is planar on $\mathbb{F}_{q^3}$ if and only if*

$$(a,b) = \left(\frac{s^{q+1}}{s^{q^2+q+1}+1}, \frac{s}{s^{q^2+q+1}+1} \right),$$

for some $s \in \mathbb{F}_{q^3}^*$ with $s^{q^2+q+1} \neq 1$.

Remark Commutative semifields (up to isotopism) over finite fields of characteristic two correspond to quadratic planar functions of characteristic two; see for instance [1, Theorem 9]. In particular, if G is a quadratic planar function over $\mathbb{F}_{2^n}$, then $(\mathbb{F}_{2^n}, +, \star)$ with multiplication $x \star y = xy + G(x+y) + G(x) + G(y)$ is a presemifield. Conversely, if $(\mathbb{F}_{2^n}, +, *)$ is a commutative presemifield, then there exist a strongly isotopic commutative presemifield $(\mathbb{F}_{2^n}, +, \star)$ and a planar function G of characteristic two such that $x \star y = xy + G(x+y) + G(x) + G(y)$. Equivalence between quadratic planar functions of characteristic two is the same as isotopism between the corresponding (pre-)semifields; see [243, Proposition 3.4]. Since the planar functions $f_{a,b}(x)$ are quadratic, the corresponding semifields' centers must contain $\mathbb{F}_q$. By the classification of semifields of order q^3 over $\mathbb{F}_q$ by Menichetti [171], such planar functions are not new.

12 APN functions

The study of APN functions is connected with differential cryptanalysis [34, 183, 184], since they have the property of being high resistant against differential attacks when they are used as substitution components of block ciphers.

Definition 12.1 A function $f : \mathbb{F}_{2^n} \to \mathbb{F}_{2^n}$ is said to be *APN* (Almost Perfect Nonlinear) or differentially 2-uniform if it has the following property

$$\forall \alpha \in \mathbb{F}_{2^n}^*, \forall \beta \in \mathbb{F}_{2^n} \implies \#\{x \in \mathbb{F}_{2^n} | f(x+\alpha) + f(x) = \beta\} \leq 2. \tag{12.1}$$

We refer to the survey [72] and references therein more details on APN functions and equivalence issues.

12.1 APN monomials

Monomial APN functions have been deeply investigated; see [45, 133, 134, 135, 113, 38].

Theorem 12.2 1. *[102, 134, 183] The function $f(x) = x^{2^j+1}$ is APN over any $\mathbb{F}_{2^n}$ with $\gcd(n, j) = 1$.*

2. *[81] The function $f(x) = x^{4^j - 2^j + 1}$ is APN over any $\mathbb{F}_{2^n}$ with $\gcd(n, j) = 1$.*

APN functions of type $f(x) = x^{2^j+1}$ are also known as APN Gold functions. The second family in the theorem above has been studied by Welch (unpublished) and they are usually called Kasami power functions [136]. As a definition, if $f(x) = x^t$ is APN over $\mathbb{F}_{2^n}$ then t is called an *APN exponent* for $\mathbb{F}_{2^n}$. Mimicking the definition of exceptional polynomials, i.e. polynomials defined over a finite field $\mathbb{F}_q$ which are permutations over infinitely many extensions $\mathbb{F}_{q^m}$ of $\mathbb{F}_q$, the term *exceptional APN function* refers to functions $f(x) \in \mathbb{F}_{2^n}[x]$ which are APN over infinitely many extensions $\mathbb{F}_{2^{mn}}$ of $\mathbb{F}_{2^n}$. Also, we say that t is an *exceptional exponent* if $f(x) = x^t$ is exceptional APN.

Conjecture 12.3 *[81, 133, 135] The only exceptional exponents are the Gold and Kasami-Welch numbers.*

Actually, this is only a special case of a more general conjecture.

Conjecture 12.4 *[7] If $f \in \mathbb{F}_q[x]$ is an exceptional APN polynomial, then $f(x)$ is CZZ-equivalent (see [48]) to a Gold or a Kasami-Welch monomial.*

The following is a well known connection between APN monomials and algebraic curves; see for instance [113].

Proposition 12.5 *The function $f_t(x) = x^t$ is APN over $\mathbb{F}_{2^n}$ if and only if the algebraic curve $\mathcal{A}_t$ defined by*

$$h_t(X, Y) := \frac{(X+1)^t + X^t + (Y+1)^t + Y^t}{(X+Y)(X+Y+1)} = 0 \tag{12.2}$$

has no affine $\mathbb{F}_{2^n}$-rational points off $X = Y$ and $X = Y + 1$.

Note that $\mathcal{A}_t$ is projectively equivalent to the curve $\mathcal{A}'_t : \frac{(X+Y+1)^t + X^t + Y^t + 1}{(X+Y)(X+1)(Y+1)} = 0$ and therefore one can investigate either $\mathcal{A}_t$ or $\mathcal{A}'_t$; see for instance [113].

The above connection with algebraic curves has been used to obtain non-existence results for exceptional APN monomials via the Hasse-Weil theorem or the Aubry-Perret bound.

Proposition 12.6 *[113, Proposition 1], [134]. If $\mathcal{A}_t$ has an absolutely irreducible $\mathbb{F}_2$-rational component then $\mathcal{A}_t$ possesses an affine $\mathbb{F}_{2^n}$-rational point (x_0, y_0) with $x_0 \neq y_0$ and $x_0 \neq y_0 + 1$ if n is sufficiently large.*

Proposition 12.7 *[135, Theorem 1 and Corollary 1.1] Let t be an odd integer, $t > 5$ and $t \neq 2^k + 1$ for any integer k. Write $t = 2^i \ell + 1$, $\ell \geq 3$ odd, $i \geq 1$. If $\gcd(\ell, 2^i - 1) < \ell$ then $\mathcal{A}_t$ has an absolutely irreducible $\mathbb{F}_2$-rational component. In particular, $f_t(x) = x^t$ is not APN over $\mathbb{F}_{2^n}$ for n large enough.*

The main ingredient in [135] is a deep investigation of the singular points of $\mathcal{A}_t$: their number and the intersection multiplicities of two putative components of $\mathcal{A}_t$ at them are considered separately; see for instance [135, Table 2]. This information is used to obtain the existence of an absolutely irreducible $\mathbb{F}_2$-rational component of $\mathcal{A}_t$ via the approach described in Subsection 4.1. Note that, in general, the curve $\mathcal{A}_t$ is singular, see e.g. [134] for the case $t = 15$, so one cannot immediately deduce the absolutely irreducibility of $\mathcal{A}_t$. Also, it is worth noting that Janwa, McGuire, and Wilson [133] conjectured that $\mathcal{A}'_t$ is absolutely irreducible if t is not Kasami or Gold. This was proved to be false in [113, Section 6], where using MAGMA [43] the authors found that $\mathcal{A}'_{205}$ is not absolutely irreducible. Finally, Conjecture 12.3 was proved completely in [113] using again the approach in Subsection 4.1, via the following result on curves $\mathcal{A}'_t$.

Theorem 12.8 *[113, Theorem 16] Let $t \mid 2^i - 1$ but $t \neq 2^i - 1$. Then $\mathcal{A}'_t$ has an absolutely irreducible $\mathbb{F}_2$-rational component. In particular, $f_t(x) = x^t$ is not APN over $\mathbb{F}_{2^n}$ for n large enough.*

12.2 APN polynomials

Although among APN functions, monomials have been mostly investigated, recently APN non-monomial functions have attracted attention; see e.g. [51, 7, 73, 193, 194].

Theorem 12.9 *[194] A polynomial $f(x) \in \mathbb{F}_{2^n}$ is APN if and only if the surface $\mathcal{S}_f$ defined by*

$$\varphi_f(X, Y, Z) := \frac{f(X) + f(Y) + f(Z) + f(X+Y+Z)}{(X+Y)(X+Z)(Y+Z)} = 0$$

has no affine $\mathbb{F}_{2^n}$-rational points off the planes $X + Y = 0$, $Y + Z = 0$, $X + Z = 0$.

Result 12.10 Note that when $f(x)$ is monomial x^t then φ_f reads

$$\frac{X^t + Y^t + Z^t + (X+Y+Z)^t}{(X+Y)(X+Z)(Y+Z)},$$

whose de-homogenization with respect to Z provides exactly the equation of the curve $\mathcal{A}'_t$. When $f(x)$ is not a monomial, $\mathcal{S}_f$ is a "real" surface.

Also, note that for a polynomial $f(x)$ of degree t, the curve $\mathcal{A}'_t$ coincides with the intersection of $\mathcal{S}_f$ with the plane at infinity.

By Theorem 12.9 and generalizations of Hasse-Weil Theorem to higher dimensions (e.g. Lang-Weil, or Ghorpade-Lachaud results [101]) Rodier [194] proved the following.

Theorem 12.11 *[194] Let $f(x) \in \mathbb{F}_{2^n}$ be a polynomial of degree $d \geq 9$. If $\mathcal{S}_f$ is absolutely irreducible or contains an absolutely irreducible $\mathbb{F}_{2^n}$-rational component and $d < 0.45 \sqrt[4]{q} + 0.5$ then $f(x)$ is not APN.*

We point out that all the non-existence results for exceptional APN functions have been obtained applying either Proposition 12.5 or Theorem 12.9. Clearly, these results can be also read as non-existence results in the $\deg(f(x)) < \sqrt[4]{q} = 2^{n/4}$ scenario. We collect in the following theorem all known results about non-existence of exceptional APN functions.

Theorem 12.12 *Let $f(x) \in \mathbb{F}_{2^n}[x]$.*

1. *[7, Theorem 2.3] If $\deg(f(x))$ is odd and different from Gold or Kasami-Welch numbers, then f is not exceptional APN.*

2. *[7, Theorem 2.4] If $\deg(f(x)) = 2e$, e odd, and $f(x)$ contains a term of odd degree then f is not exceptional APN.*

3. *[22, Proposition 1.4] If $\deg(f(x)) = 2e$, e odd, then $f(x)$ f is not exceptional APN.*

4. *[7, Theorem 3.1] If $f(x) = x^{2^k+1} + \sum_{j \leq 2^{k-1}+1} a_j x^j$ and there exists j such that $a_j \neq 0$ and $\mathcal{A}'_j$ is absolutely irreducible, then f is not exceptional APN. (The same holds if $\mathcal{A}'_j$ and $\mathcal{A}'_{2^k+1}$ do not share a component.)*

5. *[7, Theorem 3.2] If $f(x) = x^{2^k+1} + g(x)$, with $g(x)$ of degree $2^{k-1} + 2$, and $\gcd(k,n) = 1$ then f is not exceptional APN.*

6. *[73, 74, 75, 33] If $f(x) = x^{2^k+1} + g(x)$, with $\deg(g(x))$ odd, then f is not exceptional APN unless $\deg(g(x))$ is a Gold number and $\mathcal{A}'_j$ share a component with $\mathcal{A}'_{2^k+1}$ for all j such that $a_j \neq 0$ in $g(x) = \sum_j a_j x^j$.*

7. *[51, Theorem 9] If $\deg(f(x)) = 4e$, $e > 3$, and $\mathcal{A}'_e$ absolutely irreducible, then f is not exceptional APN; see also [193, Theorem 3.1].*

Other results for $f(x) = x^{2^k+1} + g(x)$, $\deg(g(x))$ even, can be found in [74, Section 2]. See [87] for partial results concerning Kasami-Welch degree polynomials.

Remark The authors in [7] use Lemma 4.15 together with Result 12.10 and investigations on the possible factorizations of $\varphi_f(X,Y,Z)$. Results in [73] are obtained considering intersections of $\mathcal{S}_f$ with specific planes. Results from [33] are obtained via a connection with permutation polynomials. The result in [22] solves one of the five pending cases listed in [72, Section 4] and strengthens [7, Theorem 2.4] essentially by removing the additional assumption that f has a term of odd degree. The proof relies on a machinery based on chains of local quadratic transformations, applied to a specific curve $\mathcal{D}_f$ obtained by intersecting $\mathcal{S}_f$ with a plane. Such a machinery shows the existence of an $\mathbb{F}_q$-rational branch centered at the origin for $\mathcal{D}_f$ which is sufficient to obtain the claim via Theorem 4.4. Actually, using this method, in the case $\deg(f(x)) \equiv 4 \pmod 8$ a partial result can be obtained: If $f(x) = \sum_{i \leq d} A_i x^i$ is exceptional APN of degree $d \equiv 4 \pmod 8$ then either $A_{d-1} = 0$ and either $A_{d-2} = A_{d-3} = 0$ or $A_d T^3 + A_{d-2} T + A_{d-3}$ has no roots in $\mathbb{F}_q$; see [13].

Acknowledgements

I am grateful to Matteo Bonini, Massimo Giulietti, Ariane Masuda, Kai-Uwe Schmidt, Marco Timpanella, Yue Zhou, Giovanni Zini, Ferdinando Zullo, for a number of valuable comments on an earlier draft.

References

[1] K. Abdukhalikov, Symplectic spreads, planar functions and mutually unbiased bases, *J. Algebr. Comb.* **41** (2015), 1055–1077.

[2] A. Akbary, D. Ghioca, Q. Wang, On permutation polynomials of prescribed shape, *Finite Fields Appl.* **15** (2009), 195–206.

[3] E. Aksoy, A. Çeşmelioğlu, W. Meidl, A. Topuzoğlu, On the Carlitz rank of a permutation polynomial, *Finite Fields Appl.* **15** (2009), 428–440.

[4] N. Anbar, A. Odžak, V. Patel, L. Quoos, A. Somoza, A. Topuzoğlu, On the difference between permutation polynomials, *Finite Fields Appl.* **49** (2018), 132–142.

[5] E. Artin, Quadratische körper im gebiet der höheren kongruenzen, I and II., *Math. Z.* **19** (1924), 153–206 and 207–246.

[6] M. Artin, *Grothendieck topologies*, Harvard University, Cambridge, Mass. (1962).

[7] Y. Aubry, G. McGuire, F. Rodier, A few more functions that are not APN infinitely often, finite fields theory and applications, *Contemporary Math.* **518** (2010), 23–31.

[8] Y. Aubry, M. Perret, A Weil theorem for singular curves, in *Arithmetic, geometry and coding theory* (eds. R. Pellikaan, M. Perret, S. G. Vladut), De Gruyter Proceedings in Mathematics, de Gruyter, Berlin (1996), pp. 1–7.

[9] T. Bai, Y. Xia, A new class of permutation trinomials constructed from Niho exponents, *Cryptogr. Commun.* **10** (2018), 1023–1036.

[10] S. Ball, A. Blokhuis, M. Lavrauw, Linear $(q+1)$-fold blocking sets in $PG(2,q^4)$, *Finite Fields Appl.* **6** (2000), 294–301.

[11] U. Bartocci, B. Segre, Ovali ed altre curve nei piani di Galois di caratteristica 2, *Acta Arith.* **18** (1971), 423–449.

[12] D. Bartoli, On a conjecture about a class of permutation trinomials, *Finite Fields Appl.* **52** (2018), 30–50.

[13] D. Bartoli, More on exceptional PN and APN functions. *In preparation* 2021.

[14] D. Bartoli, X. Hou, On a conjecture on permutation rational functions over finite fields, (2020), submitted.

[15] D. Bartoli, H. Borges, L. Quoos, Rational functions with small value set, *J. Algebra* **565** (2021), 675–690.

[16] D. Bartoli, B. Csajbók, M. Montanucci, On a conjecture about maximum scattered subspaces of $\mathbb{F}_{q^6} \times \mathbb{F}_{q^6}$, (2020), submitted.

[17] D. Bartoli, M. Giulietti, Permutation polynomials, fractional polynomials, and algebraic curves, *Finite Fields Appl.* **51** (2018), 1–16.

[18] D. Bartoli, M. Giulietti, G. Marino, O. Polverino, Maximum scattered linear sets and complete caps in Galois spaces, *Combinatorica* **38** (2018), 255–278.

[19] D. Bartoli, M. Giulietti, L. Quoos, G. Zini, Complete permutation polynomials from exceptional polynomials, *J. Number Theory* **176** (2017), 46–66.

[20] D. Bartoli, M. Giulietti, G. Zini, On monomial complete permutation polynomials, *Finite Fields Appl.* **41** (2016), 132–158.

[21] D. Bartoli, M. Montanucci, On the classification of exceptional scattered polynomials, *J. Combin. Theory Ser. A* **179** (2021), 105386.

[22] D. Bartoli, K.-U. Schmidt, Low-degree planar polynomials over finite fields of characteristic two, *J. Algebra* **535** (2019), 541–555.

[23] D. Bartoli, M. Timpanella, A family of permutation trinomials in $\mathbb{F}_{q^2}$, *Finite Fields Appl.* **70** (2021), 101781.

[24] D. Bartoli, M. Timpanella, On a generalization of planar functions, *J. Algebr. Comb.* **52** (2020), 187–213.

[25] D. Bartoli, M. Timpanella, A family of planar binomials in characteristic 2, *Finite Fields Appl.* **63** (2020), 101651.

[26] D. Bartoli, M. Timpanella, On trinomials of type $X^{n+m}(1 + AX^{m(q-1)} + BX^{n(q-1)})$, n, m odd, over $\mathbb{F}_{q^2}$, $q = 2^{2s+1}$, (2020), submitted.

[27] D. Bartoli, C. Zanella, F. Zullo, A new family of maximum scattered linear sets in $\mathrm{PG}(1, q^6)$, *Ars Math. Contemp.* **19** (2020), 125–145.

[28] D. Bartoli, Y. Zhou, Exceptional scattered polynomials, *J. Algebra* **509** (2018), 507–534..

[29] D. Bartoli, Y. Zhou, Asymptotics of Moore exponent sets, *J. Combin. Theory Ser. A* **175** (2020), 105281.

[30] D. Bartoli, K. Li, Y. Zhou, On the asymptotic classification of Kloosterman polynomials, *Proc. Amer. Math Soc.*, in press.

[31] L. A. Bassalygo, V. A. Zinoviev, On one class of permutation polynomials over finite fields of characteristic two, *Mosc. Math. J.* **15** (2015), 703–713.

[32] L. A. Bassalygo, V. A. Zinoviev, Permutation and complete permutation polynomials, *Finite Fields Appl.* **33** (2015), 198–211.

[33] T. P. Berger, A. Canteaut, P. Charpin, Y. Laigle-Chapuy, On almost perfect nonlinear functions over $\mathbb{F}_{2^n}$, *IEEE Trans. Inform. Theory* **52** (2006), 4160–4170.

[34] E. Biham, A. Shamir, Differential cryptanalysis of DES-like cryptosystems, *Contemp. Math.* **4** (1991), 3–72.

[35] S. R. Blackburn, T. Etzion, K. G. Paterson, Permutation polynomials, de Bruijn Sequences, and Linear Complexity, *J. Combin. Theory Ser. A* **76** (1996), 55–82.

[36] A. Blokhuis, M. Lavrauw, Scattered spaces with respect to a spread in $PG(n,q)$, *Geom. Dedicata* **81** (2000), 231–243.

[37] A. Blokhuis, M. Lavrauw, On two-intersection sets with respect to hyperplanes in projective spaces, *J. Combin. Theory Ser. A* **99** (2002), 377–382.

[38] C. Blondeau, K. Nyberg, Perfect nonlinear functions and cryptography, *Finite Fields Appl.* **32** (2015), 120–147.

[39] E. Bombieri, Counting points on curves over finite fields (d'après S.A. Stepanov), *Lecture Notes in Math.* **383** (1974), 234–241.

[40] E. Bombieri, H. Davenport, On two problems of Mordell, *Amer. J. Math.* **88** (1966), 61–70.

[41] H. Borges, Frobenius nonclassical components of curves with separated variables, *J. Number Theory* **159** (2016), 402–426.

[42] H. Borges, R. Conceição, On the characterization of minimal value set polynomials, *J. Number Theory* **133** (2013), 2021–2035.

[43] W. Bosma, J. Cannon, C. Playoust, The Magma algebra system. I. The user language, *J. Symbolic Comput.* **24** (1997), 235–265.

[44] A. Cafure, G. Matera, Improved explicit estimates on the number of solutions of equations over a finite field, *Finite Fields Appl.* **12** (2006), 155–185.

[45] A. Canteaut, Differential cryptanalysis of Feistel ciphers and differentially uniform mappings, in *Selected Areas on Cryptography SAC '97*, (1997), 172–184.

[46] I. Cardinali, O. Polverino, R. Trombetti, Semifield planes of order q^4 with kernel $\mathbb{F}_{q^2}$ and center $\mathbb{F}_q$, *European J. Combin.* **27** (2006), 940–961.

[47] C. Carlet, C. Ding, J. Yuan, Linear codes from perfect nonlinear mappings and their secret sharing schemes, *IEEE Trans. Inform. Theory* **51** (2005), 2089–2102.

[48] C. Carlet, P. Charpin, V. Zinoviev, Codes, bent functions and permutations suitable for DES-like cryptosystems, *Des. Codes Cryptogr.* **15** (1998), 125–156.

[49] L. Carlitz, Permutations in a finite field, *Proc. Amer. Math. Soc.* **4** (1953), 538.

[50] L. Carlitz, D. J. Lewis, W. H. Mills, E. G. Straus, Polynomials over finite fields with minimal value sets, *Mathematika* **8** (1961), 121–130.

[51] F. Caullery, A new large class of functions not APN infinitely often, *Des. Codes Cryptogr.* **73** (2014), 601–614.

[52] F. Caullery, K.-U. Schmidt, On the classification of hyperovals, *Adv. Math.* **283** (2015), 195–203.

[53] F. Caullery, K.-U. Schmidt, Y. Zhou, Exceptional planar polynomials, *Des. Codes Cryptogr.* **78** (2016), 605–613.

[54] J. S. Chahala, S. R. Ghorpade, Carlitz-Wan conjecture for permutation polynomials and Weil bound for curves over finite fields, *Finite Fields Appl.* **54** (2018), 366–375.

[55] W. Cherowitzo, Hyperovals in Desarguesian planes: an update, *Discrete Math.* **155** (1996), 31–38.

[56] S. Chowla, H. Zassenhaus, Some conjectures concerning finite fields, *Nor. Vidensk. Selsk. Forh. (Trondheim)* **41** (1968), 34–35.

[57] S. D. Cohen, The distribution of polynomials over finite fields, *Acta. Arith.* **17** (1970), 255–271.

[58] S. D. Cohen, Proof of a conjecture of Chowla and Zassenhaus on permutation polynomials, *Canad. Math. Bull.* **33** (1990), 230–234.

[59] S. D. Cohen, Permutation polynomials and primitive permutation groups, *Arch. Math. (Basel)* **57** (1991), 417–423.

[60] S. D. Cohen, G. L. Mullen, P. Jau-Shyong Shiue, The difference between permutation polynomials over finite fields, *Proc. Amer. Math. Soc.* **123** (1995), 2011–2015.

[61] S. D. Cohen, M. D. Fried, Lenstra's proof of the Carlitz–Wan conjecture on exceptional polynomials: an elementary version, *Finite Fields Appl.* **1** (1995), 372–375.

[62] S. D. Cohen, H. Niederreiter, I. E. Shparlinski, M. Zieve, Incomplete character sums and a special class of permutations, *J. Théor. Nombres Bordeaux* **13** (2001), 53–63.

[63] R. S. Coulter, R. W. Matthews, Planar functions and planes of Lenz-Barlotti class II, *Des. Codes Cryptogr.* **10** (1997), 167–184.

[64] R. S. Coulter, R. W. Matthews, Dembowski-Ostrom polynomials from Dickson polynomials, *Finite Fields Appl.* **16** (2010), 369–379.

[65] B. Csajbók, Scalar q-subresultants and Dickson matrices, *J. Algebra* **547** (2020), 116–128.

[66] B. Csajbók, G. Marino, O. Polverino, Classes and equivalence of linear sets in $PG(1,q^n)$, *J. Combin. Theory Ser. A* **157** (2018), 402–426.

[67] B. Csajbók, G. Marino, O. Polverino, C. Zanella, A new family of MRD-codes, *Linear Algebra Appl.* **548** (2018), 203–220.

[68] B. Csajbók, G. Marino, O. Polverino, Y. Zhou, MRD codes with maximum idealizers, *Discrete Math.* **343** (2020), 111985.

[69] B. Csajbók, G. Marino, O. Polverino, F. Zullo, Generalising the scattered property of subspaces, *Combinatorica*, in press.

[70] H. Davenport, D. J. Lewis, Notes on congruences (I), *Quart. J. Math.* **14** (1963), 51–60.

[71] M. De Boeck, G. Van de Voorde, The weight distributions of linear sets in $PG(1,q^5)$, https://arxiv.org/pdf/2006.04961.pdf 2020.

[72] M. Delgado, The state of the art on the conjecture of exceptional APN functions, *Note Mat.* **37** (2017), 41–51.

[73] M. Delgado, H. Janwa, On the conjecture on APN functions and absolute irreducibility of polynomials, *Des. Codes Cryptogr.* 82 (2017), 617–627.

[74] M. Delgado, H. Janwa, Some new results on the conjecture on exceptional APN functions and absolutely irreducible polynomials: the gold case, *Adv. Math. Commun.* **11** (2017), 389–395.

[75] M. Delgado, H. Janwa, Progress towards the conjecture on APN functions and absolutely irreducible polynomials, https://arxiv.org/pdf/1602.02576.pdf 2016.

[76] P. Deligne, La conjecture de Weil pour les surfaces $K3$, *Invent. Math.* **15** (1972), 206–226.

[77] P. Deligne, La conjecture de Weil I, *nst. Hautes Etudes Sci. Publ. Math.* **43** (1974), 273–307.

[78] P. Deligne, La conjecture de Weil II, *Inst. Hautes Etudes Sci. Publ. Math.* **52** (1980), 137–252.

[79] P. Dembowski, T. G. Ostrom, Planes of order n with collineation groups of order n^2, *Math. Z.* **103** (1968), 239–258.

[80] L. E. Dickson, The analytic representation of substitutions on a power of a prime number of letters with a discussion of the linear group, *Ann. Math.* **11** (1896), 65–120.

[81] J. F. Dillon, Geometry, codes and difference sets: exceptional connections, *Codes and Designs*, (eds. K. T. Arasu, A. Seress), de Gruyter, Berlin (2002), 73–85.

[82] N. Durante, R. Trombetti, Y. Zhou, Hyperovals in Knuth's binary semifield planes, *European J. Combin.* **62** (2017), 77–91.

[83] B. Dwork, On the rationality of the zeta function of an algebraic variety, *Amer J. Math.* **82** (1960), 631–648.

[84] G. L. Ebert, G. Marino, O. Polverino, R. Trombetti, Infinite families of new semifields, *Combinatorica* **29** (2009), 637–663.

[85] Y. Edel, On quadratic APN functions and dimensional dual hyperovals, *Des. Codes Cryptogr.* **57** (2010), 35–44.

[86] P. Ellingsen, P. Felke, C. Riera, P. Stănică, A. Tkachenko, C-differentials, multiplicative uniformity and (almost) perfect c-nonlinearity, *IEEE Trans. Inform. Theory* **66** (2020), 5781–5789.

[87] E. Férard, R. Oyono, F. Rodier, Some more functions that are not APN infinitely often. The case of Gold and Kasami exponents, *Contemp. Math.* **574** (2012), 27–36.

[88] N. Fernando, S. Ul Hasan, M. Pal, Reversed Dickson polynomials, Dembowski-Ostrom polynomials, and planar functions, *http://arxiv.org/abs/1905.01767v3* 2019.

[89] A. Ferraguti, G. Micheli, Full classification of permutation rational functions and complete rational functions of degree three over finite fields, *Des. Codes Cryptogr.* **88** (2020), 867–886.

[90] A. Ferraguti, G. Micheli, Exceptional scatteredness in prime degree, *J. Algebra* **565** (2021), 691–701.

[91] M. D. Fried, R. Guralnick, J. Saxl, Schur covers and Carlitz's conjecture, *Israel J. Math.* **82** (1993), 157–225.

[92] W. Fulton, *Algebraic curves*, Advanced Book Classics, Addison-Wesley Publishing Company Advanced Book Program, Redwood City, CA (1989).

[93] S. Gao, Absolute irreducibility of polynomials via Newton polytopes, *J. Algebra* **237** (2000), 501–520.

[94] S. Gao, A. Lauder, Decomposition of polytopes and polynomials, *Discrete Comput Geom* **26** (2001), 89–104.

[95] M. J. Ganley, E. Spence, Relative difference sets and quasiregular collineation groups, *J. Combin. Theory Ser. A* **19** (1975), 134–153.

[96] A. Garcia, J. F. Voloch, Wronskians and linear independence in fields of prime characteristic, *Manuscripta Math.* **59** (1987), 457–469.

[97] A. Garcia, J. F. Voloch, Fermat curves over finite fields, *J. Number Theory* **30** (1988), 345–356.

[98] J. von zur Gathen, Values of polynomials over finite fields, *Bull. Aust. Math. Soc.* **43** (1991), 141–146.

[99] G. van der Geer, Counting curves over finite fields, *Finite Fields Appl.* **32** (2015), 207–232.

[100] S. Ghorpade, G. Lachaud, Number of solutions of equations over finite fields and a conjecture of Lang and Weil, *Number Theory and Discrete Mathematics*, (eds. A. K. Agarwal, B. C. Berndt, C. F. Krattenthaler, G. L. Mullen, K. Ramachandra, M. Waldschmidt), Hindustan Book Agency, Gurgaon (2002), pp. 269–291.

[101] S. Ghorpade, G. Lachaud, Étale cohomology, Lefschetz theorem and number of points of singular varieties over finite fields, *Mosc. Math. J.* **2** (2002), 589–631.

[102] R. Gold, Maximal recursive sequences with 3-valued recursive cross-correlation functions, *IEEE Trans. Inform. Theory* **14** (1968), 154–165.

[103] S. W. Golomb, O. Moreno, On periodicity properties of Costas arrays and a conjecture on permutation polynomials, *IEEE Trans. Inform. Theory* **42** (1996), 2252–2253.

[104] J. Gomez-Calderon, A note on polynomials with minimal value set over finite fields, *Mathematika* **35** (1998), 144–148.

[105] J. Gomez-Calderon, D. J. Madden, Polynomials with small value set over finite fields, *J. Number Theory* **28** (1988), 167–188.

[106] G. Gwehenberger, Über die darstellung von permutationen durch polynome und rationale funktionen, Diss. TH, Wien, 1970.

[107] D. R. Hayes, A geometric approach to permutation polynomials over a finite field, *Duke Math. J.* **34** (1967), 293–305.

[108] R. Hartshorne, *Algebraic Geometry*, Graduate Texts in Mathematics, Springer-Verlag, New York (1977).

[109] H. Hasse, Abstrakte begründung der komplexen multiplikation und Riemannsche vermutung in funktionenkörpern, *Abh. Math. Sem. Hamburg* **10** (1934), 325–348.

[110] H. Hasse, Theorie der höheren differentiale in einem algebraischen funktionenkörper mit volkommenem konstantenkörper bei beliebiger charakteristik, I, II and III, *Journ. reine angew. Math. (Crelle)* **175** (1936), 55–62, 69–88, and 193–208.

[111] T. Helleseth, V. Zinoviev, New Kloosterman sums identities over $\mathbb{F}_{2^m}$ for all m, *Finite Fields Appl.* **9** (2003), 187–193.

[112] Ch. Hermite, Sur les fonctions de sept lettres, *C. R. Acad. Sci. Paris* **57** (1863), 750–757.

[113] F. Hernando, G. McGuire, Proof of a conjecture on the sequence of exceptional numbers classifying cyclic codes and APN functions, *J. Algebra* **343** (2011), 78–92.

[114] F. Hernando, G. McGuire, Proof of a conjecture of Segre and Bartocci on monomial hyperovals in projective planes, *Des. Codes Cryptogr.* **65** (2012), 275–289.

[115] F. Hernando, G. McGuire, F. Monserrat, On the classification of exceptional planar functions over $\mathbb{F}_p$, *Geom. Dedicata* **173** (2014), 1–35.

[116] J. W. P. Hirschfeld, *Projective geometries over finite fields (second ed.)*, Oxford Math. Monogr., Clarendon Press, Oxford University Press, New York (1998).

[117] J. W. P. Hirschfeld, G. Korchmáros, F. Torres, *Algebraic curves over a finite field*, Princeton Series in Applied Mathematics, Princeton University Press, Princeton (2008).

[118] J. W. P. Hirschfeld, L. Storme, The packing problem in statistics, coding theory and finite projective spaces: update 2001, *https://cage.ugent.be/ ls/max2000finalprocfilejames.pdf* 2001.

[119] H. D. L. Hollmann, Q. Xiang, Kloosterman sum identities over $\mathbb{F}_{2^m}$, *Discrete Math.* **279** (2004), 277–286.

[120] X. Hou, Permutation polynomials over finite fields – a survey of recent advances, *Finite Fields Appl.* **32** (2015), 82–119.

[121] X. Hou, Determination of a type of permutation trinomials over finite fields, II, *Finite Fields Appl.* **35** (2015), 16–35.

[122] X. Hou, Applications of the Hasse-Weil bound to permutation polynomials, *Finite Fields Appl.* **54** (2018), 113–132.

[123] X. Hou, On a class of permutation trinomials in characteristic 2, *Cryptogr. Commun.* **11** (2019), 1199–1210.

[124] X. Hou, On the Tu-Zeng permutation trinomial of type $(1/4, 3/4)$, *Discrete Math.* **344** (2021), 112241.

[125] X. Hou, A power sum formula by Carlitz and its applications to permutation rational functions of finite fields, *https://arxiv.org/pdf/2003.02246.pdf* 2020.

[126] X. Hou, Rational functions of degree four that permute the projective line over a finite field, https://arxiv.org/pdf/2005.07213.pdf 2020.

[127] X. Hou, A. Iezzi, An application of the Hasse-Weil bound to rational functions over finite field, *Acta Arith.* **195** (2020), 207-216.

[128] X. Hou, C. Sze, On a type of permutation rational functions over finite fields, *Finite Fields Appl.* **68** (2020), 101758.

[129] X. Hou, Z. Tu, X. Zeng, Determination of a class of permutation trinomials in characteristic three, *Finite Fields Appl.* **61** (2020), 101596.

[130] S. Hu, Sh. Li, T. Zhang, T. Feng, G. Ge, New pseudo-planar binomials in characteristic two and related schemes, *Des. Codes Cryptogr.* **76** (2015), 345–360.

[131] L. Işık, A. Topuzoğlu, A. Winterhof, Complete mappings and Carlitz rank, *Des. Codes Cryptogr.* **85** (2017), 121–128.

[132] L. Işık, A. Winterhof, Carlitz rank and index of permutation polynomials, *Finite Fields Appl.* **49** (2018), 156–165.

[133] H. Janwa, G. McGuire, R. Wilson, Double-error-correcting cyclic codes and absolutely irreducible polynomials over $GF(2)$, *J. Algebra* **178** (1995), 665–676.

[134] H. Janwa, R. M. Wilson, Hyperplane sections of Fermat varieties in P^3 in char. 2 and some applications to cyclic codes, in *Applied Algebra, Algebraic Algorithms and Error-Correcting Codes, Proceedings AAEEC-10* (eds. G. Cohen, T. Mora, O. Moreno), Springer-Verlag, New York Lecture Notes in Comput. Sci., **673** (1993).

[135] D. Jedlicka, APN monomials over $GF(2^n)$ for infinitely many n, *Finite Fields Appl.* **13** (2007), 1006–1028.

[136] T. Kasami, The weight enumerators for several classes of subcodes of the 2nd order binary Reed-Muller codes, *Inform. and Control* **18** (1971), 369–394.

[137] N. Katz, An overview of Deligne's proof of the Riemann hypothesis for varieties over finite fields, in *Mathematical Developments Arising from Hilbert Problems*, Amer. Math. Soc., Providence (1976), pp. 275–305.

[138] G. Kiss, T. Szőnyi, *Finite Geometries*, Chapman & Hall, CRC Press, Boca Raton (2020).

[139] S. L. Kleiman, Bertini and his two fundamental theorems. Studies in the history of modern mathematics, *Rend. Circ. Mat. Palermo (2)* **55** (1998), 9–37.

[140] S. Konyagin, F. Pappalardi, Enumerating permutation polynomials over finite fields by degree II, *Finite Fields Appl.* **12** (2006), 26–37.

[141] Y. Laigle-Chapuy, Permutation polynomials and applications to coding theory, *Finite Fields Appl.* **13** (2007), 58–70.

[142] S. Lang, A. Weil, Number of points of varieties in finite fields, *Amer. J. Math.* **76** (1954), 819–827.

[143] M. Lavrauw, Scattered spaces in Galois geometry, in *Contemporary Developments in Finite Fields and Applications,* World Sci. Publ., Hackensack, NJ (2016), 195–216.

[144] M. Lavrauw, G. Van de Voorde, Field reduction and linear sets in finite geometry, *Contemp. Math.* **632** (2015), 271–293.

[145] C. F. Laywine, G. L. Mullen, *Discrete mathematics using latin squares,* Wiley-Interscience Series in Discrete Mathematics and Optimization, Wiley-Interscience Publication, New York (1998).

[146] E. Leducq, Functions which are PN on infinitely many extensions of $\mathbb{F}_p$, p odd, *Des. Codes Cryptogr.* **75** (2015), 281–29.

[147] J. Levine, J. V. Brawley, Some cryptographic applications of permutation polynomials, *Cryptologia* **1** (1977), 76–92.

[148] J. Levine, R. Chandler, Some further cryptographic applications of permutation polynomials, *Cryptologia* **11** (1987), 211–218.

[149] L. Li, S. Wang, C. Li, X. Zeng, Permutation polynomials $(x - x + \delta)^{s_1} + (x - x + \delta)^{s_2} + x$ over $\mathbb{F}_{p^n}$, *Finite Fields Appl.* **51** (2018), 31–61.

[150] N. Li, On two conjectures about permutation trinomials over $\mathbb{F}_{3^{2k}}$, *Finite Fields Appl.* **47** (2017), 1–10.

[151] K. Li, L. Qu, C. Li, S. Fu, New permutation trinomials constructed from fractional polynomials, *Acta Arith.* **183** (2018), 101–116.

[152] N. Li, X. Zeng, A survey on the applications of Niho exponents, *Cryptogr. Commun.* **11** (2019), 509–548.

[153] R. Lidl, On cryptosystems based on polynomials and finite fields, *Lecture Notes in Comput. Sci.* **209** (1985), 10–15.

[154] R. Lidl, W. B. Müller, A note on polynomials and functions in algebraic cryptography, *Ars Combin.* **17** (1984), 223–229.

[155] R. Lidl, W. B. Müller, Permutation Polynomials in RSA-Cryptosystems, in *Advances in Cryptology, Proceedings of Crypto 83* (ed. D. Chaum), Springer, Boston, MA (1984), pp. 293-301.

[156] R. Lidl, H. Niederreiter, *Finite fields*, Addison-Wesley, Reading, MA (1983).

[157] A. Lipkovski, Newton polyhedra and irreducibility, *Math. Z.* **199** (1998), 119–127.

[158] L. Lombardo-Radice, Sul problema dei k-archi completi di $S_{2,q}$, *Boll. Un. Mat. Ital.* **11** (1956), 178–181.

[159] G. Longobardi, C. Zanella, A family of linear maximum rank distance codes consisting of square matrices of any even order, https://arxiv.org/pdf/2003.02246.pdf 2020.

[160] G. Lunardon, Linear k-blocking sets, *Combinatorica* **21** (2001), 571–581.

[161] G. Lunardon, MRD-codes and linear sets, *J. Combin. Theory Ser. A* **149** (2017), 1–20.

[162] G. Lunardon, G. Marino, O. Polverino, R. Trombetti, Maximum scattered linear sets of pseudoregulus type and the Segre variety $\mathcal{S}_{n,n}$, *J. Algebraic Combin.* **39** (2014), 807–831.

[163] G. Lunardon, O. Polverino, Blocking sets of size $q^t + q^{t-1} + 1$, *J. Combin. Theory Ser. A* **90** (2000), 148–158.

[164] G. Lunardon, O. Polverino, Blocking sets and derivable partial spreads, *J. Algebraic Combin.* **14** (2001), 49–56.

[165] G. Lunardon, R. Trombetti, Y. Zhou, On kernels and nuclei of rank metric codes, *J. Algebraic Combin.* **46** (2017), 313–340.

[166] G. Lunardon, R. Trombetti, Y. Zhou, Generalized twisted Gabidulin codes, *J. Combin. Theory Ser. A* **159** (2018), 79–106.

[167] J. Ma, T. Zhang, T. Feng, G. Ge, Some new results on permutation polynomials over finite fields, *Des. Codes Cryptogr.* **83** (2017), 425–44.

[168] C. R. MacCluer, On a conjecture of Davenport and Lewis concerning exceptional polynomials, *Acta Arith.* **12** (1967), 289–299.

[169] G. Marino, M. Montanucci, F. Zullo, MRD-codes arising from the trinomial $x^q + x^{q^3} + cx^{q^5} \in \mathbb{F}_{q^6}[x]$, *Linear Algebra Appl.* **591** (2020), 99–114.

[170] G. Marino, O. Polverino, R. Trombetti, Towards the classification of rank 2 semifields 6-dimensional over their center, *Des. Codes Cryptogr.* **61** (2011), 11–29.

[171] G. Menichetti, On a Kaplansky conjecture concerning three-dimensional division algebras over a finite field, *J. Algebra* **47** (1977), 400–410.

[172] W. H. Mills, Polynomials with minimal value sets, *Pacific J. Math.* **14** (1964), 225–241.

[173] M. Montanucci, C. Zanella, A class of linear sets in $\mathrm{PG}(1,q^5)$, https://arxiv.org/pdf/2006.04961.pdf 2019.

[174] K. Morrison, Equivalence for rank-metric and matrix codes and automorphism groups of Gabidulin codes, *IEEE Trans. Inform. Theory* **60** (2014), 7035–7046.

[175] G. L. Mullen, H. Niederreiter, Dickson polynomials over finite fields and complete mappings, *Canad. Math. Bull.* **30** (1987), 19–27.

[176] G. L. Mullen, D. Panario, *Handbook of finite fields*, Chapman and Hall/CRC, (2013).

[177] P. Müller, M. E. Zieve, Low-degree planar monomials in characteristic two, *J. Algebraic Combin.* **42** (2015), 695–699.

[178] W. B. Müller, W. Nöbauer, Some remarks on public-key cryptosystems, *Studia Sci. Math. Hungar.* **16** (1981),f 71–76.

[179] V. Napolitano, F. Zullo, Codes with few weights arising from linear sets, *Adv. Math. Commun.*, in press.

[180] V. Napolitano, O. Polverino, G. Zini, F. Zullo, Linear sets from projection of Desarguesian spreads, *Finite Fields Appl.* **71** (2021), 101798.

[181] H. Niederreiter, K. H. Robinson, Complete mappings of finite fields, *J. Aust. Math. Soc. Ser. A* **33** (1982), 197–212.

[182] Y. Niho, Multivalued cross-correlation functions between two maximal linear recursive sequence, Ph.D. dissertation, Univ. Southern Calif., Los Angeles, (1972),

[183] K. Nyberg, Differentially uniform mappings for cryptography, *Lecture Notes in Comput. Sci.* **765** (1994), 55–64.

[184] K. Nyberg, L. R. Knudsen, Provable security against differential attacks, *J. Cryptology* **8** (1995), 27–37.

[185] F. Oort, The Weil conjectures, *Nieuw Arch. Wiskd. (5)* **15** (2014), 211–219.

[186] Y. H. Park, J. B. Lee, Permutation polynomials and group permutation polynomials, *Bull. Aust. Math. Soc.* **63** (2001), 67–74.

[187] T. Penttila, G. F. Royle, Classification of hyperovals in PG(2,32), *J. Geom.* **50** (1994), 151–158.

[188] O. Polverino, Linear sets in finite projective spaces, *Discrete Math.* **22** (2010), 3096–3107..

[189] O. Polverino, G. Zini, F. Zullo, On certain linearized polynomials with high degree and kernel of small dimension, *J. Pure Appl. Algebra* **225** (2021), 106491.

[190] O. Polverino, F. Zullo, Connection between scattered linear sets and MRD-codes, *Bulletin of the ICA* **89** (2020), 46–74.

[191] A. Pott, Almost perfect and planar functions, *Des. Codes Cryptogr.* **78** (2016), 141–195.

[192] L. Qu, A new approach to constructing quadratic pseudo-planar functions over $\mathbb{F}_{2^n}$, *IEEE Trans. Inform. Theory* **62** (2016), 6644–6658.

[193] F. Rodier, Functions of degree 4e that are not APN infinitely often, *Cryptogr. Commun.* **3** (2011), 227–240.

[194] F. Rodier, Bornes sur le degré des polynômes presque parfaitement non-linéaires, *Contemp. Math.* **Contemp. Math.** (2009), 169–181.

[195] L. Rónyai, T. Szönyi, Planar functions over finite fields, *Combinatorica* **9** (1989), 315–320.

[196] Z. Scherr, M. E. Zieve, Some planar monomials in characteristic 2, *Ann. Comb.* **18** (2014), 723–729.

[197] K.-U. Schmidt, Y. Zhou, Planar functions over fields of characteristic two, *J. Algebraic Combin.* **40** (2014), 503–526.

[198] A. Schinzel, *Polynomials with special regard to reducibility*, Encyclopedia of Mathematics and its applications, Cambridge University press, Cambridge, Mass. (2000).

[199] F. K. Schmidt, Analytische zahlentheorie in körpern der charakteristik p, *Math. Z.* **33** (1931), 1–32.

[200] W. Schmidt, *Equations over finite fields: an elementary approach*, Lecture Notes in Math. vol. 536, Springer, New York (1976).

[201] B. Segre, Ovals in a finite projective plane, *Canad. J. Math.* **7** (1955), 414–416.

[202] B. Segre, Ovali e curve σ nei piani di Galois di caratteristica due, *Atti dell'Accad. Naz. Lincei* **32** (1962), 785–790.

[203] B. Segre, Introduction to Galois geometries (ed. by J.W.P. Hirschfeld), *Mem. Accad. Naz. Lincei (8)* **8** (1967), 133–236.

[204] J. P. Serre, Zeta and L functions, in *Arithmetical Algebraic Geometry (Proc. Conf. Purdue Univ., 1963)*, Harper & Row, New York (1965), pp. 82–92.

[205] J. P. Serre, Facteurs locaux des fonctions zêta des variétés algébriques (définitions et conjectures), *Séminaire Delange-Pisot-Poitou. Théorie des nombres* **11** (1969-1970), Exposé no. 19.

[206] J. P. Serre, Sur le nombre de points rationnels d'une courbe algébrique sur un corps fini, *C. R. Acad. Sci. Paris* **296** (1983), 397–402.

[207] R. Shaheen, A. Winterhof, Permutations of finite fields for check digit systems, *Des. Codes Cryptogr.* **57** (2010), 361–371.

[208] J. Sheekey, A new family of linear maximum rank distance codes, *Adv. Math. Commun.* **10** (2016), 475–488.

[209] J. Sheekey, MRD codes: constructions and connections, in *Combinatorics and finite fields: Difference sets, polynomials, pseudorandomness and applications* (eds. K.-U. Schmidt and A. Winterhof), *Radon Series on Computational and Applied Mathematics*, De Gruyter, Berlin (2019), pp. 255–286.

[210] J. Sheekey, G. Van de Voorde, Rank-metric codes, linear sets and their duality, *Des. Codes Cryptogr.* **88** (2020), 655–675.

[211] P. Stănică, S. Gangopadhyay, A. Chaturvedi, A. K. Gangopadhyay, S. Maitra, Investigations on bent and negabent functions via the nega-Hadamard transform, *IEEE Trans. Inform. Theory* **58** (2012), 4064–4072.

[212] H. Stichtenoth, *Algebraic function fields and codes (2nd edition)*, Graduate Texts in Mathematics, vol. *254*, Springer, Berlin (2009).

[213] K.-O. Stöhr, J. F. Voloch, Weierstrass points and curves over finite fields, *Proc. London Math. Soc.* **52** (1986), 1–19.

[214] T. Szőnyi, Small complete arcs in Galois planes, *Geom. Dedicata* **18** (1985), 161–172.

[215] O. Y. Takeshita, Permutation polynomial interleavers: an algebraic-geometric perspective, *IEEE Trans. Inform. Theory* **53** (2007), 2116–2132.

[216] A. Tietäväinen, On non-residues of a polynomial, *Ann. Univ. Turku Ser. A* **94** (1966), 6 pp.

[217] Z. Tu, X. Zeng, Two classes of permutation trinomials with Niho exponents, *Finite Fields Appl.* **53** (2018), 99–112.

[218] Z. Tu, X. Zeng, A class of permutation trinomials over finite fields of odd characteristic, *Cryptogr. Commun.* **11** (2019), 563–583.

[219] Z. Tu, X. Zeng, Y. Jiang, Two classes of permutation polynomials having the form $(x^{2^m} + x + \delta)^s + x$, *Finite Fields Appl.* **31** (2015), 12–24.

[220] Z. Tu, X. Zeng, C. Li, T. Helleseth, A class of new permutation trinomials, *Finite Fields Appl.* **50** (2018), 178–195.

[221] T. J. Tucker, M. E. Zieve, Permutation polynomials, curves without points, and latin squares, 2000.

[222] G. Turnwald, A new criterion for permutation polynomials, *Finite Fields Appl.* **1** (1995), 64–82.

[223] P. Vandendriessche, Classification of the hyperovals in PG(2, 64), *Electron. J. Combin.* **26** (2019), #P2.351.

[224] D. Wan, On a conjecture of Carlitz, *J. Austral. Math. Soc. (Series A)* **43** (1987), 375–384.

[225] D. Wan, Permutation polynomials and resolution of singularities over finite fields, *Proc. Amer. Math. Soc.* **110** (1990), 303–309.

[226] D. Wan, A generalization of the Carlitz conjecture, in *Finite fields, Coding Theory, and Advances in Communications and Computing: Proceedings of the International Conference held at the University of Nevada, Las Vegas, Nevada, August 7–10, 1991* (eds. G. L. Mullen, P. J.-S. Shiue), Lecture Notes in Pure and Applied Mathematics vol. *141*, Marcel Dekker, Inc., New York (1991), pp. 431–432.

[227] Q. Wang, Polynomials over finite fields: an index approach, *Radon Ser. Comput. Appl. Math.* **23** (2019), 319–348.

[228] A. Weil, On the Riemann hypothesis in function fields, *Proc. Nat. Acad. Sci. U.S A.* **27** (1941), 345–347.

[229] A. Weil, Numbers of solutions of equations in finite fields, *Bull. Amer. Math. Soc.* **55** (1949), 497–508.

[230] G. Weng, X. Zeng, Further results on planar DO functions and commutative semifields, *Des. Codes Cryptogr.* **63** (2012), 413–423.

[231] K. S. Williams, On extremal polynomials, *Canad. Math. Bull.* **10** (1967), 585–594.

[232] K. S. Williams, On exceptional polynomials, *Canad. Math. Bull.* **11** (1968), 179–282.

[233] A. Winterhof, Generalizations of complete mappings of finite fields and some applications, *J. Symbolic Comput.* **64** (2014), 42–52.

[234] G. Wu, N. Li, T. Helleseth, Y. Zhang, Some classes of monomial complete permutation polynomials over finite fields of characteristic two, *Finite Fields Appl.* **28** (2014), 148–165.

[235] G. Wu, N. Li, T. Helleseth, Y. Zhang, Some classes of complete permutation polynomials over $\mathbb{F}_q$, *Sci. China Math.* **58** (2015), 2081–2094.

[236] J. Yuan, C. Ding, Four classes of permutation polynomials of $\mathbb{F}_{2^m}$, *Finite Fields Appl.* **13** (2007), 869–876.

[237] J. Yuan, C. Ding, H. Wang, J. Pieprzyk, Permutation polynomials of the form $(x^p - x + \delta)^s + L(x)$, *Finite Fields Appl.* **14** (2008), 482–493.

[238] C. Zanella, A condition for scattered linearized polynomials involving Dickson matrices, *J. Geom.* **110** (2019), 50.

[239] C. Zanella, F. Zullo, Vertex properties of maximum scattered linear sets of $PG(1, q^n)$, *Discrete Math.* **343** (2020), 111800.

[240] X. Zeng, X. Zhu, L. Hu, Two new permutation polynomials with the form $(x^{2^k} + x + \delta)^s + x$ over $\mathbb{F}_{2^n}$, *Appl. Algebra Eng. Comm. Comput.* **21** (2010), 145–150.

[241] Z. Zha, L. Hu, Some classes of permutation polynomials of the form $(x^{p^m} - x + \delta)^s + x$ over $\mathbb{F}_{p^{2m}}$, *Finite Fields Appl.* **40** (2016), 150–162.

[242] D. Zheng, M. Yuan, L. Yu, Two types of permutation polynomials with special forms, *Finite Fields Appl.* **56** (2019), 1–16.

[243] Y. Zhou, $(2^n, 2^n, 2^n, 1)$-relative difference sets and their representations, *J. Combin. Des.* **21** (2013), 563–584.

[244] Y. Zhou, A. Pott, A new family of semifields with 2 parameters, *Adv. Math.* **234** (2013), 43–60.

[245] Y. Zhou, L. Qu, Constructions of negabent functions over finite fields, *Cryptogr. Commun.* **9** (2017), 165–180.

[246] M. E. Zieve, On some permutation polynomials over $\mathbb{F}_q$ of the form $x^r h(x^{(q-1)/d})$, *Proc. Amer. Math. Soc.* **137** (2009), 2209–2216.

[247] M. E. Zieve, Planar functions and perfect nonlinear monomials over finite fields, *Des. Codes Cryptogr.* **75** (2015), 71–80.

[248] M. E. Zieve, Planar polynomials over finite field. 2013.

[249] G. Zini, F. Zullo, Scattered subspaces and related codes, (2020), submitted.

[250] G. Zini, F. Zullo, On the intersection problem for linear sets in the projective line, (2020), submitted.

Dipartimento di Matematica e Informatica
Università degli Studi di Perugia
Perugia, Italy
daniele.bartoli@unipg.it

Decomposing the edges of a graph into simpler structures

Marthe Bonamy

Abstract

We will review various ways to decompose the edges of a graph into few simple substructures. We will mainly focus on variants of edge colouring, and discuss specifically the discharging method and re-colouring techniques.

All graphs considered here are simple and finite, unless specified otherwise. We assume the reader has some familiarity with elementary notions in graph theory. This chapter is meant as a walkthrough of two techniques that can be useful and that are perhaps not as widely known as they could be. The first one is the discharging method, which makes up for most of this chapter (Sections 2 to 5), and the second one is the use of re-colouring (Section 6).

1 Introduction

A recurrent theme in graph theory is how problems tend to be hard in general but easy on structures like trees – or, obviously, independent sets or matchings. Unfortunately, not all graphs are trees, let alone independent sets. We can therefore try and measure "how far" a given graph is from being "nice", in the hope that when the measure is small problems can be solved relatively fast. There are multiple ways of doing so; we focus on colouring-like problems, where the goal is to partition the vertices or the edges of the graph so that each part induces a nice structure and so that not too many different parts are involved.

In this survey, we focus mostly on edge decompositions, though vertex partitions will also occasionally be convenient to consider. The simplest and most famous way to decompose the edges of a graph is through *proper edge colouring*. For any graph G and integer k, a proper edge k-colouring of a graph G is a mapping $E(G) \to \{1, 2, \ldots, k\}$ such that any two incident edges of G receive distinct colours. Though clearly any graph G admits a proper edge $|E(G)|$-colouring, this is usually not a satisfying bound: we seek to use as few colours as possible. The smallest integer k such that G admits a proper edge k-colouring is the *chromatic index* of G, denoted $\chi'(G)$.

Before we consider upper-bounds on $\chi'(G)$, let us consider lower-bounds. For a vertex u of a graph G, we denote $d_G(u)$ the degree of u in G – we often drop the subscript when there is no ambiguity. By definition, for any graph G and any vertex u of G, we have $\chi'(G) \geq d_G(u)$. Let $\Delta(G)$ be the maximum degree of a vertex in G. The previous remark boils down to $\chi'(G) \geq \Delta(G)$. By a greedy argument, we easily obtain $\chi'(G) \leq 2\Delta(G) - 1$. However, a much stronger result was obtained by Vizing in 1964:

Theorem 1.1 *[30] For every graph G, we have $\chi'(G) \leq \Delta(G) + 1$.*

Theorem 1.1 states that given a graph G, its chromatic index $\chi'(G)$ can only be equal to $\Delta(G)$ or $\Delta(G) + 1$. The classification of graphs depending on this attracted considerable interest (for instance [27]). When trying to argue that all graphs G in some class can be coloured with $\Delta(G)$ colours, it can be convenient to consider a partial colouring (i.e. a colouring of some but not all the edges) and extend it to the whole graph. However, the extension problem is rather different from the original problem, as not all edges have the same colours already used on incident edges. A colour available for some uncoloured edge might not be available for another. The corresponding problem is called *list edge colouring*. Given an edge list assignment $L : E(G) \to \mathcal{P}(\mathbb{N})$, the graph G is *edge L-colourable* if there is a colouring $\alpha : E(G) \to \mathbb{N}$ such that $\alpha(e) \in L(e)$ for every $e \in E(G)$, and $\alpha(uv) \neq \alpha(vw)$ for any vertex v and any two neighbours u and w of v.

The behaviour of this "list version" is starkly different from what happens with vertex colouring, and the most frustratingly elusive conjecture regarding edge colouring is arguably the following:

Conjecture 1.2 *[16] For every graph G and every edge list assignment L of G such that $|L(e)| \geq \chi'(G)$ for every $e \in E(G)$, the graph G is edge L-colourable.*

While Theorem 1.1 admits a rather simple proof (see Section 6.2), the best we know about Conjecture 1.2 is that it holds asymptotically [17] (the proof relies crucially on the probabilistic method). However, there are special cases where Conjecture 1.2 is known to be true. The first notable example was that of bipartite graphs [11], though it was later generalized as follows.

Theorem 1.3 *[6] For every bipartite multigraph $G = (V, E)$ and every edge list assignment $L : E \to \mathcal{P}(\mathbb{N})$ such that $|L(u,v)| \geq max(d(u), d(v))$, the multigraph G is edge L-colourable.*

Theorem 1.3 will prove to be extremely useful for extending a partial colouring to the whole graph. However, a simple corollary which can easily be proved directly will also be extensively used:

Lemma 1.4 *For every even cycle $G = (V, E)$ and every edge list assignment $L : E \to \mathcal{P}(\mathbb{N})$ such that $|L(u,v)| \geq 2$, the graph G is edge L-colourable.*

We are now ready to discuss tools to prove decomposition results. When discussing planar graphs, the best results regarding edge colouring and list edge colouring were obtained through the so-called discharging method, which was instrumental in proving the Four Colour Theorem.

2 What is a discharging argument?

First, we would like to emphasize that there is no formal definition of what a discharging method is. A discharging proof usually follows this outline:

1. Assume we have a set S of various elements that interact in a given way (it could be the vertices, edges and faces in some planar embedding, or the n integers from 1 to n, etc), and that we want to compute a function f of S that can be expressed, for some function ω, as $f(S) = \sum_{a \in S} \omega(a)$.

2. Assign to each element a of S a weight of $\omega(a)$.

3. Design discharging rules in order to reorganize the weight along S while maintaining a constant total weight.

4. Compute, for each $a \in S$, the new weight $\omega'(a)$.

5. Observe that $\sum_{a \in S} \omega'(a)$ is easier to compute, and derive the value $f(S)$.

Mostly, we say that a proof is based on a discharging method when it relies on the idea of local counting arguments in order to derive a global formula. In particular, in a discharging proof, the local counting arguments are usually presented in terms of *discharging*. The motivation behind that is to make the calculation easier. By defining formal rules about who gives what weight to who, we ensure that the total weight is constant, and in particular that no weight is lost or counted twice.

In a way, we could say that Gauss' idea[1] about how to compute the sum of all integers from 1 to n is a primitive example of a discharging method. Here, $S = \{1, 2, \ldots, n\}$, and we want to compute $N = \sum_{a \in S} a$. Therefore, we assign to each integer its own value as a weight. Then we design a single discharging rule, that every integer a gives half its initial weight to the element $n + 1 - a$ (see Figure 1). Now, all the elements have the same[2] weight of $\frac{n+1}{2}$. It is now extremely easy to derive N as $\sum_{a \in S} a = N = n \times \frac{n+1}{2}$.

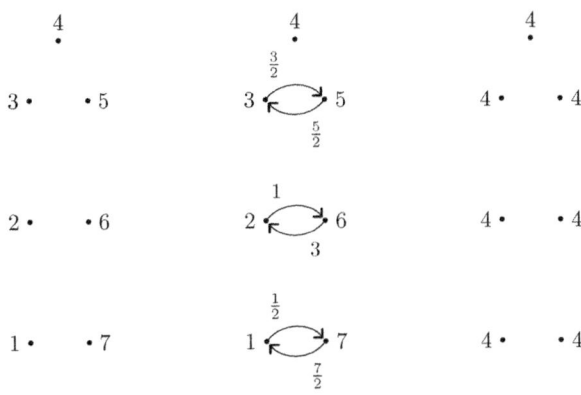

Figure 1: An illustration of Gauss' argument for $n = 7$ from a discharging perspective, from left to right.

Another nice, simple illustration of what a discharging method can be is a graphical proof of Euler's formula in planar graphs. The proof was initially presented in the more general setting of polyhedra [28]. We insist that the proof sketched here is purely for illustrational purpose, and has no pretention to rigor. Here, we have a planar embedding $\mathcal{M} = (V, E, F)$ of our favorite planar graph, and we want to compute the exact value of $|V| - |E| + |F|$, so we set $S = V \cup E \cup F$. We can pick a notion of right and left in the embedding, and without loss of generality assume

---

[1] Or so the legend has it... [12]
[2] The use of the word "discharging" makes all the more sense here: at the end the total weight is uniformly distributed, thus reaching equilibrium.

that $\mathcal{M}$ is such that no edge is perfectly horizontal. We assign to each vertex and each face (including the outer face) a weight of 1, and to each edge a weight of -1. Again, we define a single discharging rule, that each vertex and each edge gives all its weight to the face immediately to its right (see Figure 2): this is well-defined since no edge is perfectly horizontal. All vertices and edges have a final weight of 0, so we can concentrate on faces. No matter how bizarre the embedding can be, the vertices and edges immediately to the left of an inner face necessarily form a sequence of alternating vertices and edges, whose two endpoints are vertices who give their weight to another face. Thus, the number of vertices immediately to the left of an inner face will always be one less than the number of edges. It follows that the final weight of any inner face is 0. We can then safely disregard them, and look only at the final weight of the outer face. The same analysis stands in that case, except that there may be more than one sequence (if the graph is not connected, one for each connected component), and that the two endpoints of each sequence do not give their weight to another face. Consequently, each sequence has one more vertex than edges, and contributes a weight of 1 to the outer face. For $\#cc$ the number of connected components, the final weight of the outer face is thus $1 + \#cc$, hence the conclusion that $|V| - |E| + |F| = 1 + \#cc$.

For example, let us consider the same setting as in the above paragraph, up until the discharging rule. The initial weights are thus of 1 for every vertex or face, and of -1 for every edge. Here, we set a rule that every edge shares all its weight equally among its two incident vertices and two incident faces (if the edge is twice incident to the same face, the face still receives two parts of the share). It follows immediately that every vertex v has a final weight of $1 - \frac{d(v)}{4}$, and similarly for each face. Consequently, if we multiply by 4 to get rid of fractions and combine with the previous remarks, we get that

$$\sum_{v \in V}(4 - d(v)) + \sum_{f \in F}(4 - d(f)) = 4 + 4\#cc.$$

Instead of sharing equally between incident vertices and incident faces, we can choose to unbalance it in favor of vertices or faces. In fact, we can choose any ratio, and thus obtain for any $a, b \in \mathbb{N}$ that

$$\sum_{v \in V}(2(a+b) - a \times d(v)) + \sum_{f \in F}(2(a+b) - bd(f)) = 2 \times (1 + \#cc) \times (a+b).$$

For example, if we take $a = 1$ and $b = 2$, we derive

$$\sum_{v \in V}(d(v) - 6) + \sum_{f \in F}(2d(f) - 6) = -6 \times (1 + \#cc) < 0. \qquad (2.1)$$

The simple action of assigning to each vertex and each face a good weight, whose total on the graph we know to be negative, can already be informative. Here, let us set that each vertex is assigned a weight of $d(v) - 6$, and each face a weight of $2d(f) - 6$: we know the total to be negative by (2.1). Since every face of a simple planar graph has degree at least 3, no face has a negative initial weight. Therefore, at least one vertex must have a negative weight. In other words, every planar graph

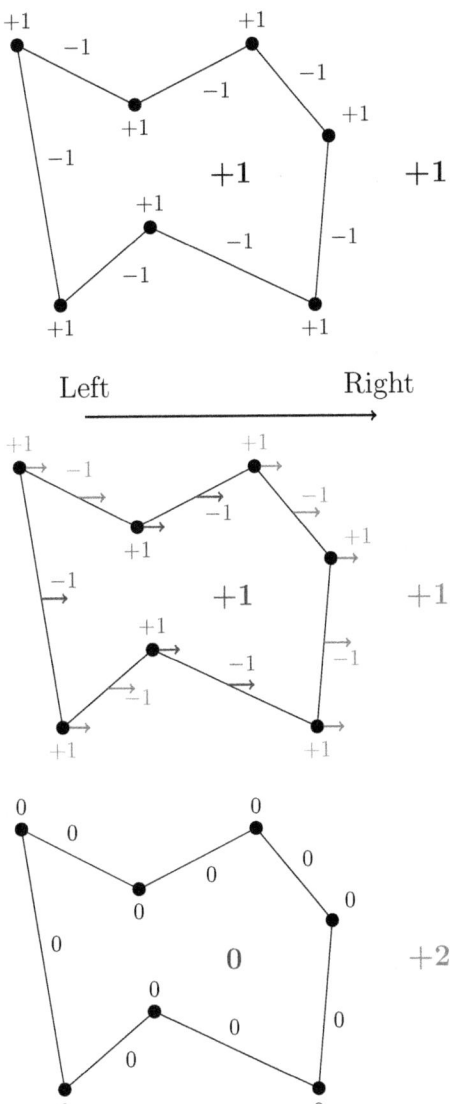

Figure 2: An illustration of the proof of Euler's formula from a discharging perspective, from top to bottom. Here we consider a simple graph with just two faces.

contains a vertex of degree at most 5. We call this an *unavoidable configuration* (a planar graph cannot avoid containing it).

We can use this fact to prove the following lemma.

Lemma 2.1 (Folklore) *Every planar graph is 6-colourable.*

Proof Assume for contradiction that some planar graphs are not 6-colourable. Among them, take G to be one with the minimum number of vertices. If G contains a vertex u of degree at most 5, we can 6-colour $G \setminus \{u\}$ by minimality. Then u has at most 5 neighbours, so there is one of the six colours that does not appear on any neighbour of u, and we colour u with it. Therefore, we can extend the 6-colouring of $G \setminus \{u\}$ to G, a contradiction. For this reason, that configuration (the graph contains a vertex of degree at most 5) is called *reducible*.

The graph G is thus a planar graph that does not contain a vertex of degree at most 5. However, we argued that this is an unavoidable configuration for planar graphs, a contradiction. Consequently, no counter-example exists and every planar graph is 6-colourable. □

This argument can be pushed a bit further to obtain that every planar graph is 5-colourable. Here no discharging at all is involved, just a good weight assignment to the vertices. Introducing even a tiny amount of discharging strengthens the conclusion a lot. Wernicke proved the first structural lemma involving discharging, as follows.

Lemma 2.2 *[33] Every planar graph G contains a vertex of degree at most 4 or a vertex of degree 5 adjacent to a vertex of degree 5 or 6.*

Proof By contradiction. Assume that G is a planar graph whose every vertex has degree at least 5, and such that the neighbours of every vertex of degree 5 are all of degree at least 7. Let $\mathcal{M} = (V, E, F)$ be a planar embedding of G. For simplicity, we only present the proof when $\mathcal{M}$ is a triangulation. We could argue that it is sufficient to prove that case, or adapt the proof to deal with the case where $\mathcal{M}$ is not a triangulation. However, our goal is merely to give an idea of the proof.

We assign an initial weight ω of $\omega(v) = d(v) - 6$ to each vertex $v \in V$, and $\omega(f) = 2d(f) - 6$ to each face $f \in F$. We know by (2.1) that the total weight of the graph is negative. We try to redistribute the weight along the graph in such a way that every vertex and every face has a non-negative final weight, a contradiction. Note that since $\mathcal{M}$ is a triangulation, every face f has an initial weight of exactly 0, so we can concentrate on the vertices. Only vertices of degree 5 have a negative weight, and every vertex of degree at least 7 has a positive weight.

We define a single rule R_1: for every vertex v of degree at least 7,

- Rule R_1 is when v is adjacent to three vertices u_1, u_2, u_3, such that (v, u_1, u_2) and (v, u_2, u_3) are faces, and $d(u_2) = 5$. Then v gives $\frac{1}{5}$ to u_2 (see Figure 3).

Let ω' be the final weight assignment on the graph, after application of Rule R_1. Let us argue that $\omega'(v) \geq 0$ for every vertex $v \in V$ and $\omega'(f) \geq 0$ for every face $f \in F$. The second holds immediately because $\omega(f) = \omega'(f)$ for every face $f \in F$.

Let u be a vertex of degree 5. By assumption, all five neighbours of u are of degree at least 7, and all give incident faces are triangles. Therefore, Rule R_1 applies fives times, and the vertex u has an initial weight of -1, receives $5 \times \frac{1}{5}$, and thus has a non-negative final weight.

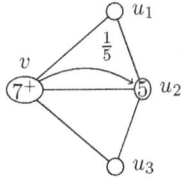

Figure 3: The discharging rule R_1.

Let u be a vertex of degree 6. By definition, Rule R_1 does not apply, and $\omega'(u) = \omega(u) = 0$, so u has a non-negative final weight.

Let u be a vertex of degree at least 7. By definition and since all incident faces are triangles, Rule R_1 applies for each neighbour of degree 5. By assumption, the vertex u cannot belong to a triangle with two vertices of degree 5. Therefore, at most $\frac{d(u)}{2}$ neighbours of u may be of degree 5. The vertex u has an initial weight of $d(u) - 6 \geq 1$, gives at most $\frac{d(u)}{2} \times \frac{1}{5}$ away, thus has a final weight of $\frac{9}{10}d(u) - 6$, which is non-negative since $d(u) \geq 7$.

□

Lemma 2.2 is a proof that these two configurations (a vertex of degree at most 4, a vertex of degree 5 adjacent to a vertex of degree 5 or 6) are unavoidable for a planar graph. Lemma 2.2 can in fact be used to obtain a linear-time algorithm to 5-colour a planar graph [26], which is asymptotically best possible.

When a colouring theorem is said to be proved through a discharging argument, it usually means that its proof has the following outline:

Assume for contradiction that the theorem is not true, and consider G a minimal counter-example (for some order $\prec$ on graphs).

1. Prove that there are some configurations $\{C_1, C_2, \ldots, C_p\}$ that G cannot contain (these are *reducible configurations*). Typically, argue that if G contains a configuration C_i, then there exists a smaller graph G' that is a counter-example to the theorem, a contradiction to the minimality of G.

2. Use a discharging method to prove the structural lemma that every graph satisfying the theorem hypotheses must contain one of $\{C_1, C_2, \ldots, C_p\}$ (this is an *unavoidable set of configurations*).

In the proof of Lemma 2.2, the discharging rules are very simple. In particular, the charge is only sent to a vertex at a bounded distance (here only to neighbours). The discharging method is based on the idea of using local counting arguments in order to derive a global formula. However, it could happen that there is a sub-structure of unbounded size on which the sum of weights is easy to compute, actually easier than through local arguments. Then, it might be interesting to design discharging rules that can send some charge arbitrarily far away. We then say that the discharging argument is *global*. This variant on the discharging method was only introduced in 2007 by Borodin, Ivanova and Kostochka [7]. When there is a single global discharging rule, we typically call it R_g (where 'g' stands for global). For the design of a global discharging rule, we are often interested in using the sub-structure without having to make everything explicit. To that purpose, it can be convenient to use the notion of *common pot*, where the sub-structure contains vertices which receive from that common pot, and vertices that give to it. To keep some information on the total weight, we need to control somehow what happens inside the common pot. Since we usually try to show that everything has a non-negative final weight, we are usually satisfied with checking that the value of the common pot at the end is non-negative, i.e. no weight was created. In that case we say the global discharging rule is *valid*. Sometimes, the global discharging rule can be designed in such a way that the weight actually does not travel arbitrarily far, but the design still depends on a sub-structure of unbounded size (thus the proof is not made of purely local arguments). We then say that the discharging argument is *semi-global*.

By increasing the number of reducible configurations to over 600 and designing involved discharging rules, the bound in Lemma 2.1 can be further lowered to 4 [26]. In the same spirit as Lemma 2.2, significant research effort has been devoted to studying unavoidable sets in planar graphs, as a source of interest regardless of whether the configurations in these sets are reducible for some colouring problems. For example, it can be proved that in a planar graph of minimum degree 4, there must be, simultaneously, a triangle, a cycle of length 5 and a cycle of length 6 [10]. More often, the goal is to prove that in a planar graph with no small vertex (i.e. no vertex of degree less than 3, 4 or 5), there is necessarily a configuration of bounded degree, which is sought as large as possible. Note that if a planar graph is allowed to have vertices of degree 2, then absolutely nothing of the kind can be said, since it suffices to artificially increase the degree of all the other vertices by adding many parallel vertices of degree 2. Lemma 2.2 states that in a planar graph of minimum degree 5, there is a configuration of two adjacent vertices of small degree (the sum of the degrees is at most 11). This bound of 11 can be proved to be optimal. The same question was studied for planar graphs of minimum degree 3 or 4. Also, what about the minimal sum of the degrees of the vertices in a triangle? In a (not necessarily induced) path of three vertices? In a (not necessarily induced) cycle of length four? These questions were largely studied, but are not the topic of this survey. We refer the reader to two nice surveys [5, 15], and from now on focus on unavoidable sets that were designed with a specific colouring problem in mind.

3 A toy problem for discharging

The presentation is necessarily biased from personal experience of discharging. However, this bias is made necessary by the fact that a full survey of various discharging arguments would hardly fit in a single chapter. We refer the reader to the nice guide to discharging by Cranston and West [9] for an overview of discharging methods used for colouring purpose. The discharging method also proves itself useful outside the area of colouring, for example in combinatorial geometry as illustrated in a paper by Radoičić and Tóth [25], but we largely disregard such applications here. In a way, some amortized analysis proofs of algorithm complexity can be said to be of the same kin as discharging proofs. This is however not the topic of this chapter.

As introduced in Section 2, a discharging proof of a colouring theorem is almost

Edge decompositions

always presented as a final product, which the authors present out of the blue in the form of a set of reducible configurations and a set of discharging rules, with the appropriate arguments about their correctness. Here we are interested in the process behind it. We consider a problem, and strive step by step to obtain the best results for it through discharging arguments. We start with standard, local arguments (see Section 4) then move on to more exotic global arguments (see Section 5), before making some more general remarks (see Section 7). The goal of this chapter is to give some intuition on the discharging method: when possible we refrain from being too formal.

We will from now on concentrate on a variant of colouring which is well-adapted for the illustration of various discharging methods. An *Adjacent Vertex-Distinguishing edge k-colouring (AVD k-colouring)* is a proper edge k-colouring such that no two neighbours are incident with the same set of colours. More formally, for any edge colouring $c : E \to \mathbb{N}$ and any vertex $u \in V$, we set $\phi_c(u) = \{c(uv) | v \in N(u)\}$. In that setting, an AVD k-colouring is a proper edge colouring $c : E \to \{1, \ldots, k\}$ such that for any edge $uv \in E$, $\phi_c(u) \neq \phi_c(v)$. Note that there is no way to AVD colour the graph K_2, as no colouring of its single edge can distinguish the two vertices. In order to avoid dealing with special cases everytime we reduce a configuration, we extend the definition so that any edge colouring of K_2 is considered to be an acceptable AVD 1-colouring. Note that this only influences results for graphs with a connected component isomorphic to K_2. We define for every graph G the AVD chromatic index χ'_{avd} as the smallest integer such that G is AVD k-colourable.

We do not strive for a full bibliography around this problem, but we present shortly some selected facts. Since an AVD colouring is a proper edge colouring, every graph G satisfies $\chi'_{avd}(G) \geq \Delta(G)$. In addition, every graph G with two adjacent vertices of degree $\Delta(G)$ satisfies $\chi'_{avd}(G) \geq \Delta(G)+1$. Note that the problem of AVD colouring is unusual in this that an AVD k-colouring of a graph does not necessarily induce an AVD k-colouring of its subgraphs, as illustrated in Figure 4.

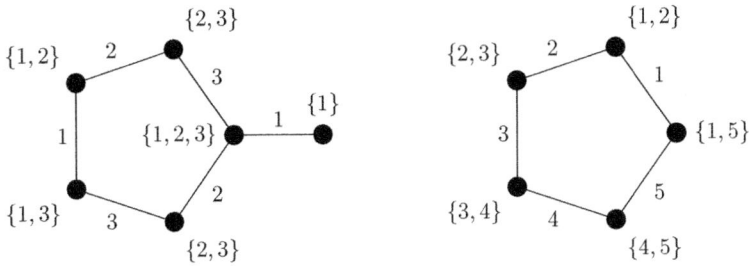

Figure 4: An example of a graph (left) which admits an AVD 3-colouring, while it admits an induced subgraph (right) which requires 5 colours.

Conjecture 3.1 *[35] Every graph G on at least 6 vertices satisfies $\chi'_{avd}(G) \leq \Delta(G) + 2$.*

For edge colouring, Theorem 1.1 ensures that the chromatic index of a graph is either $\Delta(G)$ or $\Delta(G) + 1$. For AVD colouring, Conjecture 3.1 would imply that the AVD chromatic index of a graph on at least six vertices can only have three values: $\Delta(G)$, $\Delta(G) + 1$ or $\Delta(G) + 2$. When considering a given graph class that allows two vertices of maximum degree to be adjacent, there are only two possible upper bounds: $\Delta(G) + 1$ or $\Delta(G) + 2$. Similarly, the classification of graph classes depending on this received subsequent interest. Instead of focusing on planar graphs, we focus on the density of a graph[3]. Let $\mathrm{ad}(G) = \frac{2|E(G)|}{|V(G)|}$ be the *average degree* of G, and let $\mathrm{mad}(G)$ be the *maximum average degree* of G, defined as the maximum of $\mathrm{ad}(H)$ taken over all subgraphs H of G. We consider here sufficient conditions with regards to $\Delta(G)$ and $\mathrm{mad}(G)$ for a graph G to be AVD $(\Delta(G) + 1)$-colourable.

4 Local arguments in discharging

Let $m \in \mathbb{R}^+$ and $D \in \mathbb{N}$. We seek a theorem of the form "Every graph G with $\mathrm{mad}(G) < m$ and $\Delta(G) \geq D$ is AVD $(\Delta(G) + 1)$-colourable". If we consider a minimal counter-example G to that theorem, we obtain that every proper subgraph H of G such that $\mathrm{mad}(H) < m$ and $\Delta(H) \geq D$ is AVD $(\Delta(H) + 1)$-colourable. The first condition is always true, as $\mathrm{mad}(H) \leq \mathrm{mad}(G)$ by definition. However, we can easily imagine that $\Delta(H)$ may become smaller than D, and then we cannot assume anything about the AVD colourability of H. To avoid that tricky situation, we reformulate the theorem so that the hypotheses are hereditarily satisfied: "For every integer $k \geq D$, every graph G with $\mathrm{mad}(G) < m$ and $\Delta(G) \leq k$ is AVD $(k+1)$-colourable". Note that the new statement is only stronger than the previous one.

We follow the usual outline of a discharging proof, and first look for configurations that cannot appear in a minimal counter-example. Keep in mind that our goal is to prove that the existence of a minimal counter-example is a contradiction, by showing that it cannot satisfy the theorem hypotheses. Here, we will try to prove that a minimal counter-example has large average degree. In other words, we want to argue that there cannot be too large a proportion of small vertices in the graph.

Let $k \geq D$, and let G be a minimal graph with $\Delta(G) \leq k$ that is not AVD $(k+1)$-colourable. Our goal is to prove that $\mathrm{mad}(G) \geq m$. Indeed, if every minimal graph G with $\Delta(G) \leq k$ that is not AVD $(k+1)$-colourable satisfies $\mathrm{mad}(G) \geq m$, then every graph G with $\Delta(G) \leq k$ and $\mathrm{mad}(G) < m$ is AVD $(k+1)$-colourable. We will proceed step-by-step and try to obtain the best possible lower bounds on $\mathrm{mad}(G)$, i.e. the largest possible value for m such that $\mathrm{mad}(G) \geq m$.

Note that, by the existence of C_5 which requires 5 colours, we cannot hope for any result with $D < 4$ with the stronger version of our theorem. We thus assume from now on $D \geq 4$. In particular, if G contains at most five edges, we can colour each with a different colour, thus obtaining an AVD $(k+1)$-colouring. Therefore we can assume that G contains at least six edges.

4.1 One reducible configuration, no discharging rule

We can prove that a vertex with a neighbour of degree 1 must have many other neighbours.

---
[3] Note that any planar graph G satisfies $\mathrm{ad}(G) < 6$.

Edge decompositions

Lemma 4.1 *G cannot contain a vertex u adjacent to at least one vertex of degree 1 and at most $\frac{k}{2}$ vertices of degree > 1 (see Figure 6).*

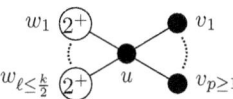

Figure 5: The configuration of Lemma 4.1.

Proof Assume for contradiction that G contains a vertex u with $p \geq 1$ neighbours $v_1, \ldots, v_p$ of degree 1 and $\ell \leq \frac{k}{2}$ vertices of degree ≥ 2. Colour by minimality $G \setminus \{v_1, \ldots, v_p\}$. We try to colour the edges (u, v_i) so as to extend the AVD $(k+1)$-colouring to the whole graph. For the colouring to be proper, every edge (u, v_i) must avoid the ℓ colours which already appear in the neighbourhood of u. That leaves $k + 1 - \ell$ possible colours. We now try to find a colouring which distinguishes u from its neighbours of same degree, which are at most ℓ. There are $\binom{k+1-\ell}{p}$ different ways of colouring the edges (u, v_i). There are at most ℓ sets of colours which are equal to $\phi_c(w)$ for some neighbour w of u. Therefore, if $\binom{k+1-\ell}{p} \geq \ell + 1$, then the colouring can be extended to G, a contradiction. Since $k - \ell \geq p \geq 1$, it suffices to have $k + 1 - \ell \geq \ell + 1$, hence the conclusion. □

Note that Lemma 4.1 implies that no vertex of degree at most $\frac{k}{2} + 1$ can have a neighbour of degree 1. Is Lemma 4.1 sufficient to obtain a theorem of the desired form? It is enough for $m = 2$.

Proposition 4.2 *The graph G satisfies $\mathrm{ad}(G) \geq 2$.*

Proof Assume for contradiction that $\mathrm{ad}(G) < 2$. A connected graph $H = (V, E)$ with a cycle satisfies $|E| \geq |V|$, hence such a graph satisfies $\mathrm{ad}(H) \geq 2$. Therefore, since G is connected by assumption, the graph G contains no cycle. Then G is a tree. We remove the leaves of the tree, and take a leaf of the resulting tree (which is non-empty since G contains more than just one edge). This vertex has at least one neighbour of degree 1 (it was not a leaf of G), and at most one neighbour of degree more than 1 (it is a leaf of the resulting tree). By Lemma 4.1, this is not possible. □

In our pursuit of a largest possible m, we therefore set $m = 2 + a$, with $a \in \mathbb{R}^+$. Note that Lemma 4.1 is not sufficient to prove the theorem with any $a > 0$, as a long cycle has maximum average degree equal to its average degree of 2, and Lemma 4.1 does not apply.

4.2 Two configurations, two discharging rules

We can prove that there cannot be a long chain of vertices of degree 2, as was done in previous works.

Lemma 4.3 ([13]) *G cannot contain a chain of length 3 (see Figure 6).*

Figure 6: The configuration of Lemma 4.3.

Proof Assume for contradiction that G contains a vertex u with two neighbours v_1 and v_2, $d(u) = d(v_1) = d(v_2) = 2$. Let w_1 and w_2 be the other neighbours of v_1 and v_2, respectively (see Figure 6). Colour by minimality $G \setminus \{u\}$. For the colouring to be proper, every edge (u, v_i) must avoid the colour that appears on (v_i, w_i). If w_1 is of degree 2, note that the single edge incident to w_1 and not to v_1 is different from (u, v_2) by assumption (otherwise G would be a C_3 and thus have less than 6 edges).

We choose for (u, v_1) a colour that is distinct from the colours of (v_1, w_1), (v_2, w_2), and from that of the single edge incident to w_1 and not to v_1 if $d(w_1) = 2$ (this last requirement to ensure we distinguish v_1 and w_1). Then we pick for (u, v_2) a colour that is distinct from the colours of (u, v_1), (v_2, w_2), (v_1, w_1), and from that of the single edge incident to w_2 and not to v_2 if $d(w_2) = 2$. Since $D \geq 4$, there are at least 5 colours and such choices are possible. Now, the vertices v_1 and w_1 are indeed distinguished, either by the fact that $d(w_1) \neq 2$ or by the colour choice of (u, v_1). Similarly, u is distinguished from v_1 by choice of (u, v_2). We symmetrically reach the same conclusions for v_2 and w_2, and u and v_2, respectively.

We thus exhibited an AVD $(k + 1)$-colouring of G, a contradiction. □

Are Lemmas 4.1 and 4.3 sufficient to obtain a theorem of the desired form with $a > 0$? We assign to each vertex u a weight of $d(u) - 2 - a$. We strive for a discharging procedure that leaves a non-negative weight on all vertices at the end. This will guarantee us that $\text{ad}(G) \geq 2 + a$ and thus $\text{mad}(G) \geq 2 + a$. Vertices of degree 1 have a negative weight of $-1 - a$, and a single neighbour. Vertices of degree 2 have a negative weight of $-a$, no neighbour of degree 1 (by Lemma 4.1) and at least one neighbour of degree more than 2 (by Lemma 4.3). Note that vertices of degree at least 3 have positive weight, and we will try to discharge the weight from them to the vertices of degree 1 or 2. Since we do not have any other piece of information about the graph, every vertex of degree at least 3 must be able to provide $1 + a$ to each of its neighbours of degree 1, and a to each of its neighbours of degree 2. By assuming $D \geq 4$ (remember that $k \geq D$), we can ensure by Lemma 4.1 that no vertex of degree 3 will have a neighbour of degree 1. However, there may be vertices of degree 3 only adjacent to vertices of degree 2 whose other neighbour is of degree 2. Therefore, a must satisfy $3 - 2 - a \geq 3 \times a$, i.e. $a \leq \frac{1}{4}$. This is in fact sufficient as soon as $D \geq 6$, and we obtain the following proposition.

Proposition 4.4 *If $D \geq 6$, the graph G satisfies $\text{ad}(G) \geq 2 + \frac{1}{4}$.*

Proof We reformulate the above arguments in a more formal way. Let us assign to each vertex u in G a weight of $\omega(u) = d(u) - 2 - \frac{1}{4}$. We design two discharging rules R_1 and R_2 (see Figure 7): the first states that every vertex u adjacent to a vertex v of degree 1 gives a charge of $1 + \frac{1}{4}$ to v, the second states that every vertex u adjacent to a vertex v of degree 2 gives a charge of $\frac{1}{4}$ to v. We apply R_1 and R_2

on G with the initial weight assignment. Let ω' be the resulting weight assignment on the vertices of G. Our goal is to prove that $\omega'(u) \geq 0$ for every vertex u of G.

Figure 7: The discharging rules R_1 (left) and R_2 (right) of Proposition 4.4.

Let u be a vertex of G. We consider different cases depending on the degree of u. Note that by Lemma 4.1, a vertex with a neighbour of degree 1 must have at least 4 neighbours of degree at least 2, and thus be itself of degree at least 5.

- *Assume $d(u) = 1$.*
  Then $\omega(u) = -1 - \frac{1}{4}$. As already noted, the neighbour of u is not of degree 1 or 2. Therefore, the vertex u gives nothing and receives $1 + \frac{1}{4}$ by R_1. Consequently, the vertex u has a non-negative final weight.

- *Assume $d(u) = 2$.*
  Then $\omega(u) = -\frac{1}{4}$. If u has a neighbour of degree 2, then they both give $\frac{1}{4}$ to the other, so it cancels out and we can pretend it doesn't happen. No neighbour of u is of degree 1, and u may not have both neighbours of degree 2 by Lemma 4.3. Therefore, the vertex u gives nothing and receives at least $\frac{1}{4}$ by R_2: it has a non-negative final weight.

- *Assume $d(u) = 3$.*
  Then $\omega(u) = \frac{3}{4}$. All the neighbours of u are of degree at least 2. The vertex u gives at most $3 \times \frac{1}{4}$ by R_2 and has a non-negative final weight.

- *Assume $4 \leq d(u) \leq 7$.*
  Then $\omega(u) = d(u) - 2 - \frac{1}{4}$. The vertex u has more than $\frac{k}{2}$, thus at least 4, neighbours of degree at least 2. Therefore, it gives at most $(d(u) - 4) \times (1 + \frac{1}{4}) + 4 \times \frac{1}{4} = d(u) \times \frac{5}{4} - 4$ by rules R_1 and R_2. It has thus a non-negative final weight, as $d(u) - 2 - \frac{1}{4} \geq d(u) \times \frac{5}{4} - 4$ when $d(u) \leq 7$.

- *Assume $d(u) \geq 8$.*
  Then $\omega(u) = d(u) - 2 - \frac{1}{4}$. The vertex u has more than $\frac{k}{2}$, thus at least $\frac{k+1}{2}$, neighbours of degree at least 2. Therefore, it gives at most $(d(u) - \frac{k+1}{2}) \times (1 + \frac{1}{4}) + \frac{k+1}{2} \times \frac{1}{4} = d(u) \times \frac{5}{4} - \frac{k+1}{2}$ by Rules R_1 and R_2. Let us argue that it has a non-negative final weight: we must have $d(u) - 2 - \frac{1}{4} \geq d(u) \times \frac{5}{4} - \frac{k+1}{2}$, i.e. $\frac{k}{2} - \frac{7}{4} \geq d(u) \times \frac{1}{4}$. Since $d(u) \leq k$, it suffices to have $\frac{k}{4} \geq \frac{7}{4}$, i.e. $k \geq 7$, which holds since $k \geq d(u) \geq 8$. Therefore, the vertex u has a non-negative final weight.

Every vertex in G has a non-negative final weight, thus $\mathrm{ad}(G) - 2 - \frac{1}{4} = \sum_{v \in V} \omega(v) = \sum_{v \in V} \omega'(v) \geq 0$, hence the conclusion. □

From now on we omit the formal case analysis: all relevant information is already contained in the forbidden configurations, rule definitions, and analysis of the bounds on a.

4.3 Three reducible configurations, two discharging rules

As noted before, Lemma 4.1 implies that no vertex of degree less than $\frac{k}{2}+1$ can have a neighbour of degree 1. One can strengthen that result.

Lemma 4.5 ([13]) *G cannot contain two adjacent vertices u and v with $d(u) \neq d(v)$ and $d(u) + d(v) \leq \frac{k}{2} + 2$ (see Figure 8).*

$$u \quad v$$
$$\circ\!\!-\!\!-\!\!-\!\!\circ$$

$$d(u) \neq d(v)$$
$$d(u) + d(v) \leq \frac{k}{2} + 2$$

Figure 8: The configuration of Lemma 4.5.

Proof Assume for contradiction that G contains two adjacent vertices u and v with $d(u) \neq d(v)$ and $d(u) + d(v) \leq \frac{k}{2} + 2$. The proof here is much more direct than that of Lemma 4.1. Colour by minimality $G \setminus \{(u,v)\}$. Since $d(u) \neq d(v)$, no effort is required to distinguish u and v. There are initially $k+1$ colours available for the edge (u, v), we possibly remove as many as $(d(u) - 1) + (d(v) - 1)$ colours to enforce the propriety of the colouring. Now, if there are at least $(d(u)-1)+(d(v)-1)+1$ choices of colours for (u, v), we know that there will be at least one that will distinguish u and v from their respective neighbours. This holds since $k+1-2(d(u)+d(v)-2) \geq 1$, hence the conclusion. □

Now, Lemma 4.5 tells us that a vertex u with $3 \leq d(u) \leq \frac{k}{2}$ has no neighbour of degree 1, nor any of degree 2. As argued before, Lemmas 4.1 and 4.3 guarantee that every vertex of degree 1 or 2 has a neighbour of degree at least 3 (assuming $D \geq 4$). Therefore, the only constraint is that vertices of degree at least $\frac{k}{2} + 1$ must afford to give weight to all their neighbours of degree 1 or 2 ($1 + a$ to each neighbour of degree 1, and a to each neighbour of degree 2). Let us consider the worst case scenario for the neighbourhood of a vertex u with $d(u) \geq \frac{k}{2} + 1$. The worst neighbours are those of degree 1. By Lemma 4.1, the vertex u cannot have more than $d(u) - \frac{k}{2} - 1$ neighbours of degree 1. Assume it has just as many. Its other neighbours are of degree at least 2, and the worst case is when they are all of degree 2. Consequently, a must satisfy $d(u) - 2 - a \geq d(u) \times a + (d(u) - \frac{k}{2} - 1) \times 1$ for every $d(u) \geq \frac{k}{2} + 1$. In other words, the constant a must satisfy $\frac{k}{2} - 1 \geq a \times (d(u) + 1)$ for every $d(u) \geq \frac{k}{2} + 1$. The strongest constraint comes from $d(u) = k$, so it suffices to have $a \leq \frac{\frac{k}{2}-1}{k+1} = \frac{1}{2} - \frac{3}{2(k+1)}$. We can take a arbitrarily close to $\frac{1}{2}$, which results in the following.

Proposition 4.6 *For every $\epsilon > 0$, if D is large enough, then $\mathrm{ad}(G) \geq 2 + \frac{1}{2} - \epsilon$.*

4.4 Shifting the density argument to a subgraph of G

There is a simple trick that drastically improves Proposition 4.6. When we try to prove that no minimal counter-example can exist, we assume one exists, and prove that its average degree is at least $2 + a$, a contradiction to the theorem hypothesis that the maximum average degree is less than $2 + a$. It could happen that the considered minimal counter-example has low average degree, but still contains a dense subgraph.

Let us sketch the consequences here. Let H be the graph obtained from G by deleting all vertices of degree 1, as was done in [13]. Again, we assign to every vertex u of H a weight of $d_H(u) - 2 - a$. By Lemma 4.1, the vertices of degree at most $\frac{k}{2}$ in H have the same degree in G. Therefore, by Lemma 4.5, vertices of degree at most $\frac{k}{2}$ in H have no neighbour of degree 2 in H. We can again concentrate on large vertices: a must satisfy $d(u) - 2 - a \geq d(u) \times a$ for every $d(u) \geq \frac{k}{2} + 1$ (by the same argumentation as for Proposition 4.6). The strongest constraint comes from $d(u) = \frac{k+1}{2}$, i.e. $\frac{k+1}{2} - 2 - a \geq \frac{k+1}{2} \times a$. In other words, we can take $a = 1 - \frac{6}{k+3}$, and thus a arbitrarily close to 1. Therefore, simply by shifting the density argument to a well-chosen subgraph of the minimal counter-example, we move up from Proposition 4.6 to the following.

Proposition 4.7 *For every $\epsilon > 0$, if D is large enough, then* $\mathrm{mad}(G) \geq 3 - \epsilon$.

4.5 Shifting the notion of minimality

There are other tricks that prove useful. For example, the considered order on graphs is traditionally the subgraph order ($H \prec G$ iff H is a proper subgraph of G). Sometimes, it is interesting to transform the graph more subtly than just by considering a subgraph. For this to be possible, we need to pick the right notion of minimality.

In our case, let us consider the lexicographic order on the sequence of the number of vertices of given degree in the graph, sorted by decreasing order. Now, if G contains a vertex u with exactly two neighbours v_1 and v_2, then G is larger than the graph obtained from G by replacing u with two vertices of degree 1, one adjacent to v_1 and the other to v_2. We call such a transformation $G \otimes \{u\}$. We can generalize it to a vertex u of any degree $p \geq 2$. In that case, $G \otimes \{u\}$ corresponds to the graph obtained from G by replacing u with p vertices of degree 1, each adjacent to a different neighbour of u in G. We can further generalize this notion to any set S of q vertices $u_1, \ldots, u_q$ of degree at least 2. We set $G \otimes S$ to be the graph obtained from G by deleting the edges in $G[S]$ then successively considering, when $d(u_i) \geq 2$, the operation $\otimes \{u_i\}$, for $1 \leq i \leq q$ (see Figure 9). Note that the order on the vertices of S has no influence on $G \otimes S$.

We do not define $G \otimes \{v\}$ when v is a vertex of degree 1, for that would result in the very same graph. As S consists only of vertices of degree at least 2, it holds that $G \otimes S \prec G$ for non-empty S.

We suddenly obtain much more information on our minimal counter-example, see Lemma 4.8 (note that a proper subgraph of G is still smaller than G in the new order, which makes the previous lemmas still valid here). Note also that

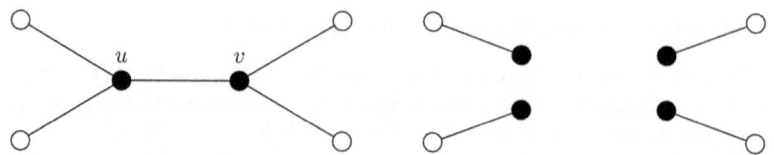

Figure 9: An example of the effect of $\otimes$: here we consider two adjacent vertices u, v of degree 3 (left), and the result of $\otimes\{u, v\}$ (right).

$\mathrm{mad}(G \otimes S) \leq \mathrm{mad}(G)$, which is in fact crucial. When considering a proper subgraph, or the result of a $\otimes$ operation on G, this is obvious and could even be omitted. However, some graph transformations, no matter how tempting (like contracting every edge between two vertices of degree 2), are made impossible by the fact that the resulting graph, though smaller, may have a larger maximum average degree and not satisfy the induction hypotheses. That constraint has to be kept in mind. On the contrary, when considering planar graphs, edge contractions are extremely convenient (multiple edges beware), while some transformations used in the setting of bounded average degree graphs might result in a non-planar graph.

Let us first make some remarks about chains of degree 2. By Lemma 4.3, in G, all maximal chains of vertices of degree 2 are of length either 1 or 2 (i.e. contain exactly one vertex of degree 2 or exactly two). Note that from an AVD point of view, a chain of length two and a chain of length one behave the same, in the sense that colouring one is equivalent to colouring the other. More formally, the graph obtained from G by contracting an edge between two vertices of degree 2 or subdividing an edge incident to a vertex of degree 2 with no neighbour of degree 2 is AVD $(k+1)$-colourable iff G is. Indeed, $D \geq 4$, and a maximal chain does not, by definition, have any neighbour of degree 2. Consider two maximal chains of degree 2, (u_1, v_1, v_2, u_2) and (u_1, v, u_2) (internal vertices of degree 2). When restricting the problem to the chain, the only constraint for the second chain is that the colours of (u_1, v) and (u_2, v) differ. The vertex v has no neighbour of degree 2, and thus does not need to be distinguished. The main constraint for the first chain is that v_1 is distinguished from v_2, i.e. that the colours of (u_1, v_1) and (u_2, v_2) differ. Then the colouring has to be proper, which means we need to find a colour for (v_1, v_2) that differs from those of (u_1, v_1) and (u_2, v_2). However, $D \geq 4$, so such a colour is available regardless of the situation, and that constraint is insignificant. Therefore, from now on, we do not make separate cases for both. We are now ready for the following lemma.

Lemma 4.8 *G cannot contain a vertex u adjacent to both a vertex of degree 1 and a vertex of degree 2 (see Figure 10).*

Figure 10: The configuration of Lemma 4.8.

Proof Assume for contradiction that G contains a vertex u adjacent to both a vertex v_1 of degree 1, and a vertex v_2 of degree 2. Let C be the maximal chain of vertices of degree 2 to which v_2 belongs, with w its other endpoint (remember $|C| \leq 2$ by Lemma 4.3). Colour by minimality $G \otimes \{C\}$. Now, by the above paragraph, the only constraint on (u, v_2) is that its colour should differ from that of the edge between w and C. Assume that both edges are of the same colour. Then we swap the colours of (u, v_2) and of (u, v_1). The set of colours incident to u remains the same, and v_1 has no neighbour to be distinguished from, but now the colour of (u, v_2) differs from that at the other end of the chain, and we can safely extend the AVD $(k+1)$-colouring to G (if $|C| = 2$, we pick a proper colour for the internal edge, which is possible since it has two constraints and there are at least 5 colours).
□

Lemma 4.8 is very convenient in the sense that, in the discharging argument, the only two troublesome neighbours are those of degree 1 and those of degree 2, which abide by different rules. This lemma guarantees us that we never have to deal with both at the same time. If a vertex has neighbours which require charge from it, then either they are all of degree 1 or all of degree 2. If we follow the same path as before, by considering the graph G where all vertices of degree 1 have been removed, we miss some of essential information gained through Lemma 4.8. The information loss might grow in the way if we are to design other lemmas similar to the last. When we shift the density argument to a proper subgraph H of the graph G on which we have structural information, it often happens that, in the discharging analysis, we have to make double considerations about the degree of the vertex in H, in G (also about the nature of its neighbours in G that do not appear in H, and about the degree in G of its neighbours in H). In a sense, while we consider H for the analysis, the actual information we need is in G. Let us consider a trick around this issue in the following section.

4.6 Ghost vertices

We introduce a way of combining the power of the previous approach (considering only a subgraph for density measures, as used for Proposition 4.7) and the information of the initial one (no vertex deletion, all information stays in the graph, as used for Proposition 4.6). The trick is to design discharging arguments that use the structure of the initial graph to prove that the considered subgraph has a high average degree.

The approach used for Proposition 4.7 was:
Dense subgraph method

- Let $V_1 \subsetneq V$, and consider $G[V_1]$.

- Every vertex u in $G[V_1]$ has an initial weight of $d_{V_1}(u) - m$.

- Can we discharge in $G[V_1]$ in such a way that every vertex in V_1 has a non-negative weight?

- If yes, then we have $\sum_{u \in V_1}(d_{V_1}(u) - m) \geq 0$, thus $\mathrm{mad}(G) \geq \mathrm{ad}(G[V_1]) \geq m$.

The new approach is:
Ghost vertices method

- Let $V_1 \cup V_2$ be a partition of V.
- Every vertex u in G has an initial weight of $d(u) - m$.
- Can we discharge in G in such a way that :
  1. Every vertex in V_1 has a non-negative weight,
  2. Every vertex u in V_2 has a final weight of at least $d(u) - m + d_{V_1}(u)$?
- If yes, then for ω' the new weight assignment, we have $\sum_{v \in V_2}(d(v) - m + d_{V_1}(v)) \leq \sum_{v \in V_2} \omega'(v)$, as well as $\sum_{v \in V} \omega(v) = \sum_{v \in V} \omega'(v)$ and $\sum_{v \in V_1} \omega'(v) \geq 0$. Therefore,

$$\sum_{v \in V_1}(d_{V_1}(v) - m)$$
$$\geq \sum_{v \in V_1}(d_{V_1}(v) - m) + \sum_{v \in V_2}(d(v) - m + d_{V_1}(v)) - \sum_{v \in V_2} \omega'(v)$$
$$\geq \sum_{v \in V_1}(d_{V_1}(v) - m) + |E(V_1, V_2)| + \sum_{v \in V_2}(d(v) - m) - \sum_{v \in V_2} \omega'(v)$$
$$\geq \sum_{v \in V_1}(d(v) - m) + \sum_{v \in V_2}(d(v) - m) - \sum_{v \in V_2} \omega'(v)$$
$$\geq \sum_{v \in V} \omega(v) - \sum_{v \in V_2} \omega'(v)$$
$$\geq \sum_{v \in V_1} \omega'(v)$$
$$\geq 0$$

We can conclude that $\mathrm{mad}(G) \geq \mathrm{ad}(G[V_1]) \geq m$.

In other words, the vertices in V_2 can be seen but, in a way, do not contribute to the sum analysis (the meaning of their final weight is essentially "this vertex has no positive contribution on the total weight of the rest of the graph"). This particularity leads us to informally refer to them as *ghost vertices*. Any result proved using ghost vertices can be proved, albeit more tediously perhaps, when deleting them completely from the graph. However, they can simplify the presentation of the discharging analysis, and this is the point of their introduction.

Similarly, the idea of ghost vertices can be translated to planar graphs, where not all vertices are mapped in the mapping considered (not all vertices have shadows). However, in that case, considering a proper subgraph cannot help: a graph may have low average degree and a subgraph of high average degree, but if a subgraph is not planar, then neither was the initial graph. The only point of introducing ghost vertices in the planar case is to reduce the number of discharging rules. That trick appears in a joint work with Jakub Przybyło on a variant of AVD colouring on planar graphs, which we do not present here.

Edge decompositions

In our problem, we consider V_2 (the set of ghost vertices) to be the set of vertices of degree 1 in G, and again assign to each vertex u in G an initial weight of $d(u)-2-a$. Vertices of degree 1 have a weight of $-1-a$, instead of needing an extra charge of $1+a$, they only need an extra charge of 1 in the latter approach. The bound on a does not change, as it still needs to be smaller than 1 but can be arbitrarily close to it (just as in the proof of Proposition 4.7, a vertex may be such that all its neighbours are of degree 2). However, this allows for a more refined bound on the corresponding d, but we do not dwell on this.

Despite this progress, we still are not able to prove anything for the class of graphs with mad < 3 in general (no ϵ inserted). We turn to a powerful variant of the discharging method, where Lemma 4.8 and the notion of ghost vertices will turn out to be useful.

5 Global arguments in discharging

We claimed that the discharging method is based on the idea of local counting arguments in order to derive a global formula. However, it could happen that there is a sub-structure of unbounded size on which the sum of weights is easy to compute, actually easier than through local arguments. Therefore, we sometimes mix local and global arguments in the discharging process. This variant on the dicharging method depends heavily on the emergence of a good sub-structure.

5.1 Alternating cycles

For example, in G, we can note that there cannot be two vertices of degree 2 with the same two neighbours. Indeed, if there are two such vertices u_1 and u_2 with the same two neighbours v_1 and v_2 (note that $d(v_1), d(v_2) \geq 3$ by Lemma 4.3), we colour $G \otimes \{u_1, u_2\}$, and consider the corresponding colouring of G (which is not necessarily an AVD colouring). Then, we switch if necessary the colours of (u_1, v_1) and (u_2, v_1) so that the colouring is proper, which results in an AVD $(k+1)$-colouring of G. In fact, this possibility to switch the colours of two incident edges can be interpreted in terms of list colouring: here the goal was to list colour a cycle on 4 vertices, where the list of colours assigned to the pair (u_i, v_j) is that of the colours of (u_1, v_j) and (u_2, v_j).

We generalize this idea by setting H to be the multigraph with vertex set $V_H = \{v \in V \mid d(v) \geq \frac{k+1}{2}\}$, and with edge set E_H the pairs of vertices in V_H that are the endpoints in G of a (non-empty) chain of vertices of degree 2. By Lemma 4.5, the set E_H is in bijection with the set of maximal chains of vertices of degree 2 in G. The previous remark implies that there is no digon in H. We can similarly prove that there is no loop, nor any triangle, etc. In fact, we can go further and claim that H is a forest.

Indeed, assume that there is a cycle C in H. Colour by minimality

$$G \otimes \{\text{Corresponding maximal chains of } C\}$$

(see Figure 11 for a case where C is a triangle). Similarly as before, it all boils down to L-colouring an even cycle, where L is an assignment of two colours to each edge.

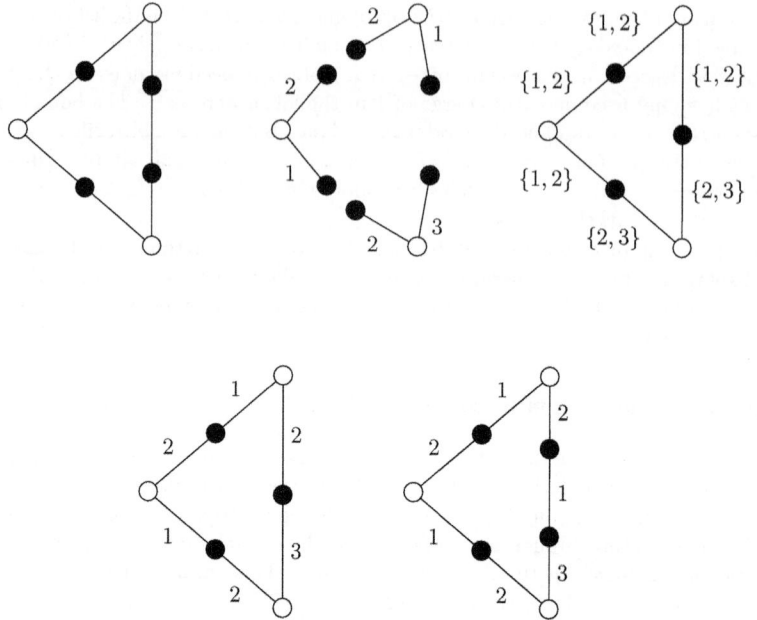

Figure 11: There is no triangle in H.

By Lemma 1.4, the even cycle is L-colourable, and G is AVD $(k+1)$-colourable. Consequently, H is a forest.

Since H is a forest, every connected component contains more vertices than edges. In other words, there are at least as many vertices of degree $\frac{k}{2}+1$ adjacent to a vertex of degree 2 than maximal chains of vertices of degree 2. We can make use of that property, as follows. We assign to each vertex u a weight of $d(u)-2-a$, and design two discharging rules. As before, we take the vertices of degree 1 to be ghost vertices, and thus set that every vertex of degree at least 2 gives a charge of 1 to every neighbour of degree 1. The second rule is what we call a *global discharging rule*, in the sense that the charge might travel arbitrarily far away. Every vertex of degree at least $\frac{k+1}{2}$ adjacent to a vertex of degree 2 gives a charge of $2a$ to a common pot, while every vertex of degree 2 receives a from the common pot. We already argued that the common pot will have a non-negative final value, which we refer to as the global discharging rule being *valid* (i.e. not used for spontaneous charge generation).

Assume $D \geq 9$. Lemma 4.1 ensures that vertices with a neighbour of degree 1 have at least 5 neighbours of degree at least 2, and no neighbour of degree 2 by Lemma 4.8. Therefore, if a vertex u has a neighbour of degree 1, then only the first discharging rule applies. The vertex u has a non-negative final weight as long as $d(u) - 2 - a \geq (d(u) - 5) \times 1$, i.e. $a \leq 3$. By Lemma 4.5, only vertices of degree at least 5 may have a neighbour of degree 2, and if they do, by Lemma 4.8, they do not have any neighbour of degree 1. Therefore, if a vertex u has a neighbour of

Edge decompositions

degree 2, then $d(u) \geq 5$ and only the second discharging rule applies. The vertex u has a non-negative final weight as long as $d(u) - 2 - a \geq 2 \times a$, i.e. $a \leq 1$. It follows that every vertex of degree 1 has a final weight of -1, and every other vertex a non-negative weight, hence the following conclusion:

Proposition 5.1 *If $D \geq 9$, then* $\mathrm{mad}(G) \geq 3$.

We defined a global discharging rule as a rule that may allow some charge to travel arbitrarily far. One may note that we could avoid that situation here. Indeed, since we know H to be a forest, we could merely pick a vertex of degree 1 in H, assign it the maximal chain which corresponds to its incident edge, and remove it. By repeating the operation until H is a stable set, evey maximal chain of vertices of degree 2 is assigned to a neighbour of it. We can then transform the global discharging rule to make it state instead that every vertex to which a chain is assigned distributes a charge of 2 equally among the vertices in the chain. Then no charge moves further away than to a vertex at distance 2. According to our definition, this should mean that the discharging proof is not global anymore. However, since the discharging rules have to take into account a structure of unbounded size in order to be defined (here, you need the full knowledge of a tree H to be able to decide to which edge a vertex in the middle gives weight to), we cannot say the method is purely local either. When a global discharging proof can be written so that no charge travels arbitrarily far away, we say that the proof is semi-global.

Consequently, this simple trick of using well-known colouring results (e.g. every even cycle is 2-choosable) so as to reduce large structures with too many small vertices enables us to reach the threshold of 3. Purely local arguments seemed too weak for such a conclusion. However, in this very special case, it turns out that a purely local argument, when combined with a right order on graphs, can be even more powerful. Indeed, by considering a yet more refined order on graphs, we can prove that there is no vertex with two neighbours of degree 2 in G. This, combined with a slightly technical argument to improve Lemma 4.5, yields a proof using purely local discharging arguments that every graph G with $\mathrm{mad}(G) < 3$ and $\Delta(G) \geq 4$ is AVD $(\Delta(G) + 1)$-colourable. This is optimal, as shows the graph in Figure 12. However, this is a purely ad hoc proof which has very little chance of being of interest in other settings, contrary to the proof presented above, so we do not go into details here. A similar global discharging argument based on Lemma 1.4 can for example be found in [8]. We do not dwell on the topic, but rather question if the threshold of 3, once close, now reached, can be broken.

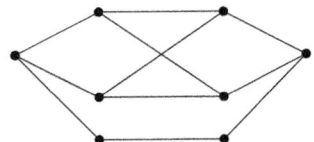

Figure 12: A graph G with $\Delta(G) = 3$ and $\mathrm{mad}(G) = \frac{11}{4} < 3$ such that $\chi'_{avd}(G) = 5$.

5.2 Beyond alternating cycles

In order to reach the threshold of 3, we simply used the classical result that an even cycle is 2-choosable, a result which proved decisive in other discharging proofs, e.g. in the case of list edge colouring [8]. There are stronger list colouring results than the fact that even cycles are 2-choosable, and the solution again came from the area of list edge colouring [6]. The following method is due to Woodall [34] who first proposed an alternative presentation of a theorem in [6] in terms of discharging, when the initial paper offered a less tell-tale sequence of equations.

We present here its transposition in the setting of AVD colouring, and choose simplicity over any improvement on the lower bound on the maximum degree.

We start with an easy structural lemma on bipartite multigraphs.

Lemma 5.2 *Let H be a bipartite multigraph with vertex set $V(H)$ bipartitioned into $A \cup B$, with $A \neq \emptyset$. For $\alpha > 0$, if for every non-empty subset $B' \subseteq B$ and $A' = N(B') \subseteq A$, there exists a vertex $u \in A'$ with $d_{B'}(u) < \alpha$, then $\alpha|A| > |B|$.*

Proof By induction on $|B|$. If $|B| < \alpha$, since $|A| \geq 1$, the conclusion holds. If $|B| \geq \alpha$, there exists $u \in A$ with $d(u) < \alpha$. We apply the induction hypothesis to the graph $H \setminus (\{u\} \cup N(u))$. It follows that $\alpha(|A| - 1) > |B| - \alpha$, hence the result. □

Similarly as for Lemma 4.5, we can prove that no small vertex has more than one small neighbour.

Lemma 5.3 *G cannot contain a vertex v adjacent to two vertices u and w with $d(u) = d(v) = d(w)$ and $d(u) \leq \frac{k}{4} + 1$.*

Proof We colour by minimality $G \setminus \{(u, v), (v, w)\}$. Each edge of (u, v) and (v, w) has at most $(d(u) - 1) + (d(u) - 2)$ colours to avoid for the colouring to be proper, and an additional $(d(u) - 1)$ colours to avoid conflicts (each neighbour of u or w might create a conflict). Therefore, each of the two edges has at least $\frac{k}{4} + 2$ available colours. Now we only have to pick the right colours among those, so that there is no conflict between v and its neighbours. We colour the edge (v, w) in such a way that there can be no conflict between u and v (this is possible as $\frac{k}{4} + 2 \geq 2$). Now we know that v cannot be in conflict with u, the edge (u, v) is the only one left uncoloured and it still has $\frac{k}{4} + 1$ possible choices if we disregard the conflicts between v and its neighbours. The vertex v has at most $\frac{k}{4}$ neighbours with whom to be in conflict, and none of them is u. We only need to remove an additional $\frac{k}{4}$ colours from the colours available for (u, v), and we colour it with one of the remaining colours (there is at least one). We thus obtain an AVD $(k + 1)$-colouring of G, a contradiction. □

We consider vertices of degree at most m to be *small*, and set D such that $m \leq \frac{k}{8}$.

By Lemma 4.5, all the neighbours of a small vertex u must either be of degree at least $\frac{3k}{8}$ or of degree exactly $d(u)$. By Lemma 5.3, the vertex u has at most one neighbour of degree $d(u)$, and thus has at least $d(u) - 1$ neighbours of degree at least $\frac{3k}{8}$. Let S be the set of vertices of degree at least 2 and at most m in G. Note that by the above remark, every vertex of S is of degree 0 or 1 in $G[S]$.

Let G' be the multigraph obtained from G by contracting any edge in $G[S]$ (we keep the multiple edges thus created, if any), and S' the set of vertices corresponding to S in G'. Note that no vertex in S' is of degree less than 2 or more than $2m$ in G'. Let B be the set of vertices of degree at least $\frac{3k}{8}$ in G. By the above remark, all the neighbours of a vertex in S are either in B or in S. Note that $B \cap S = \emptyset$ since $\frac{3k}{8} > m$. Let H be the bipartite multigraph obtained from $G'[B \cup S']$ by deleting the edges in $G'[B]$. We try to prove that $|S'| < 2m|B|$.

Assume it is not the case, and $|S'| \geq 2m|B|$.

We claim that there is a non-empty set $S'' \subsetneq S'$ such that the subgraph H'' of H obtained by considering $H[S'' \cup N(S'')]$ is such that every vertex in $B'' = N(S'')$ has degree at least $2m$ in H''. Indeed, otherwise, in every subgraph $S'' \subsetneq S'$, there is a vertex in $N(S'')$ of degree less than $2m$ in $H[S'' \cup N(S'')]$, thus Lemma 5.2 holds and $|S'| < 2m|B|$, a contradiction.

Colour by minimality $G \otimes S''$. As already noted in the case of alternating cycles, this corresponds to a list of $d_{S''}(v)$ colours assigned to each edge incident to $v \in B''$ and to a vertex in S''. Similarly, this corresponds to an edge list assignment of H'' such that $\forall u \in B'', \forall v \in S''$, if (u,v) is an edge then $|L(u,v)| = d_{H''}(u) \geq 2m \geq d_{H''}(v)$. Consequently, Theorem 1.3 applies, and H'' can be coloured. To obtain an AVD $(k+1)$-colouring, it remains to colour any edge between two vertices of degree at most m. They are already distinguished since they were contracted, so their incident sets of colours will be disjoint except for the colour of their common edge. Since $m \leq \frac{k}{8}$, it suffices to take any colour beside the $\frac{k}{4}$ that may already appear around one of the two vertices. Hence the conclusion that $|S'| < 2m|B|$.

We consider G, and assign to each vertex u a weight of $d(u) - m$. We introduce a global discharging rule, that every vertex in B gives a charge of m^2 to a common pot, from which every vertex u in S draws a charge of $d(u)$ if $d(u) \leq \frac{m}{2}$, and $m - d(u)$ if $d(u) \geq \frac{m}{2}$. Note that every small vertex draws from the common pot a charge of at most $\frac{m}{2}$. Therefore, the vertices in S' draw a total weight of at most $\frac{m}{2} \times |S'|$ from the common pot, while the vertices in B give a total weight of $m^2 \times |B|$ to the common pot. We know that $|S'| < 2m|B|$, so $\frac{m}{2} \times |S'| < m^2 \times |B|$, and the global rule is hence valid. We reintroduce the now usual rule that every vertex of degree at least $\frac{k}{2}$ gives a charge of 1 to every neighbour of degree 1. We take all vertices of degree at most $\frac{m}{2}$ to be ghosts. The two rules above ensure that every ghost vertex gives nothing and receives a weight equal to its degree, and that each small vertex that is not a ghost gives nothing and receives a weight which is just enough to get to a non-negative final weight. For large enough D, we can ensure that the vertices in B can afford to give to the common pot and to their neighbours of degree 1 (remember from Lemma 4.1 that they make for at most half the neighbours) without getting to a negative final weight. The vertices that are neither small nor in B have a final weight equal to their initial weight, which was positive by choice of "small". Therefore, the following holds.

Proposition 5.4 *If D is large enough, then* $\mathrm{mad}(G) \geq m$.

In other words, there is no threshold on the maximum average degree.

Theorem 5.5 *For every $m \in \mathbb{R}^+$, if G is a graph with $\mathrm{mad}(G) < m$ and $\Delta(G)$ large enough with regard to m, then G is AVD $(\Delta(G)+1)$-colourable.*

There are further ways to refine a discharging argument, e.g. with the so-called *potential method*, as introduced by Kostochka and Yancey [20]. There, instead of simply assuming that $\mathrm{mad}(G) < c$, i.e. $\frac{2|E(H)|}{|V(H)|} < c$ for every subgraph H of G, they consider the exact gap between $2|E|$ and $c|V|$. This allows for finer reductions, proving that in a smallest counter-example the gap is larger than given by the statement hypotheses, so that a reduced graph can be slightly denser and still fall within the requirements. We focus now on special tricks on the reducibility side.

6 Re-colouring

Not all configurations are reducible directly through a degeneracy argument or something akin to it – that is, by deleting some elements, applying induction, then picking the right order in which to colour elements and making sure that every uncoloured element still has more choices than its number of coloured neighbours. Sometimes, it makes sense to delete as few elements as possible, apply induction, then uncolour some more elements but use the knowledge that an appropriate colouring exists. This is probably rather vague, so let us proceed to a concrete example.

6.1 As an intermediary step

Say we want to argue that every graph G from a certain graph class $\mathcal{G}$ can be edge L-coloured for any edge list assignment L that assigns at least $\Delta(G)$ colours to every edge. Assume furthermore that the class $\mathcal{G}$ is hereditary, i.e. for any graph $G \in \mathcal{G}$, any induced subgraph of G belongs to $\mathcal{G}$.

Proceed by contradiction, and take a smallest counter-example G to the property. We can immediately argue some quick facts about the structure of G. The graph G is connected. There is no vertex of degree 1 in G. As far as we can tell, the graph G might contain vertices of degree 2 – but each of them must have both neighbours of degree $\Delta(G)$. Using Lemma 1.4, it is not hard to check that G does not contain two vertices of degree 2 with the same two neighbours. Replacing one of those two vertices with a vertex of degree 3 makes degeneracy arguments impossible. This is where recolouring arguments come in handy.

Lemma 6.1 *G cannot contain a vertex u of degree 2 and a vertex v of degree 3 such that u and v have two common neighbours and v has a neighbour of degree at most $\Delta(G) - 1$ (see Figure 13).*

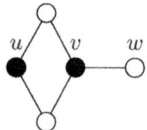

Figure 13: The configuration of Lemma 6.1 ($d(w) \leq \Delta(G) - 1$).

Proof Instead of L-colouring $G - \{u, v\}$, and failing to extend the resulting colouring to all of G, we L-colour $G - \{u\}$ by induction. Let α be the resulting colouring, and call x, y the common neighbours of u and v. Note that if α leaves at least 2 available colours for either ux or uy, the conclusion follows directly. A similar analysis holds for α restricted to $G - \{u, v\}$: if it leaves at least 3 available colours for ux, uy, vx, vy, or vw, the conclusion follows. We assume from now on that this is not the case, and with α restricted to $G - \{u, v\}$ there are exactly 2 available colours for all of ux, uy, vx, vy and vw. The readers can convince themselves that the only scenario where the colouring cannot be extended is when all of vx, vy and vw have the same two choices. However, this is impossible given the existence of α, hence the conclusion. □

This can be done much more generally, using an auxiliary graph to better analyze which elements should be recoloured and how. We refer the interested reader to [24] and in particular its Section 1.2.3 for a full overview and more intricate examples of what can be done with this recolouring approach.

Another, more famous example –though less applicable to list colouring– is that of Kempe changes. They were introduced in 1879 by Kempe in a failed attempt to prove the Four Colour Theorem [18]. Given a coloured graph, a *Kempe chain* is a maximal connected bicoloured subgraph. A *Kempe change* corresponds to selecting a Kempe chain and swapping the two colours in it. While this is not enough to prove the Four Colour Theorem, we can easily derive from it the fact that all planar graphs are five-colourable. Proceed by contradiction, and consider a smallest counter-example H. By a degeneracy argument, the graph H does not contain any vertex of degree 4 or less. By Kempe changes, we can argue a bit more.

Lemma 6.2 *H cannot contain a vertex of degree 5.*

Proof Assume for a contradiction that H contains a vertex u of degree 5, and let v_1, v_2, v_3, v_4 and v_5 be its neighbours in the clockwise order (in some planar embedding of H). By minimality of H, there is a 5-colouring α of $H - \{u\}$. The colouring can be extended to H unless α assigns five different colours to the neighbours of u. Assume that is the case, and consider without loss of generality that $\alpha(v_i) = i$ for $i \in \{1, \ldots, 5\}$. Consider the colouring α' obtained from α from doing a Kempe change on v_1 with colours 1 and 3. If $\alpha'(v_3) = 3$, then α' can be extended to H by assigning the colour 1 to u. Therefore, there is a Kempe chain with colours 1, 3 containing both v_1 and v_3. By symmetry, there is a Kempe chain with colours 2, 4 containing both v_2 and v_4. By planarity, they must intersect, which is impossible given that they are coloured disjoint colours. □

This is a standard argument for the Five Colour Theorem, though there are many alternative ways to argue it. The other standard argument consists in applying induction on H where the vertex of degree 5 has been deleted and two of its neighbours (appropriately chosen) are merged into a single vertex. Here, the notion of Kempe change is not crucial, though it offers an interesting perspective. Since its introduction in the context of 4-colouring planar graphs, much work has focused on determining which graph classes have good properties regarding Kempe changes on their vertex colourings, see e.g. [23] for a comprehensive overview or [2] for a recent result on general graphs. We refer the curious reader to the relevant chapter of a 2013 survey by Cereceda [29]. Kempe changes falls within the wider setting of combinatorial reconfiguration, which [29] is also an excellent introduction to. We consider now a case where the notion of Kempe change is indeed crucial.

6.2 As a more general tool

The proof of Theorem 1.1 relies heavily on the notion of Kempe change, introduced in the previous subsection in the context of vertex colouring but whose definition extends naturally to the realm of edge colouring. Instead of using recolourings on a local scale to argue that it suffices to apply induction on a smaller graph, the goal here is to colour the whole graph directly, albeit with too many colours, and gradually decrease the total number of colours involved. What Vizing actually proved is in fact the following.

Theorem 6.3 ([30]) *For every simple graph G, for any integer $k > \Delta(G) + 1$, for any k-edge-colouring α, there is a $(\Delta(G) + 1)$-edge-colouring that can be reached from α through a series of Kempe changes.*

Before we discuss how to prove Theorem 6.3, let us discuss further some high-level ideas around it. While some graphs need $\Delta(G) + 1$ colours, some graphs can be edge-coloured with $\Delta(G)$ colours. In the follow-up paper extending the result to multigraphs [31], and later in a more publicly available survey paper [32], Vizing asks whether an optimal colouring can always be reached through a series of Kempe changes, as follows.

Question 6.4 *[31] For every simple graph G, for any integer $k > \chi'(G)$, for any k-edge-colouring α, is there a $\chi'(G)$-edge-colouring that can be reached from α through a series of Kempe changes?*

In 2012, McDonald, Mohar and Scheide [21] answered the case $\Delta(G) = 3$ of Question 6.4 in the affirmative. In 2016, Asratian and Casselgren [1] similarly confirmed the case $\Delta(G) = 4$. We were recently able to answer Question 6.4 affirmatively in the somewhat orthogonal case where the graph is triangle-free, regardless of the value of $\Delta(G)$.

Theorem 6.5 *[3] For every triangle-free graph G, for any integer $k > \chi'(G)$, any given $\chi'(G)$-edge-colouring can be reached from any k-edge-colouring through a series of Kempe changes.*

The proof of Theorem 6.5 is rather similar in flavour to that of Theorem 6.3 – with significantly more convoluted arguments. Let us now prove Theorem 6.3.

Proof We consider a graph G, and proceed by induction on k. If $k = \Delta(G) + 1$, the statement trivially holds. Assume from now on that we are given a k-edge-colouring α of G with $k \geq \Delta(G) + 2$. Assume further that among all k-colourings reachable from α through a series of Kempe changes, α has the fewest edges coloured k. If it has none, then we can directly apply induction and conclude. Therefore, it has at least one edge coloured k. Our goal is to recolour it without creating another.

For every vertex u in $V(G)$, we select $m(u)$ to be one of the colours $\{1, \ldots, \Delta(G) + 1\}$ which α does not assign to an edge incident to u.

Consider an edge uv that is coloured k, and say we want to recolour it. If $m(u) = m(v)$, this can be done immediately without impacting the rest of the colouring, and contradicts the minimality of α. Therefore, let us consider $m(v) \neq m(u)$, and look at the obstacles around u. We define $ux_1, ux_2, \ldots, ux_p$ a maximal sequence of edges such that $\alpha(ux_1) = m(v)$ and $\alpha(ux_{i+1}) = m(x_i)$. By construction, no edge ux_i is coloured k, and in particular $v \neq x_i$ for all i.

We consider two cases depending on the value of $m(x_p)$.

- Assume $m(x_p) = m(u)$. We successively do Kempe changes with colours $m(x_i), \alpha(x_i)$ on x_i, for i from p down to 1. We then do a Kempe change with colours $k, m(v)$ on v and contradict the minimality of α.

- Assume $m(x_p) = m(x_q) = \alpha(ux_{q+1})$ for $q \in \{1, \ldots, p-2\}$. We do a Kempe change with colours $m(u), m(x_p)$ on u. At most one of x_p and x_q has now an incident edge coloured $\alpha(x_{q+1})$ and no incident edge coloured $m(u)$. Regardless of which of the two is affected, and of whether either is, we are in the first case.

□

As is often mentioned, this argument can be turned into a polynomial-time algorithm—this was formally noted by Misra and Gries in 1992 [22]. However, deciding whether a graph is $\Delta(G)$-edge-colourable is an NP-complete problem [14], even in the case of triangle-free graphs [19]. This leaves little hope for extracting a polynomial-time algorithm from the proof of Theorem 6.5. There is however no difficulty in detecting the difference between Vizing's argument and ours: we start by assuming full access to a $\Delta(G)$-edge-colouring, which is crucial in the proof.

7 Conclusion

The proof sketch behind a colouring theorem obtained with a discharging argument (with the corresponding unavoidable set $\{C_1, \ldots, C_p\}$ of reducible configurations) can equivalently be seen as proof by induction (the empty graph can be coloured, any graph obtained from a colourable graph by the reverse operation to the reduction when detecting a configuration C_i is colourable), where some space is dedicated to proving that the induction is complete (every graph can be built from the empty graph exclusively by the previous operations). The reducibility proofs could also be seen as parts of a pre-processing algorithm, which is proved by the discharging argument to always output the empty graph. Note that the reducible configurations detection can be sped up by assigning weight, applying the discharging rules, and looking around elements with a negative weight. When the discharging argument is purely local, this will very often grant a linear or quadratic colouring algorithm (see e.g. [26]).

Note that a global discharging argument relies heavily on the sub-structure that is used. We need both to have a colouring argument to deal with it when present, and a discharging argument to say that its absence has implications on the average degree. The structure of an even cycle is massively used, but Theorem 1.3 is also extremely powerful. However, there is no reason to restrain oneself from using more exotic structures, like cacti with the help of Brooks' theorem (see e.g. [4]). The main drawback to global discharging proofs is that it can significantly increase the running time of the colouring algorithm (if the global sub-structure is too hard to detect), and sometimes destroy entirely the initial hope of a colouring algorithm (if we appeal to an external, non-constructive colouring result).

Despite this warning, we should emphasize that most discharging proofs are constructive and immediately yield a polynomial colouring algorithm. We started this chapter by trying to define discharging proofs in opposition to non-discharging proofs. It could be noted that every single discharging proof can be translated in the realm of linear programming, with no discharging involved. This gives hope for at least partial automatization of the discharging proving (or checking) process.

Acknowledgements

Many thanks to the BCC committee and to the anonymous reviewer who helped improved this chapter.

References

[1] Armen S. Asratian and Carl Johan Casselgren. Solution of Vizing's problem on interchanges for the case of graphs with maximum degree 4 and related results. *Journal of Graph Theory*, 82(4):350–373, 2016.

[2] Marthe Bonamy, Nicolas Bousquet, Carl Feghali, and Matthew Johnson. On a conjecture of Mohar concerning Kempe equivalence of regular graphs. *Journal of Combinatorial Theory, Series B*, 135:179–199, 2019.

[3] Marthe Bonamy, Oscar Defrain, Tereza Klimošová, Aurélie Lagoutte, and Jonathan Narboni. On Vizing's edge colouring question. *preprint*, 2020.

[4] Marthe Bonamy, Benjamin Lévêque, and Alexandre Pinlou. Graphs with maximum degree $\delta \geq 17$ and maximum average degree less than 3 are list 2-distance $(\delta+ 2)$-colorable. *Discrete Mathematics*, 317:19–32, 2014.

[5] Oleg V Borodin. Colorings of plane graphs: A survey. *Discrete Mathematics*, 313(4):517–539, 2013.

[6] Oleg V Borodin, Alexandr V Kostochka, and Douglas R Woodall. List edge and list total colourings of multigraphs. *Journal of Combinatorial Theory, Series B*, 71(2):184–204, 1997.

[7] OV Borodin, AO Ivanova, and AV Kostochka. Oriented 5-coloring of sparse plane graphs. *Journal of Applied and Industrial Mathematics*, 1(1):9–17, 2007.

[8] Nathann Cohen and Frédéric Havet. Planar graphs with maximum degree $\Delta \geq 9$ are $(\Delta + 1)$-edge-choosable–a short proof. *Discrete Mathematics*, 310(21):3049–3051, 2010.

[9] Daniel W Cranston and Douglas B West. A guide to the discharging method. *arXiv preprint arXiv:1306.4434*, 2013.

[10] Gašper Fijavž, Martin Juvan, Bojan Mohar, and Riste Škrekovski. Planar graphs without cycles of specific lengths. *European Journal of Combinatorics*, 23(4):377–388, 2002.

[11] Fred Galvin. The list chromatic index of a bipartite multigraph. *Journal of Combinatorial Theory, Series B*, 63(1):153–158, 1995.

[12] Brian Hayes. Gauss's day of reckoning. *American Scientist*, 94(3):200–200, 2006.

[13] Hervé Hocquard and Mickaël Montassier. Adjacent vertex-distinguishing edge coloring of graphs with maximum degree Δ. *Journal of Combinatorial Optimization*, 26(1):152–160, 2013.

[14] Ian Holyer. The NP-completeness of edge-coloring. *SIAM Journal on computing*, 10(4):718–720, 1981.

[15] Stanislav Jendrol and H-J Voss. Light subgraphs of graphs embedded in the plane—a survey. *Discrete Mathematics*, 313(4):406–421, 2013.

[16] Tommy R.. Jensen and Bjarne Toft. *Graph coloring problems*. J. Wiley & sons, 1995.

[17] Jeff Kahn. Asymptotically good list-colorings. *Journal of Combinatorial Theory, Series A*, 73(1):1–59, 1996.

[18] Alfred B. Kempe. On the geographical problem of the four colours. *American journal of mathematics*, 2(3):193–200, 1879.

[19] Diamantis P. Koreas. The NP-completeness of chromatic index in triangle free graphs with maximum vertex of degree 3. *Applied mathematics and computation*, 83(1):13–17, 1997.

[20] Alexandr Kostochka and Matthew Yancey. Ore's conjecture on color-critical graphs is almost true. *Journal of Combinatorial Theory, Series B*, 109:73–101, 2014.

[21] Jessica McDonald, Bojan Mohar, and Diego Scheide. Kempe equivalence of edge-colorings in subcubic and subquartic graphs. *Journal of Graph theory*, 70(2):226–239, 2012.

[22] Jayadev Misra and David Gries. A constructive proof of Vizing's theorem. In *Information Processing Letters*. Citeseer, 1992.

[23] Bojan Mohar. Kempe equivalence of colorings. In *Graph Theory in Paris*, pages 287–297. Springer, 2006.

[24] Théo Pierron. *Induction Schemes: From Language Separation to Graph Colorings*. PhD thesis, Bordeaux, 2019.

[25] Rados Radoicic and Geza Toth. The discharging method in combinatorial geometry and the pach-sharir conjecture. *Contemporary Mathematics*, 453:319, 2008.

[26] Neil Robertson, Daniel P Sanders, Paul Seymour, and Robin Thomas. Efficiently four-coloring planar graphs. In *Proceedings of the twenty-eighth annual ACM symposium on Theory of computing*, pages 571–575, 1996.

[27] Daniel P Sanders and Yue Zhao. Planar graphs of maximum degree seven are class i. *Journal of Combinatorial Theory, Series B*, 83(2):201–212, 2001.

[28] William P Thurston. *The geometry and topology of three-manifolds*. MIT Press, 1978.

[29] Jan van den Heuvel. The complexity of change. *Surveys in combinatorics*, 409(2013):127–160, 2013.

[30] Vadim G Vizing. On an estimate of the chromatic class of a p-graph. *Diskret. Analiz*, 3(7):25–30, 1964.

[31] Vadim G. Vizing. The chromatic class of a multigraph. *Cybernetics*, 1(3):32–41, 1965.

[32] Vadim G. Vizing. Some unsolved problems in graph theory. *Russian Mathematical Surveys*, 23(6):125, 1968.

[33] Paul Wernicke. Über den kartographischen vierfarbensatz (in german). *Mathematische Annalen*, 58(3):413–426, 1904.

[34] Douglas R Woodall. The average degree of a multigraph critical with respect to edge or total choosability. *Discrete Mathematics*, 310(6):1167–1171, 2010.

[35] Zhongfu Zhang, Linzhong Liu, and Jianfang Wang. Adjacent strong edge coloring of graphs. *Applied Mathematics Letters*, 15(5):623–626, 2002.

Université de Bordeaux, LaBRI
351, cours de la Libération, CS 10004
F-33405 Talence Cedex - France
marthe.bonamy@u-bordeaux.fr

Generating graphs randomly

Catherine Greenhill

Abstract

Graphs are used in many disciplines to model the relationships that exist between objects in a complex discrete system. Researchers may wish to compare a network of interest to a "typical" graph from a family (or ensemble) of graphs which are similar in some way. One way to do this is to take a sample of several random graphs from the family, to gather information about what is "typical". Hence there is a need for algorithms which can generate graphs uniformly (or approximately uniformly) at random from the given family. Since a large sample may be required, the algorithm should also be computationally efficient.

Rigorous analysis of such algorithms is often challenging, involving both combinatorial and probabilistic arguments. We will focus mainly on the set of all simple graphs with a particular degree sequence, and describe several different algorithms for sampling graphs from this family uniformly, or almost uniformly.

1 Introduction

The modern world is full of networks, and many researchers use graphs to model real-world networks of interest. When studying a particular real-world network it is often convenient to define a family, or *ensemble*, of graphs which are similar to the network in some way. Then a random element of the ensemble provides a *null model* against which the significance of a particular property of the real-world model can be tested. For example, a researcher may observe that their network contains what looks like a large number of copies of a particular small subgraph H, also called a "motif" in network science. If this number is large compared to the average number of copies of H in some appropriate ensemble of graphs, then this provides some evidence that the high frequency of this motif may be related to the particular function of the real-world network. (For more on network motifs see for example [94].)

In this setting, the null model is a random graph model, and it may be possible to analyse the relevant properties using probabilistic combinatorics. Where this is not possible, it is very convenient to have an algorithm which provides uniformly random (or "nearly" uniformly random) graphs from the ensemble, so that the average number of copies of H can be estimated empirically. Such an algorithm should also be efficient, as a large sample may be needed. (In this survey, "efficient" means "computationally efficient".) Another motivation for the usefulness of algorithms for sampling graphs can be found in the analysis of algorithms which take graphs as input, especially when the worst-case complexity bound is suspected to be far from tight in the average case.

The aim of this survey is to describe some of the randomized algorithms which have been developed to efficiently sample graphs with certain properties, with particular focus on the problem of *uniformly sampling graphs with a given degree sequence*. We want to understand how close the output distribution is to uniform, and how the runtime of the algorithm depends on the number of vertices. Hence we restrict our attention to algorithms which have been rigorously analysed and have certain performance guarantees. In particular, statistical models such as the exponential random graph model (see for example [25, 69, 85, 118]), will not be discussed. Although we mainly restrict our attention to simple, undirected graphs, most of the ideas discussed in this survey can also be applied to bipartite graphs, directed graphs and hypergraphs, which are all extremely useful in modelling real-world networks.

It is still an open problem to find an efficient algorithm for sampling graphs with an arbitrary degree sequence. For bipartite graphs, however, the sampling problem was solved for arbitrary degree sequences by Jerrum, Sinclair and Vigoda [78], as a corollary of their breakthrough work on approximating the permanent. See Section 6.2 for more detail.

This is not a survey on random graphs. (Our focus is on *how* to randomly sample a graph from some family efficiently, from an algorithmic perspective, rather than on the properties of the resulting random graph.) However, some techniques are useful both as tools to analyse random graphs and as procedures for producing random graphs. There are many texts on random graphs [17, 52, 73], as well as Wormald's excellent survey on random regular graphs [120].

Before proceeding, we remark that in network science, the phrase "graph sampling algorithm" can refer to an algorithm for sampling vertices or subgraphs *within* a given (huge) graph (see for example [116]). For this reason, we will avoid using this phrase and will instead refer to "algorithms for sampling graphs".

2 Preliminaries and Background

2.1 Notation and assumptions

Let $[a] = \{1, 2, \ldots, a\}$ for any positive integer a.

A *multigraph* $G = (V, E)$ consists of a set of vertices V and a multiset E of edges, where each edge is an unordered pair of vertices (which are not necessarily distinct). A loop is an edge of the form $\{v, v\}$ and an edge is repeated if it has multiplicity greater than one. A *graph* is a *simple* multigraph: that is, a multigraph with no loops and no repeated edges. All graphs are finite and labelled, so V is a finite set of distinguishable vertices. A directed multigraph is defined similarly, except that edges are now ordered pairs. A directed graph is a directed multigraph which is simple, which means that it has no (directed) loops and no repeated (directed) edges.

Throughout, n will be the number of vertices of a graph, unless otherwise specified. We usually assume that the vertex set is $[n]$.

Standard asymptotic notation will be used, and asymptotics are as $n \to \infty$ unless otherwise specified. Let f, g be real-valued functions of n.

- Write $f(n) = o(g(n))$ if $\lim_{n \to \infty} f(n)/g(n) = 0$.

- Suppose that $g(n)$ is positive when n is sufficiently large. We write $f(n) = O(g(n))$ if there exists a constant C such that $|f(n)| \leq C\, g(n)$ for all n sufficiently large.

- Now suppose that $f(n)$ and $g(n)$ are both positive when n is sufficiently large. If $f = O(g)$ and $g = O(f)$ then we write $f(n) = \Theta(g(n))$.

Generating graphs randomly

We sometimes write $\approx$ to denote an informal notion of "approximately equal". In pseudocode, we write "u.a.r." as an abbreviation for *uniformly at random*.

When calculating the runtime of algorithms, we use the "Word RAM" model of computation [51, 66]. In this model, elementary operations on integers with $O(\log n)$ bits can be performed in unit time.

Randomised algorithms require a source of randomness. We assume that we have a perfect generator for random integers uniformly distributed in $\{1, 2, \ldots, N\}$ for any positive integer N. Furthermore, we assume that this perfect generator takes unit time whenever N has $O(\log n)$ bits.

2.2 Which graph families?

It is very easy to sample from some graph families:

- Let $\mathcal{S}(n)$ denote the set of all $2^{\binom{n}{2}}$ graphs on the vertex set $[n]$. We can sample from $\mathcal{G}(n)$ very easily: flip a fair coin independently for each unordered pair of distinct vertices $\{j, k\}$, and add $\{j, k\}$ to the edge set if and only if the corresponding coin flip comes up heads. Every graph on n vertices is equally likely, so this gives an exactly uniform sampling algorithm with runtime $O(n^2)$.

- Next we might consider $\mathcal{S}(n, m)$, the set of all $\binom{\binom{n}{2}}{m}$ graphs on the vertex set $[n]$ with precisely m edges. A uniformly random graph from this set can be generated edge-by-edge, starting with the vertex set n and no edges. At each step, choose a random unordered pair of distinct vertices $\{j, k\}$, without replacement, and add this edge to the graph. When the graph has m edges, it is a uniformly random element of $\mathcal{S}(n, m)$. This algorithm has runtime $O(n^2)$. Letting G_i denote the graph obtained after i edges have been added, the sequence $G_0, G_1, \ldots, G_m$ is known as the *random graph process*, with G_i a uniformly-random element of $\mathcal{S}(n, i)$ for all $i \in [m]$.

A uniform element from $\mathcal{S}(n, m)$ corresponds to the *Erdős–Rényi random graph* $\mathcal{G}(n, m)$, while the *binomial random graph model* $\mathcal{G}(n, p)$ is obtained by adapting the process for sampling from $\mathcal{S}(n)$ described above, replacing the fair coin by a biased coin which comes up heads with probability p. These two random graph models have been the subject of intense study for more than 60 years, see for example [17, 46, 52, 61, 73]. However, since polynomial-time sampling is easy for both of these families (as described above), we will say no more about them.

Instead, our focus will be on algorithms for sampling graphs with a given degree sequence. More generally, we might be interested in bipartite graphs, directed graphs or hypergraphs with a given degree sequence. Alternatively, we may want to sample graphs with a given degree sequence and some other property, such as connectedness or triangle-freeness. There are many variations, but our main focus will be on sampling from the set $G(\boldsymbol{k})$ defined below.

Definition 2.1 A graph G on vertex set $[n]$ has *degree sequence* $\boldsymbol{k} = (k_1, \ldots, k_n)$ if $\deg_G(j) = k_j$ for all $j \in [n]$, where $\deg_G(j)$ denotes the degree of j in G. Let $G(\boldsymbol{k})$ denote the set of all graphs with degree sequence $\boldsymbol{k}$. A sequence $\boldsymbol{k} = (k_1, \ldots, k_n)$ of nonnegative integers with even sum is *graphical* if $G(\boldsymbol{k})$ is nonempty. A graph with degree sequence $\boldsymbol{k}$ is a *realization* of $\boldsymbol{k}$.

We do not assume here that the elements of $\boldsymbol{k}$ are in non-ascending order, though we will usually assume that all entries of $\boldsymbol{k}$ are positive. Unlike the binomial random graph model $\mathcal{G}(n,p)$, the edges of a randomly-chosen element of $G(\boldsymbol{k})$ are not independent. This lack of independence makes sampling from $G(\boldsymbol{k})$ a non-trivial task.

If $\boldsymbol{k} = (k, k, \ldots, k)$ has every entry equal to k, then we say that $\boldsymbol{k}$ is *regular*. The set of all k-regular graphs on the vertex set $[n]$ will be denoted by $G(n, k)$ instead of $G(\boldsymbol{k})$.

We close this subsection with some more comments on graphical degree sequences. The characterisations of Erdős and Gallai [38] and Havel and Hakimi [68, 67] both give algorithms which can be used to decide, in polynomial time, whether a given sequence is graphical. The Erdős–Gallai Theorem says that if $k_1 \geq \cdots \geq k_n$ then $\boldsymbol{k}$ is graphical if and only if $\sum_{j=1}^{n} k_j$ is even and

$$\sum_{j=1}^{p} k_j \leq p(p-1) + \sum_{j=p+1}^{n} \min\{k_j, p\}$$

for all $p \in [n]$. To avoid trivialities, we will always assume that the sequence $\boldsymbol{k}$ is graphical. The Havel–Hakimi characterisation also assumes that entries of $\boldsymbol{k}$ are in non-decreasing order, and states that $\boldsymbol{k}$ is graphical if and only if

$$(k_2 - 1, \ldots, k_{k_1+1} - 1, k_{k_1+2}, \ldots, k_n)$$

has no negative entries and is graphical. This leads to a greedy algorithm to construct a realisation of $\boldsymbol{k}$ in runtime $O(n^2)$: join vertex 1 to each of vertices $2, \ldots, k_1+1$, delete vertex 1, reduce the target degree of vertices $2, \ldots, k_1+1$ by 1, sort the new degree sequence into nonincreasing order if necessary, and recurse. The runtime of this greedy algorithm is $O(n^2)$.

2.3 What kind of sampling algorithm?

To be more precise about our goals, we need some definitions. Let $(\Omega_n)_{n \in \mathcal{I}}$ be a sequence of finite sets indexed by a parameter n from some infinite index set $\mathcal{I}$, such as $\mathcal{I} = \mathbf{Z}^+$ or $\mathcal{I} = 2\mathbf{Z}^+$. Asymptotics are as n tends to infinity along elements of $\mathcal{I}$. We assume that $|\Omega_n| \to \infty$ as $n \to \infty$.

The reason that we consider a sequence of sets, rather than just one set, is that it makes no sense to say that the runtime of an algorithm is polynomial for a particular set Ω. If the runtime of an algorithm for sampling from Ω is T, then we could say that this is a *constant-time algorithm* with constant T, but then we learn nothing about how long the algorithm might take when given a different set as input. Having said that, in our notation we often drop the sequence notation and simply refer to Ω_n.

As a general rule, we say that a (uniform) sampling algorithm for Ω_n is *efficient* if its runtime is bounded above by a polynomial in $\log(|\Omega_n|)$, as it takes $\log(|\Omega_n|)$

bits to describe an element of Ω_n. However, the runtime of the algorithm may have a deterministic upper bound, or it may be a random variable, leading to the following paradigms:

- A *Monte Carlo* algorithm is a randomised algorithm which is guaranteed to terminate after some given number of steps, but has some probability of incorrect output. (The probability of incorrect output should be small.) A Monte Carlo sampling algorithm for Ω_n is efficient if its runtime is bounded above by a polynomial in $\log(|\Omega_n|)$.

- A *Las Vegas* algorithm is a randomised algorithm which is guaranteed to provide correct output with probability 1, but may have no deterministic upper bound on its running time. A Las Vegas sampling algorithm for Ω_n is efficient if its *expected* runtime is bounded above by a polynomial in $\log(|\Omega_n|)$.

Next, we focus on the output of the algorithm, and give three different definitions of "close to uniform". First we need a notion of distance for probability distributions.

Definition 2.2 Let σ and π be two probability distributions on the finite set Ω. The *total variation distance* between σ and π, denoted $d_{TV}(\sigma, \pi)$, is given by

$$d_{TV}(\sigma,\pi) = \tfrac{1}{2}\sum_{x\in\Omega}|\sigma(x)-\pi(x)| = \max_{S\subseteq\Omega}|\sigma(S)-\pi(S)|.$$

Here $\sigma(S) = \sum_{x\in S}\sigma(x)$ for any event $S \subseteq \Omega$, and similarly for $\pi(S)$.

Suppose that some sampling algorithm over Ω_n has output distribution σ_n, and let π_n denote the uniform distribution over Ω_n, for any $n \in \mathcal{I}$.

- If $\sigma_n = \pi_n$ for all $n \in \mathcal{I}$ then we say that the algorithm is a *uniform sampling algorithm* or *uniform sampler*.

- If $\lim_{n\to\infty} d_{TV}(\sigma,\pi) = 0$ then we say that the algorithm is an *asymptotically uniform sampler*. In this situation it is usually not possible to increase the accuracy by running the algorithm for longer, as the total variation distance depends only on n.

- If $d_{TV}(\sigma_n, \pi_n) < \varepsilon$ for some positive constant ε then we say that the algorithm is an *almost uniform sampler*, and that the output is ε-*close* to uniform. Often, ε is provided by the user, and higher accuracy (smaller ε) can obtained at the cost of a longer runtime.

The Markov chain approach to sampling, when successful, provides an algorithm called an FPAUS. See for example [74, Chapter 3].

Definition 2.3 Let $(\Omega_n)_{n\in\mathcal{I}}$ be a sequence of finite sets indexed by a parameter n from some infinite index set $\mathcal{I}$, such that $|\Omega_n| \to \infty$ as $n \to \infty$. A *fully-polynomial almost uniform sampler (FPAUS)* for sampling from Ω_n is an algorithm that, with probability at least $\tfrac{3}{4}$, outputs an element of Ω_n in time polynomial in $\log|\Omega_n|$ and $\log(1/\varepsilon)$, such that the output distribution is ε-*close* to the uniform distribution on Ω_n in total variation distance. (That is, $d_{TV}(\sigma_n, \pi_n) < \varepsilon$ where σ_n is the output distribution and π_n is the uniform distribution on Ω_n.)

If $\Omega_n = G(\boldsymbol{k})$ for some graphical sequence $\boldsymbol{k} = (k_1,\ldots,k_n)$ then $\log|\Omega_n| = O(M \log M)$, where M is the sum of the entries of $\boldsymbol{k}$. This can be proved using the configuration model: see (6.4). So an FPAUS for $G(\boldsymbol{k})$ must have running time bounded above by a polynomial in n and $\log(1/\varepsilon)$, since $M \leq n^2$.

Sampling and counting are closely related, and an algorithm for one problem can often be transformed into an algorithm for the other. While our focus is firmly on sampling, we will also need the following definition which describes a good approximate counting algorithm.

Definition 2.4 Let $(\Omega_n)_{n \in \mathcal{I}}$ be a sequence of finite sets indexed by a parameter n from some infinite index set $\mathcal{I}$, such that $|\Omega_n| \to \infty$ as $n \to \infty$. A *fully-polynomial randomised approximation scheme* (FPRAS) for Ω_n is an algorithm which accepts as input a parameter $\varepsilon > 0$ and outputs an estimate X for $|\Omega_n|$ such that

$$\Pr\left((1-\varepsilon)|\Omega_n| \leq X \leq (1+\varepsilon)|\Omega_n|\right) \geq \tfrac{3}{4},$$

with runtime polynomial in $\log|\Omega_n|$ and ε^{-1}.

The probability in this definition can be easily increased from $\tfrac{3}{4}$ to $1-\delta$, for any fixed $\delta \in (\tfrac{3}{4},1)$, by obtaining $O(\log \delta^{-1})$ estimates and taking the median [105, Lemma 2.1].

Before moving on, we say a little more about the connection between sampling and counting. Jerrum, Valiant and Vazirani [79] proved that for *self-reducible* problems, polynomial-time approximate counting is equivalent to polynomial-time almost-uniform sampling. Without going into too much detail, a problem is self-reducible if the solutions for a given instance can be generated recursively using a small number of smaller instances of the same problem. For example, consider the set $\mathcal{M}(G)$ of all matchings (of any size) in a graph G. Remove the edges of G one-by-one (in lexicographical order, say) to form the sequence

$$G = G_m > G_{m-1} > \cdots > G_1 > G_0 = (V, \emptyset).$$

Then $|\mathcal{M}(G_0)| = 1$ and hence $|\mathcal{M}(G)| = \prod_{j=1}^m |\mathcal{M}(G_j)|/|\mathcal{M}(G_{j-1})|$. The j'th ratio is the inverse of the probability that $e_j \notin M$, where M is a matching chosen uniformly at random from $\mathcal{M}(G_j)$ and e_j is the unique edge in $E(G_j) \setminus E(G_{j-1})$. If we can sample almost-uniformly from the sets $\mathcal{M}(G_j)$ to sufficient accuracy then estimates for these probabilities can be multiplied together to provide a good estimate for $1/|\mathcal{M}(G)|$. In this way, approximate counting can be reduced to almost-uniform sampling. See [74, Chapter 3] for full details.

For sampling graphs, the situation is a little more complicated. Erdős et al [41] showed that the problem of sampling graphs, or directed graphs, with given degrees can be made self-reducible by supplying as input a small set of forbidden edges, and sampling from the set of graphs (or bipartite graphs, or directed graphs) with specified degrees which avoid the forbidden edges. Using the fact that directed graphs can be modelled as bipartite graphs which avoid a given perfect matching, the set of forbidden edges can be taken to be a star (for undirected graphs) or the the union of a star and a perfect matching. We remark that the exact counting problem (given a graphical sequence $\boldsymbol{k}$, calculate $|G(\boldsymbol{k})|$) is not known to be #P-complete, and similarly for bipartite or directed variants.

2.4 Sampling graphs with given degrees: an overview

We assume that for all n in some infinite index set $\mathcal{I}$, we have a graphical degree sequence $\boldsymbol{k}(n) = (k_1(n), \ldots, k_n(n))$. The length of $\boldsymbol{k}(n)$ is n, and the elements of $\boldsymbol{k}(n)$ might themselves be functions of n. Our sequence $(\Omega_n)_{n \in \mathcal{I}}$ of sets (as discussed in the previous subsection) is given by taking $\Omega_n = G(\boldsymbol{k}(n))$. From now on, we simply write $\boldsymbol{k} = (k_1, \ldots, k_n)$ for the degree sequence, but we should remember that in fact we have a *sequence* of degree sequences, indexed by n.

The maximum entry in a degree sequence $\boldsymbol{k}$ is denoted $k_{\max}$. Let $M = \sum_{j \in [n]} k_j$ be the sum of the degrees. Then $m = M/2$ the number of edges of any graph in $G(\boldsymbol{k})$.

There are a few different methods for sampling graphs with given degrees (restricted to algorithms which can be rigorously analysed). We outline the main approaches here, and go into more detail in the subsequent sections.

- The *configuration model* was introduced by Bollobás [16] in 1980, as a convenient way to calculate the probability of events in random regular graphs. The model can be used as an algorithm for sampling uniformly at random from $G(\boldsymbol{k})$. However, the expected runtime is high unless the degrees are very small: specifically, $k = O(\sqrt{\log n})$ in the regular case.

- McKay and Wormald's *switchings-based algorithm* [91] from 1990 performs (exactly) uniform sampling from $G(\boldsymbol{k})$ in expected polynomial time, for a much wider range of degrees than the algorithm arising from the configuration model, namely $k_{\max} = O(M^{1/4})$. Gao and Wormald [59] and Arman, Gao and Wormald [5] have extended and improved the McKay–Wormald algorithm, allowing it to apply to a wider range of degrees and making it more efficient. These algorithms are fast, but a little complicated and difficult to implement. See the end of Section 5.1.

- Another approach is to use a *Markov chain* with uniform stationary probability over a state space $G'(\boldsymbol{k})$ which contains $G(\boldsymbol{k})$. If the Markov chain converges rapidly to its stationary distribution, and each step of the Markov chain can be implemented efficiently, then this gives an FPAUS for $G'(\boldsymbol{k})$. In 1990, Jerrum and Sinclair [76] described and analysed a Markov chain which samples from a set $G'(\boldsymbol{k})$ of graphs with degree sequence close to $\boldsymbol{k}$. The Jerrum–Sinclair chain is efficient only when $G(\boldsymbol{k})$ forms a sufficiently large fraction of $G'(\boldsymbol{k})$. When this condition on $G(\boldsymbol{k})$ holds, rejection sampling can be used to restrict the output of the chain to $G(\boldsymbol{k})$. The requirement that $|G(\boldsymbol{k})|/|G'(\boldsymbol{k})|$ is sufficiently large gives rise to a notion of *stability* of degree sequences. Another well-studied Markov chain, the *switch chain*, has state space $G(\boldsymbol{k})$ and thus avoids rejection sampling. Most proofs in this area use Sinclair's multicommodity flow method [104], and the resulting bounds, when polynomial, tend to be rather high-degree and are not believed to be tight.

- In 1999, Steger and Wormald [106] presented an algorithm for performing asymptotically uniform sampling for k-regular graphs. Their aim was to provide an algorithm which is both genuinely fast (runtime $O(k^2 n)$ when k is a small power of n) and easy to implement. While the idea for the algorithm is motivated by the bounded-degree graph process [101], the algorithm is presented as a modification of the configuration model. We also describe extensions and enhancements by Kim and Vu [82] and Bayati, Kim and Saberi [7]: in particular, Bayati et al. [7] used sequential importance sampling to provide an algorithm which is *almost* an FPAUS, when $k_{\max} = O(M^{\frac{1}{4}-\tau})$ for any constant $\tau > 0$.

The configuration model is described in Section 3, followed in Section 4 by the sequential algorithms beginning with the work of Steger and Wormald. Switchings-based algorithms are discussed in Section 5, and Markov chain (MCMC) algorithms are presented in Section 6.

3 The configuration model

The *configuration model*, introduced by Bollobás [16], is very useful in the analysis of random graphs with given degrees. It also arose in asymptotic enumeration work of Bender and Canfield [9], and was first explicitly used as an algorithm by Wormald [119].

Given a degree sequence $\boldsymbol{k}$, and recalling that $M = \sum_{j \in [n]} k_j$, we take M objects, called *points*, grouped into n *cells*, where the j'th cell contains k_j points. Each point is labelled, and hence distinguishable. (You can think of the points corresponding to vertex j as being labelled by $(j, 1), \ldots, (j, k_j)$, say. But we will not refer to these labels explicitly.)

In the network theory literature (for example [50]), points are sometimes called *stubs*, or *half-edges*, without the concept of a cell (so that k_j half-edges emanate from vertex j).

A *configuration*, also called a *pairing*, is a partition of the M points into $M/2$ pairs. This is often described as a perfect matching of the M points. Given a configuration P, shrinking each cell to a vertex and replacing each pair by an edge gives a multigraph $G(P)$ with degree sequence $\boldsymbol{k}$. The multigraph is simple if it has no loops and no repeated edges, and in this case we also say that P is simple.

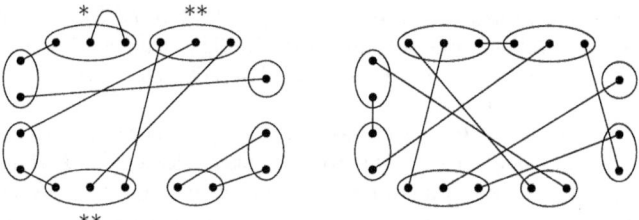

Figure 1: Two configurations with the same degree sequence

Figure 1 shows two configurations with the same degree sequence, namely $\boldsymbol{k} = (3, 3, 1, 2, 2, 3, 2, 2)$ if cells are labelled clockwise from the top-left. The small black circles represent points, which are shown inside cells, and the lines between points represent pairs. The configuration on the left is not simple, as it will produce a loop

Generating graphs randomly   141

on the vertex corresponding to the cell marked with "∗", and a repeated edge between the vertices corresponding to the two cells marked with "∗∗". The configuration on the right is simple.

Let $\mathcal{P}(\boldsymbol{k})$ be the set of all configurations corresponding to the degree sequence $\boldsymbol{k}$. The term *configuration model* typically refers to the uniform probability model over the set $\mathcal{P}(\boldsymbol{k})$. A uniformly random configuration from $\mathcal{P}(\boldsymbol{k})$ can be chosen in $O(M)$ time, as follows. Starting with all points unmatched, at each step take an arbitrary point p and pair it with a point chosen uniformly at random from the remaining unmatched points (excluding p). Once all points have been paired up, we have a configuration P and each configuration is equally likely.

The configuration model can be used as an algorithm for sampling uniformly from $G(\boldsymbol{k})$, by repeatedly sampling $P \in \mathcal{P}(\boldsymbol{k})$ uniformly at random until $G(P)$ is simple. This algorithm is displayed in Figure 2.

CONFIGURATION MODEL SAMPLING ALGORITHM
Input: graphical sequence $\boldsymbol{k}$
Output: element of $G(\boldsymbol{k})$

repeat
 choose $P \in \mathcal{P}(\boldsymbol{k})$ u.a.r.
until $G(P)$ is simple
output $G(P)$

Figure 2: The configuration model as a sampling algorithm

Observe that if G is a simple graph with degree sequence $\boldsymbol{k}$ then G corresponds to exactly $\prod_{j\in[n]} k_j!$ configurations, as there are $k_j!$ ways to assign points to the edges incident with vertex j, and these assignments can be made independently for each vertex $j \in [n]$. Hence every element of $G(\boldsymbol{k})$ is equally likely to be produced as output of the above process.

This gives a Las Vegas sampling algorithm, with expected runtime which depends linearly on the probability that a random configuration is simple. Hence, the configuration model can be used for efficient sampling when the probability that a randomly chosen configuration is simple is bounded below by $1/p(n)$, for some polynomial $p(n)$. In this case, the expected number of trials before a simple configuration is found is at most $p(n)$, and the expected runtime is $O(M\,p(n))$.

A multigraph is simple if and only if it contains no 1-cycles (loops) and no 2-cycles (arising from repeated edges). If the maximum degree is not too large compared to the number of edges, then in a uniformly random element of $\mathcal{P}(\boldsymbol{k})$, the number of 1-cycles and the number of 2-cycles are asymptotically independent Poisson random variables. In the k-regular case, Bender and Canfield [9] proved in 1978 that a uniformly random configuration is simple with probability $(1 + o(1))\,e^{-(k^2-1)/4}$. Hence the configuration model for k-regular graphs gives an expected polynomial time algorithm as long as $k = O(\sqrt{\log n})$. A very precise estimate of Pr(simple), with many significant terms, was given by McKay and Wormald [92] in 1991 under the assumption that $k_{\max}^3 = o(M)$. To prove the following result, we use the estimate (3.1) obtained by Janson [71] in 1999, which is valid for a wider range of degrees.

Theorem 3.1 [71] *Let $R = R(\boldsymbol{k})$ be defined by $R = \sum_{j\in[n]} k_j^2$. The configuration model gives a uniform sampling algorithm for $G(\boldsymbol{k})$. If $k_{\max}^2 = o(M)$ then the expected runtime of this algorithm is*

$$\Theta\bigl(M \exp\bigl(R^2/(4M^2)\bigr)\bigr)$$

when $k_{\max}^2 = o(M)$. So the expected runtime is polynomial if and only if $R = \Theta(M\sqrt{\log n})$. In particular, if $k_{\max} = O(\sqrt{\log n})$ then the expected runtime is polynomial.

Proof (Sketch.) The output is distributed uniformly as each element of $G(\boldsymbol{k})$ is simple, and hence corresponds to the same number of configurations in $\mathcal{P}(\boldsymbol{k})$. The expected number of trials required before a simple configuration is found is $1/\Pr(\text{simple})$, where $\Pr(\text{simple})$ denotes the probability that a uniformly chosen configuration from $\mathcal{P}(\boldsymbol{k})$ is simple. Janson [71] proved that if $k_{\max}^2 = o(M)$ then the probability that a random configuration is simple is

$$\Pr(\text{simple}) = \exp\left(-\frac{R^2}{4M^2} + \tfrac{1}{4}\right) + o(1). \tag{3.1}$$

Hence the expected runtime of the algorithm is $\Theta\bigl(M/\Pr(\text{simple})\bigr)$, which is bounded above by a polynomial if and only if $R = \Theta(M\sqrt{\log n})$. The last statement of the theorem follows since $R \leq k_{\max} M$. □

There are versions of the configuration model which can be used to sample bipartite graphs, directed graphs or hypergraphs with a given degree sequence. In all cases, the expected runtime is polynomial only for constant or very slowly-growing degrees.

4 Sequential algorithms and graph processes

The study of graph processes dates back to the very beginnings of the study of random graphs, in the work of Erdős and Rényi [47]. In a random graph process, edges are added to an empty graph one by one, chosen randomly from the set of all non-edges, sometimes with additional constraints. In 1979, Tinhofer [110] described such an algorithm for sampling from $G(\boldsymbol{k})$ non-uniformly. The *a posteriori* output probability could be calculated and, in theory, this could be combined with a rejection step in order to achieve uniformly distributed output. However, the runtime of the resulting algorithm (with the rejection step) is not known.

Recall that $G(n, k)$ denotes the set of all k-regular graphs on $[n]$. The bounded-degree graph process starts with the empty graph on n vertices (with no edges), and repeatedly chooses two distinct non-adjacent vertices with degree at most $k-1$, uniformly at random, and joins these two vertices by an edge. When no such pair of vertices remain, either we have a k-regular graph or the process has become stuck. (The name "bounded-degree graph process" does not mean that all the degrees are

Generating graphs randomly

$O(1)$. Rather, it means that we add edges sequentially but do not allow the degree of any vertex to exceed k, so we maintain this upper bound on all degrees.)

Ruciński and Wormald [101] proved that for any constant k, the process produces a k-regular graph with probability $1 - o(1)$. The output distribution is not uniform, and is not well understood. However, it is conjectured that the output of the bounded-degree graph process is contiguous with the uniform distribution over $G(n, k)$: see Wormald [120, Conjecture 6.1]. (Two sequences of probability spaces are *contiguous* if any event with probability which tends to 1 in one sequence must also tend to 1 in the other.)

We now turn to sequential algorithms which produce asymptotically uniform output.

4.1 The regular case

Steger and Wormald [106] described an algorithm for sampling from $G(n, k)$ using the following modification of the configuration model algorithm. Instead of choosing a configuration P uniformly at random, and then rejecting the resulting graph $G(P)$ if it is not simple, we choose one pair at a time and only keep those pairs which do not lead to a loop or a repeated edge. Specifically, we start with kn points in n cells, each with k points. Let U be the set of unpaired points, which initially contains all kn points. A set of two points $\{p, p'\} \subseteq U$ is *suitable* if p and p' belong to different cells, and no pair chosen so far contains points from the same two cells as p, p'. After repeatedly choosing pairs of suitable points, the algorithm may get stuck, or else reaches a full configuration P (with $kn/2$ pairs) and outputs the simple k-regular graph $G(P)$. The algorithm is given in pseudocode in Figure 3.

STEGER–WORMALD ALGORITHM
Input: n and k, with kn even
Output: element of $G(n, k)$

repeat
   let U be the set of all kn points
   let $P := \emptyset$
   repeat
     choose a set of two distinct points $\{p, p'\} \subseteq U$ u.a.r.
     if $\{p, p'\}$ is suitable then add $\{p, p'\}$ to P and delete $\{p, p'\}$ from U
   until U contains no suitable pairs of points
until $G(P)$ is k-regular
output $G(P)$

Figure 3: The Steger–Wormald algorithm

Though the explanation above involves the configuration model, Steger and Wormald state that their algorithm arose from adapting the bounded-degree processes. In this setting, the Steger–Wormald algorithm corresponds to choosing the vertices u, v to add at the next step with a non-uniform probability. To be specific, if $k'(x)$ denotes the current degree of vertex x in the graph formed by the edges chosen so far, then the Steger–Wormald algorithm chooses $\{u,v\}$ as the next edge with probability proportional to $(k - k'(u))(k - k'(v))$.

The following theorem is a combination of Steger and Wormald's results [106, Theorems 2.1, 2.2 and 2.3].

Theorem 4.1 [106] *Let* $\Pr(G)$ *denote the probability that a given graph* $G \in G(n, k)$ *is produced as output of the Steger–Wormald algorithm.*

(i) *If* $k = O(n^{1/28})$ *then there exists a function* $f(n, k) = o(1)$ *such that for every* $G \in G(n, k)$,
$$\left|\Pr(G) - |G(n,k)|^{-1}\right| < \frac{f(n,k)}{|G(n,k)|}.$$

(ii) *If* $k = o\bigl((n/\log^3 n)^{1/11}\bigr)$ *then there exists a function* $f(n, k) = o(1)$ *and a subset* $\mathcal{X}(n, k) \subseteq G(n, k)$ *such that*
$$\Pr(G) = (1 + O(f(n,k))) |G(n,k)|^{-1}$$
for all $G \in \mathcal{X}(n, k)$, *and* $|\mathcal{X}(n, k)| = (1 - f(n, k)) |G(n, k)|$.

(iii) *Under the same condition as (ii), the expected number of times that the outer loop of the algorithm is performed (that is, until* $G(P)$ *is regular) is* $1 + o(1)$, *and hence the runtime of the algorithm is* $O(k^2 n)$.

In particular, when $k = o\bigl((n/\log^3 n)^{1/11}\bigr)$, the output distribution of the Steger–Wormald algorithm is within $o(1)$ of uniform in total variation distance.

Kim and Vu [82] gave a new analysis of the Steger–Wormald algorithm using a concentration result of Vu [115], increasing the upper bound on the degree and confirming a conjecture of Wormald [120].

Theorem 4.2 [82] *Let* $0 < \varepsilon < \frac{1}{3}$ *be a constant. Then for any* $k \leq n^{1/3-\varepsilon}$ *and* $G \in G(n, k)$, *the probability* $\Pr(G)$ *that* G *is output by the Steger–Wormald algorithm satisfies* $\Pr(G) = (1 + o(1)) |G(n,k)|^{-1}$.

4.2 The irregular case, and an almost-FPAUS

The Steger–Wormald algorithm was generalised to irregular degree sequences in 2010 by Bayati, Kim and Saberi [7]. They stated their algorithm in terms of graphs, not configurations, and report failure (rather than restarting) if the procedure gets stuck. The pseudocode for this algorithm, which is called PROCEDURE A in [7], is given in Figure 4. Recall that $m = M/2 = \frac{1}{2} \sum_{j \in [n]} k_j$. We write $\binom{[n]}{2}$ for the set of all unordered pairs of distinct vertices in $[n]$.

This procedure is equivalent to the Steger–Wormald algorithm when $\boldsymbol{k}$ is regular, since then the factor $1 - k_i k_j/(4m)$ does not introduce any bias. For irregular degrees, this factor is chosen for the following reason. If two vertices of high degree are joined by an edge, then this choice makes it more difficult for the process to complete successfully. In [7], the authors show that the bias from edge $\{i, j\}$ is roughly $\exp(k_i k_j/(4m))$, and hence the probability $1 - k_i k_j/(4m) \approx \exp(-k_i k_j/(4m))$ is designed to cancel out this bias.

> BAYATI, KIM AND SABERI: PROCEDURE A
> *Input:*   graphical sequence $\boldsymbol{k}$
> *Output:* element of $G(\boldsymbol{k})$, or *fail*
>
> repeat
>   let $E := \emptyset$ (*set of edges, initially empty*)
>   let $\widehat{\boldsymbol{k}} := \boldsymbol{k}$ (*current degree deficit*)
>   let $a := 1$
>   repeat
>   choose an unordered pair of distinct vertices $\{i,j\} \in \binom{[n]}{2} \setminus E$
>   with probability proportional to $p_{ij} := \widehat{k}_i \widehat{k}_j \left(1 - \frac{k_i k_j}{4m}\right)$
>   let $a := a \times p_{ij}$
>   add $\{i,j\}$ to E and reduce each of $\hat{k}_i, \hat{k}_j$ by 1
>   until no more edges can be added to E
>   if $|E| = m$ then
>   output $G(P)$ and $N = (m!\, a)^{-1}$
>   else
>   report *fail* and output $N = 0$

Figure 4: Bayati, Kim and Saberi's asymptotically-uniform sampling algorithm

Bayati, Kim and Saberi [7, Theorem 1 and Theorem 2] proved the following properties of PROCEDURE A.

Theorem 4.3 [7] *Let $\boldsymbol{k}$ be a graphical degree sequence and $\tau > 0$ an arbitrary constant.*

(i) *Suppose that $k_{\max} = O(m^{1/4-\tau})$. Then* PROCEDURE A *terminates successfully with probability $1 - o(1)$ in expected runtime $O(k_{\max} m)$, and the probability $\Pr(G)$ that any given $G \in G(\boldsymbol{k})$ is output satisfies $\Pr(G) = (1+o(1))\,|G(\boldsymbol{k})|^{-1}$.*

(ii) *Now suppose that $\boldsymbol{k} = (k,\ldots,k)$, where $k = O(n^{1/2-\tau})$. Then* PROCEDURE A *has output distribution which is within distance $o(1)$ from uniform in total variation distance.*

Part (i) of this theorem extends the Kim–Vu result (Theorem 4.2) to the irregular case with essentially the same condition, since $m = kn$ when $\boldsymbol{k}$ is k-regular. Similarly, part (ii) of Theorem 4.3 generalises Theorem 4.1(ii) to irregular degree sequences with much higher maximum degree.

When successful, PROCEDURE A outputs a graph and a nonnegative number N. The value of N is not needed for asymptotically-uniform sampling, but is used to give a *fully-polynomial randomised approximation scheme* (FPRAS) for approximating $|G(\boldsymbol{k})|$, using a technique known as *sequential importance sampling* (SIS) which we outline below. Recall the definition of FPRAS from Definition 2.4.

Let $\mathcal{N}(\boldsymbol{k})$ be the set obtained by taking all possible edge-labellings of graphs $G(\boldsymbol{k})$, labelling the edges $e_1,\ldots,e_m$. Then $|\mathcal{N}(\boldsymbol{k})| = m!\,|G(\boldsymbol{k})|$. We can slightly modify PROCEDURE A so that it labels the edges in the order that they were chosen. This modified PROCEDURE A produces $H \in \mathcal{N}(\boldsymbol{k})$ with probability $P_A(H)$, denoted by a in Figure 4. The expected value of $1/P_A(\cdot)$ for an element of $\mathcal{N}(\boldsymbol{k})$ chosen according to the distribution P_A, is

$$\sum_{H \in \mathcal{N}(\boldsymbol{k})} \frac{1}{P_A(H)} P_A(H) = |\mathcal{N}(\boldsymbol{k})| = m!\,|G(\boldsymbol{k})|.$$

Therefore we can estimate $|G(\boldsymbol{k})|$ by performing r trials of PROCEDURE A and taking the average of the resulting r values of $\bigl(m!\,P_A(H_i)\bigr)^{-1}$. (Note that $m!\,P_A(H)$ is precisely the value denoted N in Figure 4 when the edge-labelled graph H is output.)

Bayati et al. prove [7, Theorem 3] that taking $r = O(\varepsilon^{-2})$ gives an FPRAS for estimating $|G(\boldsymbol{k})|$.

Finally, Bayati et al. [7] showed how to adapt the SIS approach to estimate $P_A(G)$ for (non-edge-labelled) $G \in G(\boldsymbol{k})$. This leads to an algorithm which is *almost* an FPAUS for $G(\boldsymbol{k})$, when $k_{\max} = O(m^{1/4-\tau})$ for some $\tau > 0$. The algorithm satisfies every condition from the definition of FPAUS except for the runtime: in an FPAUS the runtime must be polynomial in n and $\log(1/\varepsilon)$, but the algorithm given in [7, Section 3] has runtime which is polynomial in n and $1/\varepsilon$. In a little more detail, the Bayati–Kim–Saberi algorithm proceeds as follows:

- Given a graphical degree sequence $\boldsymbol{k}$ and parameters $\varepsilon, \delta \in (0,1)$, the FPRAS is used to obtain a sufficiently good estimate X for $|G(\boldsymbol{k})|$ (with high probability), and a random graph $G \in G(\boldsymbol{k})$ is obtained using PROCEDURE A.

- Next, we need an estimate P_G for the probability $P_A(G)$ that PROCEDURE A outputs G. This probability is estimated as follows: repeatedly choose a random ordering of the edges of G, calculate the probability that these edges were chosen *in this order* during the execution of PROCEDURE A, and take the average of these probabilities (averaged over the different orders chosen).

- Finally, G is returned as output of the almost-FPAUS with probability given by $\min\{\frac{1}{cXP_G}, 1\}$, where c is a universal constant independent of $\boldsymbol{k}$, ε, δ.

Bayati, Kim and Saberi [7, Remark 1] state their main results can be adapted to give analogous results for sampling bipartite graphs with given degrees, under the same assumptions on the maximum degree. Independently, Blanchet [13] used sequential importance sampling to give an FPRAS for counting bipartite graphs with given degrees, when the maximum degree in one part of the vertex bipartition is constant, while in the other part the maximum degree is $o(M^{1/2})$ and the sum of the squares of the degrees is $O(M)$. The arguments provided by Bayati et al. [7] and Blanchet [13] utilise concentration inequalities and Lyapunov inequalities, respectively.

Sequential importance sampling was used by Chen, Diaconis, Holmes and Liu [21] and Blitzstein and Diaconis [14] to sample graphs and bipartite graphs with given degrees, but without fully rigorous analysis. Sequential importance sampling algorithms also appear in the physics literature, for example [29, 81], again without rigorous analysis.

While sequential importance sampling algorithms may perform well in practice in many cases, Bezáková et al. [12] showed that these algorithms are provably slow in some cases.

4.3 Other graph processes

The property of having maximum degree at most k can be rephrased as the property of having no copy of the star $K_{1,k}$. More generally, for a fixed graph H, the *H-free process* proceeds from an empty graph by repeatedly choosing a random edge and adding it to the graph if it does not form a copy of H. See for example [18, 97, 117]. In particular, the *triangle-free process* is very well studied and has connections with Ramsey Theory [15, 48, 99] (we do not attempt to be comprehensive here as there is a large literature on this topic). The main focus in this area is extremal, as analysis of these processes provides a lower bound on the maximum number of edges possible in an H-free graph. This often involves application of the differential equations method, see [121]. Some pseudorandom properties of the output have been proved for these processes, see for example [15, 99]. However, it is not clear how far the output distribution varies from uniform, and so these processes may not be suitable for almost-uniform sampling.

An exception is the work of Bayati, Montanari and Saberi [8], who adapted the methods of [7] to analyse a sequential algorithm for generating graphs with a given number of edges and girth greater than ℓ (that is, no cycles of length at most ℓ), where ℓ is a fixed positive integer. As in [7, 106], the next edge is chosen non-uniformly, such that the probability that an edge e is selected is (approximately) proportional to the number of successful completions of the subgraph $G' \cup \{e\}$, where G' denotes the current graph. Bayati et al. [8] prove that the output of their algorithm is asymptotically uniform after $m = O\bigl(n^{1+1/(2\ell(\ell+3))}\bigr)$ edges have been added. The expected runtime of the algorithm is $O(n^2 m)$.

5 Switchings-based algorithms

In the sampling algorithm based on the configuration model (Figure 2), a configuration P is chosen from $\mathcal{P}(\mathbf{k})$ uniformly at random, repeatedly, until the corresponding graph $G(P)$ is simple. That is, if $G(P)$ contains any "defect" (in this case, a loop or a repeated edge) then this choice is rejected and we choose again. In 1990, McKay and Wormald [91] introduced a uniform sampling algorithm for $G(\mathbf{k})$ which begins by choosing a random element of $\mathcal{P}(\mathbf{k})$ and rejecting it only if there are "too many" defects. Once a configuration has been found with "not too many" defects, operations called *switchings* are applied, one by one, to reduce the number of defects until a simple configuration is obtained. To maintain a uniform distribution, McKay and Wormald introduce a carefully-chosen rejection probability at each step of the process.

We describe the McKay–Wormald algorithm in some detail, as this will set the scene for the significant improvements introduced by Gao and Wormald [59], to be discussed in Section 5.1. The structure of the McKay–Wormald algorithm is based on the *switching method*, introduced by McKay [89]. The switching method is used in asymptotic enumeration to obtain good approximations for the cardinality of large combinatorial sets, such as the set of all graphs with given degrees [92], when the maximum degree is not too large.

To make the phrase "not too many defects" precise, recall that a *loop* in a configuration is a pair between two points from the same cell. A *triple pair* in a configuration is a set of three distinct non-loop pairs between the same two cells, and a *double pair* is a set of two distinct non-loop pairs between the same two cells.

Definition 5.1 Let $M_2 = M_2(\mathbf{k}) = \sum_{j \in [n]} k_j(k_j - 1)$. (Note that M_2 counts the number of ways to choose an ordered pair of points from the same cell.) Define $B_1 = M_2/M$ and $B_2 = (M_2/M)^2$. Say that a configuration $P \in \mathcal{P}(\mathbf{k})$ is *good* if every cell contains at most one loop, there are no triple pairs, P contains at most B_1 loops and at most B_2 double pairs. Write $\mathcal{P}^*(\mathbf{k})$ for the set of all good configurations in $\mathcal{P}^*$.

Combining McKay and Wormald [91, Lemma 2 and Lemma 3′] with [5, Lemma 8], we can prove that if $k_{\max}^4 = O(M)$ then there exists a constant $c \in (0, 1)$ such that a uniformly-random element of $\mathcal{P}(\mathbf{k})$ is good with probability at least c.

Next, let $\mathcal{C}_{\ell,d}$ be the set of all good configurations with exactly ℓ loops and d double pairs. These sets form a partition of $\mathcal{P}^*(\mathbf{k})$. McKay and Wormald defined two switching operations, which we will refer to as *loop-switchings* and *double-switchings*. A loop-switching is used to reduce the number of loops by one, and a double-switching is used to reduce the number of double pairs by one. These switchings are illustrated in Figure 5. For example, in the loop-switching, a loop is selected together with two other pairs, such that there are 5 distinct cells involved, and performing the switching does not result in any repeated pairs. The loop-switching transforms an element of $\mathcal{C}_{\ell,d}$ to an element of $\mathcal{C}_{\ell-1,d}$. To describe a loop-switching we specify an ordered 6-tuple of points, and similarly a double-switching is specified using an ordered 8-tuple of points.

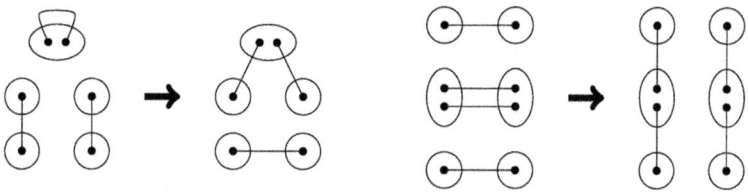

Figure 5: A loop-switching (left) and a double-switching (right)

It is possible to remove loops and double pairs using simpler switchings. In fact, McKay used simpler switching operations (as illustrated in Figure 10 below) in a very early application [90] of the switching method for asymptotic enumeration. Subsequently, McKay and Wormald found that by using the slightly more complicated switchings shown in Figure 5, they could obtain an asymptotic formula with vanishing error for a wider range of degree sequences (with a weaker bound on the maximum degree, to be precise), compared with the result of [90]. The benefits obtained by using the slightly more complicated switchings also hold here in the algorithmic setting.

The first step of McKay and Wormald's algorithm is to repeatedly choose a uniformly random element of $\mathcal{P}(\boldsymbol{k})$ until it is good. At this point, the configuration P is a uniformly random element of $\mathcal{P}^*(\boldsymbol{k})$. Next, if P contains a loop then a loop-switching is chosen uniformly at random from the set of all available options (that is, from all possible loop-switchings which could be applied to P). This switching is rejected with some probability, otherwise it is accepted and performed. The rejection probability is carefully chosen to ensure that if P has a uniform distribution over $\mathcal{C}_{\ell,d}$ then the resulting configuration has a uniform distribution over $\mathcal{C}_{\ell-1,d}$. If rejection occurs at any step then the entire algorithm restarts from the beginning.

When a configuration is reached with no loops, any double pairs are removed one by one using double-switchings, again with a rejection probability chosen to maintain uniformity. Finally, when the current configuration P belongs to $\mathcal{C}_{0,0}$ it is simple, and the algorithm outputs $G(P)$ and terminates. The algorithm is given in pseudocode in Figure 6.

McKAY–WORMALD ALGORITHM
Input: graphical sequence $\boldsymbol{k}$
Output: element of $G(\boldsymbol{k})$

repeat
    choose $P \in \mathcal{P}(\boldsymbol{k})$ u.a.r.
until P is good
\# *remove loops*
while P has at least one loop
    obtain P' from P by performing a loop-switching chosen u.a.r.
    calculate the rejection probability $q_{\text{loop}}(P, P')$
    restart with probability $q_{\text{loop}}(P, P')$; otherwise $P := P'$
\# *remove double pairs*
while P has at least one double pair
    obtain P' from P by performing a double-switching chosen u.a.r.
    calculate the rejection probability $q_{\text{double}}(P, P')$
    restart with probability $q_{\text{double}}(P, P')$; otherwise $P := P'$
output $G(P)$

Figure 6: High-level description of the McKay–Wormald algorithm

To complete the specification of the McKay–Wormald algorithm, we must define the rejection probabilities q_{loop} and q_{double}. For $P \in \mathcal{P}^*(\boldsymbol{k})$, let $f(P, X)$ denote the number of possible X-switchings $P \mapsto P'$ which may be applied to P, for $X \in \{\text{loop}, \text{double}\}$. Similarly, let $b(P', X)$ be the number of ways to produce P' using an X-switching $P \mapsto P'$, for all $P' \in \mathcal{P}^*(\boldsymbol{k})$ and $X \in \{\text{loop}, \text{double}\}$. McKay and Wormald [91, Lemma 4] gave expressions $\underline{m}(\ell, d, \text{loop})$ and $\underline{m}(d, \text{double})$, omitted here, such that for all $P \in \mathcal{C}_{\ell,d}$,

$$f(P, \text{loop}) \leq \overline{m}(\ell, \text{loop}) = 2\ell M^2, \qquad b(P, \text{loop}) \geq \underline{m}(\ell, d, \text{loop}) \qquad (5.1)$$

and for all $P \in \mathcal{C}_{0,d}$,

$$f(P, \text{double}) \leq \overline{m}(d, \text{double}) = 4d M^2, \qquad b(P, \text{double}) \geq \underline{m}(d, \text{double}). \qquad (5.2)$$

Regarding the lower bounds $\underline{m}(\cdot)$, we will only need the fact that they are positive when $k_{\max}^4 = O(M)$, for all $\ell \leq B_1$ and all $d \leq B_2$.

The upper bounds in (5.1) and (5.2) arise by counting the number of ways to choose a tuple of points (6 points for a loop-switching and 8 points for a double-switching) which satisfy some constraints of the switching and not others: typically the required pairs must be present, but we do not check that all cells involved in the switching are distinct, or that the switching does not introduce any new loops or repeated pairs. For the lower bounds, we require an upper bound on the number of bad choices of tuples, so that this may be subtracted. When the degrees get too high, the number of bad choices increases and there will be a lot of variation in this number, making the estimates less precise.

The rejection probabilities are defined by

$$\left. \begin{array}{l} q_{\text{loop}}(P, P') = 1 - \dfrac{f(P, \text{loop})\, \underline{m}(\ell - 1, d, \text{loop})}{\overline{m}(\ell, \text{loop})\, b(P', \text{loop})}, \\[2mm] q_{\text{double}}(P, P') = 1 - \dfrac{f(P, \text{double})\, \underline{m}(d - 1, \text{double})}{\overline{m}(d, \text{double})\, b(P', \text{double})} \end{array} \right\} \qquad (5.3)$$

for all $(P, P') \in \mathcal{C}_{\ell,d} \times \mathcal{C}_{\ell-1,d}$ which differ by a loop-switching, and all $(P, P') \in \mathcal{C}_{0,d} \times \mathcal{C}_{0,d-1}$ which differ by a double-switching, respectively. These probabilities are well-defined if the lower bounds $\underline{m}(\cdot)$ are positive.

Lemma 5.2 [91, Theorem 2] *If $k_{\max}^4 = O(M)$ then the output of the McKay–Wormald algorithm has uniform distribution over $G(\mathbf{k})$.*

Proof As mentioned earlier, the condition $k_{\max}^4 = O(M)$ implies that the lower bounds $\underline{m}(\cdot)$ are positive, and hence the rejection probabilities are well-defined. The initial good configuration P is distributed uniformly over $\mathcal{P}^*(\mathbf{k})$. Hence, if the initial configuration belongs to $\mathcal{C}_{\ell,d}$ then it has the uniform distribution over $\mathcal{C}_{\ell,d}$. We prove by induction that if a switching $P \mapsto P'$ is accepted and P has the uniform distribution over some set $\mathcal{C}_{\ell,d}$, then P' has the uniform distribution over the codomain of that switching. (The codomain is $\mathcal{C}_{\ell-1,d}$ if the switching is a loop-switching, while for a double-switching $\ell = 0$ and the codomain is $\mathcal{C}_{0,d-1}$.) For ease of notation we prove this for double-switchings, and note that the same argument holds for loops-switchings. For all $P' \in \mathcal{C}_{0,d-1}$, the probability that the proposed switching is not rejected and results in P' is given by

$$\Pr(P') = \sum_{\substack{P \in \mathcal{C}_{0,d} \\ P \mapsto P'}} \frac{\Pr(P)}{f(P, \text{double})} \left(1 - q_{\text{double}}(P, P')\right).$$

The sum is over all configurations $P \in \mathcal{C}_{0,d}$ such that P' can be obtained from P using a double-switching, and the factor $1/f(P, \text{double})$ is the probability that this particular double-switching is chosen to be applied to P. Substituting the value of the rejection probability from (5.3), and using the assumption that P is uniformly distributed over $\mathcal{C}_{0,d}$, we find that

$$\Pr(P') = \frac{\underline{m}(d-1, \text{double})}{|\mathcal{C}_{0,d}|\,\overline{m}(d, \text{double})} \sum_{\substack{P \in \mathcal{C}_{0,d} \\ P \mapsto P'}} \frac{1}{b(P', \text{double})}.$$

But the number of summands is precisely $b(P', \text{double})$, so the sum evaluates to 1 and we conclude that

$$\Pr(P') = \frac{\underline{m}(d-1, \text{double})}{|\mathcal{C}_{0,d}|\,\overline{m}(d, \text{double})}.$$

This depends only on d, and not on the particular configuration $P' \in \mathcal{C}_{0,d-1}$. Hence every element of $\mathcal{C}_{0,d-1}$ is equally likely to be produced after the double-switching, proving that the uniform distribution is maintained after each accepted switching step. Thus, by induction, at the end of the algorithm P is a uniformly random element of $\mathcal{C}_{0,0}$. It follows that $G(P)$ is a uniformly random element of $G(\boldsymbol{k})$, as claimed. $\square$

The previous result shows that the output of the McKay–Wormald algorithm is always correct. But what conditions on $\boldsymbol{k}$ are needed for the algorithm to be efficient? If the degrees become too large then it becomes unlikely that the randomly-chosen initial configuration is good, and there will be too much variation in the parameters $f(P, X)$, $b(P', X)$, leading to large rejection probabilities.

Theorem 5.3 [91, Theorem 3] *Suppose that $\boldsymbol{k}$ is a graphical degree sequence with $k_{\max} = O(M^{1/4})$. The McKay–Wormald algorithm for sampling from $G(\boldsymbol{k})$ can be implemented so that it has expected runtime $O(k_{\max}^2 M^2) = O(k_{\max}^4 n^2)$. If $\boldsymbol{k} = (k, k \ldots, k)$ is regular then there is an implementation with expected runtime $O(k^3 n)$, under the assumption that $k = O(n^{1/3})$.*

Proof (Sketch.) Recall that a randomly chosen element of $\mathcal{P}(\boldsymbol{k})$ is good with probability at c when $k_{\max}^4 = O(M)$, for some constant $c \in (0, 1)$. Hence it takes expected time $O(M)$ to produce a uniformly-random element of $\mathcal{P}^*(\boldsymbol{k})$. McKay and Wormald prove that the probability that there is no restart during the loop-switchings and doubles-switchings is $1 - o(1)$ when $k_{\max}^4 = o(M)$, and is bounded below by a constant when $k_{\max}^4 = \Theta(M)$.

It remains to consider the cost of performing the switching operations. Suppose that at some point in the execution of the algorithm, the current configuration is P. To choose a potential switching of the appropriate type, we can select the points of a randomly chosen loop or double pair, in a random order, and then choose the points of two other pairs, in a random order. The number of ways to make this selection is *exactly* given by the relevant upper bound from (5.1) or (5.2), and the probability that the result P' of this switching is a valid configuration in the codomain (that is, only the chosen loop/double pair has been removed, and no additional defects have been introduced) is exactly $f(P, X)/\overline{m}(x, X)$, where $(x, X) \in$

$\{(\ell, \text{loop}), (d, \text{double})\}$. This means that the value of $f(P, X)$ does not need to be calculated.

However, we do need to calculate the value $b(P', X)$ precisely for the proposed switching $P \mapsto P'$, in order to restart with the correct probability. This can be done by maintaining some information about numbers of small structures in the configuration, which is initialised before any switchings have been performed, and updated after each switching operation. The initialisation takes runtime $O(k_{\max}^2 M_2^2)$, which dominates the expected time required for the updates from each switching. See [91, Theorem 3] for more details.

In the k-regular case a further efficiency is possible, but with a much more complicated implementation, as explained in the proof of [91, Theorem 4]. $\square$

McKay and Wormald also explained how to modify their algorithm to sample bipartite graphs with given degrees, uniformly at random, see [91, Section 6]. The expected runtime of the uniform sampler for bipartite graphs with given degrees is $O(k_{\max}^4 n^2)$ when $k_{\max} = O(M^{1/4})$, where (as usual) $k_{\max}$ denotes the maximum degree.

5.1 Improvements and extensions

Starting from the McKay–Wormald algorithm, Gao and Wormald [59] introduced several new ideas which culminated in an algorithm for uniformly sampling k-regular graphs, which they called REG. The expected runtime of the Gao–Wormald algorithm is $O(k^3 n)$ when $k = o(\sqrt{n})$. This is a significant increase in the allowable range of k compared with the McKay–Wormald algorithm.

In order to handle degrees beyond $O(n^{1/3})$, Gao and Wormald must deal with triple pairs, as well as loops and double pairs. So the set of good configurations is redefined to allow "not too many" triple pairs (but still ruling out any pairs of multiplicity four or higher) and a new switching phase is performed to remove triple pairs one by one. However, it turns out that triple pairs are easily handled. In fact, the first two phases of the algorithms (removing loops and removing triples, respectively) proceed as in the McKay–Wormald algorithm. As the double pairs are the most numerous "defect", new ideas are required in phase 3, where double pairs must be removed.

The innovations introduced by Gao and Wormald in [59] are designed to reduce the probability of a rejection during a double-switching step. These ideas are described in a very general setting in [59], for ease of applications to other problems. Here, we given an overview of these ideas in the context of the double-switching, performed on configurations which have no loops, no triple pairs and at most B_2 double pairs, where $B_2 = \lfloor (1+\gamma)(k-1)^2/4 \rfloor$ for some sufficiently small constant $\gamma > 0$.

Since we now discuss only double-switchings, we drop "double" from our notation. Write $\mathcal{C}_d$ for the set of good configurations with no loops, no triple pairs and exactly d double pairs. Observe from (5.3) that the probability that a proposed switching $P \mapsto P'$ is *not* rejected is a product of a factor $f(P)/\overline{m}(d)$ which depends only on $P \in \mathcal{C}_d$, and a factor $\underline{m}(d-1)/b(P')$ which depends only on $P' \in \mathcal{C}_{d-1}$. Gao and Wormald aim to reduce both the probability of *forward rejection*, or *f-rejection*

Generating graphs randomly 153

(which depends only on P), and the probability of *backwards rejection*, or *b-rejection*, which depends only on P'.

To reduce the likelihood of an f-rejection, Gao and Wormald allow some double-switchings which would be rejected in the McKay–Wormald algorithm. The aim is to bring the values of $f(P)$ closer to the upper bound $\overline{m}(d)$. Specifically, a double-switching which introduces exactly one new double pair is allowed, as illustrated in Figure 7. The original double-switching, shown on the right of Figure 5, is known as a type I, class A switching, while switching in Figure 7 is a type I, class B switching.

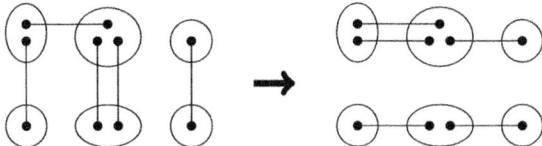

Figure 7: A type I, class B double-switching $\mathcal{C}_d \to \mathcal{C}_d$

Next, Gao and Wormald observed that some configurations in $\mathcal{C}_{d-1}$ are less likely to be produced by a (type I) double-switching than others. These configurations bring down the lower bound $\underline{m}(d-1)$ and hence increase the b-rejection probability for every element of $\mathcal{C}_{d-1}$. For this reason, Gao and Wormald introduced another new switching, called a *type II* switching, which actually increases the number of double pairs by one, as shown in Figure 8. All type II switchings have class B.

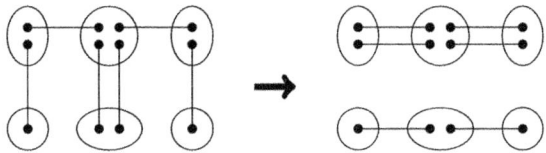

Figure 8: A type II, class B double-switching $\mathcal{C}_d \to \mathcal{C}_{d+1}$

To perform a switching step, from current configuration $P \in \mathcal{C}_d$, first the type $\tau \in \{I, II\}$ of switching is chosen, according to a probability distribution ρ (with a small restart probability if no type is chosen). Next, a type τ switching $P \mapsto P'$ is proposed, chosen randomly from all $f_\tau(P)$ type τ switchings available in P. The f-rejection probability is $1 - f_\tau(P)/\overline{m}_\tau(d)$, where $\overline{m}_\tau(d)$ is an upper bound on $f_\tau(P)$ over all $P \in \mathcal{C}_d$. Let $d' \in \{0, \ldots, B_2\}$ be the unique index such that $P' \in \mathcal{C}_{d'}$. The class $\alpha \in \{A, B\}$ of the proposed switching $P \mapsto P'$ can now be observed, and the b-rejection probability is $1 - b_\alpha(P')/\underline{m}_\alpha(d')$, where $\underline{m}_\alpha(d')$ is a lower bound on $b_\alpha(P')$ over all $P' \in \mathcal{C}_{d'}$. If there is no f-rejection or b-rejection then the proposed switching is accepted and P' becomes the current configuration. As soon as an element $P \in \mathcal{C}_0$ is reached, the algorithm stops with output $G(P)$. Here we see that the f-rejection probability depends on P and the chosen type, while the b-rejection probability depends on the outcome P' and the class α of the proposed switching from P.

Rather than maintaining a uniform distribution after each switching, as in the

McKay–Wormald algorithm, the goal in the Gao–Wormald algorithm is to ensure that the expected number of visits to each configuration $P \in \mathcal{C}_d$, over the course of (the doubles-reducing phase of) the algorithm, depends only on d and is independent of P. In particular, this guarantees that each element of $\mathcal{C}_0$ is equally likely, and hence the output of the algorithm is a uniformly random element of $G(n,k)$.

In [59, Lemma 6 and Lemma 8], Gao and Wormald gave expressions for $\overline{m}_\tau(d)$, $\underline{m}_\alpha(d)$ and showed how to choose values for $\rho_\tau(d)$ satisfying a certain system of linear equations. By [59, Lemma 5], when the parameters $\rho_\tau(d)$ satisfy this system of equations then the last element visited by the algorithm is distributed uniformly at random from $\mathcal{C}_0$, assuming that no rejection occurs. Furthermore, the solution can be chosen to satisfy $\rho_I(d) = 1 - \varepsilon > 0$ for all $1 \leq d \leq B_2$, where $\varepsilon = O(k^2/n^2)$. Since $k = o(n^{1/2})$ this means that almost every step is a "standard" double-switching (type I, class A) switching.

Having set these parameter values, the algorithm is completely specified. It remains to show that the probability of rejection during the course of the algorithm is $o(1)$, which requires careful analysis. The runtime analysis is very similar to Theorem 5.3, resulting in the following.

Theorem 5.4 *If $1 \leq k = o(\sqrt{n})$ then the Gao–Wormald algorithm REG is a uniform sampler from $G(n,k)$, and can be implemented with expected runtime $O(k^3 n)$.*

Recent work of Armand, Gao and Wormald [5] which gives an even more efficient uniform sampler for the same range of k is discussed in Section 5.1.3.

A *k-factor* is a k-regular spanning subgraph of a given graph. Gao and Greenhill [55] used the Gao–Wormald framework to give algorithms for sampling k-factors of a given graph H_n with n vertices, under various conditions on k and the maximum degree Δ of the complement $\overline{H}_n$ of H_n. The edges of the complement of H_n can be thought of as "forbidden edges", and we want to sample k-regular graphs with no forbidden edges.

Theorem 5.5 *[55, Theorem 1.1 and 1.2] Let H_n be a graph on n vertices such that $\overline{H}_n$ has maximum degree Δ.*

- *There is an algorithm which produces a uniformly random k-factor of H_n, and has expected runtime $O((k+\Delta)^3 n)$ if $(k+\Delta)k\Delta = o(n)$.*

- *Now suppose that H_n is $(n - \Delta - 1)$-regular. There is an algorithm which generates a uniformly random k-factor of H_n and has expected runtime*

$$O\big((k+\Delta)^4 (n+\Delta)^3 + (k+\Delta)^8 k^2 \Delta^2 / n + (k+\Delta)^{10} k^2 \Delta^3 / n^2\big)$$

if $k^2 + \Delta^2 = o(n)$.

In [55], the algorithms described in Theorem 5.5 are called FACTOREASY and FACTORUNIFORM, respectively. Previously the only algorithm for this problem was a rejection algorithm of Gao [54] which has expected linear runtime when $k = O(1)$ and $\overline{H}_n$ has at most a linear number of edges (but the maximum degree of $\overline{H}_n$ can be linear).

5.1.1 Asymptotically-uniform algorithms based on switchings

In [59, Theorem 3], Gao and Wormald described an algorithm REG* which performs asymptotically-uniform sampling from $G(n,k)$ in expected runtime $O(kn)$. This algorithm is obtained from REG by never performing any rejection steps and never performing any class B switchings. (So only loop-switchings, triple-switchings and type I, class A double-switchings will be used.) This is more efficient as computation of the b-rejection probabilities is the most costly part of the algorithm. The output of REG* is within total variation distance $o(1)$ of uniform when $k = o(\sqrt{n})$. Indeed, Gao and Wormald remark that the McKay–Wormald algorithm can be modified in the same way, giving an asymptotically-uniform sampling algorithm with expected runtime $O(M)$ whenever $k_{\max} = O(M^{1/4})$.

A similar performance was obtained by Zhao [122] using a slightly different approach, involving the use of a Markov chain to make local modifications starting from $G(P)$, where P is a uniformly random element of $\mathcal{P}(\boldsymbol{k})$.

Recently, Janson [72] introduced and analysed the following switching-based algorithm for asymptotically-uniform sampling from $G(\boldsymbol{k})$. Say that a pair in a configuration is *bad* if it is a loop or part of a double pair. (Recall that, as we have defined it here, "double pair" does not mean that the multiplicity of the corresponding edge is exactly two: only that the multiplicity is at least two.)

JANSON ALGORITHM
Input: graphical sequence $\boldsymbol{k}$
Output: element of $G(\boldsymbol{k})$, denoted $\widehat{G}$

choose $P \in \mathcal{P}(\boldsymbol{k})$ u.a.r.
repeat
  choose and orient a bad pair pp' in P u.a.r.
  choose and orient a distinct pair qq', u.a.r.
  delete pairs pp', qq' from P, and replace with pairs $pq, p'q'$
until $G(P)$ is simple
output $\widehat{G} = G(P)$

Figure 9: Janson algorithm, corresponding to the switched configuration model

Starting from a uniformly random configuration $P \in \mathcal{P}(\boldsymbol{k})$, if $G(P)$ is simple then we output $G(P)$. Otherwise, choose a pair uniformly at random from the set of all bad pairs in P. Next, choose a pair uniformly at random from the set of all other pairs in P. Update P by removing these two pairs and replacing them by two other pairs using the same four points, chosen uniformly at random. (In the pseudocode, this is done by randomly ordering the points in each chosen pair.) This gives a new configuration in $\mathcal{P}(\boldsymbol{k})$. At each step, the switching removes the chosen bad pair, and may cause other pairs to stop being bad or to become bad. Repeatedly apply the switching step until $G(P)$ is simple, and let $\widehat{G} = G(P)$ denote the output graph. This algorithm is shown in Figure 9. Janson calls the resulting probability space the *switched configuration model*. This is a non-uniform probability space over $G(\boldsymbol{k})$.

The switchings used in this process are illustrated in Figure 10. They were used by McKay [90] in a very early application of the switching method, and are simpler than the switchings in Figure 5. Note however that Janson does not insist that the cells involved in the switching are all distinct, or that the new pairs pq, $p'q'$ do not increase the multiplicity of an edge. So Figure 10 should be interpreted differently to Figure 5: in Figure 5, it is implied that all illustrated cells are distinct and that all new pairs lead to edges with multiplicity 1.

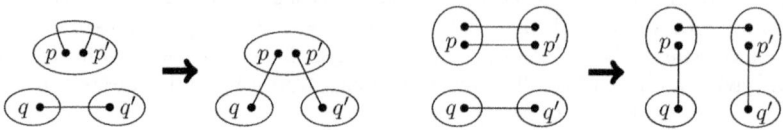

Figure 10: Possible switchings in Janson's algorithm, with chosen points labelled

Janson proved the following result [72, Theorem 2.1]. Recall the definition of $R = R(\mathbf{k})$ from Theorem 3.1.

Theorem 5.6 [72] *Suppose that $\mathbf{k}$ is a graphical degree sequence which satisfies*

$$k_{\max} = o(n^{1/2}), \qquad M = \Theta(n), \qquad R = O(n) \qquad (5.4)$$

and let $\widehat{G}$ denote the output of the switched configuration model for the degree sequence $\mathbf{k}$. Then the distribution of $\widehat{G}$ is within total variation distance $o(1)$ of uniform. With high probability, the runtime is $O(M)$ as only $O(1)$ switching steps are required.

Janson remarks that the bad pair may also be chosen deterministically according to some rule, such as lexicographically. This would lead to a slightly different distribution on the output graph, but the conclusion of Theorem 5.6 would still hold.

Furthermore, Janson [72, Corollary 2.2 and Corollary 2.3] proved that under the same conditions (5.4), statements about convergence in probability and convergence in distribution which are true for $\widehat{G}$ are also true for uniformly-random elements of $G(\mathbf{k})$.

5.1.2 Graphs with power-law degree distributions

Heavy-tailed distributions are often observed in real-world networks [24, 96], but are difficult to sample as they are far from regular and their maximum degree is too high for the sampling algorithms we have seen so far. In [60], Gao and Wormald showed how to adapt their approach to degree sequences which satisfy the following definition.

Definition 5.7 [60, Definition 1] *The degree sequence $\mathbf{k}$ is power-law distribution-bounded with parameter $\gamma > 1$ if the minimum entry in $\mathbf{k}$ is at least 1, and there is a constant $C > 0$ independent of n such that the number of entries of K which are at least i is at most $Cni^{1-\gamma}$ for all $i \geq 1$.*

Other definitions of power-law degree sequences can be found in the literature, but some only allow maximum degree $O(n^{1/\gamma})$, which is $o(n^{1/2})$ when $\gamma \in (2,3)$.

Definition 5.7 is more realistic as it allows higher degrees, as observed in real-world networks. Gao and Wormald [60, equation (4)] noted that if $\boldsymbol{k}$ is power-law distribution-bounded with parameter γ then

$$k_{\max} = O(n^{1/(\gamma-1)}), \quad M = \Theta(n), \quad M_2 = O(n^{2/(\gamma-1)}), \quad (5.5)$$

where $M_2 = M_2(\boldsymbol{k})$ is given in Definition 5.1.

The most relevant range of γ for real-world networks is $\gamma \in (2,3)$. As observed by Gao and Wormald [60], when $\gamma > 3$ it is easy to sample uniformly from $G(\boldsymbol{k})$. We provide a brief proof here.

Lemma 5.8 [60] *Suppose that $\boldsymbol{k}$ is a power-law distribution-bounded degree sequence with $\gamma > 3$. Then the configuration model (Figure 2) gives a polynomial-time uniform sampler for $G(\boldsymbol{k})$ with expected runtime $O(M)$.*

Proof It follows from (5.5) that $k_{\max}^2 = o(M)$ and $R = \sum_{j \in [n]} k_j^2 = M_2 + M = \Theta(M)$. The proof is completed by applying Theorem 3.1. □

While uniform sampling is easy when $\gamma > 3$, it is a very challenging problem when $\gamma \in (2,3)$. To cope with the very high maximum degree when $\gamma < 3$, Gao and Wormald utilised 6 different types of switching (all of the same class). First, they focussed on removing "heavy" edges or loops, where an edge is *heavy* if both its endvertices have high degree. They also introduced a new kind of rejection, called *pre-b-rejection*, which is used to equalise the number of ways to choose some additional pairs which are needed to perform some of the switchings. They described a uniform sampler PLD, and an asymptotically-uniform sampling algorithm called PLD*, obtaining the following result [60, Theorem 2, Theorem 3] when the parameter γ is a little less than 3. These are the first rigorously-analysed algorithms which can efficiently sample graphs with a realistic power-law degree distribution for some values of γ below 3.

Theorem 5.9 [60] *Suppose that $\boldsymbol{k}$ is a power-law distribution-bounded degree sequence with parameter γ such that*

$$\gamma > \tfrac{21+\sqrt{61}}{10} \approx 2.881.$$

The algorithm PLD is a uniform sampler for $G(\boldsymbol{k})$ with expected runtime $O(n^{4.081})$. The algorithm PLD* performs asymptotically-uniform sampling from $G(\boldsymbol{k})$ with expected runtime $O(n)$.

In their analysis, Gao and Wormald used a new parameter, $J(\boldsymbol{k})$, which they introduced in the context of asymptotic enumeration in [58]. We have seen that a switching argument breaks down when the number of bad choices for a given switching operation becomes too large. This often involves counting paths of length two from a given vertex. In previous work, the bound $k_{\max}^2$ was often used for this quantity. (Of course $k_{\max}(k_{\max}-1)$ is more precise but gives the same asymptotics.)

Instead, Gao and Wormald use the upper bound $J(\boldsymbol{k})$, defined as follows. First, let σ be a permutation of $[n]$ such that $k_{\sigma(1)} \geq k_{\sigma(2)} \geq \cdots \geq k_{\sigma(n)}$. Then, define

$$J(\boldsymbol{k}) = \sum_{j=1}^{k_{\max}} k_{\sigma(j)}, \qquad (5.6)$$

noting that $k_{\max} = k_{\sigma(1)}$. So $J(\boldsymbol{k})$ is the sum of the $k_{\max}$ largest entries of $\boldsymbol{k}$, and hence forms an upper bound on the number of 2-paths from an arbitrary vertex.

If $\boldsymbol{k}$ is regular then $J(\boldsymbol{k}) = k_{\max}^2$, but when $\boldsymbol{k}$ is far from regular, $J(\boldsymbol{k})$ can be significantly smaller than $k_{\max}^2$. In particular, if $\boldsymbol{k}$ is a power-law distribution-bounded degree sequence with parameter γ then $k_{\max}^2 = n^{2/(\gamma-1)}$, while

$$J(\boldsymbol{k}) = O\bigl(n^{(2\gamma-3)/(\gamma-1)^2}\bigr) = o(n^{2/(\gamma-1)}).$$

This bound on $J(\boldsymbol{k})$ is proved in [57, Lemma 5], and more briefly in [60, equation (54)].

The parameter $J(\boldsymbol{k})$ has proved very powerful when working with heavy-tailed degree distributions. As well as its use in asymptotic enumeration [58] and uniform sampling [60], it has also been used in the analysis [57] of the number of triangles and the clustering coefficient in a uniformly random element of $G(\boldsymbol{k})$, for heavy-tailed degree sequences $\boldsymbol{k}$. We will encounter $J(\boldsymbol{k})$ again in Section 6.5.

5.1.3 Incremental relaxation
Very recently, Arman, Gao and Wormald [5] introduced a new approach, called *incremental relaxation*, which allows a more efficient implementation of the b-rejection step. In incremental relaxation, the b-rejection is performed iteratively over several steps, each with its own sub-rejection probability, such that the sub-rejection probabilities are much easier to calculate than the overall probability of b-rejection. Using this idea, Arman et al. obtain improvements on the runtime of the algorithms in [59, 60, 91], and give an algorithm for uniformly sampling bipartite graphs with given degrees when the maximum degree is $O(M^{1/4})$. We collect their results together below.

Theorem 5.10 [5, Theorems 1–4] *Let $\boldsymbol{k}$ be a graphical degree sequence. There are algorithms, called* INC-GEN, INC-REG, *and* INC-POWERLAW, *respectively, which perform uniform sampling from $G(\boldsymbol{k})$ under the following assumptions on $\boldsymbol{k}$, with the stated expected runtime:*

- *If $k_{\max}^4 = O(M)$ then the expected runtime of the algorithm* INC-GEN *is $O(M)$.*

- *If $k_{\max} = (k, k, \ldots, k)$ is regular and $k = o(n^{1/2})$ then the expected runtime of the algorithm* INC-REG *is $O(kn + k^4)$.*

- *If $\boldsymbol{k}$ is a power-law distribution-bounded degree sequence with parameter $\gamma > \frac{21+\sqrt{61}}{10} \approx 2.881$ then the algorithm* INC-POWERLAW *has expected runtime $O(n)$.*

Now let $\boldsymbol{k} = (\boldsymbol{s}, \boldsymbol{t})$ be a bipartite degree sequence with $k_{\max} = \max\{\max s_j, \max t_i\}$. If $k_{\max}$ satisfies $k_{\max}^4 = O(M)$, then there is an algorithm, called INC-BIPARTITE, which has expected runtime $O(M)$ and produces a uniformly-random bipartite graph with degree sequence $\boldsymbol{s}$ on one side of the bipartition and $\boldsymbol{t}$ on the other.

Observe that incremental relaxation leads to greatly improved runtimes, for example from $O(k_{\max}^2 M^2)$ to $O(M)$ for uniform sampling from $G(\boldsymbol{k})$ when $k_{\max} = O(M^{1/4})$, and from $O(n^{4.081})$ to $O(n)$ for power-law distribution-bounded degree sequences with $\gamma > 2.882$.

At the time of writing, C code for INC-GEN and INC-REG is available from Wormald's website[1].

6 Markov chain algorithms

In this section we review sampling algorithms which use the *Markov chain Monte Carlo* (MCMC) approach. Here an ergodic Markov chain is defined with the desired stationary distribution: in our setting, the stationary distribution should be *uniform* over $G(\boldsymbol{k})$, or perhaps over a superset of $G(\boldsymbol{k})$. We refer to such algorithms as "Markov chain algorithms".

Rather than work steadily towards a particular goal, such as the sequential algorithms described in Section 4 or the McKay–Wormald algorithm discussed in Section 5, the Markov chains we consider in this section perform a random walk on $G(\boldsymbol{k})$, usually by making small random perturbations at each step. For example, the *switch chain* chooses two random edges, deletes them and replaces them with two other edges, while maintaining the degree sequence. See Figure 11. We will return to the switch chain in Section 6.4.

Figure 11: Transitions of the switch chain

A Markov chain needs a starting state: that is, we must be able to initially construct a single instance of $G(\boldsymbol{k})$. For graphs, bipartite graphs and directed graphs, if the degree sequence is graphical then a realization of that degree sequence can easily be constructed. This is done using the Havel–Hakimi algorithm [67, 68] for graphs, Ryser's algorithm [102] for bipartite graphs and an adaptation of the Havel–Hakimi algorithm for directed graphs [43]. We remark that the situation for hypergraphs is more complicated, as the existence problem ("Does a given degree sequence have a realization?") is NP-complete for 3-uniform hypergraphs [30].

In theoretical computer science, any polynomial runtime is seen as efficient. In practice, of course, an algorithm with a high-degree polynomial runtime may be too slow to use. All algorithms discussed in previous sections run until some natural stopping time is reached: that is, by looking at the current state we can tell whether or not the algorithm may successfully halt. In MCMC sampling, however, the user must specify the number of transitions T that the Markov chain will perform before producing any output. Typically T is defined to be the best-known upper bound on

---
[1] https://users.monash.edu.au/~nwormald

the mixing time $\tau(\varepsilon)$, for a suitable tolerance ε (see Definition 6.1). For this reason, loose upper bounds on the mixing time have a significant impact on the runtime.

Most Markov chain approaches to sampling from $G(\boldsymbol{k})$ have been analysed using the *multicommodity flow* method of Sinclair [104], which we describe in Section 6.3. Unfortunately, it is often very difficult to obtain tight bounds on the rate of mixing time of a Markov chain using the multicommodity flow method. In any case, it is an interesting theoretical challenge to try to characterise families of degree sequences $\boldsymbol{k}$ for which natural Markov chains on $G(\boldsymbol{k})$ have polynomial mixing time.

We now introduce some necessary Markov chain background. For more information see for example [74, 84].

6.1 Markov chain background

A time-homogenous *Markov chain* $\mathcal{M}$ on a finite state space Ω is a stochastic process $X_0, X_1, \ldots$ such that $X_t \in \Omega$ for all $t \in \mathbf{N}$, and

$$\Pr(X_{t+1} = y \mid X_0 = x_0, \ldots, X_t = x_t) = \Pr(X_{t+1} = y \mid X_t = x_t)$$

for all $t \in \mathbf{N}$ and $x_0, \ldots, x_t, y \in \Omega$. These probabilities are stored in an $|\Omega| \times |\Omega|$ matrix P, called the *transition matrix* P of $\mathcal{M}$. (This notation clashes with our earlier use of P for configurations, but we will not mention configurations in this section so this should not cause confusion.) So the (x,y) entry of P, denoted $P(x,y)$, is defined by

$$P(x, y) = \Pr(X_{t+1} = y \mid X_t = x)$$

for all $x, y \in \Omega$ and all $t \geq 0$.

A Markov chain is *irreducible* if there is a sequence of transitions which transforms x to y, for any $x, y \in \Omega$, and it is *aperiodic* if $\gcd\{t \mid P^t(x,x) > 0\} = 1$ for all $x \in \Omega$. If a Markov chain is irreducible and aperiodic then we say it is *ergodic*. The classical theory of Markov chains says that if $\mathcal{M}$ is ergodic then it has a unique stationary distribution which is the limiting distribution of the chain.

We say that the Markov chain $\mathcal{M}$ is *time-reversible* (often just called *reversible*) with respect to the distribution π on Ω if the *detailed balance* equations hold:

$$\pi(x) P(x,y) = \pi(y) P(y,x)$$

for all $x, y \in \Omega$. If a Markov chain $\mathcal{M}$ is ergodic and is time-reversible with respect to a distribution π, then π is the (unique) stationary distribution of $\mathcal{M}$. (See for example [84, Proposition 1.19].) In particular, if P is symmetric then the stationary distribution is uniform. The detailed balance equations are often used to guide the design of the transition matrix of a Markov chain, so that it has the desired stationary distribution.

Now assume that $\mathcal{M}$ is ergodic with stationary distribution π. For $x \in \Omega$ let P_x^t denote the distribution of X_t, conditioned on the event $X_0 = x$. Recall the definition of total variation distance (Definition 2.2). For any initial state $x \in \Omega$, the distance $d_{TV}(P_x^t, \pi)$ is a geometrically-decreasing function of t (see for example [84, Theorem 4.9]). This leads to the following definition.

Definition 6.1 Let $\varepsilon > 0$ be a constant. The *mixing time* of the Markov chain is the function
$$\tau(\varepsilon) = \max_{x \in \Omega} \min\{t \in \mathbf{N} \mid d_{TV}(P_x^t, \pi) < \varepsilon\}.$$

Here ε is a user-defined tolerance, which specifies how much variation from the stationary distribution is acceptable. The mixing time captures the earliest time t at which P_x^t is guaranteed to be ε-close to the stationary distribution, regardless of the starting state. Then P_x^t remains ε-close to the stationary distribution for all times $t \geq \tau(\varepsilon)$ and for every initial state $x \in \Omega$.

As usual, we are really interested in sampling from a set Ω_n, parameterised by n, where $|\Omega_n| \to \infty$, and we want to know how the runtime of the algorithm behaves as $n \to \infty$. The tolerance $\varepsilon = \varepsilon_n$ may also depend on n. We say that the Markov chain is *rapidly mixing* if the mixing time is bounded above by a polynomial in $\log |\Omega_n|$ and $\log(\varepsilon^{-1})$. Normally it is prohibitively difficult to find $\tau(\varepsilon)$ exactly, so we aim to find an upper bound T which is polynomial in $\log |\Omega_n|$ and $\log(\varepsilon^{-1})$. Then the Markov chain can be used as an FPAUS for sampling from Ω_n, in the sense of Definition 2.3, as follows: starting from a convenient initial state, run the Markov chain for T steps and output the state X_T. See for example Figure 13.

We defer introduction of the multicommodity flow method until Section 6.3.

6.2 The Jerrum–Sinclair chain

The first Markov chain algorithm for sampling from $G(\boldsymbol{k})$ was given by Jerrum and Sinclair [76] in 1990. They used Tutte's construction [111] to reduce the problem to that of sampling perfect and near-perfect matchings from an auxiliary graph $\Gamma(\boldsymbol{k})$, then applied their Markov chain from [75] to solve this problem. This resulted in a Markov chain which has uniform stationary distribution over the expanded state space $G'(\boldsymbol{k}) = \cup_{\boldsymbol{k}'} G(\boldsymbol{k}')$, where the union is taken over the set of all graphical sequences $\boldsymbol{k}' = (k'_1, \ldots, k'_n)$ such that $k'_j \leq k_j$ for all $j \in [n]$ and $\sum_{j \in [n]} |k_j - k'_j| \leq 2$.

The chain performs three types of transitions, which when mapped back to $G'(\boldsymbol{k})$ are as follows: deletion of a random edge, if the current state belongs to $G(\boldsymbol{k})$; insertion of an edge between the two distinct vertices with degree deficit one; or insertion of a random edge $\{i, j\}$ together with the deletion of a randomly-chosen neighbouring edge $\{j, \ell\}$. (We will not specify the transition probabilities precisely here.) See Figure 12, where dashed lines represent non-edges. The third type of transition is called a *hinge-flip* by Amanatidis and Kleer [3], following [25].

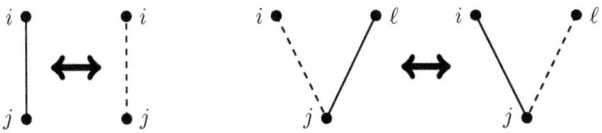

Figure 12: Transitions of the Jerrum–Sinclair chain: insertion/deletion (left) and hinge-flip (right)

We can use the Jerrum–Sinclair chain to repeatedly sample from $G'(\boldsymbol{k})$ until an element of $G(\boldsymbol{k})$ is obtained. For this to be efficient, the expected number of iterations required must be bounded above by a polynomial.

Definition 6.2 A class of degree sequences is called *P-stable* if there exists a polynomial $q(n)$ such that $|G'(\boldsymbol{k})|/|G(\boldsymbol{k})| \leq q(n)$ for every degree sequence $\boldsymbol{k} = (k_1, \ldots, k_n)$ in the class.

Jerrum and Sinclair proved the following result [76, Theorem 2.4], but did not give an explicit (polynomial) bound on the mixing time of their chain.

Theorem 6.3 [76] *There is an FPAUS for $G(\boldsymbol{k})$ for any degree sequence $\boldsymbol{k}$ which belongs to some P-stable class.*

Various classes of degree sequences are known to be P-stable, including the class of all regular sequences, and all sequences with $k_{\max}$ sufficiently small. We discuss P-stability further in Section 6.5.

6.2.1 A complete solution for the bipartite case The Jerrum–Sinclair chain for sampling perfect matchings from a given graph [75] is slow when the ratio of the number of perfect matchings to the number of near-perfect matchings is exponentially small. In 2004, Jerrum, Sinclair and Vigoda [78] described and analysed an ingenious algorithm, based on simulated annealing, which overcame this problem for bipartite graphs. Their algorithm gives an FPAUS (and hence an FPRAS) for approximately-uniformly sampling (or approximately counting) perfect matchings from a given bipartite graph. An important idea in [78] is to use a non-uniform stationary distribution over the set of all perfect and near-perfect matchings, so that the stationary probability of the set of perfect matchings is at least $1/(4n^2+1)$; that is, at most polynomially small. This is achieved by assigning weights to each "hole pattern" (for a near-perfect matching, this is the pair of vertices with deficit one, and for a perfect matching this is the empty set), as well as edge weights. Estimating good values for the weights is achieved iteratively, using simulated annealing.

As a corollary, using Tutte's construction [111], Jerrum, Sinclair and Vigoda obtained an FPAUS for sampling bipartite graphs with given degrees [78, Corollary 8.1]. In fact their result is more general: given an arbitrary bipartite subgraph H, they obtain an FPAUS for sampling subgraphs of H with a given degree sequence.

Theorem 6.4 [78, Corollary 8.1] *Given an arbitrary bipartite graph H, there is an FPAUS for the set of labelled subgraphs of H with a specified degree sequence, and there is an FPRAS for computing the number of these subgraphs.*

Bezáková, Bhatnagar and Vigoda [11] gave a more direct implementation of the algorithm from [78], which avoids Tutte's construction. This allows them to obtain faster runtime bounds compared with [78]. It follows from the proof of [11, Theorem 1] that their FPAUS is valid for any bipartite degree sequence, and has running time
$$O\big((n_1 n_2)^2 M^3 k_{\max} \log^4(n_1 n_2/\varepsilon)\big),$$
where n_1 and n_2 are the number of nodes in each part of the bipartition, and, as usual, $k_{\max}$ is the maximum degree and M is the sum of the degrees.

6.3 The multicommodity flow method

There are a few methods for bounding the mixing times of Markov chains. Before proceeding further we describe the multicommodity flow method, which has been used to analyse most MCMC algorithms for sampling graphs. For information on other methods for bounding the mixing times of Markov chains, see for example [36, 74, 84, 95].

Let $\mathcal{M}$ be a time-reversible ergodic Markov chain and let $N = |\Omega|$ be the cardinality of the state space. Then the eigenvalues of the transition matrix are real and satisfy

$$1 = \lambda_0 > \lambda_1 \geq \cdots \geq \lambda_{N-1} > -1.$$

The mixing time of the Markov chain is controlled by $\lambda_{\max} = \max\{\lambda_1, |\lambda_{N-1}|\}$. Denote the smallest stationary probability by $\pi^* = \min\{\pi(x) \mid x \in \Omega\}$. Then

$$\tau(\varepsilon) \leq (1 - \lambda_{\max})^{-1} \log\left(\frac{1}{\varepsilon \pi^*}\right), \qquad (6.1)$$

see for example [104, Proposition 1]).

If $\lambda_{\max} = |\lambda_{N-1}|$ then in particular, λ_{N-1} must be negative, in which case $1 - |\lambda_{N-1}| = 1 + \lambda_{N-1}$. For many chains we can apply a result of Diaconis and Saloff-Coste [32, p.702] (see also [64]) to establish an upper bound on $(1 + \lambda_{N-1})^{-1}$. In particular, if there is a positive probability of a null transition at any state then the following special case of that result may be useful:

$$(1 + \lambda_{N-1})^{-1} \leq \tfrac{1}{2} \max_{x \in \Omega} P(x,x)^{-1}. \qquad (6.2)$$

Another option is to work with the *lazy* version of the Markov chain $\mathcal{M}$, by replacing the transition matrix P by $(I + P)/2$. This ensures that all eigenvalues are nonnegative and hence $\lambda_{\max} = \lambda_1$. We say that a Markov chain $\mathcal{M}$ *is lazy* if $P(x,x) \geq \tfrac{1}{2}$ for all $x \in \Omega$.

Sinclair's *multicommodity flow method* [104] provides an upper bound on $(1 - \lambda_1)^{-1}$. It is a generalisation of the *canonical path* method that Jerrum and Sinclair introduced in [75].

Given a Markov chain $\mathcal{M}$ with uniform stationary distribution on a state space Ω, let $\mathcal{G}(\mathcal{M})$ be the underlying graph, where there is an edge from x to y if and only if $P(x,y) > 0$. We assume that $\mathcal{M}$ is ergodic and time-reversible with respect to the distribution π. Let $\mathcal{P}_{xy}$ be the set of all simple directed paths from x to y in $\mathcal{G}(\mathcal{M})$, and define $\mathcal{P} = \cup_{x,y} \mathcal{P}_{xy}$. A *flow* is a function $f : \mathcal{P} \to [0, \infty)$ such that for all $x, y \in \Omega$ with $x \neq y$,

$$\sum_{p \in \mathcal{P}_{xy}} f(p) = \pi(x) \pi(y).$$

(In the canonical path method, there is only one flow-carrying path from between any two pairs of states.)

If the flow can be defined so that no transition of the chain is overloaded, then the state space does not contain any "bottlenecks" and the Markov chain will be rapidly mixing. To make this precise, the total flow through a transition $e = xy$ is $f(e) = \sum_{p \ni e} f(p)$, and the *load* of e is defined by $\rho(e) = f(e)/Q(e)$, where

$Q(e) = \pi(x)P(x,y)$ is the *capacity* of the transition $e = xy$. (By time-reversibility, Q is well-defined.) Finally, the maximum load of the flow is $\rho(f) = \max_e \rho(e)$, while $\ell(f)$ denotes the length of the longest path p with $f(p) > 0$. Sinclair [104, Corollary 6'] proved that for any time-reversible Marvov chain any any flow f,

$$(1 - \lambda_1)^{-1} \leq \rho(f)\,\ell(f). \tag{6.3}$$

The next result specialises the multicommodity flow method to ergodic, time-reversible Markov chains with the uniform stationary distribution over a set Ω. It is obtained from (6.1) and (6.3), and allows two options for managing the smallest eigenvalue λ_{N-1}.

Theorem 6.5 *Let $\mathcal{M}$ be an ergodic time-reversible Markov chain with uniform stationary distribution over Ω. Define B to be 0, if it is known that $\lambda_{\max} = \lambda_1$ (for example if $\mathcal{M}$ is lazy). Otherwise, let B be an upper bound on $(1 + \lambda_{N-1})^{-1}$. Then the mixing time of the Markov chain $\mathcal{M}$ satisfies*

$$\tau(\varepsilon) \leq \max\{\rho(f)\ell(f), B\}\left(\log|\Omega| + \log(\varepsilon^{-1})\right).$$

When the multicommodity flow method is applied to the problem of sampling graphs from $G(\boldsymbol{k})$, the start and end states are graphs G, G' with degree sequence $\boldsymbol{k}$. Usually the definition of the flow is guided by the symmetric difference $H = G \triangle G'$ of G and G', and each step of a flow-bearing path is designed to make the symmetric difference smaller.

6.4 The switch chain

The *switch chain* (also called *swap chain* [93] and *Diaconis chain* [11]) is the simplest Markov chain with uniform distribution over $G(\boldsymbol{k})$. A transition of the switch chain deletes two edges and inserts two edges, while maintaining the degree sequence and without introducing any repeated edges. This is illustrated in Figure 11, at the start of Section 6. This chain was introduced by Diaconis and Gangolli [31] in 1995 in order to sample contingency tables (matrices of nonnegative integers) with given row and column sums. The transitions can be easily adapted to bipartite graphs or directed graphs. The switch chain is ergodic for graphs, and for bipartite graphs, with given degrees.

In 1999, Kannan, Tetali and Vempala [80] considered the switch chain for sampling bipartite graphs with given degrees. They used the auxiliary graph $\Gamma(\boldsymbol{k})$ obtained from Tutte's construction (modified to bipartite graphs) in order to define a multicommodity flow, and gave details only for the case of regular degrees. Unfortunately, there is a bug in their argument (specifically, in the proof of [80, Theorem 4.1]) which seems to be fatal.[2]

---

[2] The symmetric difference of two perfect matchings in $\Gamma(\boldsymbol{k})$ consists of the union of disjoint cycles. However, when mapped back to the symmetric difference of the two corresponding bipartite graphs, an alternating cycle in $\Gamma(\boldsymbol{k})$ may correspond to an alternating *walk*, which could have linearly many repeated vertices (vertices which are visited more than once on the walk). The argument of [80] does not take this into account.

> THE SWITCH CHAIN
> *Input:* graphical sequence $\bm{k}$ and positive integer $T = T(\bm{k})$
> *Output:* element of $G(\bm{k})$
>
> let G be an arbitrary initial state
> for $t = 0, \ldots, T-1$ do
>   choose two non-adjacent distinct edges $\{a,b\}, \{c,d\}$ u.a.r.
>   choose a perfect matching M of $\{a,b,c,d\}$ u.a.r.
>   if $M \cap E(G) = \emptyset$ then
>   delete the edges $\{a,b\}, \{c,d\}$ and add the edges of M
> output G_T

Figure 13: The switch chain for sampling from $G(\bm{k})$

One implementation of the switch chain for $G(\bm{k})$ is given in Figure 13. Cooper, Dyer and Greenhill [26] analysed the mixing time of the lazy version of this chain, restricted to regular degree sequences. (That is, they replaced the transition matrix P arising from the above procedure by $(I + P)/2$, which is equivalent to inserting the instruction "With probability $\frac{1}{2}$, do nothing" just inside the for-loop.) However, this is unnecessary, as the transition procedure given in Figure 13 guarantees that $(1 + \lambda_{\max})^{-1} \leq \frac{3}{2}$, by (6.2). Hence we can take $B = \frac{3}{2}$ in Theorem 6.5. Using a multicommodity flow argument, Cooper et al. established a polynomial bound on the mixing time for any regular degree sequence [26, 27].

Theorem 6.6 [26, 27] *For any $k = k(n) \geq 3$, the switch chain on $G(n, k)$ has mixing time*
$$\tau(\varepsilon) \leq k^{23} n^8 \left(kn \log(kn) + \log(\varepsilon^{-1})\right).$$

Proof (Sketch.) Cooper et al. defined a multicommodity flow for the switch chain on $G(n, k)$ and proved that maximum load $\rho(f)$ is bounded above by a polynomial in n and k. (A brief outline of the argument is given below.) The length of any flow-carrying path is at most $kn/2$. Next,
$$|G(\bm{k})| \leq \frac{M!}{2^{M/2}(M/2)! \prod_{j \in [n]} k_j!} \leq \exp(\tfrac{1}{2} M \log M), \tag{6.4}$$
where the first inequality follows from the configuration model. Applying Theorem 6.5 completes the proof. $\square$

Greenhill [63] used a similar argument to show that the switch chain for k-in, k-out (regular) directed graphs is rapidly mixing for any k with $1 \leq k = k(n) \leq n-1$ and all $n \geq 4$.

We now give some more details on the design and analysis of the multicommodity flow for the switch chain. The flow between two graphs G, G' is defined with respect to the symmetric difference $H = G \triangle G'$. Note that the symmetric difference H need not be regular, even if G, G' are both regular. Greenhill and Sfragara [65] observed that the multicommodity flow defined in [26] for regular degrees can also be used for irregular degrees. In fact, almost all parts of the analysis of the multicommodity flow also extends immediately to irregular degree sequences. For this reason, the description below is presented in the general setting of $G(\boldsymbol{k})$.

Starting from the symmetric difference $H = G \triangle G'$, Cooper et al. described how to decompose this symmetric difference into a sequence of smaller, edge-disjoint structures they called 1-circuits and 2-circuits. They identified several different ways to do this, parameterised by a set $\Psi(G, G')$. For each $\psi \in \Psi(G, G')$, the (canonical) path from G to G' indexed by ψ is denoted $\gamma_\psi(G, G')$. This path is created by "processing" each of the 1-circuits and 2-circuits, in a specified order. Processing a circuit changes the status of its edges from agreeing with G to agreeing with G', and adds some transitions to the path $\gamma_\psi(G, G')$. Finally, the flow from G to G' divided equally among these $|\Psi(G, G')|$ paths.

Once the multicommodity flow is defined, it remains to prove that no transition is too heavily loaded. Suppose that $e = (Z, Z')$ is a transition of the chain which is used on the path $\gamma_\psi(G, G')$. A common approach is to define an *encoding* L of a state Z, which records information about the symmetric difference $H = G \triangle G'$ to help us recover G and G' from the transition (Z, Z') and the encoding L. This approach will work if the set of possible encodings for Z is at most polynomially larger than $|G(\boldsymbol{k})|$, and there are at most polynomially-many options for (G, G') once Z, Z', L are all specified. In [26], encodings are defined by

$$L + Z = G + G'$$

where G, G' and Z are identified with their $n \times n$ adjacency matrices. Then L is an $n \times n$ symmetric matrix with row sums given by $\boldsymbol{k}$, and with almost all entries equal to 0 or 1. In fact, due to the careful way that 1-circuits and 2-circuits are processed, L has at most 4 entries equal to -1 or 2, and all other entries are 0 or 1. Hence L may also be thought of as a graph with most edges labelled 1, and at most four *defect edges* which may be labelled -1 or 2. The sum of all edge labels at vertex j must equal the degree k_j of j, for all $j \in [n]$. (There are some other constraints about the structure of the defect edges, stated in [26, Lemma 2].)

Given an encoding L, by removing the defect edges we obtain a graph with degree sequence which is very close to $\boldsymbol{k}$. This gives a connection between the ratio $|\mathcal{L}(Z)|/|G(\boldsymbol{k})|$ and the ratio $|G'(\boldsymbol{k})|/|G(\boldsymbol{k})|$ from the definition of P-stability (6.2). This connection is explored further in Section 6.5. In the regular case, we have the following bound.

Lemma 6.7 [26, Lemma 4] *For any $Z \in G(n, k)$, let $\mathcal{L}(Z)$ denote the set of encodings L such that every entry of $L + Z$ belongs to $\{0, 1, 2\}$. Then $|\mathcal{L}(Z)| \leq 2k^6 n^6 |G(n, k)|$.*

Proof (Sketch.) This was proved by extending the switch operation to encodings, and showing that at most three switches suffice to transform an encoding into an element of $G(n, k)$. The factor $2k^6 n^6$ is an upper bound on the number of encodings which can be transformed into an arbitrary element of $G(n, k)$. □

In fact, Greenhill and Sfragara observed that [26, Lemma 4] (restated above as Lemma 6.7) was the *only* part of the argument from [26] which relied on the

Generating graphs randomly

regularity assumption for its proof. For this reason, they called it the "critical lemma". The proof of the lemma is essentially a switching argument, used to find the relative sizes of two sets. For irregular degrees, it was no longer possible to prove that a suitable switch could always be found in any encoding.

Greenhill and Sfragara bypassed this problem by using a more powerful switching operation to prove the critical lemma. Rather than use a switch, which swaps edges for non-edges around an alternating 4-cycle, they used a operation involving an alternating 6-cycle (deleting 3 edges and inserting 3 edges at a time), illustrated in Figure 14. This operation gave them sufficient flexibility to prove the critical lemma when $k_{\max}$ is not too large.

Figure 14: Switching edges around a 6-cycle

Theorem 6.8 [65, Theorem 1.1] *Let $\boldsymbol{k}$ be a graphical degree sequence. If all entries of $\boldsymbol{k}$ are positive and $3 \leq k_{\max} \leq \frac{1}{3}\sqrt{M}$ then the mixing time of the switch chain on $G(\boldsymbol{k})$ satisfies*

$$\tau(\varepsilon) \leq k_{\max}^{14} M^9 \left(\tfrac{1}{2} M \log M + \log(\varepsilon)\right).$$

In Section 6.5 we will discuss connections between the stability of degree sequences and rapid mixing of the switch chain. First we discuss some results regarding the switch chain for bipartite graphs, directed graphs and hypergraphs, and some related topics.

6.4.1 Bipartite graphs and directed graphs We have seen in Theorem 6.4 that the algorithm of Jerrum, Sinclair and Vigoda [78] gives an FPAUS for sampling bipartite graphs with any given bipartite degree sequence. However, there is still interest in studying the switch chain for bipartite graphs, as it is a very natural and simple process.

A 1-regular bipartite graph is a permutation. Diaconis and Shahshahani [34] studied the Markov chain with state space S_n, the set of all permutations of $[n]$, and transitions defined as follows: with probability $1/n$ do nothing, and otherwise choose a transposition $(i\,j)$ uniformly at random (where $i, j \in [n]$ are distinct), and multiply the current permutation by $(i\,j)$ on the left, say. This random transposition chain is very closely related to the switch chain for a 1-regular bipartite degree sequence (the set of allowed transitions is identical, though the probability of each transition differs between the two chains). Diaconis and Shahshahani gave a very complete analysis of the random transposition chain in [34], calculating the eigenvalues and proving that the chain exhibits the "cutoff phenomenon", see [84, Chapter 18]. That is, the total variation distance to stationarity drops very quickly from $1 - o(1)$ to

$o(1)$ when the chain has taken $\frac{1}{2}n\log n + \Theta(n)$ steps (this is cutoff at $\frac{1}{2}n\log n$ with window of order n).

Other than [34], the first analysis of the switch chain for sampling bipartite graphs with given degrees was the work of Kannan et al. [80], discussed earlier.

The multicommodity flow arguments from [26, 65] can be simplified when restricted to bipartite graphs, as the symmetric difference of two bipartite graphs with the same degree sequence can be decomposed into edge-disjoint alternating cycles, and these are relatively easy to handle. The resulting bounds on the mixing time of the switch chain for bipartite graphs with given degrees were recently presented in [37, Appendix A]. These show that the switch chain is rapidly mixing for any regular bipartite degree sequence, and for arbitrary bipartite degree sequences when the maximum degree is not too large compared to the number of edges. As usual, the mixing time bounds are very high-degree polynomials.

A bipartite degree sequence is *half-regular* if all degrees on one side of the bipartition are regular. Miklos, Erdős and Soukup [93] proved that the switch Markov chain is rapidly mixing for half-regular bipartite degree sequences. Their proof also used the multicommodity flow method, but the flow is defined differently to the Cooper–Dyer–Greenhill flow described above.

For some directed degree sequences, the switch chain fails to connect the state space, as it cannot reverse the orientation of a directed 3-cycle. Rao et al. [100] observed that by the Markov chain which performs switch moves and (occasionally) reverse directed 3-cycles, is ergodic for any directed degree sequence. They noted that for many degree sequences, this additional move did not seem to be needed in order to connect the state space. This was confirmed by the work of Berger and Müller-Hannemann [10] and LaMar [83], who characterised degree sequences for which the switch chain is irreducible.

Greenhill and Sfragara [65, Theorem 1.2] adapted their argument to directed graphs, proving a similar result to Theorem 6.8. As well as an upper bound on the maximum degree, [65, Theorem 1.2] also assumes that the switch chain connects the state space. Their argument built on Greenhill's analysis [63] of the switch chain for directed graphs, replacing the proof of the "critical lemma" from [63] by one which did not require regularity.

6.4.2 The augmented switch chain and the Curveball chain

Erdős et al. [41] considered the switch chain augmented by an additional transition, namely switching the edges around an alternating 6-cycle as shown in Figure 14. They called this transition a *triple swap*. We will refer to this chain as the *augmented switch chain*. Building on the analysis from [93], Erdős et al. [41, Theorem 10] proved that the augmented switch chain for half-regular bipartite degree sequences remains rapidly mixing in the presence of set of forbidden edges given by the union of a perfect matching and a star. They also described an algorithm (similar to the Havel–Hakimi algorithm) for constructing a single realization [41, Theorem 9], to be used as the initial state. Since directed graphs can be modelled as bipartite graphs with a forbidden perfect matching, their algorithm also gives an FPAUS for directed graphs with specified in-degrees and out-degrees, where (say) the sequence of in-degrees is regular. This explains the addition of the triple swap transitions, without which the chain might not be irreducible for some directed degree sequences. By avoiding a star, the problem becomes *self-reducible* [79], which leads to an FPRAS for approximating the number of bipartite graphs with given half-regular degree sequence and some forbidden edges. As mentioned earlier, the algorithm of Jerrum, Sinclair and Vigoda [78, Corollary 8.1] can also be applied to this problem.

Erdős et al. [42] gave new conditions on bipartite and directed degree sequences which guarantee rapid mixing of the augmented switch chain. In particular, suppose that a bipartite degree sequence has degrees $s = (s_1, \ldots, s_a)$ in one part and degrees $t = (t_1, \ldots, t_b)$ in the other, where $a + b = n$. Let $s_{\max}, s_{\min}, t_{\max}, t_{\min}$ be the maximum and minimum degrees on each side. If all degrees are positive and

$$(s_{\max} - s_{\min} - 1)(t_{\max} - t_{\min} - 1) \leq \max\{s_{\min}(a - t_{\max}), t_{\min}(b - s_{\max})\}$$

then the augmented switch chain on the set of bipartite graphs with bipartite degree sequence (s, t) is rapidly mixing [42, Theorem 3]. They applied this result to the analysis of the bipartite Erdős–Rényi model $\mathcal{G}(a, b, p)$, with a vertices in one side of the bipartition, b vertices on the other and each possible edge between the two parts is included with probability p. Erdős et al. [42, Corollary 13] proved that if p is not too close to 0 or 1 then the augmented switch chain is rapidly mixing for the degree sequence arising from $\mathcal{G}(a, b, p)$, with high probability as $n \to \infty$, where $n = a + b$. They also proved analogous results for directed degree sequences [42, Theorem 4, Corollary 14]. To prove their results, they adjusted the multicommodity argument from [41, 93] and gave new proofs of the "critical lemma" for that argument.

The Curveball chain, introduced by Verhelst [114], is another Markov chain for sampling bipartite graphs with given degrees, which chooses two vertices on one side of the bipartition and randomises their neighbourhoods, without disturbing the degrees or set of common neighbours of the chosen vertices. Carstens and Kleer [20] showed that the Curveball chain is rapidly mixing whenever the switch chain is rapidly mixing.

6.4.3 New classes from old Erdős, Miklós and Toroczkai [45] described a novel way to expand the class of degree sequences (and bipartite degree sequences, and directed degree sequences) for which the switch chain is known to be rapidly mixing. Their approach utilised canonical degree sequence decompositions, introduced by Tyshkevich [112], and extended this concept to bipartite and directed degree sequences. Using the decomposition theorem from [44], if the switch chain (or augmented switch chain) is rapidly mixing on each component of this decomposition, then it is rapidly mixing on the original degree sequence.

6.4.4 Improved bounds using functional inequalities Functional inequalities can be used to give tight bounds on the convergence of of Markov chains. Suppose that $\mathcal{M}$ is a Markov chain with state space Ω, transition matrix P and stationary distribution π. The *Dirichlet form* associated to $\mathcal{M}$ is defined by

$$\mathcal{E}_{P,\pi}(f, f) = \tfrac{1}{2} \sum_{x,y \in \Omega} \big(f(x) - f(y)\big)^2 \pi(x) P(x, y)$$

for any $f : \Omega \to \mathbf{R}$. This is a weighted measure of how much f varies over pairs of states which differ by a single transition. The *variance* of f with respect to π,

defined by
$$\mathrm{Var}_\pi(f) = \tfrac{1}{2} \sum_{x,y \in \Omega} \big(f(x) - f(y)\big)^2 \pi(x)\pi(y),$$
captures the global variation of f over Ω. These functions can be used to bound the second-largest eigenvalue λ_1 of $\mathcal{M}$ as follows:
$$1 - \lambda_1 = \min_f \frac{\mathcal{E}_{P,\pi}(f,f)}{\mathrm{Var}_\pi(f)},$$
where the minimum is taken over all non-constant functions $f : \Omega \to \mathbf{R}$. See for example [84, Lemma 13.12 and Remark 13.13]. The Markov chain satisfies a *Poincaré inequality* with constant α if $\mathrm{Var}_\pi(f) \leq \alpha\, \mathcal{E}_{P,\pi}(f,f)$ for any $f : \Omega \to \mathbf{R}$. The log-Sobolev inequality has a similar (but more complicated) definition and can also be used to bound the mixing time [33].

Very recently, Tikhomirov and Youssef [109] proved[3] a sharp Poincaré inequality, and established a log-Sobolev inequality, for the switch chain for regular bipartite graphs. Using their Poincaré inequality, Tikhomirov and Youssef proved [109, Corollary 1.2] that when $3 \leq k \leq n^c$ for some universal constant c, the mixing time of the switch chain on k-regular bipartite graphs satisfies
$$\tau(\varepsilon) \leq Ckn\big(kn \log kn + \log(2\varepsilon^{-1})\big)$$
for some universal constant $C > 0$ (constants c, C not explicitly stated). This is a huge improvement on any previously-known bound. Tikhomirov and Youssef also state the following mixing time bound, obtained using their log-Sobolev inequality when $k \geq 3$ is a fixed constant:
$$\tau(\varepsilon) \leq C_k\, n \log n \left(\log n + \log\left(\frac{1}{2\varepsilon^2}\right)\right).$$

Here $C_k > 0$ is an expression which depends only on k.

The proof in [109] is long and technical, and will likely be difficult to generalise. But it is exciting to see such a low-degree polynomial bound on the mixing time of the switch chain for this non-trivial class of bipartite degree sequences.

6.4.5 Hypergraphs A hypergraph is *uniform* if every edge contains the same number of vertices. The incidence matrix of a hypergraph can be viewed as the adjacency matrix of a bipartite graph, with one part of the bipartition representing the vertices of the hypergraph and the other part representing the edges. Conversely, a bipartite graph gives rise to a simple hypergraph, by reversing this construction, if all vertices on the "edge" side of the bipartition have distinct neighbourhoods. This is needed to avoid creating a repeated edge. In the case of uniform hypergraphs, if the resulting hypergraph is simple then it arises from precisely $m!$ distinct bipartite graphs, where m is the number of edges of the hypergraph.

Hence, any algorithm for sampling bipartite graphs with a given half-regular degree sequence can be transformed into an algorithm for sampling uniform hypergraphs with given degrees, using rejection sampling. This is explored by Dyer et al.

---

[3]Subject to refereeing.

in [37]. Note that a configuration model may be defined for hypergraphs, though it is equivalent to the corresponding bipartite configuration model. The configuration model only gives polynomial-time uniform sampling when the maximum degree multiplied by the edge size is $O(\log n)$, see [37, Lemma 2.3].

Chodrow [22] introduced a Markov chain which works directly on uniform hypergraphs. A transition involves choosing two edges e, f and deleting them, then inserting two edges e', f' chosen randomly so that the degree sequence is unchanged and $e' \cap f' = e \cap f$. This transitions of this chain are analogous to the transitions of the Curveball chain [20] for sampling bipartite graphs. Chodrow proved that this chain is ergodic, but did not analyse the mixing time. It is an open problem to determine classes of degree sequences and edge sizes for which this chain is rapidly mixing.

6.5 Stability of degree sequences

Informally, a class of degree sequences is *stable* if $|G(\boldsymbol{k})|$ varies smoothly as $\boldsymbol{k}$ ranges over the class [77]. Work on the connection between the stability of degree sequences and mixing rates of Markov chains for sampling from $G(\boldsymbol{k})$ began with Jerrum and Sinclair's definition of P-stability [76], stated in Definition 6.2 above. A slightly different version of P-stability was studied by Jerrum, Sinclair and McKay [77]. Let $\|\boldsymbol{x}\|_1 = \sum_{j \in [n]} |x_j|$ denote the 1-norm of the vector $\boldsymbol{x} = (x_1, \ldots, x_n)$, and define the set $U(\boldsymbol{k})$ of all degree sequences $\tilde{\boldsymbol{k}}$ such that

$$\|\tilde{\boldsymbol{k}}\|_1 = \|\boldsymbol{k}\|_1 \quad \text{and} \quad \|\tilde{\boldsymbol{k}} - \boldsymbol{k}\|_1 = 2.$$

Jerrum, Sinclair and McKay said that a class of degree sequences is P-stable if there exists a polynomial $q(n)$ such that

$$\left| \bigcup_{\tilde{\boldsymbol{k}} \in U(\boldsymbol{k})} G(\tilde{\boldsymbol{k}}) \right| \leq q(n) |G(\boldsymbol{k})|$$

for all $\boldsymbol{k}$ in the class. If this holds then $\boldsymbol{k}$ is also P-stable in the original sense (Definition 6.2). Let $k_{\min}$ denote the smallest entry of $\boldsymbol{k}$. Jerrum et al. [77, Theorem 8.1, Theorem 8.3] gave two sufficient conditions for a degree sequence to belong to a P-stable class.

Theorem 6.9 [77] *Recall that $M = M(\boldsymbol{k})$ is the sum of entries in the degree sequence $\boldsymbol{k}$.*

(i) *The class of graphical degree sequences $\boldsymbol{k} = (k_1, \ldots, k_n)$ which satisfy*

$$(k_{\max} - k_{\min} + 1)^2 \leq 4k_{\min}(n - k_{\max} - 1)$$

is P-stable.

(ii) *The class of graphical degree sequences $\boldsymbol{k} = (k_1, \ldots, k_n)$ which satisfy*

$$(M - k_{\min} n)(k_{\max} n - M)$$
$$\leq (k_{\max} - k_{\min})\big((M - k_{\min} n)(n - k_{\max} - 1) + k_{\min}(k_{\max} n - M)\big)$$

is P-stable.

Jerrum et al. [77] listed several examples of classes of degree sequences which satisfy one of these sufficient conditions and hence are P-stable, including

- all regular sequences;
- all graphical sequences with $k_{\min} \geq 1$ and $k_{\max} \leq 2\sqrt{n} - 2$;
- all graphical sequences with $k_{\min} \geq n/9$ and $k_{\max} \leq 5n/9 - 1$.

Using (5.5) and recalling Definition 5.7, we see that the sufficient condition from Theorem 6.9(i) does not cover heavy-tailed distributions such as the power-law distribution-bounded degree sequences with $\gamma \in (2,3)$. The condition from Theorem 6.9(ii) may hold in some cases but fails whenever $M > 2k_{\min} n$. The first and third examples show that a P-stable class does not have to be sparse.

It is possible to define classes of degree sequences which are not P-stable but for which the switch chain is rapidly mixing. Jerrum, Sinclair and McKay [77] illustrated this using the degree sequence

$$\boldsymbol{k} = \boldsymbol{k}(n) = (2n-1, 2n-2, \ldots, n+1, n, n, n-1, \ldots, 2, 1)$$

on $2n$ vertices. There is a unique realisation of this degree sequence, so $|G(\boldsymbol{k})| = 1$ and the switch chain is trivially rapidly mixing on $G(\boldsymbol{k})$. However, $|G(\boldsymbol{k}')|$ is exponential in n, where

$$\boldsymbol{k}' = \boldsymbol{k}'(n) = (2n-2, 2n-2, \ldots, n+1, n, n, n-1, \ldots, 2, 2)$$

is obtained from $\boldsymbol{k}$ by decreasing the largest degree by 1 and increasing the smallest degree by 1. Hence the class $\{\boldsymbol{k}(n) \mid n \geq 2\}$ is not P-stable. Erdős et al. [40] described more general classes of degree sequences with these properties. So P-stability is not a necessary condition for the switch chain to be efficient. Rather, the standard proof techniques tend to break down when the class of degree sequences is not P-stable.

6.5.1 Strong stability Amanatidis and Kleer [3, 4] defined a new notion of stability, called *strong stability*, which is possibly stronger than P-stability. Recall that $G'(\boldsymbol{k})$ denotes the state space of the Jerrum–Sinclair chain. Say that graphs G, H are at JS-distance r if H can be obtained from G using at most r transitions of the Jerrum–Sinclair chain. Next, let $d_{JS}(\boldsymbol{k})$ denote the maximum, over all $G \in G'(\boldsymbol{k})$, of the minimum distance from G to an element of $G(\boldsymbol{k})$. Then every element of the augmented state space $G'(\boldsymbol{k})$ can be transformed into an element of $G(\boldsymbol{k})$ in at most $d_{JS}(\boldsymbol{k})$ transitions of the Jerrum–Sinclair chain.

Definition 6.10 A class of graphical degree sequences is *strongly stable* if there is a constant ℓ such that $d_{JS}(\boldsymbol{k}) \leq \ell$ for all degree sequences $\boldsymbol{k}$ in the class.

Amanatidis and Kleer proved [3, Proposition 3] that every strongly stable family is P-stable. It is not known whether the converse is also true. The main result of [3] is the following.

Theorem 6.11 [4, Proposition 2.3 and Theorem 2.4] *The switch chain is rapidly mixing for all degree sequences from a strongly stable family.*

Generating graphs randomly

The proof of Theorem 6.11 rests on the observation that it is much easier to define a good multicommodity flow for the Jerrum–Sinclair chain than for the switch chain. Next, Amanatidis and Kleer prove that when $\boldsymbol{k}$ is strongly stable, a good flow for the Jerrum–Sinclair chain can be transformed into a good flow for the switch chain.

Amanatidis and Kleer gave analogous results for bipartite degree sequences [3, Theorem 17]. Their framework provided a unified proof of many rapid mixing results for the switch chain for graphs, and bipartite graphs, with given degrees. The authors of [3] remark that their "unification of the existing results [...] is qualitative rather than quantitative, in the sense that our simpler, indirect approach provides weaker polynomial bounds for the mixing time."

6.5.2 Rapid mixing for P-stable degree classes

Erdős et al. [39] defined a new multicommodity flow for the switch chain. The symmetric difference is decomposed into *primitive* alternating circuits, such that no vertices is visited more than twice on a primitive circuit, and if a vertex is visited twice then the two occurrences are at an odd distance from each other around the circuit. Then the primitive alternating circuits are processed in a carefully-chosen order. An encoding is defined (and is called an "auxiliary matrix") such that it is at most three switches away from (the adjacency matrix of) an element of the set $G'(\boldsymbol{k})$. By definition, P-stability guarantees that $|G'(\boldsymbol{k})| \leq q(n) |G(\boldsymbol{k})|$ for some polynomial $q(n)$. Furthermore, there are a polynomial number of ways to choose each of the (at most 3) switches. Hence, when $\boldsymbol{k}$ is P-stable, we conclude that the number of encodings is at most polynomially larger than $|G(\boldsymbol{k})|$. This proves the "critical lemma" for this flow, and establishes the following.

Theorem 6.12 [39, Theorem 1.3] *The switch Markov chain is rapidly mixing on all degree sequences contained in a P-stable class.*

Erdős et al. adapted their analysis to bipartite degree sequences and directed degree sequences [39], proving the analogue of Theorem 6.12 in those settings. Hence their result extends Theorem 6.11 from strongly stable to P-stable degree classes, and includes directed degree sequences. bipartite degree sequences and directed degree sequences.

Applying Theorem 6.9(ii) and Theorem 6.12 to the degree sequence of $\mathcal{G}(n,p)$ leads to the following result.

Corollary 6.13 [39, Corollary 8.6] *When $n \geq 100$, the degree sequence of the binomial random graph $\mathcal{G}(n,p)$ satisfies the condition of Theorem 6.9(ii) with probability at least $1 - 3/n$, so long as $p, 1-p \geq \frac{5 \log n}{n-1}$. Hence the switch chain is rapidly mixing on $\mathcal{G}(n,p)$ with probability at least $1 - 3/n$.*

Indeed, applying Theorem 6.3 we can also conclude that if the conditions of Corollary 6.13 hold then with probability at least $1 - 3/n$, the Jerrum–Sinclair chain gives an FPAUS for sampling from $G(n,p)$.

6.5.3 A new notion of stability

Recently, Gao and Greenhill [56] introduced a new notion[4] of stability for classes of degree sequences.

---
[4]Called "k-stability" in [56], but here we reserve k for degrees.

Definition 6.14 Given a positive integer b and nonnegative real number α, a graphical degree sequence $\boldsymbol{k}$ is said to be (b,α)-stable if $|G(\boldsymbol{k}')| \leq M(\boldsymbol{k})^\alpha \, |G(\boldsymbol{k})|$ for every graphical degree sequence $\boldsymbol{k}'$ with $\|\boldsymbol{k}' - \boldsymbol{k}\|_1 \leq b$. Let $\mathcal{D}_{b,\alpha}$ be the set of all degree sequences that are (b,α)-stable. A family $\mathcal{D}$ of degree sequences is b-*stable* if there exists a constant $\alpha > 0$ such that $\mathcal{D} \subseteq \mathcal{D}_{b,\alpha}$.

Gao and Greenhill proved [56, Proposition 6.2] that 2-stability is equivalent to P-stability. The relationship between strong stability and 2-stability is not known.

Recall that by removing all defect edges, an encoding gives rise to a graph with degree sequence not too far from $\boldsymbol{k}$. Gao and Greenhill observe that all degree sequences $\boldsymbol{k}'$ which correspond to encodings arising from the multicommodity flow of [26, 65] satisfy $\|\boldsymbol{k}' - \boldsymbol{k}\|_1 \leq 8$. Next, assuming $(8,\alpha)$-stability they found an upper bound on the number of encodings compatible with a given graph $Z \in G(\boldsymbol{k})$: this proves the "critical lemma" and leads to the following result.

Theorem 6.15 [56, Theorem 2.1] *If the graphical degree sequence $\boldsymbol{k}$ is $(8,\alpha)$-stable for some $\alpha > 0$ then the switch chain on $G(\boldsymbol{k})$ is rapidly mixing, and*

$$\tau(\varepsilon) \leq 12 \, k_{max}^{14} \, n^6 \, M^{3+\alpha} \left(\tfrac{1}{2} M \log M + \log \varepsilon^{-1} \right).$$

Gao and Greenhill provided a sufficient condition for a degree sequence to be 8-stable, and a slightly weaker condition which guarantees P-stability and strong stability. These conditions involve the parameter $J(\boldsymbol{k})$ defined in (5.6), and have been designed to work well for heavy-tailed degree sequences.

Theorem 6.16 [56, Theorem 2.2]

(i) *Let $\boldsymbol{k}$ be a graphical degree sequence. If $M > 2J(\boldsymbol{k}) + 18k_{\max} + 56$ then $\boldsymbol{k}$ is $(8,8)$-stable.*

(ii) *Suppose that $\mathcal{D}$ is a family of degree sequences wuch that $M > 2J(\boldsymbol{k}) + 6k_{\max} + 2$ for all $\boldsymbol{k} \in \mathcal{D}$. Then $\mathcal{D}$ is both P-stable and strongly stable.*

The proof of Theorem 6.16(i) uses the switching method. Then Theorem 6.16(ii) follows using the fact, proved in [56, Lemma 4.1], that if every graphical degree sequence $\boldsymbol{k}'$ with $\|\boldsymbol{k}' - \boldsymbol{k}\|_1 \leq 6$ is $(2,\alpha)$-stable then $\boldsymbol{k}$ is $(8,4\alpha)$-stable.

Finally, Gao and Greenhill prove that various families of heavy-tailed degree sequences satisfy the condition of Theorem 6.16(i), and hence are 8-stable, strongly stable and P-stable. In particular [56, Theorem 5.3], the family of power-law distribution-bounded sequences with parameter $\gamma > 2$ is P-stable.

Gao and Greenhill gave analogous definitions and results for directed degree sequences [56, Section 7].

6.6 Restricted graph classes

We briefly describe some related work on using rapidly mixing Markov chains to sample from restricted classes of graphs with given degrees.

6.6.1 Joint degree matrices In some applications, it is desirable to be able to specify not just the degrees of a graph, but also the number of edges between vertices with given degrees. This can help to capture network properties such as *assortativity*, which is the tendency for vertices with similar degrees to be adjacent. A *joint degree matrix* [2, 98] stores the number of edges J_{ij} with one endvertex of degree i and the other of degree j, for all relevant i, j. A sequential importance sampling approach for sampling graphs with a specified joint degree matrix was given in the physics literature by Bassler et al. [6], without full analysis.

The switch chain can be adapted to sample from the set of all graphs with a given degree sequence and given joint degree matrix, by rejecting any transition which would change any entry in the joint degree matrix. Stanton and Pinar [107] gave empirical evidence that suggests that the switch chain mixes rapidly on graphs with a prescribed joint degree matrix, but there are few rigorous results. Erdős, Miklós and Toroczkai [44] proved that the switch chain is rapidly mixing on the set of all *balanced* realizations of a given joint degree matrix. Here a realization is balanced if for all vertices v with degree i, the number of neighbours w of v with degree j is within 1 of the value J_{ij}/n_i, where n_i is the number of vertices with degree i. Their proof involved a new Markov chain decomposition theorem [44, Theorem 4.3], similar to that of Martin and Randall [87].

Amanatidis and Kleer [3] showed that the switch chain is rapidly mixing on the set of all realizations of any joint degree matrix with just two degree classes. Their analysis is quite technical, and moving beyond two degree classes seems to be a challenging problem.

6.6.2 Connected graphs The switch chain may disconnect a connected graph, which can be undesirable in some applications such as communications networks. One possibility is to simply reject any proposed switch which would disconnect the graph. Gkantsidis et al. [62] investigated the performance of this restricted switch chain empirically, but without rigorous analysis. Note that the set of connected graphs with given degree sequence was shown to be connected under switches by Taylor [108] in 1981.

Mahlmann and Schindelhauer [86] proposed an alternative operation, called the k-Flipper. Here a switch is performed if the edges of the switch are at distance at most k apart in the graph. In the 1-Flipper, or *flip chain*, the switch operation takes a path of length 3 and exchanges its endvertices, as shown in Figure 15. Clearly this operation, known as a *flip*, cannot disconnect a connected graph.

Figure 15: Transitions of the flip chain

The flip chain is rapidly mixing on the set of all connected k-regular graphs, for any k. This was investigated by Feder et al. [49], with full analysis and improved mixing time bound given by Cooper et al. [28]. These proofs involve a comparison argument, where a sequence of flips is used to simulate a single switch. If a switch disconnects or connects components, then a clever "chaining" argument from [49] is used to stay within the space of connected graphs.

The flip operation can be used to re-randomise a given connected network (such as a communications network) without any risk of disconnecting the network. Expander graphs are fixed graphs which enjoy some pseudorandom properties, such as logarithmic diameter and high connectivity [70]. Allen-Zhu et al. [1, Theorem 4.2] proved that when $k \geq c \log n$ for some positive constant c, performing $O(k^2 n^2 \log n)$ randomly-chosen flips produces an expander with high probability, starting from any k-regular graph. They also applied their methods to the switch chain, showing that $O(kn)$ randomly-chosen switches suffice to produce an expander, with high probability. Hence in situations where the output does not need to be close to uniform, but where pseudorandomness is enough, the runtime of the algorithm can be much shorter.

7 Conclusion

We have discussed rigorously-analysed algorithms for sampling graphs with a given degree sequence, uniformly or approximately uniformly. Some algorithms are inefficient when the maximum degree becomes too high. For other approaches, the boundary between tractable and intractable degree sequences is not clear. Mapping out this frontier is an interesting open problem. Are there families of degree sequences for which the switch chain is provably slow? Connections with stability of degree sequences have also been discussed. As well as their theoretical interest, there are connections between the stability of degree sequences and network privacy, as investigated by Salas and Torra [103].

A challenging open problem is to find an FPRAS for counting graphs with given (arbitrary) degree sequences. The corresponding problem for bipartite graphs with solved by Jerrum, Sinclair and Vigoda [78].

There are many related sampling algorithms which are just outside the scope of this survey. One example is the use of Boltzmann samplers [35] to sample from other restricted graph classes, including planar graphs [53]. As well as providing algorithms, this approach can be used to investigate typical properties of random graphs generated in this way, see for example [88].

To close, we mention some algorithms where the degree sequence itself is a random variable. Some fuzziness in the degree sequence can be useful in some applications, perhaps to account for inaccuracies in the data, or to avoid overfitting. The excellent book by Van der Hofstad [113] is a very good reference for further reading on these topics.

- In the network theory literature, often $k_1, \ldots, k_n$ are i.i.d. random variables drawn from some fixed distribution. If the resulting sum is odd then k_n is increased by 1. More generally, a degree sequence $\boldsymbol{k}$ can be drawn at random from a given distribution, and then a graph from $G(\boldsymbol{k})$ can be sampled uniformly, or approximately uniformly, using one of the methods discussed in this survey.

- *Inhomogeneous random graphs* are similar to the binomial random graph $\mathcal{G}(n,p)$ except that different edges have different probabilities. For a sequence $\boldsymbol{w} \in \mathbf{R}^n$ of positive vertex weights and a function $f : \mathbf{R}^2 \to [0,1]$, the edge $\{i,j\}$ is included in the graph with probability $f(w_i, w_j)$, independently for each edge. An example is the *generalised random graph model* [19] with $f(w_i, w_j) = \frac{w_i w_j}{M(\boldsymbol{w}) + w_i w_j}$, where $M(\boldsymbol{w}) = \sum_{\ell \in [n]} w_\ell$ is the sum of the weights. The Chung-Lu algorithm [23] uses $f(w_i, w_j) = w_i w_j / M(\boldsymbol{w})$, under the assumption that the maximum entry of $\boldsymbol{w}$ is $o(M(\boldsymbol{w})^{1/2})$. The output of the Chung-Lu algorithm is a random graph with expected degree sequence $\boldsymbol{w}$.

- Another algorithm which produces a graph with degree sequence close to some target sequence is the *erased configuration model* [19]. First sample a uniformly random configuration, and in the corresponding graph, delete any loops and delete all but one copy of each multiple edge. Call the resulting graph $\widehat{G}$. If $k_{\max} = o(M^{1/2})$ and $R = O(M)$ then, arguing as in the proof of Theorem 3.1, with high probability only a very small number of edges were deleted, and hence the degree sequence of $\widehat{G}$ is likely to be very close to $\boldsymbol{k}$. Other variations of the configuration model are described in [113, Section 7.8], including models which are tailored to encourage other network properties, such as clustering.

Acknowledgements

The author is very grateful to the anonymous referee and to the following colleagues for their feedback, which helped improve this survey: Martin Dyer, Peter Erdős, Serge Gaspers, Pieter Kleer, Brendan McKay, Eric Vigoda and Nick Wormald. This survey is dedicated to Mike.

References

[1] Z. Allen-Zhu, A. Bhaskara, S. Lattanzi, V. Mirrokni and L. Orecchia, Expanders via local edge flips, in *Proceedings of the 27th Annual ACM-SIAM Symposium on Discrete Algorithms (SODA 2016)*, SIAM, Philadelphia (2018), pp. 259–269.

[2] G. Amanatidis, B. Green and M. Mihail, Connected realizations of joint-degree matrices, *Discrete Applied Mathematics* **250** (2018), 65–74.

[3] G. Amanatidis and P. Kleer, Rapid mixing of the switch Markov chain for strongly stable degree sequences and 2-class joint degree matrices, in *Proceedings of the 30th Annual ACM–SIAM Symposium on Discrete Algorithms (SODA 2019)*, ACM–SIAM, New York–Philadelphia (2019), pp. 966–985.

[4] G. Amanatidis and P. Kleer, Rapid mixing of the switch Markov chain for strongly stable degree sequences, *Random Structures & Algorithms* **57** (2020), 637–657.

[5] A. Arman, P. Gao and N. Wormald, Fast uniform generation of random graphs with given degree sequences, *Random Structures & Algorithms*, in press. arXiv:1905.03446.

[6] K. E. Bassler, C. I. Del Genio, P. L. Erdős, I. Miklós and Z. Toroczkai, Exact sampling of graphs with prescribed degree correlations, *New Journal of Physics* **17** (2015), #083052.

[7] M. Bayati, J.-H. Kim and A. Saberi, A sequential algorithm for generating random graphs, *Algorithmica* **58** (2010), 860–910.

[8] M. Bayati, A. Montanari and A. Saberi, Generating random networks without short cycles, *Operations Research* **66** (2018), 1227–1246.

[9] E. A. Bender and E. R. Canfield, The asymptotic number of non-negative integer matrices with given row and column sums, *Journal of Combinatorial Theory (Series A)* **24** (1978), 296–307.

[10] A. Berger and M. Müller-Hannemann, Uniform sampling of digraphs with a fixed degree sequence, in *Graph Theoretic Concepts in Computer Science, LNCS vol. 6410*, Springer, Berlin (2010), pp. 220–231.

[11] I. Bezáková, N. Bhatnagar, E. Vigoda, Sampling binary contingency tables with a greedy start, *Random Structures & Algorithms* **30** (2007), 168–205.

[12] I. Bezáková, A. Sinclair, Štefankovič and E. Vigoda, Negative examples for sequential importance sampling of binary contingency tables, *Algorithmica* **64** (2012), 606–620.

[13] J. H. Blanchet, Efficient importance sampling for binary contingency tables, *Annals of Applied Probability* **19** (2009), 949–982.

[14] J. Blitzstein and P. Diaconis, A sequential importance sampling algorithm for generating random graphs with prescribed degrees, *Internet Mathematics* **6** (2011), 489–522.

[15] T. Bohman and P. Keevash, Dynamic concentration of the triangle-free process, in *Seventh European Conference on Combinatorics, Graph Theory and Applications*, Edizioni della Normale, Pisa (2013), pp. 489–495.

[16] B. Bollobás, A probabilistic proof of an asymptotic formula for the number of labelled regular graphs, *European Journal of Combinatorics* **1(4)** (1980), 311–316.

[17] B. Bollobás, *Random Graphs (2nd edn.)*, Cambridge University Press, Cambridge (2001).

[18] B. Bollobás and O. Riordan, Constrained graph processes, *Electronic Journal of Combinatorics* **7** (2000), #R18.

[19] T. Britton, M. Deijfen and A. Martin-Löf, Generating simple random graphs with prescribed degree distribution, *Journal of Statistical Physics* **124** (2006), 1377–1397.

[20] C. J. Carstens and P. Kleer, Speeding up switch Markov chains for sampling bipartite graphs with given degree sequence, in *Approximation, Randomization and Combinatorial Optimization. Algorithms and Techniques, APPROX/RANDOM 2018, LIPIcs, vol. 116*, Schloss Dagstuhl–Leibniz-Zentrum für Informatik, Dagstuhl (2018), 36:1–36:18.

[21] Y. Chen, P. Diaconis, S.P. Holmes and J.S. Liu, Sequential Monte Carlo methods for statistical analysis of tables, *Journal of the American Statistical Association* **100** (2005), 109–120.

[22] P. S. Chodrow, Configuration models of random hypergraphs, *Journal of Complex Networks* **8** (2020), cnaa018.

[23] F. Chung and L. Lu, The average distances in random graphs with given expected degrees, *Proceedings of the National Academy of Sciences of the United States of America* **99** (2002), 15879–15882.

[24] A. Clauset, C. R. Shalizi and M. E. J. Newman, Power-law distributions in empirical data, *SIAM Review* **51** (2009), 661–703.

[25] T. Coolen, A. Annibale and E. Roberts, *Generating Random Networks and Graphs*, Oxford University Press, Oxford (2017).

[26] C. Cooper, M. Dyer and C. Greenhill, Sampling regular graphs and a peer-to-peer network, *Combinatorics, Probability and Computing* **16** (2007), 557–593.

[27] C. Cooper, M.E. Dyer and C. Greenhill, Corrigendum: Sampling regular graphs and a peer-to-peer network. `arXiv:1203.6111`

[28] C. Cooper, M. Dyer, C. Greenhill and A. Handley, The flip Markov chain for connected regular graphs, *Discrete Applied Mathematics* **254** (2019), 56–79.

[29] C. I. Del Genio, H. Kim, Z. Toroczkai and K. E. Bassler, Efficient and exact sampling of simple graphs with given arbitrary degree sequence, *PLOS ONE* **5** (2010), e10012.

[30] A. Deza, A. Levin, S. M. Meesum and S. Onn, Hypergraphic degree sequences are hard, *Bulletin of the EATCS* **127** (2019),

[31] P. Diaconis and A. Gangolli, Rectangular arrays with fixed margins, in *Discrete Probability and Algorithms* (eds. D. Aldous, P. Diaconis, J. Spencer and J. M. Steele), Springer–Verlag, New York (1995), pp. 15–41.

[32] P. Diaconis and L. Saloff-Coste, Comparison theorems for reversible Markov chains, *Annals of Applied Probability* **3** (1993), 697–730.

[33] P. Diaconis and L. Saloff-Coste, Logarithmic Sobolev inequalities for finite Markov chains, *Annals of Applied Probability* **6** (1996), 695–750.

[34] P. Diaconis and M. Shahshahani, Generating a random permutation with random transpositions, *Zeitschrift für Wahrscheinlichkeitstheorie* **57** (1981), 159–179.

[35] P. Duchon, P. Flajolet, G. Louchard and G. Schaeffer, Boltzmann samplers for the random generation of combinatorial structures, *Combinatorics, Probability and Computing* **13** (2004), 577–625.

[36] M. Dyer and C. Greenhill, Random walks on combinatorial objects, in *Surveys in Combinatorics 1999* (eds. J. D. Lamb and D. A. Preece), *London Mathematical Society Lecture Note Series vol. 267*, Cambridge University Press, Cambridge (1999), pp. 101–136.

[37] M. Dyer, C. Greenhill, P. Kleer, J. Ross and L. Stougie, Sampling hypergraphs with given degrees. arXiv:2006.12021

[38] P. Erdős and T. Gallai, Graphs with prescribed degree of vertices (in Hungarian), *Matematikai Lapok* **11** (1960), 264–274.

[39] P. L. Erdős, C. Greenhill, T. R. Mezei, I. Miklos, D. Soltész, L. Soukup, The mixing time of the switch Markov chains: a unified approach. arXiv:1903.06600

[40] P. L. Erdős, E. Győri, T. R. Mezei, I. Miklós and D. Soltész, Half-graphs, other non-stable degree sequences, and the switch Markov chain. arxiv:1909.02308

[41] P. L. Erdős, S. Z. Kiss, I. Miklós and L. Soukup, Approximate counting of graphical realizations, *PLOS ONE* **10** (2015), e0131300.

[42] P. L. Erdős, T. R. Mezei, I. Miklós and D. Soltész, Efficiently sampling the realizations of bounded, irregular degree sequences of bipartite and directed graphs, *PLOS ONE* **13** (2018), e0201995.

[43] P. L. Erdős, I. Miklós and Z. Toroczkai, A simple Havel–Hakimi type algorithm to realize graphical degree sequences of directed graphs, *Electronic Journal of Combinatorics* **17(1)** (2010), #R66.

[44] P. L. Erdős, I. Miklós and Z. Toroczkai, A decomposition based proof for fast mixing of a Markov chain over balanced realizations of a joint degree matrix, *SIAM Journal on Discrete Mathematics* **29** (2015), 481–499.

[45] P. L. Erdős, I. Miklós and Z. Toroczkai, New classes of degree sequences with fast mixing swap Markov chain sampling, *Combinatorics, Probability and Computing* **27** (2018), 186–207.

[46] P. Erdős and A. Rényi, On the evolution of random graphs. On Random Graphs I., *Publicationes Mathematicae Debrecen* **6** (1959), 290–297.

[47] P. Erdős and A. Rényi, On the evolution of random graphs, *Publications of the Mathematical Institute of the Hungarian Academy of Sciences* **5** (1960), 17–61.

[48] P. Erdős, S. Suen and P. Winkler, On the size of a random maximal graph, *Random Structures & Algorithms* **6** (1995), 309–318.

[49] T. Feder, A. Guetz, M. Mihail and A. Saberi, A local switch Markov chain on given degree graphs with application in connectivity of peer-to-peer networks, in *47th Annual IEEE Symposium on Foundations of Computer Science (FOCS'06)*, IEEE, New York (2006), pp. 69–76.

[50] B. K. Fosdick, D. B. Larremore, J. Nishimura, J. Ugander, Configuring random graph models with fixed degree sequences, *SIAM Review* **60** (2018), 315–355.

[51] M. L. Fredman and D. E. Willard, BLASTING through the information theoretic barrier with FUSION TREES, in *Proceedings of the 22nd Annual ACM Symposium on the Theory of Computing (STOC 1990)*, ACM, New York (1990), pp. 1–7.

[52] A. Frieze and M. Karoński, *Introduction to Random Graphs*, Cambridge University Press, Cambridge (2015).

[53] E. Fusy, Uniform random sampling of planar graphs in linear time, *Random Structures & Algorithms* **35** (2009), 464–521.

[54] P. Gao, Uniform generation of d-factors in dense host graphs, *Graphs and Combinatorics* **30** (2014), 581–589.

[55] P. Gao and C. Greenhill, Uniform generation of spanning regular subgraphs of a dense graph, *Electronic Journal of Combinatorics* **26(4)** (2019), #P4.28.

[56] P. Gao and C. Greenhill, Mixing time of the switch Markov chain and stable degree sequences. `arXiv:2003.08497`

[57] P. Gao, R. van der Hofstad, A. Southwell and C. Stegehuis, Counting triangles in power-law uniform random graphs, *Electronic Journal of Combinatorics* **27(3)** (2020), #P3.19.

[58] P. Gao and N. Wormald, Enumeration of graphs with a heavy-tailed degree sequence, *Advances in Mathematics* **287** (2016), 412–450.

[59] P. Gao and N. Wormald, Uniform generation of random regular graphs, *SIAM Journal on Computing* **46** (2017), 1395–1427.

[60] P. Gao and N. Wormald, Uniform generation of random graphs with power-law degree sequences, in *Proceedings of the 29th Annual ACM-SIAM Symposium on Discrete Algorithms (SODA 2018)*, SIAM, Philadelphia (2018), pp. 1741–1758.

[61] E. N. Gilbert, Random graphs, *Annals of Mathematical Statistics* **30** (1959), 1141–1144.

[62] C. Gkantsidis, M. Mihail and E. Zegura, The Markov chain simulation method for generating connected power-law random graphs, in *Proceedings of 5th Workshop on Algorithm Engineering and Experiments (ALENEX03)*, SIAM, Philadelphia (2003), pp. 16–25.

[63] C. Greenhill, A polynomial bound on the mixing time of a Markov chain for sampling regular directed graphs, *Electronic Journal of Combinatorics* **18(1)** (2011), #P234.

[64] C. Greenhill, Making Markov chains less lazy. arXiv:1203.6668

[65] C. Greenhill and M. Sfragara, The switch Markov chain for sampling irregular graphs and digraphs, *Theoretical Computer Science* **719** (2018), 1–20.

[66] T. Hagerup, Sorting and searching on the word RAM, in *Proceedings of the 15th Annual Symposium on Theoretical Aspects of Computer Science (STACS 1998)*, Springer, Berlin (1998), 366–398.

[67] S. L. Hakimi, On realizability of a set of integers as degrees of the vertices of a linear graph. I, *SIAM Journal on Applied Mathematics* **10** (1962), 496–506.

[68] V. Havel, A remark on the existence of finite graphs (in Czech), *Časopis Pro Pěstování Matematiky* **80** (1955), 477–480.

[69] P. W. Holland and S. Leinhardt, An exponential family of probability distributions for directed graphs, *Journal of the American Statistical Association* **76** (1981), 33–50.

[70] S. Hoory, N. Linial and A. Wigderson, Expander graphs and their applications, *Bulletin of the American Mathematical Society* **43** (2006), 439–561.

[71] S. Janson, The probability that a random multigraph is simple, *Combinatorics, Probability and Computing* **18** (2009), 205–225.

[72] S. Janson, Random graphs with given vertex degrees and switchings, *Random Structures & Algorithms* **57** (2020), 3–31.

[73] S. Janson, T. Łuczak and A. Ruciński, *Random Graphs,*, Wiley, New York (2000).

[74] M. Jerrum, *Counting, Sampling and Integrating: Algorithms and Complexity*, Lectures in Mathematics - ETH Zürich, Birkhäuser, Basel (2003).

[75] M. Jerrum and A. Sinclair, Approximating the permanent, *SIAM Journal on Computing* **18** (1989), 1149–1178.

[76] M. Jerrum and A. Sinclair, Fast uniform generation of regular graphs, *Theoretical Computer Science* **73** (1990), 91–100.

[77] M. Jerrum, A. Sinclair and B. D. McKay, When is a graphical sequence stable? , in *Random Graphs, Vol. 2 (Poznán, 1989)*, Wiley, New York (1992), pp. 101–115.

[78] M. Jerrum, A. Sinclair and E. Vigoda, A polynomial-time approximation algorithm for the permanent of a matrix with non-negative entries, *Journal of the ACM* **51** (2004), 671–697.

[79] M. R. Jerrum, L. G. Valiant and V. V. Vazirani, Random generation of combinatorial structures from a uniform distribution, *Theoretical Computer Science* **43** (1986), 169–188.

[80] R. Kannan, P. Tetali and S. Vempala, Simple Markov chain algorithms for generating random bipartite graphs and tournaments, *Random Structures & Algorithms* **14** (1999), 293–308.

[81] H. Kim, C. I. Del Genio, K. E. Bassler and Z. Toroczkai, Constructing and sampling directed graphs with given degree sequences, *New Journal of Physics* **14** (2012), #023012.

[82] J. H. Kim and V. H. Vu, Generating random regular graphs, *Combinatorica* **26** (2006), 683–708.

[83] M. D. LaMar, Directed 3-cycle anchored digraphs and their application in the uniform sampling of realisations from a fixed degree sequence, in *Proceedings of the 2011 Winter Simulation Conference (WSC)*, IEEE, New York (2011), pp. 3348-03359.

[84] D. A. Levin, Y. Peres and E. L. Wilmer, *Markov Chains and Mixing Times*, American Mathematical Society, Providence RI (2009).

[85] D. Lusher, J. Koskinen and G. Robins, *Exponential Random Graph Models for Social Networks: Theory, Methods and Applications*, Cambridge University Press, Cambridge (2012).

[86] P. Mahlmann and C. Schindelhauer, Peer-to-peer networks based on random transformations of connected regular undirected graphs, in *Proceedings of the 17th Annual ACM Symposium on Parallelism for Algorithms and Architectures (SPAA 2005)*, ACM Press, New York (2005), pp. 155–164.

[87] R. Martin and D. Randall, Disjoint decomposition of Markov chains and sampling circuits in Cayley graphs, *Annals of Applied Probabability* **12** (2002), 581–606.

[88] C. McDiarmid, Random graphs from a weighted minor-closed class, *Electronic Journal of Combinatorics* **20(2)** (2013), #P52.

[89] B. D. McKay, Asymptotics for 0-1 matrices with prescribed line sums, in *Enumeration and Design*, Academic Press, Canada (1984), pp. 225–238.

[90] B. D. McKay, Asymptotics for symmetric 0-1 matrices with prescribed row sums, *Ars Combinatorica* **19A** (1985), 15–25.

[91] B. D. McKay, N. C. Wormald, Uniform generation of random regular graphs of moderate degree, *Journal of Algorithms* **11(1)** (1990), 52–67.

[92] B.D. McKay and N.C. Wormald, Asymptotic enumeration by degree sequence of graphs with degrees $o(n^{1/2})$, *Combinatorica* **11** (1991), 369–383.

[93] I. Miklos, P. L. Erdős and L. Soukup, Towards random uniform sampling of bipartite graphs with given degree sequence, *Electronic Journal of Combinatorics* **20(1)** (2013),, #P16.

[94] R. Milo, S. Shen-Orr, S. Itkovitz, N. Kashtan, D. Chklovskii and U. Alon, Network motifs: simple building blocks of complex networks, *Science* **298** (2002), 825–827.

[95] R. Montenegro and P. Tetali, *Mathematical Aspects of Mixing Times in Markov Chains*, Foundations and Trends in Theoretical Computer Science, vol. *1*, Now Publishers, Hanover (2006).

[96] M. E. J. Newman, The structure and function of complex networks, *SIAM Review* **45** (2003), 167–256..

[97] D. Osthus and A. Taraz, Random maximal H-free graphs, *Random Structures & Algorithms* **18** (2001), 61–82.

[98] A. N. Patrinos and S. L. Hakimi, Relations between graphs and integer-pair sequences, *Discrete Mathematics* **15** (1976), 30–41.

[99] G. F. Pontiveros, S. Griffiths and R. Morris, *The Triangle-Free Process and the Ramsey Number $R(3,k)$*, American Mathematical Society, Providence RI (2020).

[100] A. R. Rao, R. Jana and S. Bandyopadhyay, A Markov chain Monte Carlo method for generating random $(0,1)$-matrices with given marginals, *Sankhyā: The Indian Journal of Statistics* **58** (1996), 225–242.

[101] A. Ruciński and N. C. Wormald, Random graph processes with degree restrictions, *Combinatorics, Probability and Computing* **180** (1992), 1–169.

[102] H. J. Ryser, Combinatorial properties of matrices of zeros and ones, *Canadian Journal of Mathematics* **9** (1957), 371–377.

[103] J. Salas and V. Torra, Improving the characterization of P-stability for applications in network privacy, *Discrete Applied Mathematics* **2016** (2016), 109–114.

[104] A. Sinclair, Improved bounds for mixing rates of Markov chains and multicommodity flow, *Combinatorics, Probability and Computing* **1** (1992), 351–370.

[105] A. Sinclair and M. Jerrum, Approximate counting, uniform generation and rapidly mixing Markov chains, *Information and Computation* **82** (1989), 93–133.

[106] A. Steger and N. C. Wormald, Generating random regular graphs quickly, *Combinatorics, Probability and Computing* **8** (1999), 377–396.

[107] I. Stanton and A. Pinar, Constructing and sampling graphs with a prescribed joint degree distribution, *ACM Journal of Experimental Algorithms* **17** (2011), 3-1.

[108] R. Taylor, Constrained switchings in graphs, in *Combinatorial Mathematics, VIII* (ed. K. McAvaney), Springer, Berlin (1981), pp. 314–336.

[109] K. Tikhomirov and P. Youssef, Sharp Poincaré and log-Sobolev inequalities for the switch chain on regular bipartite graphs. arXiv:2007.02729

[110] G. Tinhofer, On the generation of random graphs with given properties and known distribution, *Applied Computer Science. Berichte zur Praktischen Informatik* **13** (1979), 265–297.

[111] W. T. Tutte, A short proof of the factor theorem for finite graphs, *Canadian Journal of Mathematics* **6** (1954), 347–352.

[112] R. Tyschkevich, Canonical decomposition of a graph (in Russian), *Doklady Akademii Nauk BSSR* **24(8)** (1980), 677–679.

[113] R. van der Hofstad, *Random Graphs and Complex Networks, Cambridge Series in Statistical and Probabilistic Mathematics*, Cambridge University Press, Cambridge (2016).

[114] N. D. Verhelst, An efficient MCMC algorithm to sample binary matrices with fixed marginals, *Psychometrika* **73(4)** (2008), 705.

[115] V. H. Vu, Concentration of non-Lipschitz functions and applications, *Random Structures & Algorithms* **20** (2002), 267–316.

[116] T. Wang, Y. Chen, Z. Zhang, T. Xu, L. Jin, P. Hui, B. Deng and X. Li, Understanding graph sampling algorithms for social network analysis, in *Proceedings of the 31st International Conference on Distributed Computing Systems Workshops*, IEEE, Piscataway, NJ (2011), pp. 123–128.

[117] L. Warnke, The C_ℓ-free process, *Random Structures & Algorithms* **44** (2014), 490–526.

[118] S. Wasserman and P. E. Pattison, Logit models and logistic regression for social networks: I. An introduction to Markov graphs and p^*, *Psychometrika* **61** (1996), 401–425.

[119] N. C. Wormald, Generating random regular graphs, *Journal of Algorithms* **5** (1984), 247–280.

[120] N. C. Wormald, Models of random regular graphs, in *Surveys in Combinatorics, 1999* (eds. J.D. Lamb and D.A. Preece), *London Mathematical Society Lecture Note Series, vol. 267*, Cambridge University Press, Cambridge (1999), pp. 239–298.

[121] N. C. Wormald, The differential equations method for random graph processes and greedy algorithms, in *Lectures on Approximation and Randomized Algorithms*, PWN, Warsaw (1999), 77–155.

[122] J. H. Zhao, Expand and contract: Sampling graphs with given degrees and other combinatorial families. arXiv:1308.6627

School of Mathematics and Statistics
UNSW Sydney
Sydney NSW 2052, Australia
c.greenhill@unsw.edu.au

Recent advances on the graph isomorphism problem

Martin Grohe and Daniel Neuen

Abstract

We give an overview of recent advances on the graph isomorphism problem. Our main focus will be on Babai's quasi-polynomial time isomorphism test and subsequent developments that led to the design of isomorphism algorithms with a quasi-polynomial parameterized running time of the form $n^{\text{polylog}(k)}$, where k is a graph parameter such as the maximum degree. A second focus will be the combinatorial Weisfeiler-Leman algorithm.

1 Introduction

Determining the computational complexity of the Graph Isomorphism Problem (GI) is regarded a major open problem in computer science. Already stated as an open problem in Karp's seminal paper on the NP-completeness of combinatorial problems in 1972 [51], GI remains one of the few natural problems in NP that is neither known to be NP-complete nor known to be polynomial-time solvable. In a breakthrough five years ago, Babai [6] proved that GI is solvable in *quasi-polynomial time* $n^{\text{polylog}(n)} := n^{(\log n)^{\mathcal{O}(1)}}$, where n denotes the number of vertices of the input graphs. This was the first improvement on the worst-case running time of a general graph isomorphism algorithm over the $2^{\mathcal{O}(\sqrt{n \log n})}$ bound established by Luks in 1983 (see [10]).

Research on GI as a computational problem can be traced back to work on chemical information systems in the 1950s (e.g. [77]). The first papers in the computer science literature, proposing various heuristic approaches, appeared in the 1960s (e.g. [85]). An early breakthrough was Hopcroft and Tarjan's $\mathcal{O}(n \log n)$ isomorphism test for planar graphs [48]. Isomorphism algorithms for specific graph classes have been an active research strand ever since (e.g. [31, 39, 57, 61, 66, 76]).

Significant progress on GI was made in the early 1980s when Babai and Luks introduced non-trivial group-theoretic methods to the field [4, 11, 61]. In particular, Luks [61] introduced a general divide-and-conquer framework based on the structure of the permutation groups involved. It is the basis for most subsequent developments. Luks used it to design an isomorphism test running in time $n^{\mathcal{O}(d)}$, where d is the maximum degree of the input graphs. The divide-and-conquer framework is also the foundation of Babai's quasi-polynomial time algorithm. Primitive permutation groups constitute the difficult bottleneck of Luks's strategy. Even though we know a lot about the structure of primitive permutation groups (going back to [20] and the classification of finite simple groups), for a long time it was not clear how we can exploit this structure algorithmically. It required several groundbreaking ideas by Babai, both on the combinatorial and the group-theoretic side, to obtain a quasi-polynomial time algorithm. Starting from Luks's divide-and-conquer framework, we will describe the group-theoretic approach and (some of) Babai's new techniques in Section 4.

In subsequent work, jointly with our close collaborators Pascal Schweitzer and Daniel Wiebking, we have designed a number of *quasi-polynomial parameterized algorithms* where we confine the poly-logarithmic factor in the exponent to some parameter that can be much smaller than the number n of vertices of the input graph. The first result [41] in this direction was an isomorphism test running in time $n^{\text{polylog}(d)}$ which improves both Luks's and Babai's results (recall that d denotes the maximum degree of the input graph). After similar quasi-polynomial algorithms parameterized by tree width [87] and genus [71], the work culminated in an isomorphism test running in time $n^{\text{polylog}(k)}$ for graphs excluding some k-vertex graph as a minor [43]. In Section 5, we will present these results in a unified framework that is based on a new combinatorial closure operator and the notion of *t-CR-bounded* graphs, introduced by the second author in [71] (some ideas can be traced back as far as [67, 75]).

The combinatorial algorithm underlying this approach is a very simple partition refinement algorithm known as the Color Refinement algorithm (first proposed in [68]), which is also an important subroutine of all practical graph isomorphism algorithms. The Color Refinement algorithm can be viewed as the 1-dimensional version of a powerful combinatorial algorithm known as the Weisfeiler-Leman algorithm (going back to [86]). The strength of the Weisfeiler-Leman algorithm and its multiple, sometimes surprising connections with other areas, have been a recently very active research direction that we survey in Section 3.

It is not the purpose of this paper to cover all aspects of GI. In fact, there are many research strands that we completely ignore, most notably the work on practical isomorphism algorithms. Instead we want to give the reader at least an impression of the recent technical developments that were initiated by Babai's breakthrough. For a broader and much less technical survey, we refer the reader to [45].

2 Preliminaries

Let us review some basic notation. By $\mathbb{N}$, we denote the natural numbers (including 0), and for every positive $k \in \mathbb{N}$ we let $[k] := \{1, \ldots, k\}$. Graphs are always finite and, unless explicitly stated otherwise, undirected. We denote the vertex set and edge set of a graph G by $V(G)$ and $E(G)$, respectively. For (undirected) edges, we write vw instead of $\{v, w\}$. The *order* of a graph G is $|G| := |V(G)|$.

A *subgraph* of a graph G is a graph H with $V(H) \subseteq V(G)$ and $E(H) \subseteq E(G)$. For a set $W \subseteq V(G)$, the *induced subgraph* of G with vertex set W is the graph $G[W]$ with $V(G[W]) = W$ and $E(G[W]) := \{vw \in E(G) \mid v, w \in W\}$. Moreover, $G - W := G[V(G) \setminus W]$. The set of *neighbors* of a vertex v is $N_G(v) := \{w \mid vw \in E(G)\}$, and the *degree* of v is $\deg_G(v) := |N_G(v)|$. If G is clear from the context, we omit the subscript G. The *maximum degree* of G is $\Delta(G) := \max\{\deg(v) \mid v \in V(G)\}$.

A *homomorphism* from a graph G to a graph H is a mapping $h: V(G) \to V(H)$ that preserves adjacencies, that is, if $vw \in V(G)$ then $h(v)h(w) \in V(H)$. An *isomorphism* from G to H is a bijective mapping $\varphi: V(G) \to V(H)$ that preserves adjacencies and non-adjacencies, that is, $vw \in E(G)$ if and only if $\varphi(v)\varphi(w) \in E(H)$. We write $\varphi: G \cong H$ to indicate that φ is an isomorphism from G to H.

Sometimes, it will be convenient to work with *vertex- and arc-colored graphs*, or just *colored graphs*, $G = (V(G), E(G), \chi_V^G, \chi_E^G)$, where $(V(G), E(G))$ is a graph, $\chi_V^G: V(G) \to C_V$ is a vertex coloring, and $\chi_E^G: \{(v, w) \mid vw \in E(G)\} \to C_E$ an arc coloring. We speak of an *arc coloring* rather than edge coloring, because we color or-

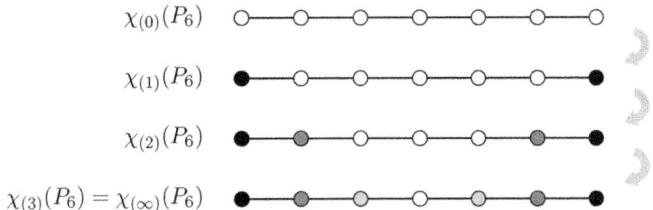

Figure 1: The iterations of Color Refinement on a path P_6 of length 6.

dered pairs (v, w) and not undirected edges vw. Homomorphisms and isomorphisms of colored graphs must respect colors.

3 The Weisfeiler-Leman Algorithm

3.1 The Color Refinement Algorithm

One of the simplest combinatorial procedures to tackle GI is the *Color Refinement algorithm*. The basic idea is to label vertices of the graph with their iterated degree sequence. More precisely, an initially uniform coloring is repeatedly refined by counting, for each color, the number of neighbors of that color.

Let $\chi_1, \chi_2 \colon V(G) \to C$ be colorings of vertices of a graph G, where C is some finite set of colors. The coloring χ_1 *refines* χ_2, denoted $\chi_1 \preceq \chi_2$, if $\chi_1(v) = \chi_1(w)$ implies $\chi_2(v) = \chi_2(w)$ for all $v, w \in V(G)$. The colorings χ_1 and χ_2 are *equivalent*, denoted $\chi_1 \equiv \chi_2$, if $\chi_1 \preceq \chi_2$ and $\chi_2 \preceq \chi_1$.

The Color Refinement algorithm takes as input a vertex- and arc-colored graph (G, χ_V, χ_E) and outputs an isomorphism-invariant coloring of the vertices. Initially, the algorithm sets $\chi_{(0)}(G) := \chi_V$. Then the coloring is iteratively refined as follows. For $i > 0$ we let $\chi_{(i)}(G)(v) := \bigl(\chi_{(i-1)}(G)(v), \mathcal{M}_i(v)\bigr)$, where

$$\mathcal{M}_i(v) := \{\!\!\{ (\chi_{(i-1)}(G)(w), \chi_E(v,w), \chi_E(w,v)) \mid w \in N_G(v) \}\!\!\}$$

is the multiset of colors for the neighbors computed in the previous round.

In each round, the algorithm computes a coloring that is finer than the one computed in the previous round, i.e., $\chi_{(i)}(G) \preceq \chi_{(i-1)}(G)$. At some point this procedure stabilizes, meaning the coloring does not become strictly finer anymore. In other words, there is an $i_\infty < n$ such that $\chi_{(i_\infty)}(G) \equiv \chi_{(i_\infty+1)}(G)$. We call $\chi_{(i_\infty)}(G)$ the *stable coloring* and denote it by $\chi_{(\infty)}(G)$.

Example 3.1 The sequence of colorings for a path of length 6 is shown in Figure 1.

The Color Refinement algorithm is very efficient, and it is used as an important subroutine in all practical isomorphism tools [22, 50, 59, 64, 65]. The stable coloring of a graph can be computed in time $\mathcal{O}((n+m)\log n)$, where n denotes the number of vertices and m the number of edges of the input graph [21] (also see [74])[1]. For a natural class of partitioning algorithms, this is best-possible [13].

3.2 Higher Dimensions

The *k-dimensional Weisfeiler-Leman algorithm (k-WL)* is a natural generalization of the Color Refinement algorithm where, instead of vertices, we color k-tuples of vertices of a graph. The 2-dimensional version, also known as the *classical Weisfeiler-Leman algorithm*, has been introduced by Weisfeiler and Leman [86], and the k-dimensional generalization goes back to Babai and Mathon (see [19]).

To introduce k-WL, we need the notion of the *atomic type* $\text{atp}(G, \bar{v})$ of a tuple $\bar{v} = (v_1, \ldots, v_\ell)$ of vertices in a graph G. It describes the isomorphism type of the labeled induced subgraph $G[\{v_1, \ldots, v_\ell\}]$, that is, for graphs G, H and tuples $\bar{v} = (v_1, \ldots, v_\ell) \in V(G)^\ell, \bar{w} = (w_1, \ldots, w_\ell) \in V(H)^\ell$ we have $\text{atp}(G, \bar{v}) = \text{atp}(H, \bar{w})$ if and only if the mapping $v_i \mapsto w_i$ is an isomorphism from $G[v_1, \ldots, v_k]$ to $H[w_1, \ldots, w_k]$. Formally, we can view $\text{atp}(G, v_1, \ldots, v_k)$ as a pair of Boolean $(k \times k)$-matrices, one describing equalities between the v_i, and one describing adjacencies. If G is a colored graph, we need one matrix for each arc color and one diagonal matrix for each vertex color.

Then k-WL computes a sequence of colorings $\chi_{(i)}^k(G)$ of $V(G)^k$ as follows. Initially, each tuple is colored by its atomic type, that is, $\chi_{(0)}^k(G)(\bar{v}) := \text{atp}(G, \bar{v})$. For $i > 0$, we define

$$\chi_{(i)}^k(G)(\bar{v}) := \left(\chi_{(i-1)}^k(G)(\bar{v}), \mathcal{M}_i^k(\bar{v})\right),$$

where

$$\mathcal{M}_i^k(\bar{v}) := \left\{\!\!\left\{ \left(\text{atp}(G, \bar{v}v), \chi_{(i-1)}^k(G)(\bar{v}[v/1]), \ldots, \chi_{(i-1)}^k(G)(\bar{v}[v/k])\right) \,\Big|\, v \in V(G) \right\}\!\!\right\}.$$

Here, for $\bar{v} = (v_1, \ldots, v_k) \in V(G)$ and $v \in V(G)$, by $\bar{v}v$ we denote the $(k+1)$-tuple $(v_1, \ldots, v_k, v)$ and, for $i \in [k]$, by $\bar{v}[v/i]$ the k-tuple $(v_1, \ldots, v_{i-1}, v, v_{i+1}, \ldots, v_k)$. Then there is an $i_\infty < n^k$ such that $\chi_{(i_\infty)}^k(G) \equiv \chi_{(i_\infty+1)}^k(G)$, and we call $\chi_{(\infty)}^k(G) := \chi_{(i_\infty)}^k(G)$ the *k-stable coloring*. The k-stable coloring of an n-vertex graph can be computed in time $\mathcal{O}(n^k \log n)$ [49].

It is easy to see that 1-WL is essentially just the Color Refinement algorithm, that is, for all graphs G, H, vertices $v \in V(G), w \in V(H)$, and $i \geq 0$ it holds that

$$\chi_{(i)}(G)(v) = \chi_{(i)}(H)(w) \iff \chi_{(i)}^1(G)(v) = \chi_{(i)}^1(H)(w). \tag{3.1}$$

In the following, we drop the distinction between the Color Refinement algorithm and 1-WL.

Example 3.2 The sequence of colorings of 2-WL on a cycle C_7 of length 7 is shown in Figure 2. In this example, the coloring is symmetric, that is, $\chi_{(i)}^2(C_7)(v, v') = \chi_{(i)}^2(C_7)(v', v)$ for all $i \geq 0$ and $v, v' \in V(C_7)$. This allows us to show the colors as undirected edges.

---

[1] To be precise, the algorithms compute the partition of the vertex set corresponding to the stable coloring, not the actual colors viewed as multisets.

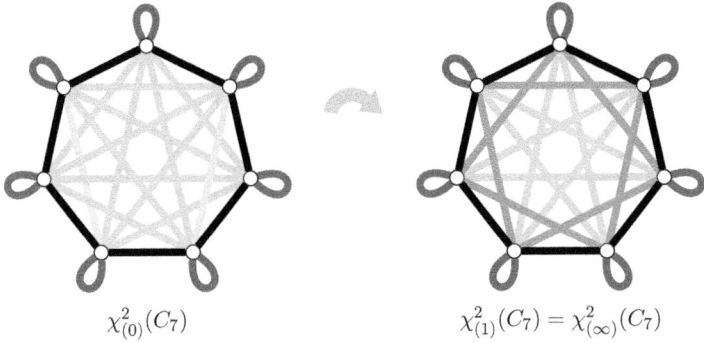

$\chi^2_{(0)}(C_7)$ $\chi^2_{(1)}(C_7) = \chi^2_{(\infty)}(C_7)$

Figure 2: The iterations of 2-WL on a cycle C_7

3.3 The Power of WL and the WL Dimension

To use the Weisfeiler-Leman algorithm as an isomorphism test, we can compare the color patterns of two given graphs. We say that k-WL *distinguishes* graphs G and H if there is a color $c \in \mathrm{rg}(\chi^k_{(\infty)}(G)) \cup \mathrm{rg}(\chi^k_{(\infty)}(H))$ such that

$$\left|\{\bar{v} \in V(G)^k \mid \chi^k_{(\infty)}(G)(\bar{v}) = c\}\right| \neq \left|\{\bar{w} \in V(H)^k \mid \chi^k_{(\infty)}(H)(\bar{w}) = c\}\right|.$$

Note that k-WL is an *incomplete* isomorphism test: if it distinguishes two graphs, we know that they are non-isomorphic, but as we will see, there are non-isomorphic graphs that k-WL does not distinguish. Let us call two graphs that are not distinguished by the k-WL k-*indistinguishable*.

Example 3.3 Any two d-regular graphs are 1-indistinguishable. The simplest example are a cycle of length 6 and the disjoint union of two triangles.

Similarly, any two strongly regular graphs with the same parameters are 2-indistinguishable. Figure 3 shows two non-isomorphic strongly regular graphs with parameters $(16, 6, 2, 2)$, which means that they have 16 vertices, each vertex has 6 neighbors, each pair of adjacent vertices has two common neighbors, and each pair of non-adjacent vertices has 2 common neighbors. The first of the graphs is the line graph of the complete bipartite graph $K_{4,4}$. The second graph is the Shrikhande graph, which can be viewed as the undirected graph underlying the Cayley graph for the group $\mathbb{Z}_4 \times \mathbb{Z}_4$ with generators $(0, 1), (1, 0), (1, 1)$. To see that they are non-isomorphic, note that the neighborhood of each vertex in the Shrikhande graph is a cycle of length 6, and the neighborhood of the each vertex in the line graph of $K_{4,4}$ is the disjoint union of two triangles.

It is not easy to find two non-isomorphic graphs that are k-indistinguishable for some $k \geq 3$. In a seminal paper, Cai, Fürer, and Immerman constructed such graphs for each k.

Theorem 3.4 (Cai, Fürer and Immerman [19]) *For every $k \geq 1$ there exist non-isomorphic 3-regular graphs G_k, H_k of order $|G_k| = |H_k| = \mathcal{O}(k)$ that are k-indistinguishable.*

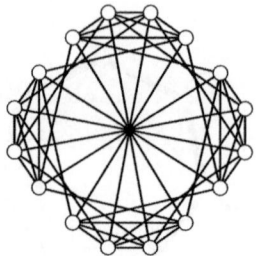 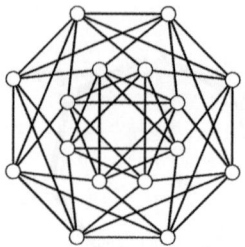

Figure 3: Two non-isomorphic strongly regular graphs with parameters $(16, 6, 2, 2)$: the line graph of $K_{4,4}$ (left) and the Shrikhande Graph (right).

The Weisfeiler-Leman algorithm colors tuples of vertices based on the isomorphism type of these tuples and the overlap with other tuples. To distinguish the so called *CFI graphs* G_k, H_k of Theorem 3.4, we need to color tuples of length linear in the order of the graphs. As the number of such tuples is $\Omega(k^k)$, this is prohibitive both in terms of running time and memory space. However, a closer analysis reveals that it suffices to color only some long tuples, to be precise, an isomorphism-invariant set of tuples of size polynomial in k (this follows from an argument due to Dawar, Richerby and Rossman [24]). In [46], the first author jointly with Schweitzer and Wiebking proposed an algorithmic framework called DEEPWL for combinatorial "coloring" algorithms that allows it to selectively color tuples within isomorphism-invariant families and thereby color tuples of length linear in the input graphs. DEEPWL can distinguish the CFI graphs in polynomial time.

While Theorem 3.4 clearly shows its limitations, the Weisfeiler-Leman algorithm still remains quite powerful. In particular, it does yield complete isomorphism tests for many natural graph classes. Let us say that k-WL *identifies* a graph G if it distinguishes G from all graphs H that are not isomorphic to G. It is long known [9] that even 1-WL (that is, the Color Refinement algorithm) identifies almost all graphs, in the sense that in the limit, the fraction of n-vertex graphs identified by 1-WL is 1 as n goes to infinity. Similarly, 2-WL identifies almost all d-regular graphs [16].

The *WL dimension* of a graph G is the least $k \geq 1$ such that k-WL identifies G. Trivially, the WL dimension of an n-vertex graph is at most n (in fact, $n-1$ for $n \geq 2$). The *WL dimension* of a class $\mathcal{C}$ of graphs is the maximum of the WL dimensions of the graphs in $\mathcal{C}$ if this maximum exists, or ∞ otherwise. It is easy to see that the WL dimension of the class of forests is 1. Building on earlier results for graphs of bounded genus and bounded tree width [33, 34, 38], the first author proved the following far-reaching result.

Theorem 3.5 (Grohe [35]) *All graph classes excluding a fixed graph as a minor have bounded WL dimension.*

Explicit bounds on the WL dimension are known for specific graph classes. Most notably, Kiefer, Ponomarenko, and Schweitzer [54] proved that the WL dimension of planar graphs is at most 3. The example of two triangles vs a cycle of length

6 shows that it is at least 2. More generally, the WL dimension of the class of all graphs embeddable in a surface of Euler genus g is at most $4g + 3$ [37]; a lower bound linear in g follows from Theorem 3.4. The second author jointly with Kiefer [53] proved that the WL dimension of the class of graphs of tree width k is at least $\lceil k/2 \rceil - 3$ and at most k.

There are also natural graph classes that do not exclude a fixed graph as a minor (and that are therefore not covered by Theorem 3.5). The WL dimension of the class of interval graphs is 2 [30] and the WL dimension of the class of graphs of rank width at most k is at most $3k+4$ [40]. Contrasting all these results, Theorem 3.4 shows that the WL dimension of the class of 3-regular graphs is ∞, even though isomorphism of 3-regular graphs can be decided in polynomial time [61] (cf. Theorem 5.4). Much more on this topic can found in Kiefer's recent survey [52].

3.4 Characterisations of WL

The relevance of the Weisfeiler-Leman algorithm is not restricted to GI. It has found applications in linear optimisation [36] and probabilistic inferences [1], and because of relations to graph kernels [83] and graph neural networks [69, 88], received attention in machine learning recently. On a theoretical level, connections to other areas are induced by several seemingly unrelated characterizations of k-indistinguishability in terms of logic, combinatorics, and algebra.

We start with a logical characterization going back to [19, 49]. Let C be the extension of first-order predicate logic using counting quantifiers of the form $\exists^{\geq \ell} x$ for $\ell \geq 1$, where $\exists^{\geq \ell} x \, \varphi(x)$ means that there are at least ℓ elements x satisfying φ. By C^k we denote the fragment of C consisting of formulas with at most k variables. For example, the following C^2-sentence φ says that all vertices of a graph have fewer than 3 neighbors of degree at least 10:

$$\varphi := \forall x \neg \exists^{\geq 3} y \big(E(x,y) \wedge \exists^{\geq 10} x \, E(y,x) \big).$$

Note that we reuse variable x for the innermost quantification to keep the number of distinct variables down to two.

Theorem 3.6 (Cai, Fürer and Immerman [19]) *Let $k \geq 1$. Then two graphs are k-indistinguishable if and only if they are C^{k+1}-equivalent, that is, they satisfy the same sentences of the logic C^{k+1}.*

This logical characterization of k-indistinguishability plays a crucial role in the proof of Theorem 3.5 and, via a game characterization of C^k-equivalence, in the proof of Theorem 3.4.

The next characterization we will discuss is combinatorial. Recall that a graph homomorphism is a mapping that preserves adjacency, and let $\hom(G, H)$ denote the number of homomorphism from G to H. An old theorem due to Lovász [60] states that two graphs G, H are isomorphic if and only if $\hom(F, G) = \hom(F, H)$ for all graphs F. If, instead of all graphs F, we only count homomorphisms from graphs of tree width k, we obtain a characterization of k-indistinguishability.

Theorem 3.7 (Dvořák [28]) *Let $k \geq 1$. Then two graphs G, H are k-indistinguishable if and only if $\hom(F, G) = \hom(F, H)$ for all graphs F of tree width at most k.*

Maybe the most surprising characterization of k-indistinguishability is algebraic. For simplicity, we focus on $k = 1$ here. Let G, H be graphs with vertex sets $V := V(G), W := V(H)$, and let $A \in \mathbb{R}^{V \times V}, B \in \mathbb{R}^{W \times W}$ be their adjacency matrices. Then G and H are isomorphic if and only if there is a permutation matrix X such that $X^{-1}AX = B$, or equivalently, $AX = XB$. We can write this as a system of linear (in)equalities:

$$\sum_{v' \in V} A_{vv'} X_{v'w} = \sum_{w' \in W} X_{vw'} B_{w'w}, \qquad (3.2)$$

$$\sum_{w' \in W} X_{vw'} = \sum_{v' \in V} X_{v'w} = 1, \qquad (3.3)$$

$$X_{vw} \geq 0, \qquad \text{for all } v \in V, w \in W. \qquad (3.4)$$

Integer solutions to this system are precisely permutation matrices $X = (X_{vw})$ satisfying $AX = XB$. Thus the system (3.2)–(3.4) has an integer solution if and only if G and H are isomorphic. Now let us drop the integrality constraints and consider rational solutions to the system, which can be viewed as doubly stochastic matrices X satisfying $AX = XB$. Let us call such solutions *fractional isomorphisms*.

Theorem 3.8 (Tinhofer [84]) *Two graphs G, H are 1-distinguishable if and only if they are fractionally isomorphic, that is, the system (3.2)–(3.4) has a rational solution.*

Atserias and Maneva [2] showed that by considering the Sherali-Adams hierarchy over the integer linear program (3.2)–(3.4), we obtain a characterization of k-indistinguishability. To be precise, solvability of the level-k Sherali-Adams (in)equalities yields an equivalence relation strictly between k-indistinguishability and $(k+1)$-indistinguishability. By combining the equations of levels $(k-1)$ and k, we obtain an exact characterization of k-indistinguishability [44, 62]. A similar, albeit looser correspondence to k-indistinguishability has been obtained for the Lassere hierarchy of semi-definite relaxations of the integer linear program (3.2)–(3.4) [3, 73] and in terms of polynomial equations and the so-called polynomial calculus [3, 14, 32].

4 The Group-Theoretic Graph Isomorphism Machinery

In this section we present recent advances on the group-theoretic graph isomorphism machinery. After giving some basics on permutation groups we first discuss Luks's algorithm for testing isomorphism of graphs of bounded degree [61]. This algorithm forms the foundation of the group-theoretic graph isomorphism machinery. Then, we give a very brief overview on Babai's quasi-polynomial-time isomorphism test [6] which also builds on Luks's algorithm and attacks the obstacle cases where the recursion performed by Luks's algorithm does not give the desired running time. Finally, we present recent extensions of Babai's techniques resulting in more efficient isomorphism tests for bounded-degree graphs [41] and hypergraphs [71].

4.1 Basics

We introduce the group-theoretic notions required in this work. For a general background on group theory we refer to [80]; background on permutation groups can be found in [26].

Permutation Groups. A *permutation group* acting on a set Ω is a subgroup $\Gamma \leq \mathrm{Sym}(\Omega)$ of the symmetric group. The size of the permutation domain Ω is called the *degree* of Γ. Throughout this section, the degree is denoted by $n := |\Omega|$. If $\Omega = [n]$, then we also write S_n instead of $\mathrm{Sym}(\Omega)$. The *alternating group* on a set Ω is denoted by $\mathrm{Alt}(\Omega)$. As before, we write A_n instead of $\mathrm{Alt}(\Omega)$ if $\Omega = [n]$. For $\gamma \in \Gamma$ and $\alpha \in \Omega$ we denote by α^γ the image of α under the permutation γ. The set $\alpha^\Gamma := \{\alpha^\gamma \mid \gamma \in \Gamma\}$ is the *orbit* of α. The group Γ is *transitive* if $\alpha^\Gamma = \Omega$ for some (and therefore every) $\alpha \in \Omega$.

For $\alpha \in \Omega$ the group $\Gamma_\alpha := \{\gamma \in \Gamma \mid \alpha^\gamma = \alpha\} \leq \Gamma$ is the *stabilizer* of α in Γ. The group Γ is *semi-regular* if $\Gamma_\alpha = \{\mathrm{id}\}$ for every $\alpha \in \Omega$ (where id denotes the identity element of the group). The *pointwise stabilizer* of a set $A \subseteq \Omega$ is the subgroup $\Gamma_{(A)} := \{\gamma \in \Gamma \mid \forall \alpha \in A \colon \alpha^\gamma = \alpha\}$. For $A \subseteq \Omega$ and $\gamma \in \Gamma$ let $A^\gamma := \{\alpha^\gamma \mid \alpha \in A\}$. The set A is Γ-*invariant* if $A^\gamma = A$ for all $\gamma \in \Gamma$. The *setwise stabilizer* of a set $A \subseteq \Omega$ is the subgroup $\Gamma_A := \{\gamma \in \Gamma \mid A^\gamma = A\}$.

For $A \subseteq \Omega$ and a bijection $\theta \colon \Omega \to \Omega'$ we denote by $\theta[A]$ the restriction of θ to the domain A. For a Γ-invariant set $A \subseteq \Omega$, we denote by $\Gamma[A] := \{\gamma[A] \mid \gamma \in \Gamma\}$ the induced action of Γ on A, i.e., the group obtained from Γ by restricting all permutations to A.

Let $\Gamma \leq \mathrm{Sym}(\Omega)$ be a transitive group. A *block* of Γ is a nonempty subset $B \subseteq \Omega$ such that $B^\gamma = B$ or $B^\gamma \cap B = \emptyset$ for all $\gamma \in \Gamma$. The trivial blocks are Ω and the singletons $\{\alpha\}$ for $\alpha \in \Omega$. The group Γ is called *primitive* if there are no non-trivial blocks. If $B \subseteq \Omega$ is a block of Γ then $\mathfrak{B} := \{B^\gamma \mid \gamma \in \Gamma\}$ builds a *block system* of Γ. Note that $\mathfrak{B}$ is an equipartition of Ω. The group $\Gamma_{(\mathfrak{B})} := \{\gamma \in \Gamma \mid \forall B \in \mathfrak{B} \colon B^\gamma = B\}$ denotes the subgroup stabilizing each block $B \in \mathfrak{B}$ setwise. Moreover, the natural action of Γ on the block system $\mathfrak{B}$ is denoted by $\Gamma[\mathfrak{B}] \leq \mathrm{Sym}(\mathfrak{B})$. Let $\mathfrak{B}, \mathfrak{B}'$ be two partitions of Ω. We say that $\mathfrak{B}$ *refines* $\mathfrak{B}'$, denoted $\mathfrak{B} \preceq \mathfrak{B}'$, if for every $B \in \mathfrak{B}$ there is some $B' \in \mathfrak{B}'$ such that $B \subseteq B'$. If additionally $\mathfrak{B} \neq \mathfrak{B}'$ we write $\mathfrak{B} \prec \mathfrak{B}'$. A block system $\mathfrak{B}$ is *minimal* if there is no non-trivial block system $\mathfrak{B}'$ such that $\mathfrak{B} \prec \mathfrak{B}'$. A block system $\mathfrak{B}$ is minimal if and only if $\Gamma[\mathfrak{B}]$ is primitive.

Let $\Gamma \leq \mathrm{Sym}(\Omega)$ and $\Gamma' \leq \mathrm{Sym}(\Omega')$. A *homomorphism* is a mapping $\varphi \colon \Gamma \to \Gamma'$ such that $\varphi(\gamma)\varphi(\delta) = \varphi(\gamma\delta)$ for all $\gamma, \delta \in \Gamma$. For $\gamma \in \Gamma$ we denote by γ^φ the φ-image of γ. Similarly, for $\Delta \leq \Gamma$ we denote by Δ^φ the φ-image of Δ (note that Δ^φ is a subgroup of Γ').

Algorithms for Permutation Groups. Next, we review some basic facts about algorithms for permutation groups. For detailed information we refer to [82].

In order to handle permutation groups computationally it is essential to represent groups in a compact way. Indeed, the size of a permutation group is typically exponential in the degree of the group which means it is not possible to store the whole group in memory. In order to allow for efficient computation, permutation groups are represented by generating sets. By Lagrange's Theorem, for each permutation group $\Gamma \leq \mathrm{Sym}(\Omega)$, there is a generating set of size $\log |\Gamma| \leq n \log n$ (recall that $n := |\Omega|$ denotes the size of the permutation domain). Actually, most algorithms are based on so-called *strong generating sets*, which can be chosen of size quadratic in the degree of the group and can be computed in polynomial time given an arbitrary generating set (see, e.g., [82]).

Theorem 4.1 (cf. [82]) *Let $\Gamma \leq \mathrm{Sym}(\Omega)$ and let S be a generating set for Γ. Then the following tasks can be performed in time polynomial in n and $|S|$:*

1. *compute the order of Γ,*
2. *given $\gamma \in \mathrm{Sym}(\Omega)$, test whether $\gamma \in \Gamma$,*
3. *compute the orbits of Γ,*
4. *compute a minimal block system of Γ (if Γ is transitive), and*
5. *given $A \subseteq \Omega$, compute a generating set for $\Gamma_{(A)}$.*

Let $\Delta \leq \mathrm{Sym}(\Omega')$ be a second group of degree $n' = |\Omega'|$. Also let $\varphi \colon \Gamma \to \Delta$ be a homomorphism given by a list of images for $s \in S$. Then the following tasks can be solved in time polynomial in n, n' and $|S|$:

6. *compute a generating set for $\ker(\varphi) := \{\gamma \in \Gamma \mid \gamma^\varphi = \mathrm{id}\}$, and*
7. *given $\delta \in \Delta$, find $\gamma \in \Gamma$ such that $\gamma^\varphi = \delta$ (if it exists).*

Observe that, by always maintaining generating sets of size at most quadratic in the degree, we can concatenate a polynomial number of subroutines from the theorem while keeping the polynomial-time bound. For the remainder of this work, we will typically ignore the role of generating sets and simply refer to groups being the input/output of an algorithm. This always means the algorithm performs computations on generating sets of size polynomial in the degree of the group.

4.2 Luks's Algorithm

The String Isomorphism Problem. In order to apply group-theoretic techniques to the Graph Isomorphism Problem, Luks [61] introduced a more general problem that allows to build recursive algorithms along the structure of the permutation groups involved. Here, we follow the notation and terminology used by Babai [6] for describing his quasi-polynomial-time algorithm for the Graph Isomorphism Problem that also employs this recursive strategy.

A *string* is a mapping $\mathfrak{x} \colon \Omega \to \Sigma$ where Ω is a finite set and Σ is also a finite set called the *alphabet*. Let $\gamma \in \mathrm{Sym}(\Omega)$ be a permutation. The permutation γ can be applied to the string $\mathfrak{x}$ by defining

$$\mathfrak{x}^\gamma \colon \Omega \to \Sigma \colon \alpha \mapsto \mathfrak{x}\left(\alpha^{\gamma^{-1}}\right). \tag{4.1}$$

Let $\mathfrak{y} \colon \Omega \to \Sigma$ be a second string. The permutation γ is an isomorphism from $\mathfrak{x}$ to $\mathfrak{y}$, denoted $\gamma \colon \mathfrak{x} \cong \mathfrak{y}$, if $\mathfrak{x}^\gamma = \mathfrak{y}$. Let $\Gamma \leq \mathrm{Sym}(\Omega)$. A Γ-*isomorphism* from $\mathfrak{x}$ to $\mathfrak{y}$ is a permutation $\gamma \in \Gamma$ such that $\gamma \colon \mathfrak{x} \cong \mathfrak{y}$. The strings $\mathfrak{x}$ and $\mathfrak{y}$ are Γ-*isomorphic*, denoted

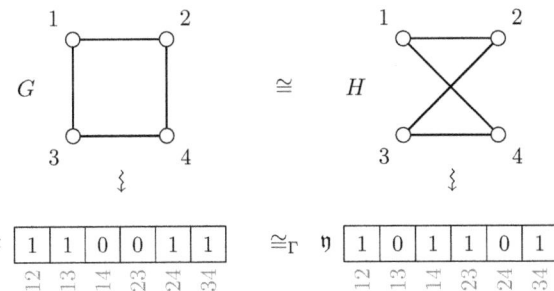

Figure 4: Reduction from the Graph Isomorphism Problem to the String Isomorphism Problem. The group Γ is the natural induced action of S_4 on $\binom{[4]}{2}$.

$\mathfrak{x} \cong_\Gamma \mathfrak{y}$, if there is a Γ-isomorphism from $\mathfrak{x}$ to $\mathfrak{y}$. The *String Isomorphism Problem* asks, given two strings $\mathfrak{x}, \mathfrak{y} \colon \Omega \to \Sigma$ and a generating set for a group $\Gamma \leq \mathrm{Sym}(\Omega)$, whether $\mathfrak{x}$ and $\mathfrak{y}$ are Γ-isomorphic. The set of Γ-isomorphisms from $\mathfrak{x}$ to $\mathfrak{y}$ is denoted by

$$\mathrm{Iso}_\Gamma(\mathfrak{x}, \mathfrak{y}) := \{\gamma \in \Gamma \mid \mathfrak{x}^\gamma = \mathfrak{y}\}. \tag{4.2}$$

The set of Γ-*automorphisms* of $\mathfrak{x}$ is $\mathrm{Aut}_\Gamma(\mathfrak{x}) := \mathrm{Iso}_\Gamma(\mathfrak{x}, \mathfrak{x})$. Observe that $\mathrm{Aut}_\Gamma(\mathfrak{x})$ is a subgroup of Γ and $\mathrm{Iso}_\Gamma(\mathfrak{x}, \mathfrak{y}) = \mathrm{Aut}_\Gamma(\mathfrak{x})\gamma := \{\delta\gamma \mid \delta \in \mathrm{Aut}_\Gamma(\mathfrak{x})\}$ for an arbitrary $\gamma \in \mathrm{Iso}_\Gamma(\mathfrak{x}, \mathfrak{y})$.

Theorem 4.2 *There is a polynomial-time many-one reduction from the Graph Isomorphism Problem to the String Isomorphism Problem.*

Proof Let G and H be two graphs and assume without loss of generality $V(G) = V(H) =: V$. Let $\Omega := \binom{V}{2}$ and $\Gamma \leq \mathrm{Sym}(\Omega)$ be the natural action of $\mathrm{Sym}(V)$ on the two-element subsets of V. Let $\mathfrak{x} \colon \Omega \to \{0,1\}$ with $\mathfrak{x}(vw) = 1$ if and only if $vw \in E(G)$. Similarly define $\mathfrak{y} \colon \Omega \to \{0,1\}$ with $\mathfrak{y}(vw) = 1$ if and only if $vw \in E(H)$ (see Figure 4). Let $\varphi \colon \mathrm{Sym}(V) \to \Gamma$ be the natural homomorphism. Then $\gamma \colon G \cong H$ if and only if $\gamma^\varphi \colon \mathfrak{x} \cong \mathfrak{y}$. □

The main advantage of the String Isomorphism Problem is that it naturally allows for algorithmic approaches based on group-theoretic techniques.

Recursion Mechanisms. The foundation for the group-theoretic approaches lays in two recursion mechanisms first exploited by Luks [61]. As before, let $\mathfrak{x}, \mathfrak{y} \colon \Omega \to \Sigma$ be two strings. For a set of permutations $K \subseteq \mathrm{Sym}(\Omega)$ and a *window* $W \subseteq \Omega$ define

$$\mathrm{Iso}_K^W(\mathfrak{x}, \mathfrak{y}) := \{\gamma \in K \mid \forall \alpha \in W \colon \mathfrak{x}(\alpha) = \mathfrak{y}(\alpha^\gamma)\}. \tag{4.3}$$

In the following, the set K is always a *coset*, i.e., $K = \Gamma\gamma := \{\gamma'\gamma \mid \gamma' \in \Gamma\}$ for some group $\Gamma \leq \mathrm{Sym}(\Omega)$ and a permutation $\gamma \in \mathrm{Sym}(\Omega)$, and the set W is Γ-invariant. In this case it can be shown that $\mathrm{Iso}_K^W(\mathfrak{x}, \mathfrak{y})$ is either empty or a coset of the group $\mathrm{Aut}_\Gamma^W(\mathfrak{x}) := \mathrm{Iso}_\Gamma^W(\mathfrak{x}, \mathfrak{x})$. Hence, the set $\mathrm{Iso}_K^W(\mathfrak{x}, \mathfrak{y})$ can be represented by a generating

set for $\mathrm{Aut}_\Gamma^W(\mathfrak{x})$ and a single permutation $\gamma \in \mathrm{Iso}_K^W(\mathfrak{x},\mathfrak{y})$. Moreover, for $K = \Gamma\gamma$, it holds that
$$\mathrm{Iso}_{\Gamma\gamma}^W(\mathfrak{x},\mathfrak{y}) = \mathrm{Iso}_\Gamma^W(\mathfrak{x},\mathfrak{y}^{\gamma^{-1}})\gamma. \tag{4.4}$$
Using this identity, it is possible to restrict to the case where K is actually a group.

With these definitions we can now formulate the two recursion mechanisms introduced by Luks [61]. For the first type of recursion suppose $K = \Gamma \leq \mathrm{Sym}(\Omega)$ is not transitive on W and let $W_1, \ldots, W_\ell$ be the orbits of $\Gamma[W]$. Then the strings can be processed orbit by orbit as follows. We iterate over all $i \in [\ell]$ and update $K := \mathrm{Iso}_K^{W_i}(\mathfrak{x},\mathfrak{y})$ in each iteration. In the end, $K = \mathrm{Iso}_\Gamma^W(\mathfrak{x},\mathfrak{y})$. This recursion mechanism is referred to as *orbit-by-orbit processing*.

The set $\mathrm{Iso}_K^{W_i}(\mathfrak{x},\mathfrak{y})$ can be computed by making one recursive call to the String Isomorphism Problem over domain size $n_i := |W_i|$. Indeed, using Equation (4.4), it can be assumed that K is a group and W_i is K-invariant. Then
$$\mathrm{Iso}_K^{W_i}(\mathfrak{x},\mathfrak{y}) = \{\gamma \in K \mid \gamma[W_i] \in \mathrm{Iso}_{K[W_i]}(\mathfrak{x}[W_i],\mathfrak{y}[W_i])\} \tag{4.5}$$
where $\mathfrak{x}[W_i]$ denotes the induced substring of $\mathfrak{x}$ on the set W_i, i.e., $\mathfrak{x}[W_i]\colon W_i \to \Sigma\colon \alpha \mapsto \mathfrak{x}(\alpha)$ (the string $\mathfrak{y}[W_i]$ is defined analogously). So overall, if Γ is not transitive, the set $\mathrm{Iso}_\Gamma^W(\mathfrak{x},\mathfrak{y})$ can be computed by making ℓ recursive calls over window size $n_i = |W_i|$, $i \in [\ell]$.

For the second recursion mechanism let $\Delta \leq \Gamma$ and let T be a *transversal*[2] for Δ in Γ. Then
$$\mathrm{Iso}_\Gamma^W(\mathfrak{x},\mathfrak{y}) = \bigcup_{\delta \in T} \mathrm{Iso}_{\Delta\delta}^W(\mathfrak{x},\mathfrak{y}) = \bigcup_{\delta \in T} \mathrm{Iso}_\Delta^W(\mathfrak{x},\mathfrak{y}^{\delta^{-1}})\delta. \tag{4.6}$$

Luks applied this type of recursion when Γ is transitive (on the window W), $\mathfrak{B}$ is a minimal block system for Γ, and $\Delta = \Gamma_{(\mathfrak{B})}$. In this case $\Gamma[\mathfrak{B}]$ is a primitive group. Let $t = |\Gamma[\mathfrak{B}]|$ be the size of a transversal for Δ in Γ. Note that Δ is not transitive (on the window W). Indeed, each orbit of Δ has size at most n/b where $b = |\mathfrak{B}|$. So by combining both types of recursion the computation of $\mathrm{Iso}_\Gamma^W(\mathfrak{x},\mathfrak{y})$ is reduced to $t \cdot b$ many instances of the String Isomorphism Problem over window size $|W|/b$. This specific combination of types of recursion is referred to as *standard Luks reduction*. Observe that the time complexity of standard Luks reduction is determined by the size of the primitive group $\Gamma[\mathfrak{B}]$.

Overall, Luks's algorithm is formulated in Algorithm 1. The running time of this algorithm heavily depends on the size of the primitive groups involved in the computation.

The Role of Primitive Groups. In order to analyse the running time of Luks's algorithm (on a specific input group Γ) the crucial step is to understand which primitive groups appear along the execution of the algorithm and to bound their size as a function of their degree. Let us first analyse the situation for GI for graphs of maximum degree d. The crucial observation is that there is a polynomial-time Turing reduction from this problem to the String Isomorphism Problem where the input group Γ is contained the class $\widehat{\Gamma}_d$ [61, 11].

---
[2] A set $T \subseteq \Gamma$ is a *transversal* for Δ in Γ if $|T| = |\Gamma|/|\Delta|$ and $\{\Delta\delta \mid \delta \in T\} = \{\Delta\delta \mid \delta \in \Gamma\}$ (such a set always exists).

Algorithm 1: Luks's Algorithm: $\mathsf{StringIso}(\mathfrak{x}, \mathfrak{y}, \Gamma, \gamma, W)$

Input : Strings $\mathfrak{x}, \mathfrak{y} \colon \Omega \to \Sigma$, a group $\Gamma \leq \mathrm{Sym}(\Omega)$, a permutation $\gamma \in \mathrm{Sym}(\Omega)$, and a Γ-invariant set $W \subseteq \Omega$

Output: $\mathrm{Iso}^W_{\Gamma\gamma}(\mathfrak{x}, \mathfrak{y})$

1 **if** $\gamma \neq 1$ **then**
2   **return** $\mathsf{StringIso}(\mathfrak{x}, \mathfrak{y}^{\gamma^{-1}}, \Gamma, 1, W)\gamma$ /* Equation (4.4) */
3 **end**
4 **if** $|W| = 1$ **then**
5   **if** $\mathfrak{x}(\alpha) = \mathfrak{y}(\alpha)$ *where* $W = \{\alpha\}$ **then**
6   **return** Γ
7   **else**
8   **return** $\emptyset$
9   **end**
10 **end**
11 **if** Γ *is not transitive on* W **then**
12   compute orbit $W' \subseteq W$
13   **return** $\mathsf{StringIso}(\mathfrak{x}, \mathfrak{y}, \mathsf{StringIso}(\mathfrak{x}, \mathfrak{y}, \Gamma, 1, W'), W \setminus W')$
14 **end**
15 compute minimal block system $\mathfrak{B}$ of the action of Γ on W
16 compute $\Delta := \Gamma_{(\mathfrak{B})}$
17 compute transversal T of Δ in Γ
18 **return** $\bigcup_{\delta \in T} \mathsf{StringIso}(\mathfrak{x}, \mathfrak{y}, \Delta, \delta, W)$

Let Γ be a group. A *subnormal series* is a sequence of subgroups $\Gamma = \Gamma_0 \trianglerighteq \Gamma_1 \trianglerighteq \cdots \trianglerighteq \Gamma_k = \{\text{id}\}$. The length of the series is k and the groups Γ_{i-1}/Γ_i are the factor groups of the series, $i \in [k]$. A *composition series* is a strictly decreasing subnormal series of maximal length. For every finite group Γ all composition series have the same family (considered as a multiset) of factor groups (cf. [80]). A *composition factor* of a finite group Γ is a factor group of a composition series of Γ.

Definition 4.3 For $d \geq 2$ let $\widehat{\Gamma}_d$ denote the class of all groups Γ for which every composition factor of Γ is isomorphic to a subgroup of S_d.

We want to stress the fact that there are two similar classes of groups that have been used in the literature both typically denoted by Γ_d. One of these is the class introduced by Luks [61] that we denote by $\widehat{\Gamma}_d$, while the other one used in [7] in particular allows composition factors that are simple groups of Lie type of bounded dimension.

Lemma 4.4 (Luks [61]) *Let* $\Gamma \in \widehat{\Gamma}_d$. *Then*

1. $\Delta \in \widehat{\Gamma}_d$ *for every subgroup* $\Delta \leq \Gamma$, *and*

2. $\Gamma^\varphi \in \widehat{\Gamma}_d$ *for every homomorphism* $\varphi \colon \Gamma \to \Delta$.

Now, the crucial advantage of $\widehat{\Gamma}_d$-groups with respect to Luks's algorithm is that the size of primitive $\widehat{\Gamma}_d$-groups is bounded polynomially in the degree for every fixed d.

Theorem 4.5 *There is a function* $f \colon \mathbb{N} \to \mathbb{N}$ *such that for every primitive permutation group* $\Gamma \in \widehat{\Gamma}_d$ *it holds that* $|\Gamma| \leq n^{f(d)}$. *Moreover, the function* f *can be chosen such that* $f(d) = \mathcal{O}(d)$.

The first part of this theorem on the existence of the function f was first proved by Babai, Cameron and Pálfy [7] implying that Luks's algorithm runs in polynomial time for every fixed number $d \in \mathbb{N}$. Later, it was observed that the function f can actually be chosen to be linear in d (see, e.g., [56]).

Theorem 4.5 allows the analysis of Luks's algorithm when the input group comes from the class $\widehat{\Gamma}_d$. Note that the class $\widehat{\Gamma}_d$ is closed under subgroups and homomorphic images by Lemma 4.4 which implies that all groups encountered during Luks's algorithm are from the class $\widehat{\Gamma}_d$ in case the input group is a $\widehat{\Gamma}_d$-group.

Corollary 4.6 *Luks's Algorithm (Algorithm 1) for groups* $\Gamma \in \widehat{\Gamma}_d$ *runs in time* $n^{\mathcal{O}(d)}$.

Proof Sketch In order to analyze the running time of Algorithm 1 let $f(n)$ denote the maximal number of leaves in a recursion tree for a $\widehat{\Gamma}_d$-group where $n := |W|$ denotes the window size. For $n = 1$ it is easy to see that $f(n) = 1$. Suppose Γ is not transitive on W and let $n_1 := |W'|$ be the size of an orbit $W' \subseteq W$. Then

$$f(n) \leq f(n_1) + f(n - n_1). \tag{4.7}$$

Finally, if Γ is transitive on W, the algorithm computes a minimal block system $\mathfrak{B}$ of $\Gamma[W]$ and performs standard Luks reduction. In this case the algorithm performs $t \cdot b$ recursive calls over window size n/b where $t = |\Gamma[\mathfrak{B}]|$. Hence,

$$f(n) \leq t \cdot b \cdot f(n/b). \tag{4.8}$$

Since $|\Gamma[\mathfrak{B}]|$ is a primitive $\widehat{\Gamma}_d$-group it holds that $t = b^{\mathcal{O}(d)}$ by Theorem 4.5. Combining the bound on t with Equation (4.7) and (4.8) gives $f(n) = n^{\mathcal{O}(d)}$. Also, each node of the recursion tree only requires computation time polynomial in n and the number of children in the recursion tree (cf. Theorem 4.1). Overall, this gives the desired bound on the running time. □

In combination with the polynomial-time Turing reduction from the Graph Isomorphism Problem for graphs of maximum degree d to the String Isomorphism Problem for $\widehat{\Gamma}_d$-groups [61, 11] the former problem can be solved in the same running time $n^{\mathcal{O}(d)}$.

Unfortunately, for the general String Isomorphism Problem, the situation is more complicated. As a very simple example, symmetric and alternating groups S_n and A_n are primitive groups of exponential size (which means that Luks's algorithm performs a brute-force search over all elements of the group). However, even in the case of arbitrary input groups, the recursion techniques of Luks are still a powerful tool. The main reason is that large primitive groups, which build the bottleneck cases for Luks's algorithm, are quite rare and well understood.

Let $m \in \mathbb{N}$, $t \leq \frac{m}{2}$ and denote $\binom{[m]}{t}$ to be the set of all t-element subsets of $[m]$. Let $S_m^{(t)} \leq \mathrm{Sym}(\binom{[m]}{t})$ denote the natural induced action of S_m on the set $\binom{[m]}{t}$. Similarly, let $A_m^{(t)} \leq \mathrm{Sym}(\binom{[m]}{t})$ denote the natural induced action of A_m on the set $\binom{[m]}{t}$. Following Babai [6], we refer to these groups as the *Johnson groups*.

The classification of large primitive groups by Cameron [20] states that a primitive permutation group of order $|\Gamma| \geq n^{1+\log n}$ is necessarily a so-called *Cameron group* which involves a large Johnson group. Here, we only state the following slightly weaker result which is sufficient for most algorithmic purposes.

Theorem 4.7 *Let $\Gamma \leq \mathrm{Sym}(\Omega)$ be a primitive group of order $|\Gamma| \geq n^{1+\log n}$ where n is greater than some absolute constant. Then there is a polynomial-time algorithm computing a normal subgroup $N \trianglelefteq \Gamma$ of index $|\Gamma : N| \leq n$ and an N-invariant equipartition $\mathfrak{B}$ such that $N[\mathfrak{B}]$ is permutationally equivalent to $A_m^{(t)}$ for some $m \geq \log n$.*

The mathematical part of this theorem follows from [20, 63] whereas the algorithmic part is resolved in [12] (see also [5, Theorem 3.2.1]). The theorem exactly characterizes the obstacle cases of Luks's algorithm and can be seen as the starting point for Babai's algorithm solving the String Isomorphism Problem in quasi-polynomial time [6].

4.3 Babai's Algorithm

Next, we give a brief overview on the main ideas of Babai's quasi-polynomial-time algorithm for the String Isomorphism Problem. By Theorem 4.2 this also gives an algorithm for the Graph Isomorphism Problem with essentially the same running time.

As already indicated above the basic strategy of Babai's algorithm is to follow standard Luks recursion until the algorithm encounters an obstacle group which, by Theorem 4.7, may be assumed to be a Johnson group. In order to algorithmically handle the case of Johnson groups Babai's algorithm utilizes several subroutines based on both group-theoretic techniques and combinatorial approaches like the Weisfeiler-Leman algorithm.

Let us start with discussing the group-theoretic subroutines, specifically the *Local Certificates Routine* which is based on two group-theoretic statements, the *Unaffected Stabilizers Theorem* and the *Affected Orbit Lemma*.

Recall that for a set M we denote by $\mathrm{Alt}(M)$ the alternating group acting with its standard action on the set M. Moreover, following Babai [6], we refer to the groups $\mathrm{Alt}(M)$ and $\mathrm{Sym}(M)$ as the *giants* where M is an arbitrary finite set. Let $\Gamma \leq \mathrm{Sym}(\Omega)$. A *giant representation* is a homomorphism $\varphi\colon \Gamma \to S_k$ such that $\Gamma^\varphi \geq A_k$. Observe that Johnson groups naturally admit giant representations and thus, the obstacle cases from Theorem 4.7 also have a giant representation (this is a main feature of an obstacle exploited algorithmically).

Given a string $\mathfrak{x}\colon \Omega \to \Sigma$, a group $\Gamma \leq \mathrm{Sym}(\Omega)$ and a giant representation $\varphi\colon \Gamma \to S_k$, the aim of the Local Certificates Routine is to determine whether $(\mathrm{Aut}_\Gamma(\mathfrak{x})^\varphi) \geq A_k$ and to compute a meaningful certificate in either case. To achieve this goal the central tool is to split the set Ω into *affected* and *non-affected* points.

Definition 4.8 (Affected Points, Babai [6]) Let $\Gamma \leq \mathrm{Sym}(\Omega)$ be a group and $\varphi\colon \Gamma \to S_k$ a giant representation. Then an element $\alpha \in \Omega$ is *affected by φ* if $\Gamma_\alpha^\varphi \not\geq A_k$.

We remark that, if $\alpha \in \Omega$ is affected by φ, then every element in the orbit α^Γ is affected by φ. Hence, the set α^Γ is called an *affected orbit* (with respect to φ).

Let $\Delta \leq \Gamma$ be a subgroup. We define $\mathrm{Aff}(\Delta,\varphi) := \{\alpha \in \Omega \mid \Delta_\alpha^\varphi \not\geq A_k\}$. Observe that, for $\Delta_1 \leq \Delta_2 \leq \Gamma$ it holds that $\mathrm{Aff}(\Delta_1,\varphi) \supseteq \mathrm{Aff}(\Delta_2,\varphi)$.

The Local Certificates Routine is described in Algorithm 2. The algorithm computes a sequence of groups $\Gamma = \Gamma_0 \geq \Gamma_1 \geq \cdots \geq \Gamma_i$ as well as a sequence of windows $\emptyset = W_0 \subseteq W_1 \subseteq \cdots \subseteq W_i$. Throughout, the algorithm maintains the property that $\Gamma_i = \mathrm{Aut}_\Gamma^{W_i}(\mathfrak{x})$, i.e., the algorithm tries to "approximate" the automorphism group $\mathrm{Aut}_\Gamma(\mathfrak{x})$ taking larger and larger substrings into account. Observe that W_{i+1} is Γ_i-invariant and thus, $\mathrm{Aut}_\Gamma(\mathfrak{x}) \leq \Gamma_{i+1}$.

When growing the window W_i the algorithm adds those points $\alpha \in \Omega$ that are affected (with respect to the current group Γ_i) and computes the group Γ_{i+1} using recursion. The while-loop is terminated as soon as $\Gamma_i^\varphi \not\geq A_k$ or the window W_i stops growing. In the first case the algorithm returns $\Lambda := \Gamma_i^\varphi$ which clearly gives the desired outcome. In the second case the algorithm returns $\Delta := (\Gamma_i)_{(\Omega \setminus W_i)}$. It is easy to verify that $\Delta \leq \mathrm{Aut}_\Gamma(\mathfrak{x})$ since all points inside the window W_i are taken into account by the definition of group Γ_i and all points outside of W_i are fixed. The key insight is that Δ contains a large number of automorphisms, more specifically, $\Delta^\varphi \geq A_k$. This is guaranteed by the *Unaffected Stabilizers Theorem*, one of the main conceptual contributions of Babai's algorithm.

Algorithm 2: LocalCertificates($\mathfrak{x}, \Gamma, \varphi$)

Input : A string $\mathfrak{x}\colon \Omega \to \Sigma$, a group $\Gamma \leq \mathrm{Sym}(\Omega)$, and a giant representation $\varphi\colon \Gamma \to S_k$ such that $k \geq \max\{8, 2+\log_2 n\}$

Output: A non-giant $\Lambda \leq S_k$ with $(\mathrm{Aut}_\Gamma(\mathfrak{x}))^\varphi \leq \Lambda$ or $\Delta \leq \mathrm{Aut}_\Gamma(\mathfrak{x})$ with $\Delta^\varphi \geq A_k$.

1 $W_0 := \emptyset$
2 $\Gamma_0 := \Gamma$
3 $i := 0$
4 **while** $\Gamma_i^\varphi \geq A_k$ **and** $W_i \neq \mathrm{Aff}(\Gamma_i, \varphi)$ **do**
5 $\quad W_{i+1} := \mathrm{Aff}(\Gamma_i, \varphi)$
6 $\quad N := \ker(\varphi|_{\Gamma_i})$
7 $\quad \Gamma_{i+1} := \emptyset$
8 $\quad$ **for** $\gamma \in \Gamma_i^\varphi$ **do**
9 $\quad\quad$ compute $\bar\gamma \in \varphi^{-1}(\gamma)$
10 $\quad\quad \Gamma_{i+1} := \Gamma_{i+1} \cup \mathrm{Aut}_{N\bar\gamma}^{W_{i+1}}(\mathfrak{x})$
11 $\quad$ **end**
12 $\quad i := i+1$
13 **end**
14 **if** $\Gamma_i^\varphi \not\geq A_k$ **then**
15 $\quad$ **return** Γ_i^φ
16 **else**
17 $\quad$ **return** $(\Gamma_i)_{(\Omega \setminus W_i)}$
18 **end**

Theorem 4.9 (Unaffected Stabilizers Theorem, Babai [6]) *Let $\Gamma \leq \operatorname{Sym}(\Omega)$ be a permutation group of degree n and let $\varphi \colon \Gamma \to S_k$ be a giant representation such that $k > \max\{8, 2 + \log n\}$. Let $D \subseteq \Omega$ be the set of elements not affected by φ.*

Then $(\Gamma_{(D)})^\varphi \geq A_k$. In particular $D \neq \Omega$, that is, at least one point is affected by φ.

Overall, this gives the correctness of the Local Certificates Routine. So let us analyse the running time. Here, the crucial property is that the orbits of $N[W]$ are small which guarantees efficient standard Luks reduction when computing Γ_{i+1}.

Lemma 4.10 (Affected Orbit Lemma, Babai [6]) *Let $\Gamma \leq \operatorname{Sym}(\Omega)$ be a permutation group and suppose $\varphi \colon \Gamma \to S_k$ is a giant representation for $k \geq 5$. Suppose $A \subseteq \Omega$ is an affected orbit of Γ (with respect to φ). Then every orbit of $\ker(\varphi)$ in A has size at most $|A|/k$.*

Overall, this means that the Local Certificates Routine runs in time $k! n^{\mathcal{O}(1)}$ and performs at most $k! n$ many recursive calls to the String Isomorphism Problem over domain size at most n/k. Unfortunately, this not fast enough if k is significantly larger than $\log n$. The solution to this problem is to not run the Local Certificates Routine for the entire set $[k]$, but only for *test sets* $T \subseteq [k]$ of size $t = \mathcal{O}(\log n)$.

Let $T \subseteq [k]$ be a test set of size $|T| = t$. We extend the notion of point- and setwise stabilizers to the image of φ and define $\Gamma_{(T)} := \varphi^{-1}((\Gamma^\varphi)_{(T)})$ and $\Gamma_T := \varphi^{-1}((\Gamma^\varphi)_T)$. The test set T is called *full* if $((\operatorname{Aut}_{\Gamma_T}(\mathfrak{x}))^\varphi)[T] \geq \operatorname{Alt}(T)$. A *certificate of fullness* is a group $\Delta \leq \operatorname{Aut}_{\Gamma_T}(\mathfrak{x})$ such that $(\Delta^\varphi)[T] \geq \operatorname{Alt}(T)$. A *certificate of non-fullness* is a non-giant group $\Lambda \leq \operatorname{Sym}(T)$ such that $((\operatorname{Aut}_{\Gamma_T}(\mathfrak{x}))^\varphi)[T] \leq \Lambda$. Given a test set $T \subseteq [k]$ we can use the Local Certificates Routine to determine whether T is full and, depending on the outcome, compute a certificate of fullness or non-fullness (simply run the Local Certificates Routine with input $(\mathfrak{x}, \Gamma_T, \varphi_T)$ where $\varphi_T(\gamma) = (\gamma^\varphi)[T]$ for all $\gamma \in \Gamma_T$). Observe that, for $t = \mathcal{O}(\log n)$, the recursion performed by the Local Certificate Routine for test sets of size t only results in quasi-polynomial running time.

Going back to the original problem of testing isomorphism of strings, the Local Certificates Routine is applied for all test sets of size t as well as both input strings $\mathfrak{x}, \mathfrak{y}$. Based on the results, one can achieve one of the following two outcomes:

1. subsets $M_\mathfrak{x}, M_\mathfrak{y} \subseteq [k]$ of size at least $\frac{3}{4}k$ and a group $\Delta \leq \operatorname{Aut}_{\Gamma_{M_\mathfrak{x}}}(\mathfrak{x})$ such that $\Delta^\varphi[M_\mathfrak{x}] \geq \operatorname{Alt}(M_\mathfrak{x})$ and $(M_\mathfrak{x})^\varphi(\gamma) = M_\mathfrak{y}$ for all $\gamma \in \operatorname{Iso}_\Gamma(\mathfrak{x}, \mathfrak{y})$, or

2. two families of $r = k^{\mathcal{O}(1)}$ many t-ary relational structures $(\mathfrak{A}_{\mathfrak{x},j})_{j \in [r]}$ and $(\mathfrak{A}_{\mathfrak{y},j})_{j \in [r]}$, where each relational structure has domain $[k]$ and only few symmetries, such that

$$\{\mathfrak{A}_{\mathfrak{x},1}, \ldots, \mathfrak{A}_{\mathfrak{x},r}\}^{\varphi(\gamma)} = \{\mathfrak{A}_{\mathfrak{y},1}, \ldots, \mathfrak{A}_{\mathfrak{y},r}\}$$

for all $\gamma \in \operatorname{Iso}_\Gamma(\mathfrak{x}, \mathfrak{y})$.

Here, we shall not formally specify what it means for a structure to have few symmetries, but roughly speaking, this implies that the automorphism group is exponentially smaller than the full symmetric group.

Intuitively speaking, the first option can be achieved if there are many test sets which are full. In this case, the certificates of fullness (for the string $\mathfrak{x}$) can be combined to obtain the group Δ. Otherwise, there are many certificates of non-fullness which can be combined into the relational structures.

Now suppose the first option is satisfied. For simplicity assume that $M_{\mathfrak{x}} = M_{\mathfrak{y}} = [k]$ and $\Delta^{\varphi} = S_k$. Then $\mathrm{Iso}_{\Gamma}(\mathfrak{x}, \mathfrak{y}) \neq \emptyset$ if and only if $\mathrm{Iso}_{\ker(\varphi)}(\mathfrak{x}, \mathfrak{y}) \neq \emptyset$. Hence, it suffices to recursively determine whether $\mathrm{Iso}_{\ker(\varphi)}(\mathfrak{x}, \mathfrak{y}) \neq \emptyset$. Since $\ker(\varphi)$ is significantly smaller than Γ this gives sufficient progress for the recursion to obtain the desired running time.

In the other case the situation is far more involved and here, Babai's algorithm builds on combinatorial methods to achieve further progress.

For simplicity, suppose the algorithm is given a single pair of isomorphism-invariant relational structures $\mathfrak{A} := \mathfrak{A}_{\mathfrak{x},1}$ and $\mathfrak{B} := \mathfrak{A}_{\mathfrak{y},1}$ (the general case reduces to this case by *individualizing* one of the relational structures). Recall that $\mathfrak{A}$ and $\mathfrak{B}$ only have few symmetries. On a high level, the aim of the combinatorial subroutines is to compute "simpler" isomorphism-invariant relational structures $\mathfrak{A}^*$ and $\mathfrak{B}^*$ such that $\mathfrak{A}^*$ and $\mathfrak{B}^*$ still have few symmetries and all isomorphisms between $\mathfrak{A}^*$ and $\mathfrak{B}^*$ can be computed in polynomial time. Roughly speaking, this allows the algorithm to make sufficient progress by computing $\Lambda\lambda := \mathrm{Iso}(\mathfrak{A}^*, \mathfrak{B}^*)$ and updating $\Gamma\gamma := \varphi^{-1}(\Lambda\lambda)$, i.e., the algorithm continues to compute $\mathrm{Iso}_{\Gamma\gamma}(\mathfrak{x}, \mathfrak{y})$. Actually, as before, we compute small families of relational structures $\mathfrak{A}_1^*, \ldots, \mathfrak{A}_m^*$ and $\mathfrak{B}_1^*, \ldots, \mathfrak{B}_m^*$ for some number m which is quasi-polynomial in k. This does not impose any additional problems since, once again, it is possible to individualize a single relational structure.

To achieve the goal, Babai's algorithm builds on two subroutines, the *Design Lemma* and the *Split-or-Johnson Routine*. The *Design Lemma* first turns the t-ary relational structures into (a small family of) graphs (without changing the domain). For this task, the *Design Lemma* mainly builds on the t-dimensional Weisfeiler-Leman algorithm. The output is passed to the *Split-or-Johnson Routine* which produces one of the following:

1. an isomorphism-invariant coloring where each color class C has size $|C| \leq \frac{3}{4}k$, or

2. an isomorphism-invariant non-trivial equipartition of a subset $M \subseteq [k]$ of size $|M| \geq \frac{3}{4}k$, or

3. an isomorphism-invariant non-trivial Johnson graph defined on a subset $M \subseteq [k]$ of size $|M| \geq \frac{3}{4}k$.

Here, the Johnson graph $J(m,t)$ is the graph with vertex set $V(J(m,t)) := \binom{[m]}{t}$ and edge set $E(J(m,t)) := \{XY \mid |X \setminus Y| = 1\}$. We remark that $\mathrm{Aut}(J(m,t)) = S_m^{(t)}$.

Once again, the *Split-or-Johnson Routine* actually produces a family of such objects. Note that, for each possible outcome, isomorphisms between objects can be computed efficiently and the number of isomorphisms is exponentially smaller than the entire symmetric group. This provides the desired progress for the recursive algorithm.

The *Split-or-Johnson Routine* again builds on the Weisfeiler-Leman algorithm as well as further combinatorial tools. We omit any details here and refer the interested reader to [5, 47]. Overall, Babai's algorithm results in the following theorem.

Theorem 4.11 (Babai [6]) *The String Isomorphism Problem can be solved in time* $n^{\operatorname{polylog}(n)}$.

4.4 Faster Certificates for Groups with Restricted Composition Factors

In the following two subsections we discuss extensions of Babai's algorithm for the String Isomorphism Problem. We start by considering again the String Isomorphism Problem for $\widehat{\Gamma}_d$-groups. For such an input group, Babai's algorithm never encounters an obstacle case (assuming $d = \mathcal{O}(\log n)$) and thus, it simply follows Luks's algorithm which solves the problem in time $n^{\mathcal{O}(d)}$ (see also Corollary 4.6). However, in light of the novel group-theoretic subroutines employed in Babai's algorithm, it seems plausible to also hope for improvements in case the input group is a $\widehat{\Gamma}_d$-group.

Looking at Babai's algorithm from a high level, there are two major hurdles that one needs to overcome. First, we need to analyse the obstacle cases for Luks's algorithm, i.e., one needs to classify primitive $\widehat{\Gamma}_d$-groups of size larger than $n^{\mathcal{O}(\log d)}$. Without going into any details, let us just state that by a deep analysis of primitive $\widehat{\Gamma}_d$-groups [41], one can prove that such groups are, once again, essentially Johnson groups in a similar manner as in Theorem 4.7.

The second crucial hurdle is the adaptation of the Local Certificates Routine. Recall that Babai's algorithm applies the Local Certificates Routine to test sets T of size $|T| = t = \mathcal{O}(\log n)$. The routine runs in time $t!n^{\mathcal{O}(1)}$ and performs $t!n$ many recursive calls to the String Isomorphism Problem over domain size at most n/t. The central issue is that, in order to prove correctness building on the Unaffected Stabilizers Theorem, the Local Certificates Routine requires that $t > \max\{8, \log n + 2\}$. However, to apply the Local Certificates Routine in the setting of $\widehat{\Gamma}_d$-groups, we wish to choose $t = \mathcal{O}(\log d)$ to attain the desired running time. Hence, a main obstacle for a faster algorithm for $\widehat{\Gamma}_d$-groups is to obtain a suitable variant of the Unaffected Stabilizers Theorem. Unfortunately, it is not difficult to see that the natural variant of the Unaffected Stabilizers Theorem (where we restrict the input group to be in $\widehat{\Gamma}_d$ and update the bound on t as desired) does not hold (see, e.g., [70, Example 5.5.10]).

The main idea to circumvent this problem [41] is to first normalize the input to ensure suitable restrictions on the input group Γ that enable us to prove the desired variant of the Unaffected Stabilizers Theorem. In the following, we only describe the main idea for the normalization. Given a normalized input, one can provide a suitable variant of the Local Certificates Routine and, building on a more precise analysis of the recursion, extend Babai's algorithm to the setting of $\widehat{\Gamma}_d$-groups. This results in an algorithm solving the String Isomorphism Problem for $\widehat{\Gamma}_d$-groups in time $n^{\operatorname{polylog}(d)}$. We remark that the algorithm does not exploit the combinatorial subroutines directly, but one can simply rely on using Babai's algorithm as a black box instead. In particular, this also simplifies the analysis of the entire algorithm.

To describe the normalization procedure, we follow the framework developed in the second author's PhD thesis [70]. A *rooted tree* is a pair (T, v_0) where T is a directed tree and $v_0 \in V(T)$ is the root of T (all edges are directed away from the root). Let $L(T)$ denote the set of leaves of T, i.e., vertices $v \in V(T)$ without outgoing edges. For $v \in V(T)$ we denote by T^v the subtree of T rooted at vertex v.

Let $\Gamma \leq \text{Sym}(\Omega)$ be a permutation group. A *structure tree* for Γ is a rooted tree (T, v_0) such that $L(T) = \Omega$ and $\Gamma \leq (\text{Aut}(T))[\Omega]$, i.e., all $\gamma \in \Gamma$ preserve the structure of T.

Lemma 4.12 *Let $\Gamma \leq \text{Sym}(\Omega)$ be a transitive group and (T, v_0) a structure tree for Γ. For every $v \in V(T)$ the set $L(T^v)$ is a block of Γ. Moreover, $\{L(T^w) \mid w \in v^{\text{Aut}(T)}\}$ forms a block system of the group Γ.*

Let $\Gamma \leq \text{Sym}(\Omega)$ be a transitive group. The last lemma implies that every structure tree (T, v_0) gives a sequence of Γ-invariant partitions $\{\Omega\} = \mathfrak{B}_0 \succ \cdots \succ \mathfrak{B}_k = \{\{\alpha\} \mid \alpha \in \Omega\}$ (for each level $h \geq 0$, the blocks associated with the vertices at distance h from the root form one of the partitions). On the other hand, every such sequence of partitions gives a structure tree (T, v_0) with

$$V(T) = \Omega \cup \bigcup_{i=0,\ldots,k-1} \mathfrak{B}_i$$

and

$$E(T) = \{(B, B') \mid B \in \mathfrak{B}_{i-1}, B' \in \mathfrak{B}_i, B' \subseteq B\} \cup \{(B, \alpha) \mid B \in \mathfrak{B}_{k-1}, \alpha \in B\}.$$

The root is $v_0 = \Omega$.

The following definition describes the desired structure of a normalized group. For $\mathfrak{B}$ a partition of Ω and $S \subseteq \Omega$ we denote by $\mathfrak{B}[S] := \{B \cap S \mid B \in \mathfrak{B}, B \cap S \neq \emptyset\}$ the induced partition on S.

Definition 4.13 (Almost d-ary Sequences of Partitions) Let $\Gamma \leq \text{Sym}(\Omega)$ be a group and let $\{\Omega\} = \mathfrak{B}_0 \succ \cdots \succ \mathfrak{B}_k = \{\{\alpha\} \mid \alpha \in \Omega\}$ be a sequence of Γ-invariant partitions. The sequence $\mathfrak{B}_0 \succ \cdots \succ \mathfrak{B}_k$ is *almost d-ary* if for every $i \in [k]$ and $B \in \mathfrak{B}_{i-1}$ it holds that

1. $|\mathfrak{B}_i[B]| \leq d$, or

2. $\Gamma_B[\mathfrak{B}_i[B]]$ is semi-regular.

If the first option is always satisfied the sequence $\mathfrak{B}_0 \succ \cdots \succ \mathfrak{B}_k$ is called *d-ary*. Similarly, a structure tree (T, v_0) for Γ is *(almost) d-ary* if the corresponding sequence of partitions is.

Let Γ be a group for which there is an almost d-ary structure tree (T, v_0). Now, we can execute Luks's algorithm along the given structure tree (T, v_0) (by always picking the next partition when performing standard Luks reduction). Intuitively speaking, this allows us to restrict the primitive groups that are encountered during Luks's algorithm. For a d-ary structure tree all primitive groups are subgroups of S_d. For an almost d-ary structure tree we additionally need to be able to handle semi-regular groups. However, handling such groups is simple since they have size at most n where n denotes the size of the permutation domain.

The goal of the *Normalization Routine* is, given an instance $(\Gamma, \mathfrak{x}, \mathfrak{y})$ of the String Isomorphism Problem where $\Gamma \in \widehat{\Gamma}_d$, to compute a normalized equivalent instance

$(\Gamma^*, (T, v_0), \mathfrak{x}^*, \mathfrak{y}^*)$ such that (T, v_0) forms an almost d-ary structure tree for Γ^*. The main tool to achieve the normalization are tree unfoldings of certain *structure graphs*.

Let (G, v_0) be a rooted acyclic directed graph. For $v \in V(G)$ define $N^+(v) := \{w \in V(G) \mid (v, w) \in E(G)\}$ to be the set of outgoing neighbors of v. The *forward degree* of v is $\deg^+(v) := |N^+(v)|$. A vertex is a *leaf* of G if it has no outgoing neighbors, i.e., $\deg^+(v) = 0$. Let $L(G) = \{v \in V(G) \mid \deg^+(v) = 0\}$ denote the set of leaves of G.

Let $\Gamma \leq \mathrm{Sym}(\Omega)$ be a permutation group. A *structure graph* for Γ is a triple (G, v_0, φ) where (G, v_0) is a rooted acyclic directed graph such that $L(G) = \Omega$ and $\Gamma \leq (\mathrm{Aut}(G))[\Omega]$ and $\varphi \colon \Gamma \to \mathrm{Aut}(G)$ is a homomorphism such that $(\gamma^\varphi)[\Omega] = \gamma$ for all $\gamma \in \Gamma$.

Note that each structure tree can be viewed as a structure graph (for trees the homomorphism φ is uniquely defined and can be easily computed). As indicated above the strategy to normalize the action is to consider the tree unfolding of a suitable structure graph. The permutation domain of the normalized action then corresponds to the leaves of the tree unfolding for which there is a natural action of the group Γ.

Let (G, v_0) be a rooted acyclic directed graph. A *branch* of (G, v_0) is a sequence $(v_0, v_1, \ldots, v_k)$ such that $(v_{i-1}, v_i) \in E(G)$ for all $i \in [k]$. A branch $(v_0, v_1, \ldots, v_k)$ is *maximal* if it cannot be extended to a longer branch, i.e., if v_k is a leaf of (G, v_0). Let $\mathrm{Br}(G, v_0)$ denote the set of branches of (G, v_0) and $\mathrm{Br}^*(G, v_0)$ denote the set of maximal branches. Note that $\mathrm{Br}^*(G, v_0) \subseteq \mathrm{Br}(G, v_0)$. Also, for a maximal branch $\bar{v} = (v_0, v_1, \ldots, v_k)$ let $L(\bar{v}) := v_k$. Note that $L(\bar{v}) \in L(G)$.

For a rooted acyclic directed graph (G, v_0) the *tree unfolding* of (G, v_0) is defined to be the rooted tree $\mathrm{Unf}(G, v_0)$ with vertex set $\mathrm{Br}(G, v_0)$ and edge set

$$E(\mathrm{Unf}(G, v_0)) = \{((v_0, \ldots, v_k), (v_0, \ldots, v_k, v_{k+1})) \mid (v_0, \ldots, v_{k+1}) \in \mathrm{Br}(G, v_0)\}.$$

Note that $L(\mathrm{Unf}(G, v_0)) = \mathrm{Br}^*(G, v_0)$, i.e., the leaves of the tree unfolding of (G, v_0) are exactly the maximal branches of (G, v_0).

Example 4.14 Let $m \leq d$, $t \leq \frac{m}{2}$ and consider the Johnson group $\Gamma = A_m^{(t)}$. Then a structure graph (G, v_0, φ) for $\Gamma \leq \mathrm{Sym}(\binom{[m]}{t})$ can be constructed as follows. The vertices of the graph are all subsets of $[m]$ of size at most t, i.e.,

$$V(G) := \binom{[m]}{\leq t} = \{X \subseteq [m] \mid |X| \leq t\}.$$

Two vertices X and Y are connected by an edge if Y is the extension of X by a single element, i.e.,

$$E(G) := \{(X, Y) \mid X \subseteq Y \land |Y \setminus X| = 1\}.$$

The root of the structure graph is $v_0 := \emptyset$. Note that $L(G) = \binom{[m]}{t}$ as desired. Intuitively speaking, (G, v_0) corresponds to the first $t+1$ levels of the subset lattice of the set $[m]$. An example is given in Figure 5.

For $\gamma \in A_m$ let $\gamma^{(t)}$ be the element obtained from the natural action of γ on $\binom{[m]}{t}$. So $\Gamma = \{\gamma^{(t)} \mid \gamma \in A_m\}$. Let $\varphi \colon \Gamma \to \mathrm{Aut}(G)$ be defined by $X^{\varphi(\gamma^{(t)})} = X^\gamma$ where $X \in V(G)$. Then (G, v_0, φ) is a structure graph for Γ.

Recent advances on the graph isomorphism problem

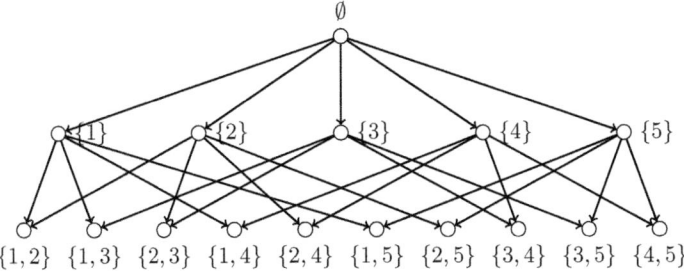

Figure 5: A structure graph for the Johnson group $A_5^{(2)}$

Now consider the tree unfolding $\mathrm{Unf}(G, v_0)$ of the graph (G, v_0). A maximal branch $\bar{v} \in \mathrm{Br}^*(G, v_0)$ is a sequence $\bar{v} = (X_0, X_1, \ldots, X_t)$ where $X_{i-1} \subsetneq X_i$ and $|X_i| = i$ for all $i \in [t]$. This gives an ordering of the elements in X_t. Indeed, it is not difficult to see that there is a one-to-one correspondence between the maximal branches $\mathrm{Br}^*(G, v_0)$ and the set $[m]^{\langle t \rangle}$ of ordered t-tuples over the set $[m]$ with pairwise distinct elements.

The group A_m also acts naturally on the set $[m]^{\langle t \rangle}$. Let $A_m^{\langle t \rangle}$ be the permutation group obtained from this action. Now it can be observed that $\mathrm{Unf}(G, v_0)$ gives a structure tree for $A_m^{\langle t \rangle}$. Also, the degree of the permutation group $A_m^{\langle t \rangle}$ is only slightly larger than the degree of $A_m^{(t)}$. Indeed,

$$\left| \binom{[m]}{t} \right|^{\log m} = \binom{m}{t}^{\log m} \geq \left(\frac{m}{t}\right)^{t \log m} \geq 2^{t \log m} = m^t \geq |[m]^{\langle t \rangle}|. \quad (4.9)$$

In order to generalize this example to obtain a normalization for all groups one can first observe that a group Γ naturally acts on the set of maximal branches of a structure graph and additionally, the tree unfolding of the structure graph forms a structure tree for this action.

Lemma 4.15 *Let $\Gamma \leq \mathrm{Sym}(\Omega)$ be a permutation group and let (G, v_0, φ) be a structure graph for Γ. Then there is an action $\psi \colon \Gamma \to \mathrm{Sym}(\mathrm{Br}^*(G))$ on the set of maximal branches of (G, v_0) such that*

1. *$L(\bar{v}^{\psi(\gamma)}) = (L(\bar{v}))^\gamma$ for all $\bar{v} \in \mathrm{Br}^*(G)$ and $\gamma \in \Gamma$, and*

2. *$\mathrm{Unf}(G, v_0)$ forms a structure tree for Γ^ψ.*

Moreover, given the group Γ and the structure graph (G, v_0, φ), the homomorphism ψ can be computed in time polynomial in $|\mathrm{Br}^*(G, v_0)|$.

Hence, for the normalization, the strategy is to first compute a suitable structure graph (G, v_0, φ) for Γ. Then we update the group by setting $\Gamma^* := \Gamma^\psi$ to be the *standard action on the set of maximal branches* described in Lemma 4.15. Here, the tree unfolding $\mathrm{Unf}(G, v_0)$ should form an almost d-ary structure tree for Γ^*. Towards this end, we say that (G, v_0, φ) is an *almost d-ary* structure graph for Γ if $\mathrm{Unf}(G, v_0)$ forms an almost d-ary structure tree for Γ^*. Moreover, we also update the strings $\mathfrak{x}, \mathfrak{y} \colon \Omega \to \Sigma$ by setting $\mathfrak{x}^* \colon \operatorname{Br}^*(G) \to \Sigma \colon \bar{v} \mapsto \mathfrak{x}(L(\bar{v}))$ and similarly $\mathfrak{y}^* \colon \operatorname{Br}^*(G) \to \Sigma \colon \bar{v} \mapsto \mathfrak{y}(L(\bar{v}))$. One can easily show that

$$(\operatorname{Iso}_\Gamma(\mathfrak{x}, \mathfrak{y}))^\psi = \operatorname{Iso}_{\Gamma^*}(\mathfrak{x}^*, \mathfrak{y}^*).$$

Hence, to complete the normalization, it only remains to find an almost d-ary structure graph for which the number of maximal branches is not too large.

Theorem 4.16 ([41, 70]) *Let* $\Gamma \leq \operatorname{Sym}(\Omega)$ *be a* $\widehat{\Gamma}_d$*-group. Then there is an almost d-ary structure graph* (G, v_0, φ) *for* Γ *such that* $|\operatorname{Br}(G, v_0)| \leq n^{\mathcal{O}(\log d)}$. *Moreover, there is an algorithm computing such a structure graph in time polynomial in the size of* G.

The proof is the theorem is far from trivial and builds on the characterization of large primitive $\widehat{\Gamma}_d$-groups (see [41]).

Corollary 4.17 *There is a Turing-reduction from the String Isomorphism Problem for* $\widehat{\Gamma}_d$*-groups to the String Isomorphism Problem for groups equipped with an almost d-ary structure tree running in time* $n^{\mathcal{O}(\log d)}$.

As already indicated above, having normalized the input to the String Isomorphism Problem, one can prove an adaptation of the Unaffected Stabilizers Theorem and extend Babai's algorithm to the setting of $\widehat{\Gamma}_d$-groups.

Theorem 4.18 (Grohe, Neuen and Schweitzer [41]) *The String Isomorphism Problem for* $\widehat{\Gamma}_d$*-groups can be solved in time* $n^{\operatorname{polylog}(d)}$.

Recall that there is a polynomial-time Turing reduction from the Graph Isomorphism Problem for graphs of maximum degree d to the String Isomorphism Problem for $\widehat{\Gamma}_d$-groups [61, 11]. Hence, the Graph Isomorphism Problem for graphs of maximum degree d can be solved in time $n^{\operatorname{polylog}(d)}$. Actually, we reprove this result in Section 5 using slightly different tools.

4.5 From Strings to Hypergraphs

Next, we present another extension of the presented methods taking more general input structures into account. Specifically, up to this point, we only considered strings for the input structures which are quite restrictive. In this subsection, we briefly describe how to extend the methods to isomorphism testing for hypergraphs.

A hypergraph is a pair $\mathcal{H} = (V, \mathcal{E})$ where V is a finite vertex set and $\mathcal{E} \subseteq 2^V$ (where 2^V denotes the power set of V). As for strings, we are interested in the isomorphism problem between hypergraphs where we are additionally given a permutation group that restricts possible isomorphisms between the two given hypergraphs. More precisely, in this work, the *Hypergraph Isomorphism Problem* takes as input two hypergraphs $\mathcal{H}_1 = (V, \mathcal{E}_1)$ and $\mathcal{H}_2 = (V, \mathcal{E}_2)$ over the same vertex set and a permutation group $\Gamma \leq \operatorname{Sym}(V)$ (given by a set of generators), and asks whether is some $\gamma \in \Gamma$ such that $\gamma \colon \mathcal{H}_1 \cong \mathcal{H}_2$ (i.e., γ is an isomorphism from $\mathcal{H}_1$ to $\mathcal{H}_2$).

For the purpose of designing an algorithm it is actually more convenient to consider the following equivalent problem. The *Set-of-Strings Isomorphism Problem* takes as input two sets $\mathfrak{X} = \{\mathfrak{x}_1, \ldots, \mathfrak{x}_m\}$ and $\mathfrak{Y} = \{\mathfrak{y}_1, \ldots, \mathfrak{y}_m\}$ where $\mathfrak{x}_i, \mathfrak{y}_i \colon \Omega \to \Sigma$

are strings, and a group $\Gamma \leq \mathrm{Sym}(\Omega)$, and asks whether there is some $\gamma \in \Gamma$ such that $\mathfrak{X}^\gamma := \{\mathfrak{x}_1^\gamma, \ldots, \mathfrak{x}_m^\gamma\} = \mathfrak{Y}$.

Theorem 4.19 *The Hypergraph Isomorphism Problem for $\widehat{\Gamma}_d$-groups is polynomial-time equivalent to the Set-of-Strings Isomorphism Problem for $\widehat{\Gamma}_d$-groups under many-one reductions.*

Proof Sketch A hypergraph $\mathcal{H} = (V, \mathcal{E})$ can be translated into a set $\mathfrak{X} := \{\mathfrak{x}_E \mid E \in \mathcal{E}\}$ over domain $\Omega := V$ where $\mathfrak{x}_E$ is the characteristic function of E, i.e., $\mathfrak{x}(v) = 1$ if $v \in E$ and $\mathfrak{x}(v) = 0$ if $v \notin E$.

In the other direction, a set of strings $\mathfrak{X}$ over domain Ω and alphabet Σ can be translated into a hypergraph $\mathcal{H} = (V, \mathcal{E})$ where $V := \Omega \times \Sigma$ and $\mathcal{E} := \{E_\mathfrak{x} \mid \mathfrak{x} \in \mathfrak{X}\}$ where $E_\mathfrak{x} := \{(\alpha, \mathfrak{x}(\alpha)) \mid \alpha \in \Omega)\}$. The group $\Gamma \leq \mathrm{Sym}(\Omega)$ is translated into the natural action of Γ on V defined via $(\alpha, a)^\gamma := (\alpha^\gamma, a)$.

It is easy to verify that both translations preserve isomorphisms. □

For the purpose of building a recursive algorithm, we consider a slightly different problem that crucially allows us to modify instances in a certain way.

Let $\Gamma \leq \mathrm{Sym}(\Omega)$ be a group and let $\mathfrak{P}$ be a Γ-invariant partition of the set Ω. A $\mathfrak{P}$-*string* is a pair $(P, \mathfrak{x})$ where $P \in \mathfrak{P}$ and $\mathfrak{x}\colon P \to \Sigma$ is a string over a finite alphabet Σ. For $\sigma \in \mathrm{Sym}(\Omega)$ the string $\mathfrak{x}^\sigma$ is defined by $\mathfrak{x}^\sigma \colon P^\sigma \to \Sigma\colon \alpha \mapsto \mathfrak{x}(\alpha^{\sigma^{-1}})$. A permutation $\sigma \in \mathrm{Sym}(\Omega)$ is a Γ-*isomorphism* from $(P, \mathfrak{x})$ to a second $\mathfrak{P}$-string $(Q, \mathfrak{y})$ if $\sigma \in \Gamma$ and $(P^\sigma, \mathfrak{x}^\sigma) = (Q, \mathfrak{y})$.

The *Generalized String Isomorphism Problem* takes as input a permutation group $\Gamma \leq \mathrm{Sym}(\Omega)$, a Γ-invariant partition $\mathfrak{P}$ of Ω, and $\mathfrak{P}$-strings $(P_1, \mathfrak{x}_1), \ldots, (P_m, \mathfrak{x}_m)$ and $(Q_1, \mathfrak{y}_1), \ldots, (Q_m, \mathfrak{y}_m)$, and asks whether there is some $\gamma \in \Gamma$ such that

$$\{(P_1^\gamma, \mathfrak{x}_1^\gamma), \ldots, (P_m^\gamma, \mathfrak{x}_m^\gamma)\} = \{(Q_1, \mathfrak{y}_1), \ldots, (Q_m, \mathfrak{y}_m)\}.$$

We denote $\mathfrak{X} = \{(P_1, \mathfrak{x}_1), \ldots, (P_m, \mathfrak{x}_m)\}$ and $\mathfrak{Y} = \{(Q_1, \mathfrak{y}_1), \ldots, (Q_m, \mathfrak{y}_m)\}$. Additionally, $\mathrm{Iso}_\Gamma(\mathfrak{X}, \mathfrak{Y})$ denotes the set of Γ-isomorphisms from $\mathfrak{X}$ to $\mathfrak{Y}$ and $\mathrm{Aut}_\Gamma(\mathfrak{X}) := \mathrm{Iso}_\Gamma(\mathfrak{X}, \mathfrak{X})$.

It is easy to see that the Set-of-Strings Isomorphism Problem forms a special case of the Generalized String Isomorphism Problem where $\mathfrak{P}$ is the trivial partition consisting of one block.

For the rest of this subsection we denote by $n := |\Omega|$ the size of the domain, and m denotes the size of $\mathfrak{X}$ and $\mathfrak{Y}$ (we always assume $|\mathfrak{X}| = |\mathfrak{Y}|$, otherwise the problem is trivial). The goal is to sketch an algorithm that solves the Generalized String Isomorphism Problem for $\widehat{\Gamma}_d$-groups in time $(n + m)^{\mathrm{polylog}(d)}$.

As a starting point it was already observed by Miller [67] that Luks's algorithm can be easily extended to hypergraphs resulting in an isomorphism test running in time $(n + m)^{\mathcal{O}(d)}$ for $\widehat{\Gamma}_d$-groups. Similar to the previous subsection, the main obstacle we are facing to obtain a more efficient algorithm is the adaptation of the Local Certificates Routine.

Let $\Gamma \leq \mathrm{Sym}(\Omega)$ be a $\widehat{\Gamma}_d$-group, $\mathfrak{P}$ a Γ-invariant partition of Ω, and $\mathfrak{X}, \mathfrak{Y}$ two sets of $\mathfrak{P}$-strings. Recall that, in a nutshell, the Local Certificates Routine considers a Γ-invariant window $W \subseteq \Omega$ such that $\Gamma[W] \leq \mathrm{Aut}(\mathfrak{X}[W])$ (i.e., the group Γ respects $\mathfrak{X}$ restricted to the window W) and aims at creating automorphisms of the entire structure $\mathfrak{X}$ (from the local information that Γ respects $\mathfrak{X}$ on the window W). In order to create these automorphisms the Local Certificates Routine considers the group $\Gamma_{(\Omega\setminus W)}$ fixing every point outside of W. Remember that the Unaffected Stabilizers Theorem (resp. the variant suitable for $\widehat{\Gamma}_d$-groups) guarantees that the group $\Gamma_{(\Omega\setminus W)}$ is large.

For the String Isomorphism Problem it is easy to see that $\Gamma_{(\Omega\setminus W)}$ consists only of automorphisms of the input string (assuming Γ respects the input string $\mathfrak{x}$ on the window W) since there are no dependencies between the positions within the window W and outside of W. However, for the Generalized String Isomorphism Problem, this is not true anymore. As a simple example, suppose the input is a graph on vertex set Ω (which, in particular, can be interpreted as a hypergraph and translated into a set of strings) and the edges between W and $\Omega\setminus W$ form a perfect matching. Then a permutation $\gamma \in \Gamma_{(\Omega\setminus W)}$ is not necessarily an automorphism of G even if it respects $G[W]$, since it may not preserve the edges between W and $\Omega\setminus W$. Actually, since the edges between W and $\Omega\setminus W$ form a perfect matching, the only automorphism of G in the group $\Gamma_{(\Omega\setminus W)}$ is the identity mapping.

In other words, in order to compute $\text{Aut}(\mathfrak{X})$, it is not possible to consider $\mathfrak{X}[W]$ and $\mathfrak{X}[\Omega \setminus W]$ independently as is done by the Local Certificates Routine.

The solution to this problem is guided by the following simple observation. Suppose that $\mathfrak{X}[W]$ is *simple*, i.e.,

$$m_{\mathfrak{X}[W]}(P) := |\{\mathfrak{x}[W \cap P] \mid (P,\mathfrak{x}) \in \mathfrak{X}, W \cap P \neq \emptyset\}| = 1$$

for all $P \in \mathfrak{P}[W]$ (in other words, when restricted to W, each block P only contains one string). In this case it is possible to consider $\mathfrak{X}[W]$ and $\mathfrak{X}[\Omega \setminus W]$ independently since there can be no more additional dependencies between W and $\Omega \setminus W$.

Let $W \subseteq \Omega$ be a Γ-invariant set such that $\Gamma[W] \leq \text{Aut}(\mathfrak{X}[W])$ (this is the case during the Local Certificates Routine). In order to solve the problem described above in general, the basic idea is to introduce another normalization procedure, referred to as *Simplification Routine*, modifying the instance in such a way that $\mathfrak{X}[W]$ becomes simple. Eventually, this allows us to extend the Local Certificates Routine to the setting of the Generalized String Isomorphism Problem.

In the following we briefly describe the *Simplification Routine* which exploits the specific definition of the Generalized String Isomorphism Problem. Consider a set $P \in \mathfrak{P}$. In order to "simplify" the instance we define an equivalence relation on the set $\mathfrak{X}[[P]] := \{\mathfrak{x} \mid (P,\mathfrak{x}) \in \mathfrak{X}\}$ of all strings on P. Two $\mathfrak{P}$-strings $(P,\mathfrak{x}_1)$ and $(P,\mathfrak{x}_2)$ are W-equivalent if they are identical on the window W, i.e., $\mathfrak{x}_1[W\cap P] = \mathfrak{x}_2[W \cap P]$. For each equivalence class we create a new block P' containing exactly the strings from the equivalence class. Since the group Γ respects the induced sub-instance $\mathfrak{X}[W]$ it naturally acts on the equivalence classes. This process is visualized in Figure 6 and formalized below.

Let $\Omega' := \bigcup_{P\in\mathfrak{P}} P \times \{\mathfrak{x}[W\cap P] \mid (P,\mathfrak{x}) \in \mathfrak{X}\}$ and $\mathfrak{P}' := \{P \times \{\mathfrak{x}[W\cap P]\} \mid (P,\mathfrak{x}) \in \mathfrak{X}\}$. Also define

$$\mathfrak{X}' := \left\{ \left(P \times \{\mathfrak{x}[W \cap P]\}, \mathfrak{x}'\right) \;\middle|\; (P,\mathfrak{x}) \in \mathfrak{X} \right\}$$

where $\mathfrak{x}' \colon P \times \{\mathfrak{x}[W \cap P]\} \to \Sigma \colon (\alpha, \mathfrak{x}[W \cap P]) \mapsto \mathfrak{x}(\alpha)$. The set $\mathfrak{Y}'$ is defined similarly for the instance $\mathfrak{Y}$. Note that $\mathfrak{X}'$ and $\mathfrak{Y}'$ are sets of $\mathfrak{P}'$-strings. Finally, the group Γ faithfully acts on the set Ω' via $(\alpha, \mathfrak{z})^\gamma = (\alpha^\gamma, \mathfrak{z}^\gamma)$ yielding an injective

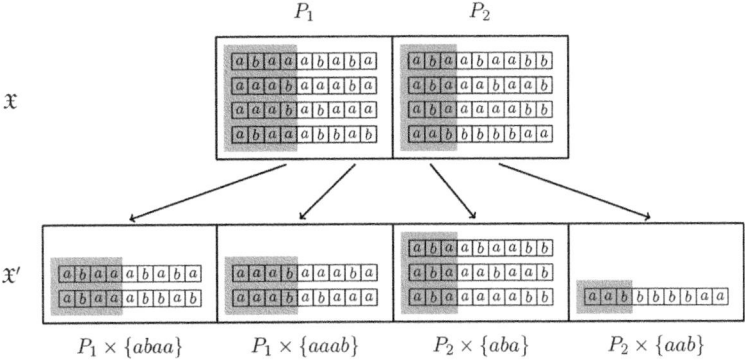

Figure 6: A set $\mathfrak{X}$ of $\mathfrak{P}$-strings is given in the top and the "simplified" instance $\mathfrak{X}'$ is given below. The window W is marked in gray. Note that $\mathfrak{X}'[W']$ is simple where W' denotes the window marked in gray in the bottom part of the figure.

homomorphism $\psi \colon \Gamma \to \mathrm{Sym}(\Omega')$. Define $\Gamma' := \Gamma^\psi$. It can be easily checked that the updated instance is equivalent to the original instance. Also, $\mathfrak{X}'[W']$ is simple where $W' := \{(\alpha, \mathfrak{z}) \in \Omega' \mid \alpha \in W\}$.

While this simplification allows us to treat $\mathfrak{X}'[W']$ and $\mathfrak{X}'[W' \setminus \Omega']$ independently and thus solves the above problem, it creates several additional issues that need to be addressed. First, this modification may destroy the normalization property (i.e., the existence of an almost d-ary sequence of partitions). As a result, the Local Certificates Routine constantly needs to re-normalize the input instances which requires a precise analysis of the increase in size occurring from the re-normalization. In turn, the re-normalization of instances creates another problem. In the *Aggregation of Local Certificates* (where the local certificates are combined into a relational structure), the outputs of the Local Certificates Routine are compared with each other requiring the outputs to be isomorphism-invariant. However, the re-normalization procedure is not isomorphism-invariant. The solution to this problem is to run the Local Certificates Routine in parallel on all pairs of test sets compared later on. This way, one can ensure that all instances are normalized in the same way.

Overall, these ideas allow us to obtain the following theorem.

Theorem 4.20 (Neuen [71]) *The Generalized String Isomorphism Problem for $\widehat{\Gamma}_d$-groups can be solved in time $(n + m)^{\mathrm{polylog}(d)}$ where n denotes the size of the domain and m the number of strings in the input sets.*

Note that this gives an algorithm for the Set-of-Strings Isomorphism Problem as well as the Hypergraph Isomorphism Problem for $\widehat{\Gamma}_d$-groups with the same running time. In particular, the Hypergraph Isomorphism Problem without any input group can be solved in time $(n + m)^{\mathrm{polylog}(n)}$. This is the fasted known algorithm for the problem. A different algorithm with the same running time was obtained by Wiebking [87].

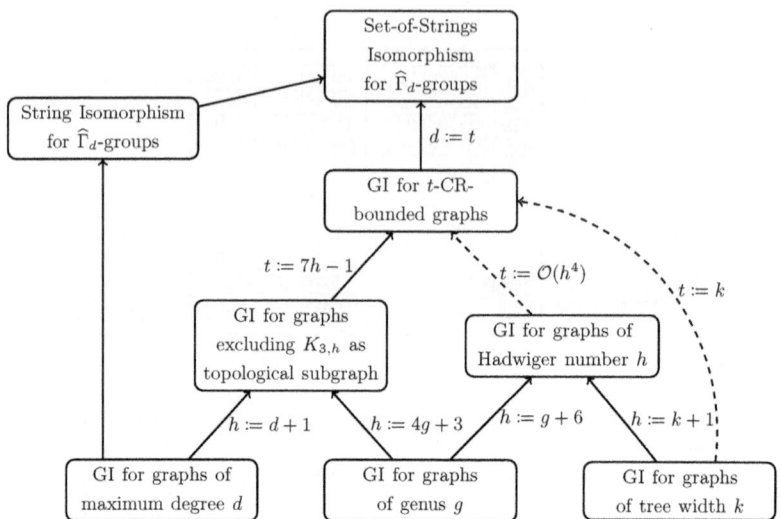

Figure 7: Dependencies between isomorphism problems considered in this paper. Each solid edge represents a polynomial-time Turing reduction where the parameter is set according to the edge label. The dashed edges represent fpt reductions.

5 Quasi-Polynomial Parameterized Algorithms for Isomorphism Testing

In this section we present several applications of the results presented above for isomorphism testing on restricted classes of graphs. Towards this end, we first introduce the notion of t-CR-$bounded$ graphs recently introduced in [71] and build an isomorphism test for such graphs based on Theorem 4.20. It turns out that the notion of t-CR-bounded graphs forms a powerful tool when it comes to the task of designing isomorphism tests for restricted classes of graphs. In this direction we build a series of reductions for the isomorphism problem for well-known parameterized classes of graphs to the isomorphism problem for t-CR-bounded graphs leading to the most efficient algorithms for isomorphism testing for mentioned classes. An overview on the reductions can be found in Figure 7.

5.1 Allowing Color Refinement to Split Small Classes

As indicated above, the central tool for designing more efficient isomorphism tests based on the results of the previous section is the notion of t-CR-bounded graphs. Intuitively speaking, a vertex-colored graph (G, χ) is t-CR-$bounded$ if the vertex-coloring can be transformed into a discrete coloring (i.e., a coloring where each vertex has its own color) by repeatedly applying the following operations:

1. applying the Color Refinement algorithm, and

2. picking a color class $[v]_\chi \coloneqq \{w \in V(G) \mid \chi(w) = \chi(v)\}$ for some vertex v such that $|[v]_\chi| \leq t$ and assigning each vertex in the class its own color.

The next definition formalizes this intuition. For formal reason, it turns out to be more convenient to consider vertex- and arc-colored graphs.

Definition 5.1 A vertex- and arc-colored graph (G, χ_V, χ_E) is *t-CR-bounded* if the sequence $(\chi_i)_{i \geq 0}$ reaches a discrete coloring where $\chi_0 := \chi_V$,

$$\chi_{2i+1} := \chi_{(\infty)}(G, \chi_{2i}, \chi_E)$$

and

$$\chi_{2i+2}(v) := \begin{cases} (v, 1) & \text{if } |[v]_{\chi_{2i+1}}| \leq t \\ (\chi_{2i+1}(v), 0) & \text{otherwise} \end{cases}$$

for all $i \geq 0$.

Also, for the minimal $i_\infty \geq 0$ such that $\chi_{i_\infty} \equiv \chi_{i_\infty+1}$, we refer to χ_{i_∞} as the *t-CR-stable* coloring of G and denote it by $\chi_{t\text{-CR}}(G)$.

In order to give an efficient isomorphism test for t-CR-bounded graphs we present a polynomial-time reduction to the Set-of-Strings Isomorphism Problem discussed in Section 4.5.

Theorem 5.2 ([71]) *There is a polynomial-time Turing reduction from the Graph Isomorphism Problem for t-CR-bounded graphs to the Set-of-Strings Isomorphism Problem for $\widehat{\Gamma}_t$-groups.*

Proof Sketch Let G be a t-CR-bounded graph and let $(\chi_i)_{i \geq 0}$ be the corresponding sequence of colorings. For simplicity we assume the arc-coloring of G is trivial and restrict ourselves to computing a generating set for the automorphism group $\text{Aut}(G)$. Let $\mathfrak{P}_i := \pi(\chi_i)$ be the partition of the vertex set into the color classes of the coloring χ_i. Observe that $\mathfrak{P}_i$ is defined in an isomorphism-invariant manner with respect to G and $\mathfrak{P}_{i+1} \preceq \mathfrak{P}_i$ for all $i \geq 0$.

The idea for the algorithm is to iteratively compute $\widehat{\Gamma}_t$-groups $\Gamma_i \leq \text{Sym}(\mathfrak{P}_i)$ such that $(\text{Aut}(G))[\mathfrak{P}_i] \leq \Gamma_i$ for all $i \geq 0$. First note that this gives an algorithm for computing the automorphism group of G. Since G is t-CR-bounded there is some $i_\infty \geq 0$ such that $\mathfrak{P}_{i_\infty}$ is the discrete partition. Hence, $\text{Aut}(G) \leq \Gamma_{i_\infty}$[3] where Γ_{i_∞} is a $\widehat{\Gamma}_t$-group and the automorphism group of G can be computed using Theorems 4.19 and 4.20.

For the sequence of groups the algorithm sets $\Gamma_0 := \{\text{id}\}$ to be the trivial group (this is correct since χ_0 is the vertex-coloring of the input graph G). Hence, let $i > 0$ and suppose the algorithm already computed a generating set for Γ_{i-1}.

If i is even $\mathfrak{P}_i$ is obtained from $\mathfrak{P}_{i-1}$ by splitting all sets $P \in \mathfrak{P}_{i-1}$ of size at most t. Towards this end, the algorithm first updates $\Gamma'_{i-1} := \text{Aut}_{\Gamma_{i-1}}(\{\mathfrak{x}_{i-1}\})$ where $\mathfrak{x}_{i-1}(P) = |P|$ for all $P \in \mathfrak{P}_{i-1}$. Then, for each orbit $\mathfrak{B}$ of Γ'_{i-1} there is some number $p \in \mathbb{N}$ such that $|P| = p$ for all $P \in \mathfrak{B}$. Now Γ_i is obtained from Γ'_{i-1} by taking the wreath product with the symmetric group S_p for all orbits where $p \leq t$. Clearly, Γ_i is still a $\widehat{\Gamma}_t$-group.

---

[3]Formally, this is not correct since $\text{Aut}(G) \leq \text{Sym}(V(G))$ and $\Gamma_{i_\infty} \leq \text{Sym}(\{\{v\} \mid v \in V(G)\})$. However, the algorithm can simply identify v with the singleton set $\{v\}$ to obtain the desired supergroup.

If i is odd $\mathfrak{P}_i$ is obtained from $\mathfrak{P}_{i-1}$ by performing the Color Refinement algorithm. Consider a single iteration of the Color Refinement algorithm. More precisely, let $\mathfrak{P}$ be an invariant partition (with respect to $\operatorname{Aut}(G)$) and $\Gamma \leq \operatorname{Sym}(\mathfrak{P})$ be a $\widehat{\Gamma}_t$-group such that $(\operatorname{Aut}(G))[\mathfrak{P}] \leq \Gamma$. Let $\mathfrak{P}'$ be the partition obtained from $\mathfrak{P}$ by performing a single iteration of the Color Refinement algorithm. We argue how to compute a $\widehat{\Gamma}_t$-group $\Gamma' \leq \operatorname{Sym}(\mathfrak{P}')$ such that $(\operatorname{Aut}(G))[\mathfrak{P}'] \leq \Gamma'$. Repeating this procedure for all iterations of the Color Refinement algorithm then gives the desired group Γ_i.

In order to compute the group Γ' consider the following collection of strings. For every $v \in V(G)$ define $\mathfrak{x}_v \colon \mathfrak{P} \to \mathbb{N}^2$ via $\mathfrak{x}_v(P) = (1, |N(v) \cap P|)$ if $v \in P$ and $\mathfrak{x}_v(P) = (0, |N(v) \cap P|)$ otherwise. Observe that, by the Definition of the Color Refinement algorithm, $\mathfrak{x}_v = \mathfrak{x}_w$ if and only if there is some $P' \in \mathfrak{P}'$ such that $v, w \in P'$. Hence, there is a natural one-to-one correspondence between $\mathfrak{X} := \{\mathfrak{x}_v \mid v \in V(G)\}$ and $\mathfrak{P}'$. Now define Γ' be the induced action of $\operatorname{Aut}_\Gamma(\mathfrak{X})$ on the set $\mathfrak{P}'$ (obtained from this correspondence). It is easy to verify that $(\operatorname{Aut}(G))[\mathfrak{P}'] \leq \Gamma'$.

Clearly, all steps of the algorithm can be performed in polynomial time using an oracle to the Set-of-Strings Isomorphism Problem for $\widehat{\Gamma}_t$-groups. $\square$

We remark that the proof of the last theorem also implies that $\operatorname{Aut}(G) \in \widehat{\Gamma}_t$ for every t-CR-bounded graph G. In combination with Theorem 4.20 we obtain an efficient isomorphism test for t-CR-bounded graphs.

Corollary 5.3 *The Graph Isomorphism Problem for t-CR-bounded graphs can be solved in time $n^{\operatorname{polylog}(t)}$.*

For the remainder of this section we shall exploit the algorithm from the corollary to design efficient isomorphism tests for a number of graph classes. Towards this end, we typically build on another standard tool for isomorphism testing which is individualization of single vertices. Intuitively, this allows us to break potential regularities in the input graph (for example, on a d-regular graph, the t-CR-stable coloring achieves no refinement unless $t \geq n$) and identify a "starting point" for analyzing the t-CR-stable coloring.

Let G be a graph and let $X \subseteq V(G)$ be a set of vertices. Let $\chi_V^* \colon V(G) \to C$ be the vertex-coloring obtained from individualizing all vertices in the set X, i.e., $\chi_V^*(v) := (v, 1)$ for $v \in X$ and $\chi_V^*(v) := (0, 0)$ for $v \in V(G) \setminus X$. Let $\chi := \chi_{t\text{-CR}}(G, \chi_V^*)$ denote the t-CR-stable coloring with respect to the input graph (G, χ_V^*). We define the t-*closure* of the set X (with respect to G) to be the set

$$\operatorname{cl}_t^G(X) := \{v \in V(G) \mid |[v]_\chi| = 1\}.$$

Observe that $X \subseteq \operatorname{cl}_t^G(X)$. For $v_1, \ldots, v_\ell$ we use $\operatorname{cl}_t^G(v_1, \ldots, v_\ell)$ as a shorthand for $\operatorname{cl}_t^G(\{v_1, \ldots, v_\ell\})$.

If $\operatorname{cl}_t^G(X) = V(G)$ then there is an isomorphism test for G running in time $n^{|X|+\operatorname{polylog}(t)}$. An algorithm first individualizes all vertices from X creating $n^{|X|}$ many instances of GI for t-CR-bounded graphs each of which can be solved using Corollary 5.3. This provides us a generic and powerful method for obtaining polynomial-time isomorphism tests for various classes of graphs. As a first, simple example we argue that isomorphism for graphs of maximum degree d can be tackled this way.

Theorem 5.4 *Let G be a connected graph of maximum degree d and let $v \in V(G)$. Then $\mathrm{cl}_d^G(v) = V(G)$.*

In particular, there is a polynomial-time Turing reduction from the Graph Isomorphism Problem for graphs of maximum degree d to the Graph Isomorphism Problem for d-CR-bounded graphs.

Proof Let $\chi := \chi_{d\text{-CR}}(G, \chi_V^*)$ denote the d-CR-stable coloring where χ_V^* is the coloring obtained from individualizing v. For $i \geq 0$ let $V_i := \{w \in V(G) \mid \mathrm{dist}_G(v,w) \leq i\}$. We prove by induction on $i \geq 0$ that $V_i \subseteq \mathrm{cl}_d^G(v)$. Since $V_n = V(G)$ this implies that $\mathrm{cl}_d^G(v) = V(G)$.

The base case $i = 0$ is trivial since $V_0 = \{v\} \subseteq \mathrm{cl}_d^G(v)$ as already observed above. So suppose $i \geq 0$ and let $w \in V_{i+1}$. Then there is some $u \in V_i$ such that $uw \in E(G)$. Moreover, $u \in \mathrm{cl}_d^G(v)$ by the induction hypothesis. This means that $|[u]_\chi| = 1$. Since χ is stable with respect to the Color Refinement algorithm $[w]_\chi \subseteq N(u)$. So $|[w]_\chi| \leq \deg(u) \leq d$. Hence, $|[w]_\chi| = 1$ because χ is d-CR-stable. □

This gives an isomorphism test for graphs of maximum degree d running in time $n^{\mathrm{polylog}(d)}$. Observe that the algorithm is obtained via a reduction to the Set-of-Strings Isomorphism Problem for $\widehat{\Gamma}_d$-groups. We remark that, using similar, but slightly more involved ideas, there is also a polynomial-time Turing reduction from the isomorphism problem for graphs of maximum degree d to the String Isomorphism Problem for $\widehat{\Gamma}_d$-groups [61, 11] (see also Section 4.2). Recall that the String Isomorphism Problem is a simpler problem which can be seen as a special case of the Set-of-Strings Isomorphism Problem (where the set of strings contains only one element). Indeed, the original $n^{\mathrm{polylog}(d)}$ isomorphism test for graphs of maximum degree d from [41] is obtained via this route. However, considering the remaining applications of the isomorphism test for t-CR-bounded graphs in this section, such a behaviour seems to be an exception and reductions to the String Isomorphism Problem for $\widehat{\Gamma}_d$-groups are usually not known. This highlights the significance of the Set-of-Strings Isomorphism Problem in comparison to the String Isomorphism Problem.

5.2 Graphs of Small Genus

Next, we turn to the isomorphism problem for graphs of bounded Euler genus. Recall that a graph has Euler genus at most g if it can be embedded on a surface of Euler genus g. We omit a formal definition of the genus of a graph and instead only rely on the following basic property. A graph H is a *minor* of graph G if H can be obtained from G by deleting vertices and edges as well as contracting edges. A graph G *excludes H as a minor* if it has no minor isomorphic to H. It is well known that graphs of Euler genus g exclude $K_{3,4g+3}$ (the complete bipartite graph with 3 vertices on the left and $4g + 3$ vertices on the right) as a minor [78]. The next lemma connects graphs that exclude $K_{3,h}$ as a minor to t-CR-bounded graphs.

Recall that a graph G is *3-connected* if $G-X$ is connected for every set $X \subseteq V(G)$ of size $|X| \leq 2$.

Lemma 5.5 *Let (G, χ) be a 3-connected, vertex-colored graph that excludes $K_{3,h}$ as a minor and suppose $V_1 \uplus V_2 = V(G)$ such that*

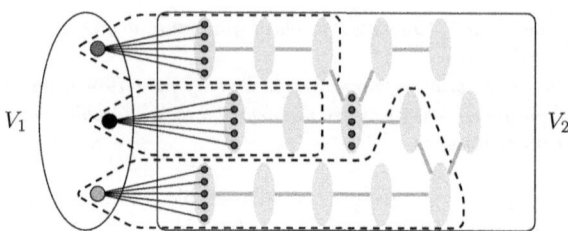

Figure 8: Visualization of the spanning tree T described in the proof of Lemma 5.5. Contracting the dashed regions into single vertices gives a subgraph isomorphic to $K_{3,h}$.

1. $|[v]_\chi| = 1$ for all $v \in V_1$,
2. χ is stable with respect to the Color Refinement algorithm, and
3. $|V_1| \geq 3$.

Then there exists $u \in V_2$ such that $|[u]_\chi| \leq h - 1$.

Proof Let $C := \chi(V(G))$, $C_1 := \chi(V_1)$ and $C_2 := \chi(V_2)$. Also define H to be the graph with vertex set $V(H) := C$ and edge set

$$E(H) = \{c_1c_2 \mid \exists v_1 \in \chi^{-1}(c_1), v_2 \in \chi^{-1}(c_2) \colon v_1v_2 \in E(G)\}.$$

Let $C' \subseteq C_2$ be the vertex set of a connected component of $H[C_2]$. Then $|N_H(C')| \geq 3$ since each $v \in V_1$ forms a singleton color class with respect to χ and G is 3-connected.

Now let $c_1, c_2, c_3 \in N_H(C')$ be distinct and also let $v_i \in \chi^{-1}(c_i)$ for $i \in [3]$. Also let T be a spanning tree of $H[C' \cup \{c_1, c_2, c_3\}]$ such that $c_1, c_2, c_3 \in L(T)$ where $L(T)$ denotes the set of leaves of T (see Figure 8). Moreover, let T' be the subtree of T obtained from repeatedly removing all leaves $c \in C'$. Hence, $L(T') = \{c_1, c_2, c_3\}$. Then there is a unique color c such that $\deg_{T'}(c) = 3$. Also, for $i \in [3]$, define C'_i to be the set of internal vertices on the unique path from c_i to c in the tree T'.

Since $|\chi^{-1}(c_i)| = 1$ and χ is stable with respect to the Color Refinement algorithm it holds that

$$G\left[\chi^{-1}(C'_i \cup \{c_i\})\right]$$

is connected. Let $U_i := \chi^{-1}(C'_i \cup \{c_i\})$, $i \in [3]$. Also let $U = \chi^{-1}(c)$ and suppose that $|U| \geq h$. Then $N(U_i) \cap U \neq \emptyset$ by the definition of the tree T. Moreover, this implies $U \subseteq N(U_i)$ since χ is stable with respect to the Color Refinement algorithm. Hence, G contains a minor isomorphic to $K_{3,h}$. □

Corollary 5.6 ([71]) Let (G, χ_V, χ_E) be a 3-connected, vertex- and arc-colored graph that excludes $K_{3,h}$ as a minor and let $v_1, v_2, v_3 \in V(G)$ be distinct vertices. Then $\mathrm{cl}^G_{h-1}(v_1, v_2, v_3) = V(G)$.

Recent advances on the graph isomorphism problem   219

Proof Let χ_V^* be the coloring obtained from χ_V after individualizing v_1, v_2 an v_3. Also let $\chi := \chi_{(h-1)\text{-CR}}(G, \chi_V^*, \chi_E)$ be the $(h-1)$-CR-stable coloring with respect to the graph (G, χ_V^*, χ_E).

Suppose towards a contradiction that χ is not discrete (i.e., not every color class is a singleton). Let $V_2 := \{v \in V(G) \mid |[v]_\chi| > 1\}$ and let $V_1 := N_G(V_2)$. Then $|V_1| \geq 3$ since $|V(G) \setminus V_2| \geq 3$ and G is 3-connected. Also note that $\chi|_{V_1 \cup V_2}$ is a stable coloring for the graph $G[V_1 \cup V_2]$. Hence, by Lemma 5.5, there is some vertex $u \in V_2$ such that $|[u]_\chi| \leq h - 1$. Also $|[u]_\chi| > 1$ by the definition of V_2. But this contradicts the definition of an $(h-1)$-CR-stable coloring. □

Note that Corollary 5.6 only provides an isomorphism test for 3-connected graphs excluding $K_{3,h}$ as a minor. We can easily remedy this problem by using the well-known fact that, for a minor-closed class of graph, it suffices to solve the isomorphism problem on vertex- and arc-colored 3-connected graphs exploiting the standard decomposition into 3-connected components [48]. This gives us an isomorphism test for graphs of Euler genus at most g running in time $n^{\text{polylog}(g)}$.

5.3 Graphs of Small Tree Width

After discussing two fairly simple applications of the isomorphism test for t-CR-bounded graphs we now turn to the first more involved application considering graphs of small tree width. In the following subsection we will extend some of the basic ideas discussed for graphs of small tree width further to deal with the more general case of excluding the complete graph K_h as a minor.

We assume that the reader is familiar with the notion of tree width (see, e.g., [25, Chapter 12], [23, Chapter 7]). For the sake of completeness, let us nevertheless give the basic definitions. Let G be a graph. A *tree decomposition* of G is a pair (T, β) where T is a rooted tree and $\beta \colon V(T) \to 2^{V(G)}$, where $2^{V(G)}$ denotes the powerset of $V(G)$, such that

(T.1) for every edge $vw \in E(G)$ there is some $t \in V(T)$ such that $v, w \in \beta(t)$, and

(T.2) for every $v \in V(G)$ the set $\beta^{-1}(v) := \{t \in V(T) \mid v \in \beta(t)\}$ is non-empty and connected in T, that is, the induced subgraph $T[\beta^{-1}(v)]$ is a subtree.

The sets $\beta(t)$, $t \in V(T)$, are the *bags* of the decomposition. Also, the sets $\beta(t) \cap \beta(s)$, $st \in E(T)$, are the *adhesion sets* of the decomposition. The *width* of a tree decomposition is defined as the maximum bag size minus one, i.e., $\text{wd}(T, \beta) := \max_{t \in V(T)} |\beta(t)| - 1$. The *tree width* of a graph G, denoted $\text{tw}(G)$, is the minimal width of any tree decomposition of G.

Our goal for this section is to design efficient isomorphism tests for graphs of small tree width. Once again, we want to build on the isomorphism test for t-CR-bounded graphs. However, compared to graphs of small degree or genus, identifying parts of graphs of bounded tree width that are t-CR-bounded (after fixing a constant number of vertices) is more challenging. This comes from the fact that the automorphism group of a graph of tree width k may contain arbitrarily large symmetric groups. For example, let $T_{d,h}$ denote a complete d-ary tree of height h, where we should think of d as being unbounded and much larger than k. Also, let $K_k \otimes T_{d,h}$ denote the graph obtained from $T_{d,h}$ by replacing each vertex with a clique of size k and two cliques being completely connected whenever there is an edge in $T_{d,h}$. Then $\operatorname{tw}(K_k \otimes T_{d,h}) \leq 2k$, the graph $K_k \otimes T_{d,h}$ does not contain any $(k-1)$-separator, and the automorphism group contains various symmetric groups S_d.

To circumvent this problem we exploit *clique-separator decompositions* of improved graphs which already form an essential part of the first fixed-parameter tractable isomorphism test for graphs of bounded tree width [58], running in time $2^{\mathcal{O}(k^5 \log k)} n^5$. We first require some additional notation. Let G be a graph. A pair (A, B) where $A \cup B = V(G)$ is called a *separation* if $E(A \setminus B, B \setminus A) = \emptyset$ (here, $E(X, Y) \coloneqq \{vw \in E(G) \mid v \in X, w \in Y\}$ denotes the set of edges with one endpoint in X and one endpoint in Y). In this case we call $A \cap B$ a *separator*. A separation (A, B) is a called a *clique separation* if $A \cap B$ is a clique and $A \setminus B \neq \emptyset$ and $B \setminus A \neq \emptyset$. In this case we call $A \cap B$ a *clique separator*.

Definition 5.7 ([15, 58]) The *k-improvement* of a graph G is the graph G^k obtained from G by adding an edge between every pair of non-adjacent vertices v, w for which there are more than k pairwise internally vertex disjoint paths connecting v and w. We say that a graph G *is k-improved* when $G^k = G$.

A graph is *k-basic* if it is k-improved and does not have any separating cliques.

Note that a k-basic graph is 2-connected. We summarize several structural properties of G^k.

Lemma 5.8 ([58]) *Let G be a graph and $k \in \mathbb{N}$.*

1. *The k-improvement G^k is k-improved, i.e., $(G^k)^k = G^k$.*

2. *Every tree decomposition of G of width at most k is also a tree decomposition of G^k.*

3. *There exists an algorithm that, given G and k, runs in $\mathcal{O}(k^2 n^3)$ time and either correctly concludes that $\operatorname{tw}(G) > k$, or computes G^k.*

Since the construction of G^k from G is isomorphism-invariant, the concept of the improved graph can be exploited for isomorphism testing. Given a graph, we can compute the k-improvement and assign a special color to all added edges to distinguish them from the original edges. Hence, it suffices to solve the isomorphism problem for k-improved graphs. In order to further reduce isomorphism testing to the case of k-basic graphs, we build on a decomposition result of Leimer [55] which provides an isomorphism-invariant tree decomposition of a graph into clique-separator free parts.

Theorem 5.9 ([55, 29]) *For every connected graph G there is an isomorphism-invariant tree decomposition (T, β) of G, called clique separator decomposition, such that*

1. *for every $t \in V(T)$ the graph $G[\beta(t)]$ is clique-separator free, and*

2. *each adhesion set of (T, β) is a clique.*

Moreover, given a graph G, the clique separator decomposition of G can be computed in polynomial time.

Observe that applying the theorem to a k-improved graph results in a tree decomposition (T, β) such that $G[\beta(t)]$ is k-basic for every $t \in V(T)$. Using standard dynamic programming arguments, this decomposition allows us to essentially restrict to the case of k-basic graphs at the cost of an additional factor $2^{\mathcal{O}(k \log k)} n^{\mathcal{O}(1)}$ in the running time. Now, a key insight is that k-basic graphs are once again k-CR-bounded after individualizing a single vertex. The next lemma provides us the main tool for proving this. Recall that, for a graph G and sets $A, B \subseteq V(G)$, we denote $E(A, B) \coloneqq \{vw \in E(G) \mid v \in A, w \in B\}$.

Lemma 5.10 *Let G be a graph and let χ be a coloring that is stable with respect to the Color Refinement algorithm. Let $X_1, \ldots, X_m$ be distinct color classes of χ and suppose that $E(X_i, X_{i+1}) \neq \emptyset$ for all $i \in [m-1]$. Then there are $\min_{i \in [m]} |X_i|$ many vertex-disjoint paths from X_1 to X_m.*

To gain some intuition on the lemma, consider the simple case where $\ell \coloneqq |X_i| = |X_j|$ for all $i, j \in [m]$. Since χ is stable we conclude that the bipartite graph $H_i \coloneqq (X_i \cup X_{i+1}, E(X_i, X_{i+1}))$ is non-empty and biregular. Hence, by Hall's Marriage Theorem, there is a perfect matching M_i in H_i of size $|M_i| = \ell$. Taking the disjoint union of all sets M_i, $i \in [m-1]$, gives ℓ vertex-disjoint paths from X_1 to X_m. The general case can be proved using similar arguments. The details can be found in [43].

Lemma 5.11 *Let G be a k-basic graph. Also let $v \in V(G)$ such that $\deg(v) \leq k$. Then $\mathrm{cl}_k^G(v) = V(G)$.*

Proof Let χ denote the k-CR-stable coloring of G after individualizing v. First note that $N(v) \subseteq \mathrm{cl}_k^G(v)$ since $\deg(v) \leq k$. Now suppose towards a contradiction that $\mathrm{cl}_k^G(v) \neq V(G)$ and let W be the vertex set of a connected component of $G - \mathrm{cl}_k^G(v)$. Then $v \notin N(W)$. Since G does not have any separating cliques there exist $w_1, w_2 \in N(W)$ such that $w_1 w_2 \notin E(G)$. Let $u_0, u_1, \ldots, u_m, u_{m+1}$ be a path from $w_1 = u_0$ to $w_2 = u_{m+1}$ such that $u_i \in W$ for all $i \in [m]$. Let $X_i \coloneqq [u_i]_\chi$ and note that $|X_i| \geq k + 1$. Also note that $X_1 \subseteq N(w_1)$ and $X_m \subseteq N(w_2)$ since $w_1, w_2 \in \mathrm{cl}_k^G(v)$ and χ is stable. Hence, by Lemma 5.10, there are $k + 1$ internally vertex-disjoint paths from w_1 to w_2. But this contradicts the fact that G is k-improved. $\square$

Overall, this gives an isomorphism test for graphs of tree width k running in time $2^{\mathcal{O}(k \log k)} n^{\mathrm{polylog}(k)}$ as follows. First, the algorithm computes the clique-separator decomposition of the k-improved graph using Lemma 5.8 and Theorem 5.9. The isomorphism problem for the k-basic parts can be solved in time $n^{\mathrm{polylog}(k)}$ by Lemma 5.11 and Corollary 5.3. Finally, the results for the bags can be joined using dynamic programming along the clique separator decomposition. (Formally, this requires the algorithm to solve a more general problem for k-basic graphs where information obtained by dynamic programming on the adhesion sets to the children of a bag are incorporated. But this can be easily achieved within the group-theoretic framework; see e.g. [43].)

Using further advanced techniques (which are beyond the scope of this article) this result can be strengthened into two directions. First, building isomorphism-invariant decompositions of k-basic graphs, it is possible to build an improved fixed-parameter tractable isomorphism test for graphs of tree width k running in time $2^k \operatorname{polylog}(k) n^3$ [42] (improving on the $2^{O(k^5 \log k)} n^5$ algorithm by Lokshtanov et al. [58]).

And second, isomorphism of graphs of tree width k can also be tested in time $n^{\operatorname{polylog}(k)}$. Here, the crucial step is to speed up to dynamic programming procedure along the clique separator decomposition. Let D be the vertex set of the root bag of the clique separator decomposition and let $Z_1, \ldots, Z_\ell$ be connected components of $G - D$ such that $S := N_G(Z_i) = N_G(Z_j)$ for all $i, j \in [\ell]$. Observe that $|S| \leq k + 1$ since it is a clique by the properties of the clique separator decomposition. After recursively computing representations for the sets of isomorphisms between $(G[Z_i \cup S], S)$ and $(G[Z_j \cup S], S)$ for all $i, j \in [\ell]$ the partial solutions have to be joined into the automorphism group of $(G[Z_1 \cup \cdots \cup Z_\ell \cup S], S)$. A simple strategy is to iterate through all permutations $\gamma \in \operatorname{Sym}(S)$ and check in polynomial time which permutations can be extended to an automorphism. This gives a running time $2^{O(k \log k)} \ell^{O(1)}$ as already exploited above. Giving another extension of Babai's quasi-polynomial time algorithm, Wiebking [87] presented an algorithm solving this problem in time $\ell^{\operatorname{polylog}(k)}$. This enables us to achieve the running time $n^{\operatorname{polylog}(k)}$ for isomorphism testing of graphs of tree width k.

5.4 Graphs Excluding Small (Topological) Minors

In the previous two subsections we have seen isomorphism tests running in time $n^{\operatorname{polylog}(g)}$ for graphs of genus g and $n^{\operatorname{polylog}(k)}$ for graphs of tree width k. Both algorithms follow a similar pattern of computing an isomorphism-invariant tree-decomposition of small adhesion-width such that (the torso of) each bag is t-CR-bounded after individualizing a constant number of vertices. Recall that the same strategy also works for graphs of bounded degree (taking the trivial tree decomposition with a single bag). This suggests that a similar approach might work in a much more general setting considering graphs that exclude an arbitrary h-vertex graph as a minor or even as a topological subgraph.

Let G, H be graphs. Recall that H is a minor of G if H can be obtained from G by deleting vertices and edges as well as contracting edges. Moreover, H is a *topological subgraph* of G if H can be obtained from G by deleting vertices and edges as well as dissolving degree-two vertices (i.e., deleting the vertex and connecting its two neighbors). Note that, if H is a topological subgraph of G, then H is also a minor of G (in general, this is not true in the other direction). A graph G *excludes* H *as a (topological) minor* if it has no (topological) minor isomorphic to H.

A common argument for the correctness of the algorithms considered so far is the construction of forbidden subgraphs from t-CR-stable colorings that are not discrete. In order to generalize the existing algorithms we first introduce a tool that allows us to construct large (topological) minors from t-CR-stable colorings which are not discrete, but still contain some individualized vertices. Let G be a graph, let $\chi \colon V(G) \to C$ be a vertex-coloring and let T be a tree with vertex set $V(T) = C$. A subgraph $H \subseteq G$ *agrees* with T if $\chi|_{V(H)} \colon H \cong T$, i.e., the coloring χ induces an isomorphism between H and T. Observe that each $H \subseteq G$ that agrees with a tree T is also a tree.

For a tree T we define $V_{\leq i}(T) := \{t \in V(T) \mid \deg(t) \leq i\}$ and $V_{\geq i}(T) := \{t \in V(T) \mid \deg(t) \geq i\}$. It is well known that for trees T it holds that $|V_{\geq 3}(T)| < |V_{\leq 1}(T)|$.

Lemma 5.12 ([43]) *Let G be a graph and let $\chi\colon V(G) \to C$ be a vertex-coloring that is stable with respect to the Color Refinement algorithm. Also let T be a tree with vertex set $V(T) = C$ and assume that for every $c_1 c_2 \in E(T)$ there are $v_1 \in \chi^{-1}(c_1)$ and $v_2 \in \chi^{-1}(c_2)$ such that $v_1 v_2 \in E(G)$. Let $m := \min_{c \in C} |\chi^{-1}(c)|$ and let $\ell := 2|V_{\leq 1}(T)| + |V_{\geq 3}(T)|$.*

Then there are (at least) $\lfloor \frac{m}{\ell} \rfloor$ pairwise vertex-disjoint trees in G that agree with T.

Intuitively speaking, assuming there is a stable coloring in which all color classes are large, the lemma states that one can find a large number of vertex-disjoint trees with predefined color patterns. Observe that Lemma 5.12 generalizes Lemma 5.10 which provides such a statement for paths instead of arbitrary trees (ignoring the additional factor of two for the number of vertex-disjoint paths in comparison to the minimum color class size).

The proof of this lemma is quite technical and we omit any details here. As a first, simple application we can provide a strengthened version of Lemma 5.5.

Lemma 5.13 *Let (G, χ) be a 3-connected, vertex-colored graph that excludes $K_{3,h}$ as a topological subgraph and suppose $V_1 \uplus V_2 = V(G)$ such that*

1. *$|[v]_\chi| = 1$ for all $v \in V_1$,*

2. *χ is stable with respect to the Color Refinement algorithm, and*

3. *$|V_1| \geq 3$.*

Then there exists $u \in V_2$ such that $|[u]_\chi| \leq 7h - 1$.

In comparison with Lemma 5.5 the graph G is only required to exclude $K_{3,h}$ as a topological subgraph instead of a minor. Hence, Lemma 5.13 covers a much larger class of graphs. Observe that the bound on the size of the color class $[u]_\chi$ is slightly worse which however is not relevant for most algorithmic applications since the additional factor disappears in the $\mathcal{O}$-notation.

The proof of Lemma 5.13 is similar to the proof of Lemma 5.5. Recall that in the proof of Lemma 5.5 we constructed a tree T' with three leaves representing three vertices $v_1, v_2, v_3 \in V_1$. Instead of using T' to construct a minor $K_{3,h}$ this tree can also be used to construct a topological subgraph $K_{3,h}$ building on Lemma 5.12. Note that $2|V_{\leq 1}(T')| + |V_{\geq 3}(T')| \leq 7$ which gives the additional factor on the bound of $|[u]_\chi|$.

Following the same arguments as in Section 5.2 we obtain a polynomial-time Turing reduction from GI for graphs excluding $K_{3,h}$ as a topological subgraph to GI for $(7h-1)$-CR-bounded graphs.

Theorem 5.14 *The Graph Isomorphism Problem for graphs excluding $K_{3,h}$ as a topological subgraph can be solved in time $n^{\text{polylog}(h)}$.*

This generalizes the previous algorithms for graphs of bounded degree as well as graphs of bounded genus (see Figure 7).

Next, we wish to generalize the algorithms for graphs of bounded genus and graphs of bounded tree width. More precisely, we aim for an algorithm that decides isomorphism of graphs that exclude an arbitrary h-vertex graph as a minor in time $n^{\text{polylog}(h)}$. Towards this end, first observe that a graph G excludes an arbitrary h-vertex graph as a minor if and only if G excludes K_h as a minor. The maximum h such that K_h is a minor of G is also known as the *Hadwiger number* of G. Note that a graph of Hadwiger number h excludes K_{h+1} as a minor. So in other words, the goal is to give an isomorphism test for graphs of Hadwiger number h running in time $n^{\text{polylog}(h)}$.

As a first basic tool we argue that, given a set $X \subseteq V(G)$, the t-closure of X only stops to grow at a separator of small size. Note that similar properties are implicitly proved for graphs of small genus and graphs of small tree width in the previous two subsections.

Lemma 5.15 ([43]) *Let $t \geq 3h^3$. Let G be a graph that excludes K_h as a topological subgraph. Also let $X \subseteq V(G)$ and let Z be the vertex set of a connected component of $G - \text{cl}_t^G(X)$. Then $|N(Z)| < h$.*

The strategy for the proof is very similar to the previous arguments building on Lemma 5.12. We assume towards a contradiction that $|N(Z)| \geq h$ for some connected component Z of $G - \text{cl}_t^G(X)$ and construct a topological subgraph K_h. Towards this end, it suffices to find vertex-disjoint paths between all pairs of vertices of a subset $Y \subseteq N(Z)$ of size at least h. Using Lemma 5.12 we can actually construct a large number of vertex-disjoint trees such that each $v \in Y$ has a neighbor in all of those trees. In particular, this gives the desired set of vertex-disjoint paths.

Now, similar to the previous examples, the central step is to obtain a suitable isomorphism-invariant decomposition of the input graph so that each part of the decomposition is t-CR-bounded after individualizing a constant number of vertices. For graphs of bounded genus as well as graphs of bounded tree width we could rely on existing decompositions from the literature. However, for graphs excluding K_h as a minor, such results do not seem to be known. In particular, neither of the decompositions applied for graphs of bounded genus and graphs of bounded tree width are suitable for graph classes that only exclude K_h as a minor.

The central idea to resolve this problem is to use the t-closure of a set itself as a way of defining a graph decomposition. We have already established in the last lemma that the t-closure can only stop to grow at a separator of small size. In particular, this enables us to bound the adhesion-width of the resulting tree decomposition (i.e., the maximum size of an adhesion set). The second, more involved challenge is to obtain a decomposition which is isomorphism-invariant. Recall that, in all previous examples, we individualized a constant number of vertices to "initiate the growth of the t-closure". In order to still obtain an isomorphism-invariant decomposition we have to ensure that $\text{cl}_t^G(v)$ is independent of the specific choice of v. Towards this end, we want to argue that there is an isomorphism-invariant set $X \subseteq V(G)$ such that $\text{cl}_t^G(v) = \text{cl}_t^G(w)$ for all $v, w \in X$.

This basic idea can indeed be realized, but the concrete formulation turns out to be slightly more complicated. More precisely, in order to incorporate information not "visible" to the Color Refinement algorithm, the t-closure is computed on an extended graph which allows us to explicitly encode certain structural information of the input graph.

Lemma 5.16 ([43]) *Let* $t = \Omega(h^4)$. *There is a polynomial-time algorithm that, given a connected vertex-colored graph G, either correctly concludes that G has a minor isomorphic to K_h or computes a vertex-colored [4] graph (G', χ') and a set $\emptyset \neq X \subseteq V(G')$ such that*

1. $X = \{v \in V(G') \mid \chi'(v) = c\}$ *for some color c,*

2. $X \subseteq \mathrm{cl}_t^{(G', \chi')}(v)$ *for every $v \in X$, and*

3. $X \subseteq V(G)$.

Moreover, the output of the algorithm is isomorphism-invariant with respect to G.

The proof of the lemma is rather technical and builds on the 2-dimensional Weisfeiler-Leman algorithm. For details we refer the reader to [43].

Now, an algorithm for testing isomorphism of graphs excluding K_h as a minor can proceed as follows. Given a graph G, we first compute an isomorphism-invariant set $X \subseteq V(G)$ using Lemma 5.16 as well as its closure $D := \mathrm{cl}_t^G(X)$. Then $|N_G(Z)| < h$ for every connected component Z of the graph $G - D$ by Lemma 5.15. Actually, by proceeding recursively on all connected components of $G - D$ we obtain an isomorphism-invariant tree decomposition (T, β) of adhesion-width at most $h - 1$. Moreover, for each bag, we can consider the corresponding extended graph from Lemma 5.16 to obtain a structure associated with the bag that is t-CR-bounded for $t = \Omega(h^4)$ (after individualizing a single vertex). Hence, using similar techniques as for graphs of bounded tree width in the previous subsection, we obtain the following result.

Theorem 5.17 (Grohe, Neuen and Wiebking [43]) *The Graph Isomorphism Problem for graphs excluding K_h as a minor can be solved in time $n^{\mathrm{polylog}(h)}$.*

From the algorithm described above we can also infer some information on the structure of automorphism groups of graphs excluding K_h as a minor.

Theorem 5.18 *Let G be a graph that excludes K_h as a minor. Then there is an isomorphism-invariant tree decomposition (T, β) of G such that*

1. *the adhesion-width of (T, β) is at most $h - 1$, and*

2. *for all $t \in V(T)$ there exists $v \in \beta(t)$ such that $(\mathrm{Aut}(G))_v[\beta(t)] \in \widehat{\Gamma}_d$ for $d = \mathcal{O}(h^4)$.*

---

[4]In [43, Theorem IV.2], this graph is pair-colored, i.e., a color is assigned to each pair of vertices. By adding gadgets for the pair colors such a graph can easily be transformed into a vertex-colored graph while preserving all required properties.

6 Concluding Remarks

We survey recent progress on the Graph Isomorphism Problem. Our focus is on Babai's quasi-polynomial time algorithm and a series of subsequent quasi-polynomial parameterized algorithms. In particular, we highlight the power of a new technique centered around the notion of t-CR-boundedness. Actually, upon completion of this survey article, the second author showed that these techniques can be pushed even further obtaining an isomorphism test running in time $n^{\text{polylog}(h)}$ for graphs only excluding a graph of order h as a topological subgraph [72]. More precisely, building on a very lengthy and technical analysis, it is possible to generalize Lemma 5.16 to graphs only excluding K_h as a topological subgraph. Following the same strategy outlined above, this implies a faster isomorphism test graphs excluding K_h as a topological subgraph.

Research on the isomorphism problem is still very active, and interesting questions remain open. Of course, the most important question remains whether isomorphism of graphs can be tested in polynomial time. Let us also mention a few more specific questions that seem to be within reach.

1. Can isomorphism of hypergraphs be tested in time $n^{\text{polylog}(n)} m^{\mathcal{O}(1)}$ where n denotes the number of vertices and m the number of hyperedges of the input?

2. Can the techniques based on t-CR-boundedness be exploited to design faster isomorphism tests for dense graphs? For example, is there an isomorphism algorithm running in time $n^{\text{polylog}(k)}$ for graphs of rank width at most k?

3. Is GI parameterized by the Hadwiger number fixed-parameter tractable, that is, is there an isomorphism algorithm running in time $f(k)n^{\mathcal{O}(1)}$ for graphs excluding a graph of order k as a minor, for some arbitrary function f? The same question can be asked for topological subgraphs.

4. Is GI parameterized by maximum degree number fixed-parameter tractable?

The fourth question and the third for topological subgraphs seem to be hardest. The question whether GI parameterized by maximum degree is fixed parameter tractable has already been raised more than twenty years ago in Downey and Fellow's monograph on parameterized complexity [27].

Other interesting research directions include the Group Isomorphism Problem (see, e.g., [8, 17, 18, 79]) and the Tournament Isomorphism Problem (see, e.g., [11, 81]).

References

[1] Babak Ahmadi, Kristian Kersting, Martin Mladenov, and Sriraam Natarajan, *Exploiting symmetries for scaling loopy belief propagation and relational training*, Mach. Learn. **92** (2013), no. 1, 91–132.

[2] Albert Atserias and Elitza N. Maneva, *Sherali-Adams relaxations and indistinguishability in counting logics*, SIAM J. Comput. **42** (2013), no. 1, 112–137.

[3] Albert Atserias and Joanna Ochremiak, *Definable ellipsoid method, sums-of-squares proofs, and the isomorphism problem*, Proceedings of the 33rd Annual ACM/IEEE Symposium on Logic in Computer Science, LICS 2018, Oxford, UK, July 09-12, 2018 (Anuj Dawar and Erich Grädel, eds.), ACM, 2018, pp. 66–75.

[4] László Babai, *Monte Carlo algorithms in graph isomorphism testing*, Tech. Report 79-10, Université de Montréal, 1979.

[5] _____, *Graph isomorphism in quasipolynomial time*, CoRR **abs/1512.03547** (2015).

[6] _____, *Graph isomorphism in quasipolynomial time [extended abstract]*, Proceedings of the 48th Annual ACM SIGACT Symposium on Theory of Computing, STOC 2016, Cambridge, MA, USA, June 18-21, 2016 (Daniel Wichs and Yishay Mansour, eds.), ACM, 2016, pp. 684–697.

[7] László Babai, Peter J. Cameron, and Péter P. Pálfy, *On the orders of primitive groups with restricted nonabelian composition factors*, J. Algebra **79** (1982), no. 1, 161–168.

[8] László Babai, Paolo Codenotti, and Youming Qiao, *Polynomial-time isomorphism test for groups with no abelian normal subgroups - (extended abstract)*, Automata, Languages, and Programming - 39th International Colloquium, ICALP 2012, Warwick, UK, July 9-13, 2012, Proceedings, Part I (Artur Czumaj, Kurt Mehlhorn, Andrew M. Pitts, and Roger Wattenhofer, eds.), Lecture Notes in Computer Science, vol. 7391, Springer, 2012, pp. 51–62.

[9] László Babai, Paul Erdös, and Stanley M. Selkow, *Random graph isomorphism*, SIAM J. Comput. **9** (1980), no. 3, 628–635.

[10] László Babai, William M. Kantor, and Eugene M. Luks, *Computational complexity and the classification of finite simple groups*, 24th Annual Symposium on Foundations of Computer Science, Tucson, Arizona, USA, 7-9 November 1983, IEEE Computer Society, 1983, pp. 162–171.

[11] László Babai and Eugene M. Luks, *Canonical labeling of graphs*, Proceedings of the 15th Annual ACM Symposium on Theory of Computing, 25-27 April, 1983, Boston, Massachusetts, USA (David S. Johnson, Ronald Fagin, Michael L. Fredman, David Harel, Richard M. Karp, Nancy A. Lynch, Christos H. Papadimitriou, Ronald L. Rivest, Walter L. Ruzzo, and Joel I. Seiferas, eds.), ACM, 1983, pp. 171–183.

[12] László Babai, Eugene M. Luks, and Ákos Seress, *Permutation groups in NC*, Proceedings of the 19th Annual ACM Symposium on Theory of Computing, 1987, New York, New York, USA (Alfred V. Aho, ed.), ACM, 1987, pp. 409–420.

[13] Christoph Berkholz, Paul S. Bonsma, and Martin Grohe, *Tight lower and upper bounds for the complexity of canonical colour refinement*, Theory Comput. Syst. **60** (2017), no. 4, 581–614.

[14] Christoph Berkholz and Martin Grohe, *Limitations of algebraic approaches to graph isomorphism testing*, Automata, Languages, and Programming - 42nd International Colloquium, ICALP 2015, Kyoto, Japan, July 6-10, 2015, Proceedings, Part I (Magnús M. Halldórsson, Kazuo Iwama, Naoki Kobayashi, and Bettina Speckmann, eds.), Lecture Notes in Computer Science, vol. 9134, Springer, 2015, pp. 155–166.

[15] Hans L. Bodlaender, *Necessary edges in k-chordalisations of graphs*, J. Comb. Optim. **7** (2003), no. 3, 283–290.

[16] Béla Bollobás, *Distinguishing vertices of random graphs*, Ann. Discrete Math. **13** (1982), 33–49.

[17] Jendrik Brachter and Pascal Schweitzer, *On the Weisfeiler-Leman dimension of finite groups*, LICS '20: 35th Annual ACM/IEEE Symposium on Logic in Computer Science, Saarbrücken, Germany, July 8-11, 2020 (Holger Hermanns, Lijun Zhang, Naoki Kobayashi, and Dale Miller, eds.), ACM, 2020, pp. 287–300.

[18] Peter A. Brooksbank, Joshua A. Grochow, Yinan Li, Youming Qiao, and James B. Wilson, *Incorporating Weisfeiler-Leman into algorithms for group isomorphism*, CoRR **abs/1905.02518** (2019).

[19] Jin-yi Cai, Martin Fürer, and Neil Immerman, *An optimal lower bound on the number of variables for graph identification*, Comb. **12** (1992), no. 4, 389–410.

[20] Peter J. Cameron, *Finite permutation groups and finite simple groups*, Bull. London Math. Soc. **13** (1981), no. 1, 1–22.

[21] Alain Cardon and Maxime Crochemore, *Partitioning a graph in $O(|A|\log_2 |V|)$*, Theor. Comput. Sci. **19** (1982), no. 1, 85–98.

[22] Paolo Codenotti, Hadi Katebi, Karem A. Sakallah, and Igor L. Markov, *Conflict analysis and branching heuristics in the search for graph automorphisms*, 25th IEEE International Conference on Tools with Artificial Intelligence, ICTAI 2013, Herndon, VA, USA, November 4-6, 2013, IEEE Computer Society, 2013, pp. 907–914.

[23] Marek Cygan, Fedor V. Fomin, Lukasz Kowalik, Daniel Lokshtanov, Dániel Marx, Marcin Pilipczuk, Michal Pilipczuk, and Saket Saurabh, *Parameterized algorithms*, Springer, 2015.

[24] Anuj Dawar, David Richerby, and Benjamin Rossman, *Choiceless polynomial time, counting and the Cai-Fürer-Immerman graphs*, Ann. Pure Appl. Log. **152** (2008), no. 1-3, 31–50.

[25] Reinhard Diestel, *Graph theory*, 5th ed., Springer Verlag, 2016.

[26] John D. Dixon and Brian Mortimer, *Permutation groups*, Graduate Texts in Mathematics, vol. 163, Springer-Verlag, New York, 1996.

[27] Rodney G. Downey and Michael R. Fellows, *Parameterized complexity*, Monographs in Computer Science, Springer, 1999.

[28] Zdenek Dvorák, *On recognizing graphs by numbers of homomorphisms*, J. Graph Theory **64** (2010), no. 4, 330–342.

[29] Michael Elberfeld and Pascal Schweitzer, *Canonizing graphs of bounded tree width in logspace*, ACM Trans. Comput. Theory **9** (2017), no. 3, 12:1–12:29.

[30] Sergei Evdokimov, Ilia N. Ponomarenko, and Gottfried Tinhofer, *Forestal algebras and algebraic forests (on a new class of weakly compact graphs)*, Discret. Math. **225** (2000), no. 1-3, 149–172.

[31] I. S. Filotti and Jack N. Mayer, *A polynomial-time algorithm for determining the isomorphism of graphs of fixed genus (working paper)*, Proceedings of the 12th Annual ACM Symposium on Theory of Computing, April 28-30, 1980, Los Angeles, California, USA (Raymond E. Miller, Seymour Ginsburg, Walter A. Burkhard, and Richard J. Lipton, eds.), ACM, 1980, pp. 236–243.

[32] Erich Grädel, Martin Grohe, Benedikt Pago, and Wied Pakusa, *A finite-model-theoretic view on propositional proof complexity*, Log. Methods Comput. Sci. **15** (2019), no. 1, 4:1–4:53.

[33] Martin Grohe, *Fixed-point logics on planar graphs*, Thirteenth Annual IEEE Symposium on Logic in Computer Science, Indianapolis, Indiana, USA, June 21-24, 1998, IEEE Computer Society, 1998, pp. 6–15.

[34] _____, *Isomorphism testing for embeddable graphs through definability*, Proceedings of the Thirty-Second Annual ACM Symposium on Theory of Computing, May 21-23, 2000, Portland, OR, USA (F. Frances Yao and Eugene M. Luks, eds.), ACM, 2000, pp. 63–72.

[35] Martin Grohe, *Descriptive complexity, canonisation, and definable graph structure theory*, Lecture Notes in Logic, vol. 47, Cambridge University Press, 2017.

[36] Martin Grohe, Kristian Kersting, Martin Mladenov, and Erkal Selman, *Dimension reduction via colour refinement*, Algorithms - ESA 2014 - 22th Annual European Symposium, Wroclaw, Poland, September 8-10, 2014. Proceedings (Andreas S. Schulz and Dorothea Wagner, eds.), Lecture Notes in Computer Science, vol. 8737, Springer, 2014, pp. 505–516.

[37] Martin Grohe and Sandra Kiefer, *A linear upper bound on the Weisfeiler-Leman dimension of graphs of bounded genus*, 46th International Colloquium on Automata, Languages, and Programming, ICALP 2019, July 9-12, 2019, Patras, Greece (Christel Baier, Ioannis Chatzigiannakis, Paola Flocchini, and Stefano Leonardi, eds.), LIPIcs, vol. 132, Schloss Dagstuhl - Leibniz-Zentrum für Informatik, 2019, pp. 117:1–117:15.

[38] Martin Grohe and Julian Mariño, *Definability and descriptive complexity on databases of bounded tree-width*, Database Theory - ICDT '99, 7th International Conference, Jerusalem, Israel, January 10-12, 1999, Proceedings (Catriel Beeri and Peter Buneman, eds.), Lecture Notes in Computer Science, vol. 1540, Springer, 1999, pp. 70–82.

[39] Martin Grohe and Dániel Marx, *Structure theorem and isomorphism test for graphs with excluded topological subgraphs*, SIAM J. Comput. **44** (2015), no. 1, 114–159.

[40] Martin Grohe and Daniel Neuen, *Canonisation and definability for graphs of bounded rank width*, 34th Annual ACM/IEEE Symposium on Logic in Computer Science, LICS 2019, Vancouver, BC, Canada, June 24-27, 2019, IEEE, 2019, pp. 1–13.

[41] Martin Grohe, Daniel Neuen, and Pascal Schweitzer, *A faster isomorphism test for graphs of small degree*, 59th IEEE Annual Symposium on Foundations of Computer Science, FOCS 2018, Paris, France, October 7-9, 2018 (Mikkel Thorup, ed.), IEEE Computer Society, 2018, pp. 89–100.

[42] Martin Grohe, Daniel Neuen, Pascal Schweitzer, and Daniel Wiebking, *An improved isomorphism test for bounded-tree-width graphs*, ACM Trans. Algorithms **16** (2020), no. 3, 34:1–34:31.

[43] Martin Grohe, Daniel Neuen, and Daniel Wiebking, *Isomorphism testing for graphs excluding small minors*, 61st IEEE Annual Symposium on Foundations of Computer Science, FOCS 2020, November 16-19, 2020 (Virtual Conference) (Sandy Irani, ed.), IEEE Computer Society, 2020, pp. 625–637.

[44] Martin Grohe and Martin Otto, *Pebble games and linear equations*, J. Symb. Log. **80** (2015), no. 3, 797–844.

[45] Martin Grohe and Pascal Schweitzer, *The graph isomorphism problem*, Commun. ACM **63** (2020), no. 11, 128–134.

[46] Martin Grohe, Pascal Schweitzer, and Daniel Wiebking, *Deep Weisfeiler Leman*, Proceedings of the 32nd ACM-SIAM Symposium on Discrete Algorithms, 2021.

[47] Harald Andrés Helfgott, Jitendra Bajpai, and Daniele Dona, *Graph isomorphisms in quasi-polynomial time*, CoRR **abs/1710.04574** (2017).

[48] John E. Hopcroft and Robert Endre Tarjan, *Isomorphism of planar graphs*, Proceedings of a symposium on the Complexity of Computer Computations, held March 20-22, 1972, at the IBM Thomas J. Watson Research Center, Yorktown Heights, New York, USA (Raymond E. Miller and James W. Thatcher, eds.), The IBM Research Symposia Series, Plenum Press, New York, 1972, pp. 131–152.

[49] Neil Immerman and Eric Lander, *Describing graphs: A first-order approach to graph canonization*, Complexity Theory Retrospective (Alan L. Selman, ed.), Springer New York, New York, NY, 1990, pp. 59–81.

[50] Tommi A. Junttila and Petteri Kaski, *Conflict propagation and component recursion for canonical labeling*, Theory and Practice of Algorithms in (Computer) Systems - First International ICST Conference, TAPAS 2011, Rome, Italy, April 18-20, 2011. Proceedings (Alberto Marchetti-Spaccamela and

Michael Segal, eds.), Lecture Notes in Computer Science, vol. 6595, Springer, 2011, pp. 151–162.

[51] Richard M. Karp, *Reducibility among combinatorial problems*, Proceedings of a symposium on the Complexity of Computer Computations, held March 20-22, 1972, at the IBM Thomas J. Watson Research Center, Yorktown Heights, New York, USA (Raymond E. Miller and James W. Thatcher, eds.), The IBM Research Symposia Series, Plenum Press, New York, 1972, pp. 85–103.

[52] Sandra Kiefer, *The Weisfeiler-Leman algorithm: An exploration of its power*, ACM SIGLOG News **7** (2020), no. 3, 5–27.

[53] Sandra Kiefer and Daniel Neuen, *The power of the Weisfeiler-Leman algorithm to decompose graphs*, 44th International Symposium on Mathematical Foundations of Computer Science, MFCS 2019, August 26-30, 2019, Aachen, Germany (Peter Rossmanith, Pinar Heggernes, and Joost-Pieter Katoen, eds.), LIPIcs, vol. 138, Schloss Dagstuhl - Leibniz-Zentrum für Informatik, 2019, pp. 45:1–45:15.

[54] Sandra Kiefer, Ilia Ponomarenko, and Pascal Schweitzer, *The Weisfeiler-Leman dimension of planar graphs is at most 3*, J. ACM **66** (2019), no. 6, 44:1–44:31.

[55] Hanns-Georg Leimer, *Optimal decomposition by clique separators*, Discret. Math. **113** (1993), no. 1-3, 99–123.

[56] Martin W. Liebeck and Aner Shalev, *Simple groups, permutation groups, and probability*, J. Amer. Math. Soc. **12** (1999), no. 2, 497–520.

[57] Steven Lindell, *A logspace algorithm for tree canonization (extended abstract)*, Proceedings of the 24th Annual ACM Symposium on Theory of Computing, May 4-6, 1992, Victoria, British Columbia, Canada (S. Rao Kosaraju, Mike Fellows, Avi Wigderson, and John A. Ellis, eds.), ACM, 1992, pp. 400–404.

[58] Daniel Lokshtanov, Marcin Pilipczuk, Michal Pilipczuk, and Saket Saurabh, *Fixed-parameter tractable canonization and isomorphism test for graphs of bounded treewidth*, SIAM J. Comput. **46** (2017), no. 1, 161–189.

[59] José Luis López-Presa, Luis F. Chiroque, and Antonio Fernández Anta, *Novel techniques to speed up the computation of the automorphism group of a graph*, J. Appl. Math. **2014** (2014), 934637:1–934637:15.

[60] László Lovász, *Operations with structures*, Acta Math. Acad. Sci. Hungar. **18** (1967), 321–328.

[61] Eugene M. Luks, *Isomorphism of graphs of bounded valence can be tested in polynomial time*, J. Comput. Syst. Sci. **25** (1982), no. 1, 42–65.

[62] Peter N. Malkin, *Sherali-Adams relaxations of graph isomorphism polytopes*, Discret. Optim. **12** (2014), 73–97.

[63] Attila Maróti, *On the orders of primitive groups*, J. Algebra **258** (2002), no. 2, 631–640.

[64] Brendan D. McKay, *Practical graph isomorphism*, Congr. Numer. **30** (1981), 45–87.

[65] Brendan D. McKay and Adolfo Piperno, *Practical graph isomorphism, II*, J. Symb. Comput. **60** (2014), 94–112.

[66] Gary L. Miller, *Isomorphism testing for graphs of bounded genus*, Proceedings of the 12th Annual ACM Symposium on Theory of Computing, April 28-30, 1980, Los Angeles, California, USA (Raymond E. Miller, Seymour Ginsburg, Walter A. Burkhard, and Richard J. Lipton, eds.), ACM, 1980, pp. 225–235.

[67] _____, *Isomorphism of graphs which are pairwise k-separable*, Inf. Control. **56** (1983), no. 1/2, 21–33.

[68] Howard L. Morgan, *The generation of a unique machine description for chemical structures-a technique developed at chemical abstracts service.*, J. Chem. Doc. **5** (1965), no. 2, 107–113.

[69] Christopher Morris, Martin Ritzert, Matthias Fey, William L. Hamilton, Jan Eric Lenssen, Gaurav Rattan, and Martin Grohe, *Weisfeiler and Leman go neural: Higher-order graph neural networks*, The Thirty-Third AAAI Conference on Artificial Intelligence, AAAI 2019, Honolulu, Hawaii, USA, January 27 - February 1, 2019, AAAI Press, 2019, pp. 4602–4609.

[70] Daniel Neuen, *The power of algorithmic approaches to the graph isomorphism problem*, Ph.D. thesis, RWTH Aachen University, Aachen, Germany, 2019.

[71] _____, *Hypergraph isomorphism for groups with restricted composition factors*, 47th International Colloquium on Automata, Languages, and Programming, ICALP 2020, July 8-11, 2020, Saarbrücken, Germany (Virtual Conference) (Artur Czumaj, Anuj Dawar, and Emanuela Merelli, eds.), LIPIcs, vol. 168, Schloss Dagstuhl - Leibniz-Zentrum für Informatik, 2020, pp. 88:1–88:19.

[72] _____, *Isomorphism testing for graphs excluding small topological subgraphs*, CoRR **abs/2011.14730** (2020).

[73] Ryan O'Donnell, John Wright, Chenggang Wu, and Yuan Zhou, *Hardness of robust graph isomorphism, Lasserre gaps, and asymmetry of random graphs*, Proceedings of the Twenty-Fifth Annual ACM-SIAM Symposium on Discrete Algorithms, SODA 2014, Portland, Oregon, USA, January 5-7, 2014 (Chandra Chekuri, ed.), SIAM, 2014, pp. 1659–1677.

[74] Robert Paige and Robert E. Tarjan, *Three partition refinement algorithms*, SIAM J. Comput. **16** (1987), no. 6, 973–989.

[75] Ilia N. Ponomarenko, *Polynomial isomorphism algorithm for graphs which do not pinch to $K_{3,g}$*, J. Math. Sci. **34** (1986), 1819–1831.

[76] _____, *The isomorphism problem for classes of graphs that are invariant with respect to contraction*, Zap. Nauchn. Sem. Leningrad. Otdel. Mat. Inst. Steklov. (LOMI) **174** (1988), no. Teor. Slozhn. Vychisl. 3, 147–177, 182.

[77] Louis C. Ray and Russell A. Kirsch, *Finding chemical records by digital computers*, Science **126** (1957), no. 3278, 814–819.

[78] Gerhard Ringel, *Das geschlecht des vollständigen paaren graphen*, Abh. Math. Semin. Univ. Hambg. **28** (1965), no. 3, 139–150.

[79] David J. Rosenbaum, *Beating the generator-enumeration bound for solvable-group isomorphism*, ACM Trans. Comput. Theory **12** (2020), no. 2, 12:1–12:18.

[80] Joseph J. Rotman, *An introduction to the theory of groups*, fourth ed., Graduate Texts in Mathematics, vol. 148, Springer-Verlag, New York, 1995.

[81] Pascal Schweitzer, *A polynomial-time randomized reduction from tournament isomorphism to tournament asymmetry*, 44th International Colloquium on Automata, Languages, and Programming, ICALP 2017, July 10-14, 2017, Warsaw, Poland (Ioannis Chatzigiannakis, Piotr Indyk, Fabian Kuhn, and Anca Muscholl, eds.), LIPIcs, vol. 80, Schloss Dagstuhl - Leibniz-Zentrum für Informatik, 2017, pp. 66:1–66:14.

[82] Ákos Seress, *Permutation group algorithms*, Cambridge Tracts in Mathematics, vol. 152, Cambridge University Press, Cambridge, 2003.

[83] Nino Shervashidze, Pascal Schweitzer, Erik Jan van Leeuwen, Kurt Mehlhorn, and Karsten M. Borgwardt, *Weisfeiler-Lehman graph kernels*, J. Mach. Learn. Res. **12** (2011), 2539–2561.

[84] Gottfried Tinhofer, *A note on compact graphs*, Discret. Appl. Math. **30** (1991), no. 2-3, 253–264.

[85] Stephen H. Unger, *GIT - a heuristic program for testing pairs of directed line graphs for isomorphism*, Commun. ACM **7** (1964), no. 1, 26–34.

[86] Boris Y. Weisfeiler and Andrei A. Leman, *The reduction of a graph to canonical form and the algebra which appears therein*, NTI, Series 2 (1968), English transalation by G. Ryabov available at https://www.iti.zcu.cz/wl2018/pdf/wl_paper_translation.pdf.

[87] Daniel Wiebking, *Graph isomorphism in quasipolynomial time parameterized by treewidth*, 47th International Colloquium on Automata, Languages, and Programming, ICALP 2020, July 8-11, 2020, Saarbrücken, Germany (Virtual Conference) (Artur Czumaj, Anuj Dawar, and Emanuela Merelli, eds.), LIPIcs, vol. 168, Schloss Dagstuhl - Leibniz-Zentrum für Informatik, 2020, pp. 103:1–103:16.

[88] Keyulu Xu, Weihua Hu, Jure Leskovec, and Stefanie Jegelka, *How powerful are graph neural networks?*, 7th International Conference on Learning Representations, ICLR 2019, New Orleans, LA, USA, May 6-9, 2019, OpenReview.net, 2019.

RWTH Aachen University
Ahornstr. 55
52074 Aachen
Germany
grohe@informatik.rwth-aachen.de

CISPA Helmholtz Center for Information Security
Stuhlsatzenhaus 5
66123 Saarbruecken
Germany
daniel.neuen@cispa.saarland

Extremal aspects of graph and hypergraph decomposition problems

Stefan Glock   Daniela Kühn   Deryk Osthus

Abstract

We survey recent advances in the theory of graph and hypergraph decompositions, with a focus on extremal results involving minimum degree conditions. We also collect a number of intriguing open problems, and formulate new ones.

1 Introduction

The problem of decomposing large objects into (simple) smaller ones pervades many areas of Mathematics. This is particularly true for Combinatorics. Here, we focus on graphs and hypergraphs: given two graphs F and G, an *F-decomposition of G* is a collection of subgraphs of G, each isomorphic to F, such that every edge of G is used exactly once.

Graph decomposition problems have a long history, dating back to the work of Euler on orthogonal Latin squares. In 1847, Kirkman [72] proved that the complete graph K_n has a K_3-decomposition if and only if $n \equiv 1, 3 \mod 6$. A related problem posed by Kirkman asks for a K_3-decomposition of K_n such that the triangles can be organised into edge-disjoint K_3-factors, where a K_3-*factor* is a set of vertex-disjoint triangles covering all vertices. Also in the 19th century, Walecki proved the existence of decompositions of the complete graph K_n into edge-disjoint Hamilton cycles (for odd n) and Hamilton paths (for even n). Note here that the assumptions on the parity of n are necessary. Indeed, K_n has $n(n-1)/2$ edges, and a Hamilton cycle has n edges, so a decomposition into edge-disjoint Hamilton cycles has to consist of $(n-1)/2$ such cycles, implying that n is odd. Similar 'necessary divisibility conditions' can be observed for essentially every decomposition problem, and we will encounter many of these throughout this survey.

Classical graph decomposition results have mostly been obtained based on the symmetry of the underlying structures, thus often involving algebraic techniques. Recently, much progress has been made in the area of decompositions using probabilistic techniques. This has gone hand in hand with the realisation that for many questions, it is not necessary that the underlying structure to be decomposed is 'complete' or 'highly symmetric'. This leads to the consideration of extremal aspects of such questions: *Are there natural (density) conditions which ensure (subject to the divisibility conditions) the existence of such decompositions?*

For Hamilton cycles and perfect matchings, the above question was resolved in [30], proving the so-called Hamilton decomposition conjecture and the 1-factorization conjecture for large n: the former states that for $d \geq \lfloor n/2 \rfloor$, every d-regular n-vertex graph G has a decomposition into Hamilton cycles and at most one perfect matching, the latter states that the corresponding threshold for decompositions into perfect matchings is $d \geq 2\lceil n/4 \rceil - 1$. Here a decomposition into perfect matchings is often called a 1-*factorization*. Similarly, Kelly's conjecture on Hamilton decompositions of regular tournaments not only turned out to be correct, but such a decomposition already exists in regular oriented graphs of minimum semi-degree at least $(3/8 + o(1))n$ [75]. For triangle decompositions (i.e. Steiner triple systems), the corresponding question translates to a conjecture by Nash-Williams, which we introduce at the end of this section.

Minimum degree versions of decomposition problems have a natural application to 'completion problems': For instance, the 1-factorization conjecture has an interpretation in terms of scheduling round-robin tournaments (where n players play all of each other in $n-1$ rounds): one can schedule the first half of the rounds arbitrarily before one needs to plan the remainder of the tournament. More generally, there are applications to completions of partial designs, to hypergraph Euler tours and to the completion of Latin squares, to mention only a few. We will discuss some of these applications in this survey.

Another feature of minimum degree conditions is that they provide a large class of graphs (or hypergraphs) whose membership is easy to verify algorithmically. In general, the question of whether a given (hyper-)graph G has some decomposition into a given class of subgraphs is NP-complete [35]. Thus it is natural to seek simple sufficient conditions for the existence of such decompositions. Indeed, the minimum degree requirement yields attractive results and conjectures in this respect. Another very fruitful notion is that of quasirandomness – we will briefly discuss this here too (mainly in the guise of 'typicality'), but will omit a detailed discussion.

We will focus mainly on decompositions into small subgraphs. (We will also discuss related topics such as decompositions into 2-factors and spanning trees. For Hamilton decompositions, we refer e.g. to [76]). The main conjecture in this area is due to Nash-Williams, which we introduce now: Observe that if a graph G admits a K_3-decomposition, then the number of edges of G must be divisible by 3, and all the vertex degrees of G must be even. We say that G is K_3-*divisible* if it has these properties. Clearly, not every K_3-divisible graph has a K_3-decomposition (e.g. C_6). In fact, to decide whether a given graph has a K_3-decomposition is NP-hard [35]. However, the following beautiful conjecture of Nash-Williams suggests that if the minimum degree of G is sufficiently large, then the existence of a K_3-decomposition hinges only on the necessary divisibility condition.

Conjecture 1.1 ([85]) *For sufficiently large n, every K_3-divisible graph G on n vertices with $\delta(G) \geq 3n/4$ has a K_3-decomposition.*

The following class of extremal examples shows that the bound on the minimum degree would be best possible.

Example 1.2 Given any $k \in \mathbb{N}$, let G_1 and G_2 be vertex-disjoint $(6k+2)$-regular graphs with $|G_1| = |G_2| = 12k+6$ and let G_3 be the complete bipartite graph between $V(G_1)$ and $V(G_2)$. Let $G := G_1 \cup G_2 \cup G_3$. (In the standard construction, each of G_1 and G_2 is a union of two disjoint cliques of size $6k+3$, see Figure 1.) It is straightforward to check that $\delta(G) = 3|G|/4 - 1$ and that G is K_3-divisible. However, every triangle in G contains at least one edge from $G_1 \cup G_2$. Since $2e(G_1 \cup G_2) < e(G_3)$, G cannot have a K_3-decomposition.[1]

---

[1]This type of extremal example is called a 'space barrier'. There are simply not enough edges in $G_1 \cup G_2$ for a triangle decomposition. In other constructions, the obstacle might be a 'divisibility barrier'.

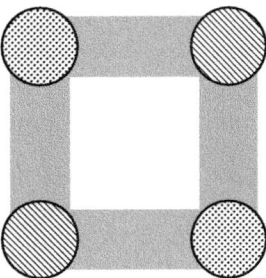

Figure 1: The standard example is a blown-up C_4, where G_1 and G_2 each consist of two disjoint cliques.

Conjecture 1.1 is still open. Nevertheless, it has already inspired very fruitful research, and will serve us as a thread running through this survey.

1.1 Organisation of this survey

In Section 1.2, we collect some basic notation. In Section 2, we consider approximate versions of Conjecture 1.1 and its generalisations, and in Section 3 we discuss how such approximate results can be turned into exact ones. In Section 4, we introduce the decomposition problem for hypergraphs. Subsequently, in Section 5 we present an application of hypergraph decompositions to Euler tours. Section 6 is devoted to the Oberwolfach problem. Finally, in Section 7, we close by briefly mentioning some further open problems.

1.2 Notation

Let us briefly agree on some general notation. A *hypergraph* G consists of a set of vertices $V(G)$ and a set of edges $E(G)$, where each edge is a subset of the vertex set. The hypergraph is called *k-uniform*, or just a *k-graph*, if every edge has size k. Hence, a 2-graph is simply a graph. We let $|G|$ denote the number of vertices of G and $e(G)$ the number of edges.

Let G be a k-graph. For a set $S \subseteq V(G)$ with $|S| = k-1$, we let $N_G(S)$ denote the *neighbourhood of S in G*, that is, the set of vertices which form an edge in G together with S. For a set $S \subseteq V(G)$ with $0 \leq |S| \leq k-1$, the *degree* $d_G(S)$ *of S* is the number of edges in G containing S. We let $\delta(G)$ and $\Delta(G)$ denote the minimum and maximum $(k-1)$-degree of G, respectively, that is, the minimum/maximum value of $d_G(S)$ over all $S \subseteq V(G)$ of size $k-1$.

Some of the results we state apply for 'quasirandom' (hyper-)graphs, by which we mean the following notion.

Definition 1.3 (Typicality) A k-graph G on n vertices is called (c, h, p)-*typical* if for any set A of $(k-1)$-subsets of $V(G)$ with $|A| \leq h$, we have $|\bigcap_{S \in A} N_G(S)| = (1 \pm c)p^{|A|}n$.

We let K_n^k denote the complete k-graph on n vertices. Finally, $[n]$ denotes the set $\{1,\ldots,n\}$.

2 Approximate and fractional decompositions

As Conjecture 1.1 turned out to be hard, one might try to prove at least an approximate version. For instance, one could attempt to obtain a collection of edge-disjoint triangles which do not cover all edges of G, but may leave $o(|G|^2)$ edges uncovered. We refer to such types of 'almost' decompositions as *approximate decompositions*. Another relaxation is to find only a 'fractional' decomposition. As we shall see, these two concepts are closely related, and played a pivotal role in recent progress towards Conjecture 1.1 and many related problems.

2.1 From fractional to approximate decompositions

Fractional relaxations of graph parameters have been studied extensively in recent decades. Often, it turns out that these fractional relaxations give good approximations for the original parameter. This can also have important algorithmic ramifications.

A *fractional F-decomposition* of a graph G is a function ω that assigns to each copy of F in G a value in $[0,1]$ such that for all $e \in E(G)$,

$$\sum_{F':\, e\in E(F')} \omega(F') = 1,$$

where the sum is over all copies F' of F which contain e.

Thus, an F-decomposition is a fractional F-decomposition taking values in $\{0,1\}$. Note that it can be much easier to obtain a fractional F-decomposition rather than an F-decomposition. For example, K_n always has a fractional F-decomposition (assuming $n \geq |F|$ of course), by giving every copy of F the same weight.

The following theorem of Haxell and Rödl [59] allows us to turn a fractional decomposition into an approximate decomposition.

Theorem 2.1 ([59]) *If an n-vertex graph G has a fractional F-decomposition, then all but $o(n^2)$ edges of G can be covered with edge-disjoint copies of F.*

The proof is based on Szemerédi's regularity lemma and a theorem of Frankl and Rödl [45]. The latter result allows one to obtain an approximate F-decomposition of a dense graph G whenever every edge of G is contained in roughly the same number of copies of F. This is a special case of a more general theorem on almost perfect matchings in (almost) regular hypergraphs with small maximum codegree, proved using the celebrated 'Rödl nibble'. Roughly speaking, the main idea in the proof of Theorem 2.1 is to partition G into edge-disjoint subgraphs, each of which has the above regularity condition. Such a partition is obtained with the help of Szemerédi's regularity lemma, using the assumption that G has a fractional F-decomposition. One can then apply the Frankl–Rödl result to all these subgraphs individually to obtain the desired approximate F-decomposition.

Given that there are numerous algorithmic applications, it would be very interesting to obtain a proof of Theorem 2.1 which does not rely on Szemerédi's regularity lemma.

Theorem 2.1 has been generalised to hypergraphs by Rödl, Schacht, Siggers and Tokushige [88].

2.2 Fractional decomposition thresholds

Motivated by this, we could aim to prove the fractional version of Conjecture 1.1. Observe that Example 1.2 also shows that the same minimum degree condition would be optimal for this relaxed version.

We define the *fractional F-decomposition threshold* δ_F^* to be the infimum of all $\delta \in [0,1]$ with the following property: there exists $n_0 \in \mathbb{N}$ such that any F-divisible graph G on $n \geq n_0$ vertices with $\delta(G) \geq \delta n$ has a fractional F-decomposition. We remark that it is somehow unnatural to assume divisibility here since a graph can have a fractional F-decomposition without being F-divisible. The sole purpose of this assumption is to have the inequality $\delta_F^* \leq \delta_F$ hold trivially, where δ_F is the threshold for F-decompositions which we will define in Section 3. This underpins our motivation that fractional decompositions are a relaxation of the original decomposition problem.[2]

Until recently, the best bound on $\delta_{K_3}^*$ was obtained by Dross [36], who showed that $\delta_{K_3}^* \leq 0.9$, using an elegant approach based on the max-flow-min-cut theorem. This improved earlier work of Yuster [99], as well as Dukes [37, 38] and Garaschuk [46]. Delcourt and Postle [34] hold the current record (a slightly weaker bound was obtained simultaneously by Dukes and Horsley [39]). An obvious open problem is to obtain further significant improvements.

Theorem 2.2 ([34]) $\delta_{K_3}^* \leq (7 + \sqrt{21})/14 \approx 0.82733$.

The best current bound on the fractional decomposition threshold of larger cliques was proved by Montgomery [82].

Theorem 2.3 ([82]) *For each $r \geq 4$, we have $\delta_{K_r}^* \leq 1 - 1/(100r)$.*

This result is best possible up to the constant 100. Indeed, a natural generalisation of Example 1.2 (see [99]) shows that the following conjecture (which is implicit in [54]) would be optimal.

Conjecture 2.4 *For each $r \geq 3$, we have $\delta_{K_r}^* \leq 1 - 1/(r+1)$.*

So far, we have only stated bounds for the fractional decomposition threshold of cliques. Yuster [100] showed that these bounds can be used to obtain approximate F-decompositions for general r-chromatic graphs F.

---

[2]Having said this, we point out that all the results in this section on δ_F^* also hold without the divisibility assumption, that is, none of the proofs showing existence of fractional F-decompositions make any use of the divisibility assumption, and we suspect that the value of δ_F^* is the same for both variants.

Theorem 2.5 ([100]) *For any fixed graph F with $r = \chi(F)$, any sufficiently large n-vertex graph G with $\delta(G) \geq (\delta^*_{K_r} + o(1))n$ contains edge-disjoint copies of F covering all but $o(n^2)$ edges of G.*

2.3 Bandwidth theorem for approximate decompositions

Not only does $\delta^*_{K_r}$ yield minimum degree bounds for approximate decompositions into a fixed r-chromatic graph F, but even for spanning r-chromatic graphs F, provided that F has bounded degree, the chromatic number of F is 'essentially' equal to r and F is 'path-like'. More precisely, a graph F is (r, η)-*chromatic* if the graph F', obtained from F by deleting isolated vertices, can be properly coloured with $r + 1$ colours such that one colour class has size at most $\eta|F'|$. For instance, the cycle C_ℓ is $(2, 1/\ell)$-chromatic. Moreover, F is η-*separable* if there exists a set S of at most $\eta|F|$ vertices such that each component of $F \setminus S$ has size at most $\eta|F|$. For bounded degree graphs, being separable is equivalent to having sublinear bandwidth. Examples of separable graphs include cycles, powers of cycles, trees and planar graphs.

The following result by Condon, Kim, Kühn and Osthus [29] provides a degree condition which ensures that a regular graph G has an approximate decomposition into $\mathcal{F}$ for any collection $\mathcal{F}$ of (r, η)-chromatic η-separable graphs of bounded degree. The degree condition is best possible in general (unless one has additional information about the graphs in $\mathcal{F}$). The 'original' bandwidth theorem of Böttcher, Schacht and Taraz [17] involves a similar degree condition for embedding a single spanning subgraph.

Given a collection $\mathcal{F}$ of graphs, we let $e(\mathcal{F}) = \sum_{F \in \mathcal{F}} e(F)$. Moreover, we say that $\mathcal{F}$ *packs* into a graph G if the graphs from $\mathcal{F}$ can be embedded edge-disjointly into G.

Theorem 2.6 ([29]) *For all $\Delta, r \geq 2$ and $\varepsilon > 0$, there exist $\eta > 0$ and $n_0 \in \mathbb{N}$ such that the following holds for all $n \geq n_0$. Assume that $\mathcal{F}$ is a collection of (r, η)-chromatic, η-separable n-vertex graphs with maximum degree at most Δ. Assume that G is a d-regular n-vertex graph with $d \geq (\max\{\delta^*_{K_r}, 1/2\} + \varepsilon)n$. If $e(\mathcal{F}) \leq (1-\varepsilon)e(G)$, then $\mathcal{F}$ packs into G.*

The proof of Theorem 2.6 is based on the blow-up lemma for approximate decompositions developed in [71], a subsequent shorter proof of which can be found in [41].

Since $\delta^*_{K_2} = 0$ and trees are separable, Theorem 2.6 has the following immediate consequence for approximate decompositions into bounded degree trees.

Corollary 2.7 ([29]) *For all $\varepsilon, \Delta > 0$, the following holds for sufficiently large n. Let $\mathcal{T}$ be a collection of trees on at most n vertices with maximum degree at most Δ. Assume that G is a d-regular n-vertex graph with $d \geq (1/2 + \varepsilon)n$. If $e(\mathcal{T}) \leq (1 - \varepsilon)e(G)$, then $\mathcal{T}$ packs into G.*

Similarly, 2-regular graphs are separable and have chromatic number at most 3. Recall that $\delta^*_{K_3}$ is conjectured to be $3/4$, though the currently best known upper bound is roughly 0.83.

Corollary 2.8 ([29]) *For all $\varepsilon > 0$, the following holds for sufficiently large n. Assume that $\mathcal{F}$ is a collection of 2-regular n-vertex graphs. Assume that G is a d-regular n-vertex graph with $d \geq (\delta^*_{K_3} + \varepsilon)n$. If $e(\mathcal{F}) \leq (1-\varepsilon)e(G)$, then $\mathcal{F}$ packs into G.*

This corollary was instrumental in the resolution of the Oberwolfach problem (see Section 6).

Moreover, if a 2-regular graph has only long cycles, say of length at least ℓ, then it is $(2, 1/\ell)$-chromatic.

Corollary 2.9 ([29]) *For all $\varepsilon > 0$, there exist $\ell, n_0 \in \mathbb{N}$ such that the following holds for all $n \geq n_0$. Assume that $\mathcal{F}$ is a collection of 2-regular n-vertex graphs with girth at least ℓ. Assume that G is a d-regular n-vertex graph with $d \geq (1/2 + \varepsilon)n$. If $e(\mathcal{F}) \leq (1-\varepsilon)e(G)$, then $\mathcal{F}$ packs into G.*

Later in Section 6, we will discuss possible 'exact' versions (i.e. 'full' decompositions) of Corollaries 2.8 and 2.9 in the context of the Oberwolfach problem.

3 Decomposition thresholds for fixed graphs F

Recall that a graph G is K_3-divisible if $3 \mid e(G)$ and $2 \mid d_G(v)$ for all $v \in V(G)$, and that any graph with a K_3-decomposition must be K_3-divisible. Now, we generalise this to arbitrary graphs. Let F be a fixed graph. Define $\gcd(F)$ as the greatest common divisor of all vertex degrees of F. We say that a graph G is F-*divisible* if $e(F) \mid e(G)$ and $\gcd(F) \mid \gcd(G)$. It is easy to see that a necessary condition for G to have an F-decomposition is to be F-divisible.

In 1976, Wilson [93, 94] proved the following fundamental result.

Theorem 3.1 ([93, 94]) *Given any graph F, for sufficiently large n, the complete graph K_n has an F-decomposition whenever it is F-divisible.*

Gustavsson [54] showed that the same is true with K_n replaced by an 'almost-complete' graph G, where $\delta(G) \geq (1-\varepsilon)|G|$ for some tiny $\varepsilon > 0$.[3]

We define the F-*decomposition threshold* δ_F to be the infimum of all $\delta \in [0,1]$ with the following property: there exists $n_0 \in \mathbb{N}$ such that any F-divisible graph G on $n \geq n_0$ vertices with $\delta(G) \geq \delta n$ has an F-decomposition.

Thus, Conjecture 1.1 would imply that $\delta_{K_3} \leq 3/4$ (which would be tight by Example 1.2). Of course, it is interesting to investigate the decomposition threshold of arbitrary graphs, not just triangles.

Problem 3.2 *Determine the F-decomposition threshold for every graph F.*

The remainder of this section is devoted to this problem. In [11], a general method was developed that turns an approximate decomposition into an exact decomposition. This approach was refined in [49] to yield the following result, which gives a general upper bound on the decomposition threshold.

---

[3]This proof has not been without criticism.

Theorem 3.3 ([49]) *For any graph F, we have*

$$\delta_F \leq \max\{\delta_F^*, 1 - 1/(\chi(F)+1)\}.$$

Theorem 3.3 improves a bound of $\delta_F \leq \max\{\delta_F^*, 1 - 1/3r\}$ proved in [11] for r-regular graphs F. Also, the cases where $F = K_3$ or C_4 were already proved in [11].

Since $\delta_{K_r}^* \geq 1 - 1/(r+1)$ (see Section 2.2), Theorem 3.3 implies that the decomposition threshold for cliques equals its fractional relaxation.

Corollary 3.4 ([49]) *For all $r \geq 3$, $\delta_{K_r} = \delta_{K_r}^*$.*

An affirmative answer to Conjecture 2.4 would, via (a variant of) Theorem 3.3 and Theorem 2.5, also imply the following general upper bound on the F-decomposition threshold.

Conjecture 3.5 ([49]) *For every graph F, we have $\delta_F \leq 1 - 1/(\chi(F)+1)$.*

This would be tight in general. However, a natural question is whether the bound can be improved for certain graphs F. This will be the topic of Sections 3.2 and 3.3. Before, we discuss very roughly the idea behind the proof of Theorem 3.3.

3.1 Turning approximate decompositions into exact ones

Let G be a large F-divisible graph. We always assume the minimum degree to be at least $(\delta_F^* + o(1))|G|$. By definition of δ_F^*, this gives us a fractional F-decomposition of G for free, and Theorem 2.1 then provides us with an approximate F-decomposition. The challenge is to deal with the leftover of such an approximate decomposition. For this, the following concept is crucial.

Definition 3.6 An *F-absorber for a graph L* is a graph A which is edge-disjoint from L such that both A and $A \cup L$ have an F-decomposition.

(Here, one can think of both A and L being subgraphs of a given host graph, but this host graph plays no role in the definition.) This motivates the following naive strategy: For any 'possible leftover' L, find an F-absorber A_L in G such that all these absorbers are edge-disjoint. Let A be the union of these 'exclusive' absorbers. Then obtain an approximate decomposition of $G - E(A)$. Now, we are guaranteed that the leftover will be one of the graphs L for which we found an absorber A_L. We can use the fact that $A_L \cup L$ has an F-decomposition by Definition 3.6. Similarly, all other absorbers also have an F-decomposition themselves, which gives in total an F-decomposition of G.

The astute reader will already have noticed (at least) one major problem with this approach. The number of possible leftovers L is gigantic. Although we know that $e(L) = o(|G|^2)$, there are still exponentially many possibilities, and so there is no hope to find edge-disjoint(!) absorbers for each of them.

However, by using an 'iterative absorption' process, one can successively obtain approximate decompositions of smaller and smaller pieces of G, combined with some partial absorption steps, until finally the number of possible leftovers is so small that the above naive strategy works as the last step. We do not go into more detail for this. Ultimately, the remaining problem one has to solve is the following: Given G

Extremal aspects of graph and hypergraph decomposition problems   243

and a small subgraph $L \subseteq G$ (we can even assume that $|L|$ is bounded), find an F-absorber for L in G. This can be achieved if $\delta(G) \geq (1 - 1/(\chi(F)+1) + o(1))|G|$, which together with our initial assumption (needed for approximate decompositions) yields the bound in Theorem 3.3.

The expository article [9] contains a short proof, with all details, of Theorem 3.3 in the case when F is a triangle.

3.2 Bipartite graphs

One big advantage when dealing with bipartite graphs is that we do not have to worry about approximate decompositions. Indeed, if F is bipartite, then its Turán density vanishes, which means that we can iteratively pull out copies of F from a graph G until at most $o(|G|^2)$ edges remain.

In [49], the decomposition threshold δ_F was determined for every bipartite graph F. Before stating the general result, let us mention preceding work. Yuster [95] showed that $\delta_F = 1/2$ if F is a tree, and later generalised his own result by drawing the same conclusion if F is connected and contains a vertex of degree one [98]. Bryant and Cavenagh [20] proved that $\delta_{C_4} \leq 31/32$. Barber, Kühn, Lo and Osthus [11] showed that $\delta_{K_{r,r}} \leq 1 - 1/(r+1)$ and $\delta_{C_\ell} = 1/2$ for all even $\ell \geq 6$. Note that their first result implies $\delta_{C_4} \leq 2/3$. The following extremal example due to Taylor (see [11]) shows that this is tight, giving C_4 a special role among the even cycles.

Example 3.7 Consider the following graph G: Let A, B, C be sets of size roughly $n/3$ where $G[A], G[C]$ are complete, B is independent and $G[A, B]$ and $G[B, C]$ are complete bipartite. Clearly, $\delta(G) \approx 2n/3$. It is easy to see that any C_4 in G contains an even number of edges from A. Thus, if $e_G(A)$ is odd, then G cannot have a C_4-decomposition. By slightly altering the sizes of A, B, C, it is not difficult to ensure this while also ensuring that G is C_4-divisible.

From Theorem 3.3, we have that $\delta_F \leq 2/3$ for all bipartite graphs F.[4] We have seen that this is tight for C_4, so a natural question is whether we can show this for other graphs too. Example 3.7 motivates the following definition. A set $X \subseteq V(F)$ is called C_4-supporting in F if there exist distinct $a, b \in X$ and $c, d \in V(F) \setminus X$ such that $ac, bd, cd \in E(F)$. Observe that if F is a copy of C_4 in G as in Example 3.7, then $V(F) \cap A$ is not C_4-supporting in F. We define

$$\tau(F) := \gcd\{e(F[X]) : X \subseteq V(F) \text{ is not } C_4\text{-supporting in } F\}.$$

For instance, $\tau(C_4) = 2$. Whenever $\tau(F) > 1$, we can adapt Example 3.7 as follows: ensure that $e_G(A)$ is not divisible by $\tau(F)$, then G cannot have an F-decomposition by the same logic. At the same time, ensure that G is F-divisible. (This needs some additional adjustment, but is a mere technicality, so we omit the details.) We note that $\tau(F) \mid \gcd(F)$ (consider all sets X consisting of a vertex and its neighbourhood).

If $\tau(F) = 1$, which happens for instance if there exists an edge in F that is not contained in any C_4, or if $\gcd(F) = 1$, then the construction fails. In fact, it turns

---
[4]Note that Theorem 3.3 is formulated in terms of the fractional decomposition threshold. However, for bipartite graphs, the detour via fractional decompositions is not necessary as explained at the beginning of this section.

out that in this case, the special structure of F can be exploited to design absorbers which show that $\delta_F \leq 1/2$. This is tight if F is connected and $e(F) \geq 2$. In order to deal with disconnected graphs, we further define

$$\tilde{\tau}(F) := \gcd\{e(C) : C \text{ is a component of } F\}.$$

If $\tilde{\tau}(F) > 1$, we can use as an extremal example the disjoint union of two cliques on roughly $n/2$ vertices (modulo some simple modifications to ensure divisibility). Moreover, if $\tilde{\tau}(F) = 1$, by adding one 'cross-edge' between the cliques, it is clear that there cannot be an F-decomposition if every edge of F is contained in a cycle.

Altogether, we have the following complete picture.

Theorem 3.8 ([49]) *Let F be a bipartite graph. Then*

$$\delta_F = \begin{cases} 2/3 & \text{if } \tau(F) > 1; \\ 0 & \text{if } \tilde{\tau}(F) = 1 \text{ and } F \text{ has a bridge}; \\ 1/2 & \text{otherwise.} \end{cases}$$

In particular, this implies that if F is a bipartite connected graph with at least two edges, then $\delta_F = 1/2$ if $\gcd(F) = 1$ or F has an edge that is not contained in any C_4. Moreover, since $\tau(K_{s,t}) = \gcd(s,t)$, we have $\delta_{K_{s,t}} = 1/2$ if s and t are coprime and $\delta_{K_{s,t}} = 2/3$ otherwise.

Given that Theorem 3.8 asymptotically determines the decomposition threshold for bipartite graphs, the next question is of course whether one can determine the exact threshold. This was achieved for even cycles (with the exception of 6-cycles) by Taylor [89] and for trees (for infinitely many n) by Yuster [97]. It would be particularly interesting to solve the cases when F is complete bipartite and when F is a 6-cycle.

3.3 A discretisation result

Recall that $\delta_F \leq \max\{\delta_F^*, 1 - 1/(\chi(F) + 1)\}$ holds, and probably $\delta_F \leq 1 - 1/(\chi(F) + 1)$ for every graph F. In the previous subsection, we saw that if $\chi := \chi(F) = 2$, then $\delta_F \in \{1 - 1/(\chi - 1), 1 - 1/\chi, 1 - 1/(\chi + 1)\}$. The question suggests itself whether a similar phenomenon occurs in general. (Note that we have $\delta_F \geq 1 - 1/(\chi - 1)$ as shown by the Turán graph.) Progress towards this has been made in [49] for graphs of chromatic number at least 5.

Theorem 3.9 ([49]) *Let F be a graph with $\chi := \chi(F) \geq 5$. Then we have $\delta_F \in \{\delta_F^*, 1 - 1/\chi, 1 - 1/(\chi + 1)\}$.*

The proof of this result has the nice feature that the assumption $\delta_F < 1 - 1/(\chi+1)$ is used indirectly to obtain better absorbers. Very roughly speaking, the idea is as follows: Suppose that $\delta_F < 1 - 1/(\chi + 1)$. By the definition of δ_F, this means that a graph which is very close to a large balanced complete $(\chi + 1)$-partite graph has an F-decomposition if it is F-divisible. One can use such graphs as absorbers, or more precisely, as building blocks of absorbers. Although they might be very large in size, the fact that they are $(\chi + 1)$-partite allows us to find them in a given graph G already if $\delta(G) \geq (1 - 1/\chi + o(1))|G|$, which we can use (modulo other parts of

Extremal aspects of graph and hypergraph decomposition problems   245

the argument) to show that $\delta_F \leq 1 - 1/\chi$. Similarly, assuming $\delta_F < 1 - 1/\chi$ allows one to find absorbers if $\delta(G) \geq (1 - 1/(\chi - 1) + o(1))|G|$. The complete proof of Theorem 3.9 involves a number of reductions for which the assumption $\chi \geq 5$ is needed.

It would be interesting to investigate the cases $\chi(F) \in \{3, 4\}$ further. Perhaps a good starting point are odd cycles.

Problem 3.10 Determine $\delta_{C_{2\ell+1}}$.

It is shown in [11, Proposition 12.1] that $\delta_{C_{2\ell+1}} \geq \frac{1}{2} + \frac{1}{4\ell}$ (and the same construction gives the same bound for $\delta^*_{C_{2\ell+1}}$). In the first version of this survey, we posed the problem of at least showing that $\delta_{C_{2\ell+1}} \to 1/2$ as $\ell \to \infty$. By the results in [11], it suffices to prove this for the fractional threshold. This was confirmed very recently by Joos and Kühn [66]. Moreover, they obtain an analogous result for fractional decompositions of hypergraphs.

3.4 Decompositions of partite graphs and Latin squares

A *Latin square* of order n is an $n \times n$ array of cells, each containing a symbol from $[n]$, where no symbol appears twice in any row or column. A Latin square corresponds to a K_3-decomposition of the complete tripartite graph $K_{n,n,n}$ with vertex classes consisting of the rows, columns and symbols.

More generally, two Latin squares R (red) and B (blue) drawn in the same $n \times n$ array of cells are *orthogonal* if no two cells contain the same combination of red symbol and blue symbol. As above, it is easy to see that a pair of orthogonal Latin squares corresponds to a K_4-decomposition of $K_{n,n,n,n}$. Even more generally, there is a bijection between sequences of $r - 2$ *mutually orthogonal* Latin squares (where every pair from the sequence are orthogonal) and K_r-decompositions of balanced complete r-partite graphs.

A partial Latin square is defined similarly as a Latin square, except that cells are allowed to be empty. Daykin and Häggkvist [32] made the following conjecture:

Conjecture 3.11 ([32]) *Given a partial Latin square L of order n where each row, column and symbol is used at most $n/4$ times, it is possible to complete L into a Latin square of order n.*

This can be seen as a natural analogue of the conjecture of Nash-Williams (Conjecture 1.1). Indeed, it turns out that they are closely related: Barber, Kühn, Lo, Osthus and Taylor [12] proved an analogue of Corollary 3.4 for r-partite graphs. Complemented by results of Bowditch and Dukes [18] on fractional K_3-decompositions of balanced tripartite graphs and results of Montgomery [81] for larger cliques, this implies that a (sequence of mutually orthogonal) partial Latin squares can be completed provided that no row, column or coloured symbol has already been used too often (see [12] for a precise statement).

In particular, the case $r = 3$ shows that Conjecture 3.11 holds with $1/25 - o(1)$ instead of $1/4$.

Theorem 3.12 ([12, 18]) *Given a partial Latin square L of order n where each row, column and symbol is used at most $(\frac{1}{25} - o(1))n$ times, it is possible to complete L into a Latin square of order n.*

It would be very interesting to improve these bounds (recall that the obvious 'bottleneck' consists of finding improved bounds for the fractional decomposition problem).

4 F-decompositions of hypergraphs

So far, we have only considered decomposition problems for graphs, but of course the same type of questions can also be asked for hypergraphs. In fact, most definitions and questions translate almost verbatim to hypergraphs. However, the known results are much more limited.

Let G and F be k-graphs. An *F-decomposition of G* is a collection of copies of F in G such that every edge of G is contained in exactly one of these copies. As for graphs, the existence of an F-decomposition necessitates certain divisibility conditions. For instance, we surely need $e(F) \mid e(G)$. More generally, define

$$d_F(i) := \gcd\{d_F(S) : S \subseteq V(F), |S| = i\}$$

for all $0 \leq i \leq k - 1$. Note that $d_F(0) = e(F)$. Moreover, $d_F(1) = \gcd(F)$ for $k = 2$. Now, G is called *F-divisible* if $d_F(i) \mid d_G(i)$ for all $0 \leq i \leq k - 1$. It is easy to see that G must be F-divisible in order to admit an F-decomposition.

In 1853, Steiner asked for which $n > r > k$ there exists a K_r^k-decomposition of K_n^k. Clearly, the above divisibility conditions need to be satisfied. The folklore 'Existence conjecture' postulated that these conditions are also sufficient, at least when r and k are fixed and n is sufficiently large. This was proved in a recent breakthrough by Keevash [67].

Theorem 4.1 ([67]) *For fixed r, k and sufficiently large n, K_n^k has a K_r^k-decomposition if it is K_r^k-divisible.*

In fact, Keevash proved a more general result that holds also if K_n^k is replaced by a typical k-graph G.

An obvious question (e.g. asked by Keevash) is whether K_r^k can be replaced by any k-graph F. For instance, Hanani [58] settled the problem if F is an octahedron (viewed as a 3-uniform hypergraph).

Recently, the authors together with Lo solved the general problem [51], thus extending Wilson's theorem (Theorem 3.1) to hypergraphs. Moreover, the proof method in [51] is quite different from the one developed in [67], so this gives also a new proof of the Existence conjecture.

Theorem 4.2 ([51]) *For all $k \in \mathbb{N}$, $p \in [0, 1]$ and any k-graph F, there exist $c > 0$ and $h, n_0 \in \mathbb{N}$ such that the following holds. Suppose that G is a (c, h, p)-typical k-graph on at least n_0 vertices. Then G has an F-decomposition whenever it is F-divisible.*

Keevash [68] later proved an even more general result. In particular, he obtained a partite version of Theorem 4.2. This has the following nice application to *resolvable* decompositions. For simplicity, we only consider graphs. An F-decomposition of G is *resolvable* if it can be partitioned into F-factors (where an F-factor is a vertex-disjoint union of copies of F covering al vertices of G). Recall that Kirkman's famous schoolgirl problem asked for a resolvable K_3-decomposition of K_{15}. Now, suppose we seek a resolvable F-decomposition of some graph G. The number of F-factors in such a decomposition should clearly be $t := \frac{e(G)|F|}{|G|e(F)}$. Define an auxiliary graph G' by adding a set T of t new vertices and all edges between G and T. Moreover, let F' be the graph obtained from F by adding a new vertex x' and all edges to $V(F)$. Now, suppose we can find an F'-decomposition of G' with the additional property that the copy of x' always lies in T ([68] contains a general framework for this). Then every copy of F' gives a copy of F in G. Moreover, for a fixed vertex in T, the copies of F' containing it induce an F-factor. Thus, we have found a resolvable F-decomposition of G. This result, for $F \in \{C_3, C_4, C_5\}$, was instrumental in the resolution of the Oberwolfach problem (see Section 6).

4.1 Minimum degree versions

As Wilson's theorem (Theorem 3.1) has been extended to graphs of large enough minimum degree, one can also ask the same for hypergraphs. The definition of the decomposition threshold straightforwardly generalises. For a k-graph F, define δ_F as the infimum of all $\delta \in [0,1]$ with the property that any sufficiently large F-divisible k-graph G with $\delta(G) \geq \delta|G|$ has an F-decomposition. (Recall that $\delta(G)$ denotes the minimum $(k-1)$-degree of G.)

Explicit bounds for the parameters c and h in Theorem 4.2 were obtained in [51]. Since every k-graph with sufficiently large minimum degree is typical, this gives explicit upper bounds for δ_F. However, these bounds are very weak. This is partly due to a reduction in Theorem 4.2 that reduces the problem for general F to such F which are 'weakly regular'. For weakly regular F (see [51] for the definition), the method in [51] gives much better results. In particular, we have the following bound on the decomposition threshold of cliques.

Theorem 4.3 ([51]) *For all $k < r$, we have $\delta_{K_r^k} \leq 1 - \frac{k!}{3 \cdot 14^k r^{2k}}$.*

We remark that even for fractional decompositions, the best known bounds are only slightly better. More precisely (for simplicity we regard the uniformity k as fixed and consider asymptotics in r), it is shown in [10] that $\delta^*_{K_r^k} \leq 1 - \Omega_k(r^{-2k+1})$, where δ^*_F is the fractional F-decomposition threshold defined in the obvious way.

Note that Theorem 4.3 implies the following result on completions of partial Steiner systems: for all $r > k$, there is an n_0 so that whenever X is a 'partial' Steiner system, consisting of a set of edge-disjoint K_r^k on n vertices and $n^* \geq \max\{n_0, \frac{3 \cdot 14^k r^{2k}}{k!}n\}$ satisfies the necessary divisibility conditions, then X can be extended to a K_r^k-decomposition of $K_{n^*}^k$. In other words, X can be extended into a so-called (n^*, r, k)-Steiner system. For the case of Steiner triple systems (i.e. $r = 3$ and $k = 2$), Bryant and Horsley [22] showed that one can take $n^* = 2n + 1$, which proved a conjecture of Lindner. It would be interesting to extend this exact result to other parameter values.

It is not clear what the correct value of $\delta_{K_r^k}$ should be. As observed in [51], a construction from [73] can be modified to obtain (for fixed $k < r$) infinitely many k-graphs G with $\delta(G) \geq (1 - \mathcal{O}_k(\frac{\log r}{r^{k-1}}))|G|$ which are even K_r^k-free. Modulo the divisibility of such examples, this seems to suggest that $\delta_{K_r^k} \geq 1 - \mathcal{O}_k(\frac{\log r}{r^{k-1}})$. In view of the case $k = 2$, perhaps the following is true.

Conjecture 4.4 $\delta_{K_r^k} = 1 - \Theta_k(r^{-k+1})$.

Moreover, the following generalisation of Corollary 3.4 might be true.

Conjecture 4.5 *For all $r > k \geq 2$, $\delta_{K_r^k} = \delta^*_{K_r^k}$.*

Prior to [51], the only explicit result for hypergraph decomposition thresholds was due to Yuster [96], who showed that if T is a linear k-uniform hypertree, then every T-divisible k-graph G on n vertices with minimum vertex degree at least $(\frac{1}{2^{k-1}} + o(1))\binom{n}{k-1}$ has a T-decomposition. This is asymptotically best possible for nontrivial T. Moreover, the result implies that $\delta_T \leq 1/2^{k-1}$.

5 Euler tours in hypergraphs

Finding an *Euler tour* in a graph is a problem as old as graph theory itself: Euler's negative resolution of the Seven Bridges of Königsberg problem in 1736 is widely considered the first theorem in graph theory. Euler observed that if a (multi-)graph contains a closed walk which traverses every edge exactly once, then all the vertex degrees are even. Hence, he observed a *necessary divisibility condition* for the existence of such a closed walk, which we now call an *Euler tour*. Is this divisibility condition also sufficient? In general, no, since the graph might be disconnected but still fulfil the divisibility condition. However, if a graph is connected, then it contains an Euler tour if and only if the divisibility condition is satisfied. This fact was already stated by Euler, and its proof is often attributed to Hierholzer and Wiener.

In this section, we consider Euler tours in hypergraphs. There are several ways of generalising the concept of paths/cycles, and similarly Euler trails/tours, to hypergraphs. We focus here on the 'tight' regime. Given a k-graph G, a sequence of vertices $\mathcal{W} = x_1 x_2 \ldots x_\ell$ is a *(tight self-avoiding) walk* in G if $\{x_i, x_{i+1}, \ldots, x_{i+k-1}\} \in E(G)$ for all $i \in [\ell - k + 1]$, and no edge of G appears more than once in this way. Similarly, we say that $\mathcal{W}$ is a *closed walk* if $\{x_i, x_{i+1}, \ldots, x_{i+k-1}\} \in E(G)$ for all $i \in [\ell]$, with indices modulo ℓ, and no edge of G appears more than once in this way. We let $E(\mathcal{W})$ denote the set of edges appearing in $\mathcal{W}$.

Definition 5.1 An *Euler tour of G* is a closed walk $\mathcal{W}$ in G with $E(\mathcal{W}) = E(G)$.

Clearly, if G is 2-graph, then this coincides with the usual definition of an Euler tour, and a necessary condition for the existence of such a tour is that every vertex degree is even. Can we formulate an analogous condition for k-graphs? To do so, assume that a k-graph G has an Euler tour $\mathcal{W}$. Fix any vertex v. Note that v might appear several times in the sequence $\mathcal{W}$, however, for every such appearance, it is contained in exactly k edges. Since every edge appears exactly once in $\mathcal{W}$, we can conclude that the degree $d_G(v)$ of v is divisible by k. Now, the question is again: is this condition also sufficient for the existence of an Euler tour? Again, the answer is no, as the given hypergraph might be divisible (i.e. satisfy this degree condition) but consist of disjoint pieces. Moreover, the problem of deciding whether a given 3-graph has an Euler tour has been shown to be NP-complete [78], thus when $k > 2$, there is probably no simple characterisation of k-graphs having an Euler tour. Surprisingly, until recently, it was not even known whether the complete k-graph has an Euler tour if it is divisible. Using the language of *universal cycles*, this was formulated as a conjecture by Chung, Diaconis and Graham [27, 28] in 1989. More precisely, they conjectured that for every fixed $k \in \mathbb{N}$ and sufficiently large n, there exists an Euler tour in K_n^k whenever k divides $\binom{n-1}{k-1}$.

Clearly, this is true for $k = 2$. Numerous partial results have been obtained. In particular, Jackson proved the conjecture for $k = 3$ [64] and for $k \in \{4, 5\}$ (unpublished), and Hurlbert [63] confirmed the cases $k \in \{3, 4, 6\}$ if n and k are coprime (see also [79]). Various approximate versions of the conjecture have been obtained in [14, 31, 33, 79].

Recently, the conjecture was proven for all k by the authors and Joos [48]. In fact, the result is more general and applies to quasirandom k-graphs in the sense of Definition 1.3. We state a simplified version here which applies for almost complete k-graphs.

Theorem 5.2 ([48]) *For all $k \in \mathbb{N}$ there exists $\varepsilon > 0$ such that any sufficiently large k-graph G with $\delta(G) \geq (1-\varepsilon)|G|$ has an Euler tour if $k \mid d_G(v)$ for every $v \in V(G)$.*

We conjecture that the minimum degree condition can be significantly improved. We discuss this in more detail in Section 5.2.

5.1 Euler tours: Proof sketch

The proof of Theorem 5.2 relies on Theorem 4.2 in order to complete a suitable partial Euler tour into a 'full' one. More precisely, the proof proceeds as follows. We call a walk $\mathcal{W}$ in a k-graph G *spanning* if every ordered $(k-1)$-set of vertices appears consecutively in $\mathcal{W}$ at least once. The motivation behind this definition is as follows. Suppose $\mathcal{W}$ is a closed spanning walk in G and $\mathcal{W}'$ is some other closed walk which is edge-disjoint from $\mathcal{W}$. Then we can 'insert' $\mathcal{W}'$ into $\mathcal{W}$ as follows: take any $(k-1)$-tuple which appears in $\mathcal{W}'$. Since $\mathcal{W}$ is spanning, we know that this ordered tuple appears in $\mathcal{W}$ too, so we can follow $\mathcal{W}$ until we reach an appearance of this tuple, then follow $\mathcal{W}'$ until we reach this tuple again, and then continue with $\mathcal{W}$. It is easy to see that this yields a new closed spanning walk which uses precisely the edges of $\mathcal{W}$ and $\mathcal{W}'$. Hence, we have the following:

Observation 5.3 *If G can be decomposed into closed walks such that one of them is spanning, then G has an Euler tour.*

This approach breaks the proof into two parts: First, we need to find a spanning walk. Note that there are $\Theta(n^{k-1})$ ordered $(k-1)$-sets of vertices, so in order to be spanning, our walk $\mathcal{W}$ needs to have length $\Omega(n^{k-1})$. Since G has $\Theta(n^k)$ edges, there is at least enough room. Moreover, if we construct $\mathcal{W}$ in some random fashion, say using $n^{k-1} \log^2 n$ edges, then we might hope that every $(k-1)$-set only appears in

$\log^3 n$ edges of $\mathcal{W}$. That is, the subgraph formed by the edges of $\mathcal{W}$ has very small maximum degree, so removing these edges leaves G essentially unchanged.

Now, we come to the second part. We only need to decompose the remainder $G - E(\mathcal{W})$ into any number of closed walks. In particular, a decomposition into tight cycles would be sufficient. This we can achieve using the F-decomposition result from Section 4, with F being a tight cycle. Let C_ℓ^k denote the tight k-uniform cycle of length ℓ, that is, the vertices of C_ℓ^k are $v_1, \ldots, v_\ell$, and the edges are all the k-tuples of the form $\{v_i, v_{i+1}, \ldots, v_{i+k-1}\}$, with indices modulo ℓ.

We want to apply Theorem 4.2 with $F = C_{2k}^k$. Conveniently, a k-graph G is C_{2k}^k-divisible whenever $2k \mid e(G)$ and $k \mid d_G(v)$ for all $v \in V(G)$, that is, there is no divisibility condition for i-sets with $i > 1$.[5] That the vertex degrees in $G - E(\mathcal{W})$ are divisible by k follows automatically from the divisibility of the initial graph G and since we removed a closed walk. Moreover, by removing greedily a few copies of C_{2k+1}^k, say, we can make the number of (remaining) edges divisible by $2k$. Theorem 4.2 then does the rest for us.

Let us say a few more words about finding the spanning walk. Essentially, we show that a random self-avoiding walk of length $n^{k-1}\log^2 n$ has the desired properties with high probability.[6] Fix any $k-1$ vertices $X_1, \ldots, X_{k-1}$ as a start tuple. Now, in each step, with the current walk being $X_1, \ldots, X_{i-1}$, choose a vertex X_i uniformly at random from all vertices that form an edge together with the last $k-1$ vertices $X_{i-1}, \ldots, X_{i-k+1}$ of the current walk and this edge has not been used previously by the walk. If no such vertex exists, then stop.

In order to analyse this random walk, fix vertices $v_1, \ldots, v_{k-1}$. We say that the walk *visits* (these vertices) at step i if $X_{i-k+2} = v_1, \ldots, X_i = v_{k-1}$. Now, assume that in some step i, the walk visits vertices $v'_1, \ldots, v'_{k-1}$. We want to ask ourselves: what is the probability that the walk visits $v_1, \ldots, v_{k-1}$ at step $i+k$? For simplicity, ignore the condition that the walk ought to be self-avoiding, and also assume that $v'_1, \ldots, v'_{k-1}, v_1, \ldots, v_{k-1}$ are distinct. In each of the following k steps, the walk has clearly at most n choices for the next vertex, so the total number of choices for the walk is at most n^k. Crucially, using the minimum degree assumption, we can check that there are $\Omega(n)$ vertices v^* such that $v'_1, \ldots, v'_{k-1}, v^*, v_1, \ldots, v_{k-1}$ is a tight walk and thus an admissible choice for the next k steps. Hence, the (conditional) probability that the walk visits $v_1, \ldots, v_{k-1}$ at step $i+k$ is $\Theta(n^{-k+1})$. Thus, if we let the walk continue for about $n^{k-1}\log^2 n$ steps and consider the steps i which are multiples of k, then the expected number of visits is roughly $(\log^2 n)/k$. Moreover, since the stated bound for the probability holds for any outcome of previous such steps, a Chernoff–Hoeffding type inequality applies, and we can infer that the probability of the walk not visiting at all is tiny. A union bound over all ordered $(k-1)$-tuples shows that the walk is spanning with high probability. A similar argument shows that the walk is unlikely to have maximum degree larger than $\log^3 n$. This justifies the above analysis also for the self-avoiding walk, since the number of admissible 'link' vertices v^* is still $\Omega(n)$. (Technically, we 'stop' the walk as soon as some $(k-1)$-set has too large degree, and then analyse this stopped walk. We refer to [48] for the remaining details.)

---

[5]We have $d_{C_{2k}^k}(\{v_1, \ldots, v_{i-1}, v_k\}) = 1$ and hence $d_{C_{2k}^k}(i) = 1$.

[6]We ignore here that the walk should be closed in the end. This can be easily achieved afterwards in $\mathcal{O}(1)$ deterministic steps.

5.2 Open problems on hypergraph decompositions and Euler tours

The following conjecture would provide a 'genuine' minimum degree version of Theorem 5.2.

Conjecture 5.4 *For all $k > 2$ and $\varepsilon > 0$, every sufficiently large k-graph G with $\delta(G) \geq (1 - \frac{1}{k} + \varepsilon)|G|$ has a tight Euler tour if all vertex degrees are divisible by k.*

It seems possible that the approach for Theorem 5.2 can be extended to attain Conjecture 5.4: recall that the proof consisted of two steps. First, we found a spanning walk with small maximum degree. For this, we analysed a random self-avoiding walk. The crucial property was that, given any two disjoint $(k-1)$-tuples $(v_1, \ldots, v_{k-1}), (v_{k+1}, \ldots, v_{2k-1})$, there are $\Omega(n)$ vertices v_k such that $v_i v_{i+1} \ldots v_{i+k-1}$ is an edge for all $i \in [k]$. This property is already satisfied if $\delta(G) \geq (1 - \frac{1}{k} + \varepsilon)|G|$. In fact, one can even make the argument work if only $\delta(G) \geq (1/2 + \varepsilon)|G|$, by considering longer paths connecting the tuples $(v_1, \ldots, v_{k-1})$ and $(v_{k+1}, \ldots, v_{2k-1})$. Amongst other things, this latter fact motivated the conjecture in [48] (restated in an earlier version of this survey) that $1/2$ could be the right threshold for Euler tours for any uniformity k. However, this was apparently too optimistic. Very recently, Piga and Sanhueza-Matamala [86] provided a counterexample in the case $k = 3$, showing that the threshold needs to be at least $2/3$. Moreover, they proved that this is the right threshold (that is, Conjecture 5.4 for $k = 3$).

The bottleneck is the second step, where we decomposed the remaining k-graph into tight cycles. In the proof of Theorem 5.2, we applied the F-decomposition theorem from Section 4 to obtain a decomposition into tight cycles of length $2k$, but clearly the length of the cycles does not matter. In order to improve the minimum degree assumption, it is probably better to use longer cycles. (In fact, it would be enough to decompose into closed walks.) The following conjecture, if true, would complement the random walk analysis sketched above, and thus imply Conjecture 5.4. It would also be significant in its own right.[7]

Conjecture 5.5 *Any k-graph with $\delta(G) \geq (1 - \frac{1}{k} + o(1))|G|$ can be decomposed into tight cycles, provided that all vertex degrees are divisible by k.*

Note that an approximate decomposition is easy to obtain. Indeed, since C_{2k}^k is k-partite and thus has Turán density 0 by a well-known result of Erdős [42], we can iteratively pull out copies of C_{2k}^k until $o(|G|^k)$ edges remain.

6 Oberwolfach problem

The Oberwolfach problem, posed by Ringel in 1967, asks for a decomposition of the complete graph K_n into edge-disjoint copies of a given 2-factor. Clearly, this can only be possible if n is odd.

Problem 6.1 (Oberwolfach problem, Ringel, 1967) *Let $n \in \mathbb{N}$ and let F be a 2-regular graph on n vertices. For which (odd) n and F does K_n have an F-decomposition?*

---

[7]In an earlier version, we conjectured that the threshold is $1/2$. However, this also was disproved in [86] in the case $k = 3$.

The problem is named after the Mathematical Research Institute of Oberwolfach, where Ringel formulated it as follows: assume n conference participants are to be seated around circular tables for $\frac{n-1}{2}$ meals, where the total number of seats is equal to n, but the tables may have different sizes. Is it possible to find a seating chart such that every person sits next to any other person exactly once?

Note that when F consists of only one cycle (that is, there is one large table), then we seek a decomposition of K_n into Hamilton cycles, which is possible by Walecki's theorem from 1892. In the other extreme, if all tables have only size 3, then we seek a decomposition of K_n into triangle factors. This was Kirkman's (generalised) schoolgirl problem from 1850, eventually solved by Ray-Chaudhuri and Wilson [87] and independently by Lu.

Over the years, the Oberwolfach problem and its variants have received enormous attention, with more than 100 research papers produced. Most notably, Bryant and Scharaschkin [24] proved it for infinitely many n. Traetta [91] solved the case when F consists of two cycles only, Alspach, Schellenberg, Stinson and Wagner [4] solved the case when all cycles have equal length, and Hilton and Johnson [60] solved the case when all but one cycle have equal length.

An approximate solution to the Oberwolfach problem was obtained by Kim, Kühn, Osthus and Tyomkyn [71] and Ferber, Lee and Mousset [44]. More precisely, it follows from their (much more general) results that K_n contains $n/2 - o(n)$ edge-disjoint copies of any given 2-factor F.

A related conjecture of Alspach stated that for all odd n the complete graph K_n can be decomposed into any collection of cycles of length at most n whose lengths sum up to $\binom{n}{2}$. This was solved by Bryant, Horsley, and Pettersson [23].

Very recently, the Oberwolfach problem was solved by the authors together with Joos and Kim [47]. More precisely, they showed that for all odd $n \geq n_0$, there is a solution for any given 2-factor F. The remaining cases could (in theory) be decided by exhaustive search, but this is not practically possible as n_0 is rather large. It would be very interesting to complete the picture. Perhaps there are not many exceptions (currently, there are four known exceptions).

We state the result in the following slightly more general way, where K_n can be replaced by an almost-complete graph. Recall that this means one can obtain a solution to the Oberwolfach problem even if the first $o(n)$ copies of F are chosen greedily.

Theorem 6.2 ([47]) *There exists $\varepsilon > 0$ such that for all sufficiently large n, the following holds: Let F be any 2-regular graph on n vertices, and let G be a d-regular graph on n vertices for some even $d \in \mathbb{N}$. If $d \geq (1 - \varepsilon)n$, then G has an F-decomposition.*

As mentioned earlier, the proof relies on Corollary 2.8 (to obtain a suitable approximate decomposition) and the results on resolvable cycle decompositions in [68] (as part of an absorbing approach).

In the spirit of this survey, the obvious question is of course: can the minimum degree assumption in Theorem 6.2 be weakened? We discuss this further in Section 6.1.

An immediate consequence of Theorem 6.2 is that if n is even, then K_n can be decomposed into one perfect matching and otherwise copies of F. For this variant

Extremal aspects of graph and hypergraph decomposition problems 253

of the Oberwolfach problem as well, many partial results were previously obtained (see e.g. [21, 61, 62]).

Another natural extension is the following 'generalised Oberwolfach problem'. Suppose $F_1, \ldots, F_{(n-1)/2}$ are (possibly distinct) 2-factors on n vertices. Is it possible to decompose K_n into $F_1, \ldots, F_{(n-1)/2}$? In the special case where the list contains only two distinct 2-factors, this is known as the Hamilton–Waterloo problem, which was also solved in [47] (for sufficiently large n). In fact, Theorem 6.2 holds in this general setting provided that some 2-factor appears linearly many times in the list.

Improving on this, Keevash and Staden [69] recently solved the generalised Oberwolfach problem. Their result applies in the setting of dense typical graphs, and they also prove an appropriate version of this for directed graphs.

Theorem 6.3 ([69]) *For every $p > 0$ there exist $c > 0$ and $h \in \mathbb{N}$ such that the following holds. Any sufficiently large (c, h, p)-typical graph G which is d-regular for some even $d \in \mathbb{N}$ can be decomposed into any $d/2$ given 2-factors.*

6.1 Open problems related to the Oberwolfach problem

6.1.1 Minimum degree thresholds As we have already seen, the problem of decomposing a graph G into a given 2-factor not only makes sense if G is complete. Of course, we should assume that G is regular with even degrees.

Conjecture 6.4 ([47]) *For all $\varepsilon > 0$, the following holds for sufficiently large n. Let F be any 2-regular graph on n vertices, and let G be a d-regular graph on n vertices for some even $d \in \mathbb{N}$. If $d \geq (3/4 + \varepsilon)n$, then G has an F-decomposition.*

If F is a triangle-factor, then the 'threshold' $3/4$ would be optimal. On the other hand, if F is a Hamilton cycle, then it can be lowered to $1/2$ ([30]). It would be interesting to 'interpolate' between these extremal cases. More specifically, it could be true that if all cycles in F have even length, then $2/3$ is sufficient, and if C_4 is excluded in addition, then the threshold is $1/2$. Similarly, if the girth of F is sufficiently large compared to $1/\varepsilon$, then $1/2$ should also be sufficient. Note that Corollaries 2.8 and 2.9 give some partial approximate results in this direction.

6.1.2 Hypergraphs It seems natural to ask for an analogue of the Oberwolfach problem for hypergraphs. Bailey and Stevens [5] conjectured that K_n^k has a decomposition into tight Hamilton cycles if and only if k divides $\binom{n-1}{k-1}$. This is still open and would generalise Walecki's theorem to hypergraphs. Clearly, the condition $k \mid \binom{n-1}{k-1}$ is necessary since every Hamilton cycle contains k edges at any fixed vertex. Moreover, it also implies that the total number of edges of K_n^k is divisible by n, the number of edges in one Hamilton cycle. We conjecture that the same divisibility condition guarantees a decomposition into any tight cycle factor.

Conjecture 6.5 *For fixed k, the following holds for sufficiently large n. Let F be the vertex-disjoint union of tight k-uniform cycles, each of length at least $2k-1$, with n vertices in total. Then K_n^k has an F-decomposition if k divides $\binom{n-1}{k-1}$.*

To the best of our knowledge, this has not been explicitly asked before. We note that the somewhat generous assumption that each cycle has length at least $2k - 1$

ensures that there are no divisibility obstructions for i-sets with $i > 1$. We also note that the very general result of Ehard and Joos [40] concerning approximate decompositions of quasirandom hypergraphs into bounded degree subgraphs yields an approximate solution to the above conjecture, in that K_n^k contains any collection of $(1 - o(1))\binom{n-1}{k-1}/k$ edge-disjoint tight cycle factors.

Another problem which is related to the conjecture of Bailey and Stevens was made by Baranyai as well as Katona. First, recall Baranyai's theorem [6] stating that K_n^k has a 1-factorization whenever k divides n. As in the case of graphs, a 1-*factor* (or *perfect matching*) is a set of disjoint edges covering all the vertices, and a 1-*factorization* is a set of edge-disjoint 1-factors covering all the edges. Baranyai [7] and Katona conjectured an extension of Baranyai's theorem to the case when the divisibility condition $k \mid n$ is not satisfied: instead of decomposing into 1-factors, the aim is to decompose into *wreaths*. Here, given a cyclically ordered set of n vertices, a wreath is obtained by greedily choosing hyperedges as follows: the first hyperedge consists of k consecutive vertices and in each step the next hyperedge consists of the k vertices which come directly after the vertices in the previous hyperedge. This process stops as soon as one obtains a regular hypergraph. So for instance, if $k = 4$ and $n = 6$, then $\{1234, 5612, 3456\}$ is a wreath. If $k \mid n$, then a wreath is a perfect matching. If n and k are co-prime, then a wreath is a tight Hamilton cycle. The *wreath decomposition conjecture* postulates that K_n^k can always be decomposed into wreaths.

While this problem is still open for the complete hypergraph, we propose the following minimum degree version of Baranyai's theorem.

Conjecture 6.6 *For fixed k and $\varepsilon > 0$, the following holds for all sufficiently large n. An n-vertex k-graph G with $\delta(G) \geq (1/2 + \varepsilon)n$ can be decomposed into perfect matchings if and only if $k \mid n$ and G is vertex-regular.*

6.1.3 Decompositions into r-factors
Finally, we restate the following conjecture formulated in [47], which can be viewed as a far-reaching generalisation of the Oberwolfach problem from 2-regular to regular graphs of arbitrary degrees.

Conjecture 6.7 ([47]) *For all $\Delta \in \mathbb{N}$, there exists an $n_0 \in \mathbb{N}$ so that the following holds for all $n \geq n_0$. Let $F_1, \ldots, F_t$ be n-vertex graphs such that F_i is r_i-regular for some $r_i \leq \Delta$ and $\sum_{i \in [t]} r_i = n - 1$. Then there is a decomposition of K_n into $F_1, \ldots, F_t$.*

This conjecture is clearly extremely challenging. So it would be interesting to prove it for restricted families, such as graphs which are separable or have high girth. An approximate version of the above conjecture was proved by Kim, Kühn, Osthus and Tyomkyn [71].

7 Related decomposition problems

In this final section, we briefly mention some further decomposition problems. We also remark that, for all the questions we discussed, it is interesting to ask for algorithmic variants (can a decomposition of a dense hypergraph be found in

Extremal aspects of graph and hypergraph decomposition problems   255

polynomial time?), counting problems (how many different decompositions of a graph exist?) and many other directions which we did not cover here.

7.1 Weighted decompositions into triangles and edges

Recall Kirkman's theorem that K_n has a triangle decomposition whenever it is divisible. We might ask what happens if we ignore divisibility? Can we decompose into triangles and a few edges? For a decomposition of an n-vertex graph G into e edges and t triangles, we define $2e + 3t$ as the weight of the decomposition, and the aim is to find a decomposition of minimum weight, denoted by $\pi_3(G)$.[8] Clearly, we always have $\pi_3(G) \geq e(G)$. In particular,

$$\pi_3(K_n) \geq \binom{n}{2} = (1 + o(1))n^2/2.$$

Similarly, if G is triangle-free, then $\pi_3(G) = 2e(G)$. In particular,

$$\pi_3(K_{\lceil n/2 \rceil, \lfloor n/2 \rfloor}) = 2 \cdot \lceil n/2 \rceil \cdot \lfloor n/2 \rfloor = (1 + o(1))n^2/2.$$

Define $\pi_3(n)$ as the maximum of $\pi_3(G)$ over all n-vertex graphs G. The problem of determining $\pi_3(n)$ was first considered by Györi and Tuza [56]. Král', Lidický, Martins and Pehova [74] resolved this problem asymptotically, by showing that

$$\pi_3(n) = (1/2 + o(1))n^2.$$

Blumenthal, Lidický, Pehova, Pfender, Pikhurko and Volec [15] were able to strengthen this to an exact bound. It turns out that K_n and $K_{\lceil n/2 \rceil, \lfloor n/2 \rfloor}$ are the only extremal examples. A crucial tool in the proof of these results was the triangle case of Theorem 3.3, proved in [11].

An immediate consequence of the above results is that every n-vertex graph with $n^2/4 + k$ edges contains $2k/3 - o(n^2)$ edge-disjoint triangles. A problem of Tuza [92] would generalise the latter bound to arbitrary cliques.

Problem 7.1 ([92]) Does every n-vertex graph with $\frac{r-2}{2r-2}n^2 + k$ edges contain $\frac{2}{r}k - o(n^2)$ edge-disjoint copies of K_r?

A minimum degree version of this problem was considered by Yuster [101]: what is the largest number of edge-disjoint copies of K_r one can find in a graph of given minimum degree? Again, this question is still open.

7.2 Packing and covering number

For graphs F and G, the F-*packing number* of G, denoted $P(F, G)$, is the maximum number of edge-disjoint copies of F in G. The 'dual' notion is the F-*covering number*, denoted $C(F, G)$, which is the minimum number of copies of F

---
[8]More generally, suppose a fixed set $\mathcal{H}$ of graphs and a weight function w on $\mathcal{H}$ are given. For a graph G which is decomposed into $H_1, \ldots, H_s \in \mathcal{H}$, define the weight of this decomposition as $\sum_{i=1}^{s} w(H_i)$. One can then ask for the minimum weight of such a decomposition of G. The case when $\mathcal{H}$ is the set of all cliques and $w(K_r) = r$ has received much attention.

in G that cover every edge of G. Clearly, G has an F-decomposition if and only if $P(F,G) = C(F,G) = e(G)/e(F)$.

Assume F is a fixed graph and G is a sufficiently large and dense graph. Note that a collection of edge-disjoint copies of F in G forms an F-decomposition of some subgraph of G. Hence, a natural way for proving lower bounds on $P(F,G)$ is to delete as few edges as possible from G to obtain an F-divisible graph, and then to apply a decomposition result such as Gustavsson's theorem (see Section 3). Using this approach, Caro and Yuster [25, 26] determined $P(F, K_n)$ and $C(F, K_n)$ exactly for all sufficiently large n. Moreover, Alon, Caro and Yuster [3] proved that when G is large and very dense ($\delta(G) \geq (1 - o(1))|G|$), then $P(F, G)$ and $C(F, G)$ can be computed in polynomial time. The minimum degree threshold in this result can probably be significantly improved, perhaps a natural guess is that $\delta(G) \geq (\delta_F + o(1))|G|$ suffices.

7.3 Decomposing highly connected graphs into trees

Essentially all decomposition results we discussed in this survey apply only for dense graphs with linear minimum degree. It would be very interesting to investigate different conditions which ensure that a given F-divisible graph G has an F-decomposition.

One such example is a beautiful conjecture of Barát and Thomassen [8] on decompositions into a fixed tree T. Recall from Section 3.2 that the decomposition threshold of T is $1/2$. Moreover, since $\gcd(T) = 1$, the only necessary divisibility condition for G to have a T-decomposition is $e(T) \mid e(G)$. The reason why the minimum degree threshold cannot be lowered is that G could consist of two equal-sized vertex-disjoint cliques (with a few edges removed), such that the total number of edges is divisible by $e(T)$, but the number of edges in each clique is not. However, this example is not very robust. Just adding a constant number of edges across would allow us to find a T-decomposition. In particular, if G is highly connected, then it seems hard to construct any such example. Barát and Thomassen [8] conjectured that in fact this is impossible. This was proved recently by Bensmail, Harutyunyan, Le, Merker and Thomassé [13] via probabilistic methods, but also involving tools based on nowhere-zero flows [80, 90].

Theorem 7.2 ([13]) *For any tree T, there exists a constant k_T such that any graph G which is k_T-edge-connected and satisfies $e(T) \mid e(G)$ has a T-decomposition.*

The value of k_T needed for their proof is quite large, and it would be interesting to improve it.

7.4 Tree packings

We now discuss some further results on decompositions into trees. Since the term 'tree decomposition' is already reserved for another graph-theoretical concept, this problem is usually referred to as 'tree packing'. The main open problem in the area is the so-called 'tree packing conjecture' due to Gyárfás and Lehel [55].

Conjecture 7.3 ([55]) *For every n, the complete graph K_n can be decomposed into any sequence of trees $T_1, \ldots, T_n$ where $|T_i| = i$.*

Joos, Kim, Kühn and Osthus [65] proved this for bounded degree trees. Slightly earlier, Allen, Böttcher, Hladky and Piguet [2] proved an approximate version for trees whose maximum degree is allowed to be as large as $o(n/\log n)$. Very recently, Allen, Böttcher, Clemens and Taraz [1] showed that the tree packing conjecture holds for almost all sequences of trees. Each of these results applies in a considerably more general setting than stated here, and there are many more results which we do not mention here.

Another famous question on tree packings was formulated by Ringel in 1963, who asked whether K_{2n+1} can be decomposed into any tree with n edges. Very recently, Montgomery, Pokrovskiy and Sudakov [84] and Keevash and Staden [70] solved Ringel's conjecture for large enough n.

Theorem 7.4 ([70, 84]) *For sufficiently large n, K_{2n+1} can be decomposed into any tree with n edges.*

The proof in [84] is based on finding a single rainbow copy of the desired tree T in a suitably edge coloured K_{2n+1}. The approach in [70] builds on results in [68]. One crucial ingredient in both papers is to consider three cases according to the structure of the given tree, which was developed in [83] to prove an approximate version of Ringel's conjecture.

The following conjecture of Graham and Häggkvist [57] generalises Ringel's conjecture to arbitrary regular graphs.

Conjecture 7.5 ([57]) *For any tree T, any $2e(T)$-regular graph G has a T-decomposition.*

The main result in [70] implies that this is true if the host graph G is in addition dense and quasirandom. Moreover, Corollary 2.7 gives an approximate version if $|T| \geq (1/4 + o(1))|G|$ and T has bounded maximum degree. It would be interesting to settle this 'dense' case exactly.

7.5 Sparse decompositions of dense graphs: Erdős meets Nash-Williams

Recall Kirkman's theorem that every K_3-divisible complete graph has a K_3-decom-position. Much of the content of this survey has been inspired by Conjecture 1.1, which would be a far-reaching generalisation of Kirkman's theorem, and which was posed by Nash-Williams in 1970. Around the same time, Erdős proposed another beautiful extension of Kirkman's theorem. Define the *girth* of a set $\mathcal{T}$ of triangles to be the smallest $g \geq 4$ for which there is a set of g vertices spanning at least $g - 2$ triangles from the set $\mathcal{T}$. Note that any K_3-decomposition has girth at least 6. Erdős [43] conjectured that there are Steiner triple systems (i.e. K_3-decompositions of K_n) of arbitrarily large girth. (Decompositions with high girth are also called 'locally sparse' since any set of $4 \leq j < g$ vertices contains at most $j - 3$ triangles.)

Conjecture 7.6 ([43]) *For every fixed g, any sufficiently large K_3-divisible K_n has a K_3-decomposition with girth at least g.*

This conjecture has been proved exactly only for the first non-trivial case, namely $g = 7$, in a series of papers [19, 52, 53, 77]. Recently, it was solved approximately for all fixed g [16, 50]. A generalisation of Conjecture 7.6 to Steiner systems with arbitrary parameters was formulated in [50].

We are tempted to propose the following combination of the conjectures of Erdős and Nash-Williams.

Conjecture 7.7 *For every fixed g, any sufficiently large K_3-divisible graph G with $\delta(G) \geq 3|G|/4$ has a K_3-decomposition with girth at least g.*

Of course, given that both conjectures themselves are still open, this seems very challenging. In view of this, it would even be interesting to obtain approximate decompositions of large girth in a sufficiently dense graph, for instance, to show that any sufficiently large graph G with $\delta(G) \geq 0.9|G|$, say, has an approximate K_3-decomposition with arbitrarily high girth.

References

[1] P. Allen, J. Böttcher, D. Clemens, and A. Taraz, *Perfectly packing graphs with bounded degeneracy and many leaves*, arXiv:1906.11558 (2019).

[2] P. Allen, J. Böttcher, J. Hladký, and D. Piguet, *Packing degenerate graphs*, Adv. Math. **354** (2019), Art. 106739.

[3] N. Alon, Y. Caro, and R. Yuster, *Packing and covering dense graphs*, J. Combin. Des. **6** (1998), no. 6, 451–472.

[4] B. Alspach, P. J. Schellenberg, D. R. Stinson, and D. Wagner, *The Oberwolfach problem and factors of uniform odd length cycles*, J. Combin. Theory Ser. A **52** (1989), 20–43.

[5] R. F. Bailey and B. Stevens, *Hamiltonian decompositions of complete k-uniform hypergraphs*, Discrete Math. **310** (2010), no. 22, 3088–3095.

[6] Zs. Baranyai, *On the factorization of the complete uniform hypergraph*, Infinite and Finite Sets I (A. Hajnal, R. Rado, and V. T. Sós, eds.), Colloq. Math. Soc. János Bolyai, vol. 10, North-Holland, Amsterdam, 1975, pp. 91–108.

[7] _____, *The edge-coloring of complete hypergraphs. I*, J. Combin. Theory Ser. B **26** (1979), no. 3, 276–294.

[8] J. Barát and C. Thomassen, *Claw-decompositions and Tutte-orientations*, J. Graph Theory **52** (2006), no. 2, 135–146.

[9] B. Barber, S. Glock, D. Kühn, A. Lo, R. Montgomery, and D. Osthus, *Minimalist designs*, Random Structures Algorithms **57** (2020), 47–63.

[10] B. Barber, D. Kühn, A. Lo, R. Montgomery, and D. Osthus, *Fractional clique decompositions of dense graphs and hypergraphs*, J. Combin. Theory Ser. B **127** (2017), 148–186.

[11] B. Barber, D. Kühn, A. Lo, and D. Osthus, *Edge-decompositions of graphs with high minimum degree*, Adv. Math. **288** (2016), 337–385.

[12] B. Barber, D. Kühn, A. Lo, D. Osthus, and A. Taylor, *Clique decompositions of multipartite graphs and completion of Latin squares*, J. Combin. Theory Ser. A **151** (2017), 146–201.

[13] J. Bensmail, A. Harutyunyan, T.-N. Le, M. Merker, and S. Thomassé, *A proof of the Barát-Thomassen conjecture*, J. Combin. Theory Ser. B **124** (2017), 39–55.

[14] S. R. Blackburn, *The existence of k-radius sequences*, J. Combin. Theory Ser. A **119** (2012), no. 1, 212–217.

[15] A. Blumenthal, B. Lidický, Y. Pehova, F. Pfender, O. Pikhurko, and J. Volec, *Sharp bounds for decomposing graphs into edges and triangles*, Combin. Probab. Comput. (to appear).

[16] T. Bohman and L. Warnke, *Large girth approximate Steiner triple systems*, J. Lond. Math. Soc. **100** (2019), 895–913.

[17] J. Böttcher, M. Schacht, and A. Taraz, *Proof of the bandwidth conjecture of Bollobás and Komlós*, Math. Ann. **343** (2009), no. 1, 175–205.

[18] F. C. Bowditch and P. J. Dukes, *Fractional triangle decompositions of dense 3-partite graphs*, J. Combin. **10** (2019), 255–282.

[19] A. E. Brouwer, *Steiner triple systems without forbidden subconfigurations*, Tech. Report ZW 104/77, Mathematisch Centrum Amsterdam, 1977.

[20] D. Bryant and N. J. Cavenagh, *Decomposing graphs of high minimum degree into 4-cycles*, J. Graph Theory **79** (2015), no. 3, 167–177.

[21] D. Bryant and P. Danziger, *On bipartite 2-factorizations of $K_n - I$ and the Oberwolfach problem*, J. Graph Theory **68** (2011), 22–37.

[22] D. Bryant and D. Horsley, *A proof of Lindner's conjecture on embeddings of partial Steiner triple systems*, J. Combin. Des. **17** (2009), no. 1, 63–89.

[23] D. Bryant, D. Horsley, and W. Pettersson, *Cycle decompositions V: Complete graphs into cycles of arbitrary lengths*, Proc. Lond. Math. Soc. **108** (2014), 1153–1192.

[24] D. Bryant and V. Scharaschkin, *Complete solutions to the Oberwolfach problem for an infinite set of orders*, J. Combin. Theory Ser. B **99** (2009), no. 6, 904–918.

[25] Y. Caro and R. Yuster, *Packing graphs: the packing problem solved*, Electron. J. Combin. **4** (1997), no. 1, Research Paper 1, 7 pages.

[26] _____, *Covering graphs: the covering problem solved*, J. Combin. Theory Ser. A **83** (1998), no. 2, 273–282.

[27] F. Chung, P. Diaconis, and R. Graham, *Universal cycles for combinatorial structures*, Proceedings of the Twentieth Southeastern Conference on Combinatorics, Graph Theory, and Computing (Florida, 1989), Congr. Numer. 70–74, Utilitas Math., 1990.

[28] _____, *Universal cycles for combinatorial structures*, Discrete Math. **110** (1992), 43–59.

[29] P. Condon, J. Kim, D. Kühn, and D. Osthus, *A bandwidth theorem for approximate decompositions*, Proc. Lond. Math. Soc. **118** (2019), 1393–1449.

[30] B. Csaba, D. Kühn, A. Lo, D. Osthus, and A. Treglown, *Proof of the 1-factorization and Hamilton decomposition conjectures*, Mem. Amer. Math. Soc. **244** (2016), monograph 1154, 164 pages.

[31] D. Curtis, T. Hines, G. Hurlbert, and T. Moyer, *Near-universal cycles for subsets exist*, SIAM J. Discrete Math. **23** (2009), 1441–1449.

[32] D. E. Daykin and R. Häggkvist, *Completion of sparse partial Latin squares*, Graph Theory and Combinatorics (Cambridge, 1983), Academic Press, London, 1984, pp. 127–132.

[33] M. Dębski and Z. Lonc, *Universal cycle packings and coverings for k-subsets of an n-set*, Graphs Combin. **32** (2016), 2323–2337.

[34] M. Delcourt and L. Postle, *Progress towards Nash-Williams' conjecture on triangle decompositions*, J. Combin. Theory Ser. B **146** (2021), 382–416.

[35] D. Dor and M. Tarsi, *Graph decomposition is NP-complete: a complete proof of Holyer's conjecture*, SIAM J. Comput. **26** (1997), no. 4, 1166–1187.

[36] F. Dross, *Fractional triangle decompositions in graphs with large minimum degree*, SIAM J. Discrete Math. **30** (2016), no. 1, 36–42.

[37] P. Dukes, *Rational decomposition of dense hypergraphs and some related eigenvalue estimates*, Linear Algebra Appl. **436** (2012), no. 9, 3736–3746.

[38] _____, *Corrigendum to "Rational decomposition of dense hypergraphs and some related eigenvalue estimates" [Linear Algebra Appl. 436 (9) (2012) 3736–3746]*, Linear Algebra Appl. **467** (2015), 267–269.

[39] P. J. Dukes and D. Horsley, *On the minimum degree required for a triangle decomposition*, SIAM J. Discrete Math. (2020), 597–610.

[40] S. Ehard and F. Joos, *Decompositions of quasirandom hypergraphs into hypergraphs of bounded degree*, arXiv:2011.05359 (2020).

[41] _____, *A short proof of the blow-up lemma for approximate decompositions*, arXiv:2001.03506 (2020).

[42] P. Erdős, *On extremal problems of graphs and generalized graphs*, Israel J. Math. **2** (1964), 183–190.

[43] _____, *Problems and results in combinatorial analysis*, Colloquio Internazionale sulle Teorie Combinatorie (Rome, 1973), Accad. Naz. Lincei, 1976, pp. 3–17.

[44] A. Ferber, C. Lee, and F. Mousset, *Packing spanning graphs from separable families*, Israel J. Math. **219** (2017), no. 2, 959–982.

[45] P. Frankl and V. Rödl, *Near perfect coverings in graphs and hypergraphs*, European J. Combin. **6** (1985), no. 4, 317–326.

[46] K. Garaschuk, *Linear methods for rational triangle decompositions*, Ph.D. thesis, University of Victoria, 2014.

[47] S. Glock, F. Joos, J. Kim, D. Kühn, and D. Osthus, *Resolution of the Oberwolfach problem*, arXiv:1806.04644 (2018).

[48] S. Glock, F. Joos, D. Kühn, and D. Osthus, *Euler tours in hypergraphs*, Combinatorica **40** (2020), 679–690.

[49] S. Glock, D. Kühn, A. Lo, R. Montgomery, and D. Osthus, *On the decomposition threshold of a given graph*, J. Combin. Theory Ser. B **139** (2019), 47–127.

[50] S. Glock, D. Kühn, A. Lo, and D. Osthus, *On a conjecture of Erdős on locally sparse Steiner triple systems*, Combinatorica **40** (2020), 363–403.

[51] _____, *The existence of designs via iterative absorption: hypergraph F-designs for arbitrary F*, Mem. Amer. Math. Soc. (to appear).

[52] M. J. Grannell, T. S. Griggs, and C. A. Whitehead, *The resolution of the anti-Pasch conjecture*, J. Combin. Des. **8** (2000), no. 4, 300–309.

[53] T. S. Griggs, J. Murphy, and J. S. Phelan, *Anti-Pasch Steiner triple systems*, J. Comb. Inf. Syst. Sci. **15** (1990), no. 1-4, 79–84.

[54] T. Gustavsson, *Decompositions of large graphs and digraphs with high minimum degree*, Ph.D. thesis, Stockholm University, 1991.

[55] A. Gyárfás and J. Lehel, *Packing trees of different order into K_n*, Combinatorics (Proc. Fifth Hungarian Colloq., Keszthely, 1976), Vol. I, Colloq. Math. Soc. János Bolyai, vol. 18, North-Holland, Amsterdam-New York, 1978, pp. 463–469.

[56] E. Győri and Zs. Tuza, *Decompositions of graphs into complete subgraphs of given order*, Studia Sci. Math. Hungar. **22** (1987), no. 1-4, 315–320.

[57] R. Häggkvist, *Decompositions of complete bipartite graphs*, Surveys in combinatorics, 1989 (Norwich, 1989) (J. Siemons, ed.), Lond. Math. Soc. Lecture Note Ser., vol. 141, Cambridge Univ. Press, Cambridge, 1989, pp. 115–147.

[58] H. Hanani, *Decomposition of hypergraphs into octahedra*, Ann. New York Acad. Sci. **319** (1979), no. 1, 260–264.

[59] P. E. Haxell and V. Rödl, *Integer and fractional packings in dense graphs*, Combinatorica **21** (2001), no. 1, 13–38.

[60] A. J. W. Hilton and M. Johnson, *Some results on the Oberwolfach problem*, J. London Math. Soc. **64** (2001), no. 3, 513–522.

[61] D. G. Hoffman and P. J. Schellenberg, *The existence of C_k-factorizations of $K_{2n} - F$*, Discrete Math. **97** (1991), 243–250.

[62] C. Huang, A. Kotzig, and A. Rosa, *On a variation of the Oberwolfach problem*, Discrete Math. **27** (1979), 261–277.

[63] G. Hurlbert, *On universal cycles for k-subsets of an n-set*, SIAM J. Discrete Math. **7** (1994), 598–604.

[64] B. W. Jackson, *Universal cycles of k-subsets and k-permutations*, Discrete Math. **117** (1993), 141–150.

[65] F. Joos, J. Kim, D. Kühn, and D. Osthus, *Optimal packings of bounded degree trees*, J. Eur. Math. Soc. **21** (2019), 3573–3647.

[66] F. Joos and M. Kühn, *Fractional cycle decompositions in hypergraphs*, arXiv:2101.05526 (2021).

[67] P. Keevash, *The existence of designs*, arXiv:1401.3665 (2014).

[68] ———, *The existence of designs II*, arXiv:1802.05900 (2018).

[69] P. Keevash and K. Staden, *The generalised Oberwolfach problem*, arXiv:2004.09937 (2020).

[70] ———, *Ringel's tree packing conjecture in quasirandom graphs*, arXiv:2004.09947 (2020).

[71] J. Kim, D. Kühn, D. Osthus, and M. Tyomkyn, *A blow-up lemma for approximate decompositions*, Trans. Amer. Math. Soc. **371** (2019), no. 7, 4655–4742.

[72] T. P. Kirkman, *On a problem in combinatorics*, Cambridge Dublin Math. J. **2** (1847), 191–204.

[73] A. Kostochka, D. Mubayi, and J. Verstraëte, *On independent sets in hypergraphs*, Random Structures Algorithms **44** (2014), no. 2, 224–239.

[74] D. Král', B. Lidický, T. L. Martins, and Y. Pehova, *Decomposing graphs into edges and triangles*, Combin. Probab. Comput. **28** (2019), no. 3, 465–472.

[75] D. Kühn and D. Osthus, *Hamilton decompositions of regular expanders: A proof of Kelly's conjecture for large tournaments*, Adv. Math. **237** (2013), 62–146.

[76] D. Kühn and D. Osthus, *Hamilton cycles in graphs and hypergraphs: an extremal perspective*, Proceedings of the International Congress of Mathematicians—Seoul 2014. Vol. IV, Kyung Moon Sa, Seoul, 2014, pp. 381–406.

[77] A. C. H. Ling, C. J. Colbourn, M. J. Grannell, and T. S. Griggs, *Construction techniques for anti-Pasch Steiner triple systems*, J. Lond. Math. Soc. **61** (2000), no. 3, 641–657.

[78] Z. Lonc, P. Naroski, and P. Rząrzewski, *Tight Euler tours in uniform hypergraphs—computational aspects*, Discrete Math. Theor. Comput. Sci. **19** (2017), 13 pages.

[79] Z. Lonc, T. Traczyk, and M. Truszczyński, *Optimal f-graphs for the family of all k-subsets of an n-set*, Data base file organization (Warsaw, 1981), Notes Rep. Comput. Sci. Appl. Math., vol. 6, pp. 247–270.

[80] M. Merker, *Decomposing highly edge-connected graphs into homomorphic copies of a fixed tree*, J. Combin. Theory Ser. B **122** (2017), 91–108.

[81] R. Montgomery, *Fractional clique decompositions of dense partite graphs*, Combin. Probab. Comput. **26** (2017), 911–943.

[82] _____, *Fractional clique decompositions of dense graphs*, Random Structures Algorithms **54** (2019), 779–796.

[83] R. Montgomery, A. Pokrovskiy, and B. Sudakov, *Embedding rainbow trees with applications to graph labelling and decomposition*, J. Eur. Math. Soc. **22** (2020), 3101–3132.

[84] _____, *A proof of Ringel's Conjecture*, arXiv:2001.02665 (2020).

[85] C. St. J. A. Nash-Williams, *An unsolved problem concerning decomposition of graphs into triangles*, Combinatorial Theory and its Applications III (P. Erdős, A. Rényi, and V.T. Sós, eds.), North Holland, 1970, pp. 1179–1183.

[86] S. Piga and N. Sanhueza-Matamala, *Cycle decompositions in 3-uniform hypergraphs*, arXiv:2101.12205 (2021).

[87] D. K. Ray-Chaudhuri and R. M. Wilson, *Solution of Kirkman's schoolgirl problem*, Proc. Sympos. Pure Math. **19** (1971), 187–203.

[88] V. Rödl, M. Schacht, M. H. Siggers, and N. Tokushige, *Integer and fractional packings of hypergraphs*, J. Combin. Theory Ser. B **97** (2007), no. 2, 245–268.

[89] A. Taylor, *On the exact decomposition threshold for even cycles*, J. Graph Theory **90** (2019), 231–266.

[90] C. Thomassen, *Decomposing a graph into bistars*, J. Combin. Theory Ser. B **103** (2013), no. 4, 504–508.

[91] T. Traetta, *A complete solution to the two-table Oberwolfach problems*, J. Combin. Theory Ser. A **120** (2013), 984–997.

[92] Zs. Tuza, *Unsolved Combinatorial Problems, Part I*, BRICS Lecture Series (2001).

[93] R. M. Wilson, *An existence theory for pairwise balanced designs III. Proof of the existence conjectures*, J. Combin. Theory Ser. A **18** (1975), 71–79.

[94] _____, *Decompositions of complete graphs into subgraphs isomorphic to a given graph*, Proceedings of the Fifth British Combinatorial Conference (Aberdeen, 1975), Congr. Numer. 15, Utilitas Math., 1976, pp. 647–659.

[95] R. Yuster, *Tree decomposition of graphs*, Random Structures Algorithms **12** (1998), no. 3, 237–251.

[96] _____, *Decomposing hypergraphs into simple hypertrees*, Combinatorica **20** (2000), no. 1, 119–140.

[97] _____, *Packing and decomposition of graphs with trees*, J. Combin. Theory Ser. B **78** (2000), no. 1, 123–140.

[98] _____, *The decomposition threshold for bipartite graphs with minimum degree one*, Random Structures Algorithms **21** (2002), no. 2, 121–134.

[99] _____, *Asymptotically optimal K_k-packings of dense graphs via fractional K_k-decompositions*, J. Combin. Theory Ser. B **95** (2005), no. 1, 1–11.

[100] _____, *H-packing of k-chromatic graphs*, Mosc. J. Comb. Number Theory **2** (2012), no. 1, 73–88.

[101] _____, *Edge-disjoint cliques in graphs with high minimum degree*, SIAM J. Discrete Math. **28** (2014), no. 2, 893–910.

Stefan Glock
Institute for Theoretical Studies
ETH Zürich
Clausiusstrasse 47
8092 Zürich
Switzerland
dr.stefan.glock@gmail.com

Daniela Kühn
School of Mathematics
University of Birmingham
Edgbaston
B15 2TT Birmingham
United Kingdom
d.kuhn@bham.ac.uk

Deryk Osthus
School of Mathematics
University of Birmingham
Edgbaston
B15 2TT Birmingham
United Kingdom
d.osthus@bham.ac.uk

Borel combinatorics of locally finite graphs

Oleg Pikhurko[1]

Abstract

We provide a gentle introduction, aimed at non-experts, to Borel combinatorics that studies definable graphs on topological spaces. This is an emerging field on the borderline between combinatorics and descriptive set theory with deep connections to many other areas.

After giving some background material, we present in careful detail some basic tools and results on the existence of Borel satisfying assignments: Borel versions of greedy algorithms and augmenting procedures, local rules, Borel transversals, etc. Also, we present the construction of Andrew Marks of acyclic Borel graphs for which the greedy bound $\Delta + 1$ on the Borel chromatic number is best possible.

In the remainder of the paper we briefly discuss various topics such as relations to LOCAL algorithms, measurable versions of Hall's marriage theorem and of the Lovász Local Lemma, applications to equidecomposability, etc.

1 Introduction

Borel combinatorics, also called *measurable combinatorics* or *descriptive (graph) combinatorics*, is an emerging and actively developing field that studies graphs and other combinatorial structures on topological spaces that are "definable" from the point of view of descriptive set theory. It is an interesting blend of descriptive set theory and combinatorics that also has deep connections to measure theory, probability, group actions, ergodic theory, etc. We refer the reader to the survey by Kechris and Marks [36].

Many recent advances in this field came from adopting various proofs and concepts of finite combinatorics to the setting of descriptive set theory. However, this potentially very fruitful interplay is not yet fully explored. One of the reasons is that a fairly large amount of background is needed in order to understand the proofs (or even the statements) of some results of this field. Thus the purpose of this paper is give a gentle introduction to some basic concepts and tools, aimed at non-experts, as well as to provide some pointers to further reading for those who would like to learn more.

In order to give quickly a flavour of what types of objects and questions we will consider, here is one simple but illustrative example.

Definition 1.1 (Irrational Rotation Graph $\mathcal{R}_\alpha$) Let $\alpha \in \mathbb{R} \setminus \mathbb{Q}$ be irrational. Let $\mathcal{R}_\alpha$ be the graph whose vertex set is the half-open interval $[0,1)$ of reals, where $x, y \in [0,1)$ are adjacent if their difference is equal to $\pm \alpha$ modulo 1.

Another way to define $\mathcal{R}_\alpha$ is to consider the transformation $T_\alpha : [0,1) \to [0,1)$ which maps x to $x + \alpha \pmod 1$; then the edge set consists of all unordered pairs $\{x, T_\alpha(x)\}$ over $x \in [0,1)$. Thus $\mathcal{R}_\alpha$ can be viewed as the graph coming from an action of the group $\mathbb{Z}$ on $[0,1)$ with the generator $1 \in \mathbb{Z}$ acting via T_α. The

---
[1]Supported by Leverhulme Research Project Grant RPG-2018-424.

name comes from identifying the interval $[0,1)$ with the unit circle $\mathbb{S}^1 \subseteq \mathbb{R}^2$ via $x \mapsto (\cos(2\pi x), \sin(2\pi x))$ where T_α corresponds to the rotation by angle $2\pi\alpha$. Since α is irrational, $\mathcal{R}_\alpha$ is a 2-regular graph with each component being a *line* (a doubly-infinite path).

Under the Axiom of Choice, $\mathcal{R}_\alpha$ is combinatorially as trivial as a single line. For example, if we want to properly 2-colour its vertices then we can take a *transversal* S (a set containing exactly one vertex from each component) and colour every $x \in [0,1)$ depending on the parity of the distance from x to S in $\mathcal{R}_\alpha$. However, this proof is non-constructive.

So what is the chromatic number of $\mathcal{R}_\alpha$ if we want each colour class, as a subset of the real interval $[0,1)$, be "definable"? Of course, we have to agree first which sets are "definable". We would like the family of such sets to be closed under Boolean operations and under countable unions/intersections (i.e. to be a σ-algebra) so that various constructions, including some that involve passing to a limit after countably many iterations, do not take us outside the family. There are three important σ-algebras on $V = [0,1)$.

One, denoted by $\mathcal{B}$, consists of *Borel* sets and is, by definition, the smallest σ-algebra that contains all open sets, which for $[0,1)$ is equivalent to containing all intervals or just ones with rational endpoints. We can build Borel sets by starting with open sets and iteratively adding complements and countable unions of already constructed sets. Then each Borel set appears after β-many iterations for some countable ordinal β and thus can be "described" with countably many bits of information (motivating the name of descriptive set theory).

The other two σ-algebras can be built by taking all Borel sets which can be additionally modified by adding or removing any "negligible" set of elements. For example, from the topological point of view any *nowhere dense* set X (that is, a set whose closure has empty interior) is "negligible": every non-empty open set U contains a "substantial" part (namely, some non-empty open $W \subseteq U$) that avoids X completely. We also consider *meager* sets (that is, countable unions of nowhere dense sets) as "negligible". Now, the corresponding σ-algebra $\mathcal{T}$ (of all sets X with the symmetric difference $X \triangle E$ being meager for some Borel E) consists precisely of the sets that have the *property of Baire* and thus may be regarded as "topologically nice".

The definition of our third σ-algebra $\mathcal{B}_\mu$ of μ-*measurable* sets depends on a measure μ on Borel sets (which we take to be the Lebesgue measure in our Example 1.1). A set X is called μ-*measurable* (or just *measurable* when μ is understood) if there is a Borel set E such that the symmetric difference $X \triangle E$ is *null* (that is, is contained in a Borel set of measure 0). For $[0,1)$ and the Lebesgue measure, a set is null if and only if, for every $\varepsilon > 0$, it can be covered by countably many intervals whose sum of lengths is at most ε. The measure μ extends in the obvious way to a measure on $\mathcal{B}_\mu \supseteq \mathcal{B}$.

Clearly, $\mathcal{B}$ is a sub-family of $\mathcal{T}$ and of $\mathcal{B}_\mu$ but the last two σ-algebras are not compatible, with the notions of a "negligible" set being quite strikingly different. For example, one can partition the interval $[0,1)$ into two Borel sets one of which is meager and the other has Lebesgue measure 0, see e.g. [69, Theorem 1.6].

As it turns out, it is impossible to find a proper 2-colouring $[0,1) = X_0 \cup X_1$ of the graph $\mathcal{R}_\alpha$ from Example 1.1 with $X_0, X_1 \in \mathcal{B}_\mu$. Indeed, since the colours must alternate on each line, we have that the measure-preserving map T_α swaps X_0 and X_1 so these sets, if measurable, have Lebesgue measure 1/2 each. However, by $T_\alpha(T_\alpha(X_0)) = X_0$ this would contradict the fact that the composition $T_\alpha \circ T_\alpha = T_{2\alpha}$ is *ergodic*, meaning that every invariant measurable set has measure 0 or 1. (For two different proofs of the last property, via Fourier analysis and via Lebesgue's density theorem, see e.g. [80, Proposition 4.2.1].) Thus, although there are no edges between the lines of $\mathcal{R}_\alpha$, we have to be careful with what we do on different lines so that the combined outcome is measurable.

A similar argument using the *generic ergodicity* of $T_{2\alpha}$ (namely, that every invariant set with the property of Baire is meager or has meager complement) shows that $\mathcal{R}_\alpha$ cannot be 2-coloured with both parts in $\mathcal{T}$, see e.g. [79, Example 21.6] for a derivation.

On the other hand, it is easy to show that we can properly 3-colour $\mathcal{R}_\alpha$ with Borel colour classes X_0, X_1 and X_2. For example, we can let $X_2 := [0, c)$ for any $c > 0$ which makes X_2 independent (namely, $c \leqslant \min\{\alpha \pmod 1, 1 - \alpha \pmod 1\}$ is enough) and then colour every $x \in [0, 1) \setminus X_2$ by the parity of the minimum integer $n \geqslant 1$ with $T_\alpha^n(x) \in X_2$. (Such n exists by the irrationality of α.) If Y_n denotes the set of vertices for which the n-th iterate of T_α is the first one to hit X_2, then $Y_0 = X_2$ and, for each $n \geqslant 1$, we have that $Y_n = T_\alpha^{-n}(X_2) \setminus (Y_0 \cup \cdots \cup Y_{n-1})$ is a finite union of half-open intervals by induction on n and thus is Borel. We see that, for each of the σ-algebras $\mathcal{B}$, $\mathcal{B}_\mu$ and $\mathcal{T}$, the minimum number of colours in a "definable" proper colouring of $\mathcal{R}_\alpha$ happens to be the same, namely 3.

To keep this paper of reasonable size, we will concentrate on results, where the assignments that we construct have to be Borel. Furthermore, except a few places where it is explicitly stated otherwise, we will restrict ourselves to *locally finite* graphs, where every vertex has finitely many neighbours (but we usually do not require that the degrees are uniformly bounded by some constant). This is already a very rich area. Also, results of this type often form the proof basis for other settings. For example, if we want to find a proper colouring whose classes are measurable (resp. have the property of Baire) then one common approach is to build a Borel colouring and then argue that the set of vertices in connectivity components with at least one conflict (two adjacent vertices of the same colour) is null (resp. meager); in such situations, we are allowed to fix such components in an arbitrary fashion, e.g. by applying the Axiom of Choice. (Everywhere in this paper we assume that the Axiom of Choice holds.)

The following quotation of Lubotzky [53, Page xi] applies almost verbatim to this paper: *"Generally speaking, I tried to write it in a form of something I wish had existed when, eight years ago, I made my first steps into these subjects without specific background in any of them."*

This paper is organised as follows. Some notation that is frequently used in the paper is collected in Section 2. Then Section 3 lists some basic facts about Borel sets. We assume that the reader is familiar with the fundamentals of topology. However, we try to carefully state all required, even rather basic results on Borel sets and functions, with references to complete proofs. Section 4 defines Borel graphs, our main object of study.

Section 5 presents some basic results for locally finite Borel graphs as follows. Section 5.1 contains the proofs of some classical results of Kechris, Solecki and

Todorcevic [38] on Borel chromatic numbers and maximal independent sets. In Section 5.2, we prove that each "locally defined" labelling is a Borel function, a well-known and extremely useful result. Section 5.3 presents one application of this result, namely that if one can pick exactly one vertex from each connectivity component in a Borel way then every locally checkable labelling problem (LCL for short) that is satisfiable admits a global Borel solution. Section 5.4 present another important application namely that, for every LCL, we can carry all augmentations supported on vertex sets of size at most r in a Borel way so that none remains.

The above mentioned results from Sections 5.2–5.4 are frequently used and well-known to experts. However, detailed and accessible proofs of these results are hard to find in the literature. So, in order to fill this gap, the author carefully states and proves rather general versions of these results, deviating from the philosophy of the rest of this paper of presenting just a simple case that conveys main ideas. The reader may prefer to skip the proofs from Sections 5.2–5.4 and move to Sections 6–13 that discuss more interesting results.

Section 6 presents the surprising result of Marks [58] that the upper bound $\Delta+1$ on the Borel chromatic number from Section 5.1 (that comes from an easy greedy algorithm) is in fact best possible even for acyclic graphs.

At this point we choose to give brief pointers to some other areas, namely, Borel equivalence relations (Section 7), assignments with the property of Baire (Section 8), and μ-measurable assignments (Section 9).

With this background, however brief, we can discuss various results, in particular those that connect Borel combinatorics to measures and the property of Baire. Section 10 presents the recent result of Bernshteyn [7] that efficient LOCAL algorithms can be used to find Borel satisfying assignments of the corresponding LCLs. Section 11 discusses some "purely Borel" existence results where measures come up in the proofs (but not in the statements). Section 12 discusses the class of graphs with subexponential growth for which one can prove some very general Borel results. Finally, Section 13 presents some applications of descriptive combinatorics to *equidecomposability* (where we try to split two given sets into congruent pieces).

2 Notation

Here we collect some notation that is used in the paper. In order to help the reader to get into a logician's mindset we use some conventions from logic and set theory.

We identify a non-negative integer k with the set $\{0,\ldots,k-1\}$, following the recursive definition of natural numbers as $0 := \emptyset$ being the empty set and $k+1 := k \cup \{k\}$. Thus $i \in k$ is a convenient shorthand for $i \in \{0,\ldots,k-1\}$. The set of natural numbers is denoted by $\omega := \{0,1,\ldots\}$ and our integer indices usually start from 0.

Let π_i denote the projection from a product $\prod_{j \in r} X_j$ of sets to its $(i+1)$-st coordinate X_i. We identify a function $f : X \to Y$ with its graph $\{(x, f(x)) : x \in X\} \subseteq X \times Y$. Thus the *domain* and the *image* of f can be written respectively as $\pi_0(f)$ and $\pi_1(f)$. Also, the restriction of a function $f : X \to Y$ to $Z \subseteq X$ is $f \restriction Z := f \cap (Z \times Y)$, which is a function from Z to Y. For sets X and Y, the set of all functions $X \to Y$ is denoted by Y^X. For functions $f_i : X \to Y_i$, $i \in 2$, let

$(f_0, f_1) : X \to Y_0 \times Y_1$ denote the function that maps $x \in X$ to $(f_0(x), f_1(x))$.

Let X be a set. The *diagonal* Diag_X is the set $\{(x,x) : x \in X\} \subseteq X^2$. We identify the elements of 2^X with the subsets of X, where a function $f : X \to 2$ corresponds to the preimage $f^{-1}(1) \subseteq X$. A *total order* on X is a partial order on X in which every two elements are compatible, that is, a transitive and antisymmetric subset $\preccurlyeq$ of X^2 such that for every $x, y \in X$ it holds that $x \preccurlyeq y$ or $y \preccurlyeq x$. The set X is *countable* if it admits an injective map into ω (in particular, every finite set is countable); then its cardinality $|X|$ is the unique $k \in \omega \cup \{\omega\}$ such that there is a bijection between X and k.

Let G be a *graph* by which we mean a pair (V, E), where V is a set and E is a subset of $V^2 \setminus \mathrm{Diag}_V$ which is *symmetric*, i.e. for every $x, y \in V$, $(x, y) \in E$ if and only if $(y, x) \in E$. When it is convenient, we may work with unordered pairs, with $\{x, y\} \in E$ translating into $(x, y), (y, x) \in E$, etc. Note that we do not allow multiple edge nor loops. Of course, when we talk about matchings, edge colourings, etc, we mean subsets of E, functions on E, etc, which are symmetric (in the appropriate sense). When the graph G is understood, we use V and E by default to mean its vertex set and its edge set, and often remove the reference to G from notation (except in the statements of theorems and lemmas, all of which we try to state fully).

The usual definitions of graph theory apply. A *walk* (of length $i \in \omega$) is a sequence $(x_0, \ldots, x_i) \in V^{i+1}$ such that $(x_j, x_{j+1}) \in E$ for every $j \in i$. A *path* is a walk in which vertices do not repeat. Note that the length of a walk (or a path) refers to the number of edges. For $x, y \in V$, their *distance* $\mathrm{dist}_G(x, y)$ is the shortest length of a path connecting x to y in G (and is ω if no such path exists). A *rooted graph* is a pair (G, x) where G is a graph and x is a vertex of G.

The *neighbourhood* of a set $A \subseteq V$ is

$$N_G(A) := \{y \in V : \exists x \in A \ (x, y) \in E\}.$$

We abbreviate $N_G(\{x\})$ to $N_G(x)$. Note that $N_G(x)$ does not include x. We denote the degree of x by $\deg_G(x) := |N_G(x)|$. The graph G is *locally finite* (resp. *locally countable*) if the neighbourhood of every vertex is finite (resp. countable).

For $r \in \omega$, the *r-th power* G^r of G is the graph on the same vertex set V where two distinct vertices are adjacent if they are at distance at most r in G. Also, the *r-ball* around A,

$$N_G^{\leqslant r}(A) := N_{G^r}(A) \cup A,$$

consists of those vertices of G that are at distance at most r from A. A set of vertices $X \subseteq V$ is called *r-sparse* if the distance between any two distinct vertices of X is strictly larger than r (or, equivalently, if X is an independent set in G^r).

The subgraph *induced* by $X \subseteq V$ in G is $G \upharpoonright X := (X, E \cap X^2)$. We call a set $X \subseteq V$ *connected* if the induced subgraph $G \upharpoonright X$ is connected, that is, every two vertices of X are connected by a path with all vertices in X. By a *(connectivity) component* of G we mean a maximal connected set $X \subseteq V$. Let

$$\mathcal{E}_G := \{(x, y) \in V^2 : \mathrm{dist}(x, y) < \omega\}, \tag{2.1}$$

denote the *connectivity relation* of G. In other words, $\mathcal{E}_G$ is the transitive closure of $E \cup \mathrm{Diag}_V$. Clearly, it is an equivalence relation whose classes are the components of G. The *saturation* of a set $A \subseteq V$ is

$$[A]_G := \{x \in V : \text{dist}(x, A) < \omega\},$$

that is, the union of all components intersecting A. In particular, the component of a vertex $x \in V$ is $[x]_G := [\{x\}]_G$.

Finally, we will need the following generalisation of Hall's marriage theorem to (infinite) graphs by Rado [70], whose proof (that relies on the Axiom of Choice) can also be found in e.g. [78, Theorem C.2].

Theorem 2.1 (Rado [70]) *Let G be a bipartite locally finite graph. If $|N(X)| \geqslant |X|$ for every finite set X inside a part then G has a perfect matching.* □

3 Some Background and Standard Results

Here we present some basic results from analysis and descriptive set theory that we will use in this paper. We do not try to give any historic account of these results. Instead we just refer to their modern proofs (using sources which happen to be most familiar to the author).

3.1 Algebras and σ-Algebras

An *algebra* on a set X is a non-empty family $\mathcal{A} \subseteq 2^X$ of subsets of X which is closed under Boolean operations inside X (for which it is enough to check that $A \cup B, X \setminus A \in \mathcal{A}$ for every $A, B \in \mathcal{A}$). In particular, the empty set and the whole set X belong to $\mathcal{A}$.

A σ-*algebra* on set X is an algebra on X which also is closed under countable unions. It follows that it is also closed under countable intersections. Also, if we have a countable sequence of sets $A_0, A_1, \ldots$ in a σ-algebra $\mathcal{A}$, then $\liminf_n A_n$ (resp. $\limsup_n A_n$), the set of elements that belong to all but finitely sets A_n (resp. belong to infinitely many sets A_n), is also in $\mathcal{A}$. Indeed, we have, for example, that

$$\liminf_n A_n = \cup_{n \in \omega} \cap_{m=n}^{\infty} A_m,$$

belongs to $\mathcal{A}$ as the countable union of $\cap_{m=n}^{\infty} A_m \in \mathcal{A}$.

For an arbitrary family $\mathcal{F} \subseteq 2^X$, let $\sigma_X(\mathcal{F})$ denote the σ-algebra on X generated by $\mathcal{F}$, that is, $\sigma_X(\mathcal{F})$ is the intersection of all σ-algebras on X containing $\mathcal{F}$. (Note that the intersection is taken over a non-empty set since 2^X is an example of such a σ-algebra.)

3.2 Polish Spaces

Although the notion of a Borel set could be defined for general topological spaces, the theory becomes particularly nice and fruitful when the underlying space is Polish. Let us discuss this class of spaces first.

By a *topological* space we mean a pair (X, τ), where τ is a topology on a set X (specifically, we view $\tau \subseteq 2^X$ as the collection of all open sets). When the topology τ on X is understood, we usually just write X. The space X is *separable* if there is a countable subset $Y \subseteq X$ which is *dense* (that is, every non-empty $U \in \tau$

intersects Y). Also, X is *metrizable* if there is a metric d which is *compatible* with τ (namely, $U \in \tau$ if and only if for every $x \in U$ there is real $r > 0$ with the *r-ball* $\{y \in X : d(x,y) < r\}$ lying inside U). Furthermore, if d can be chosen to be *complete* (i.e. every Cauchy sequence converges to some element of X) then we call X *completely metrizable*. We call a topological space *Polish* if it is separable and completely metrizable. For more details on Polish spaces, we refer to, for example, Cohn [14, Chapter 8.1], Kechris [34, Chapter 3], or Tserunyan [79, Part 1].

This class of spaces was first extensively studied by Polish mathematicians (Sierpiński, Kuratowski, Tarski and others), hence the name. It has many nice properties, in particular being closed under many topological operations (such as passing to closed or more generally G_δ subsets, or taking various constructions like countable products or countable disjoint unions, etc). Also, Polish spaces satisfy various results crucial in descriptive set theory (such as the Baire Category Theorem). So this class is the primary setting for Borel combinatorics. All topological spaces that we will consider in this survey are assumed to be Polish. As it is customary in this field, we do not fix a metric.

Two basic examples of Polish spaces are the integers $\mathbb{Z}$ (with the *discrete topology* where every set is open) and the real line $\mathbb{R}$ (with the usual topology, where a set is open if and only if it is a union of open intervals). Many other Polish spaces can be obtained from these; in fact, every Polish space is homeomorphic to a closed subset of $\mathbb{R}^\omega$ ([34, Theorem 4.17]).

Also, it can be proved (without assuming the Continuum Hypothesis) that every uncountable Polish space has the same cardinality as e.g. the set of reals:

Theorem 3.1 *Each Polish space is either countable or has continuum many points.*

Proof See e.g. [34, Corollary 6.5] or [79, Corollary 4.6]. □

3.3 The Borel σ-Algebra of a Polish Space

Let $X = (X, \tau)$ be a Polish space. Its *Borel σ-algebra* is $\mathcal{B}(X, \tau) := \sigma_X(\tau)$, the σ-algebra on X generated by all open sets. When the meaning is clear, we will usually just write $\mathcal{B}(X)$ or $\mathcal{B}$ instead of $\mathcal{B}(X, \tau)$.

The definition of the Borel σ-algebra behaves well with respect to various operations on Polish spaces. For example, the Borel σ-algebra of the product space $X \times Y$ is equal to the product of σ-algebras $\mathcal{B}(X) \times \mathcal{B}(Y)$:

$$\mathcal{B}(X \times Y) = \mathcal{B}(X) \times \mathcal{B}(Y), \tag{3.1}$$

see e.g. [14, Proposition 8.1.7] (and [14, Section 5.1] for an introduction to products of σ-algebras).

A function $f : X \to Y$ is *Borel (measurable)* if the preimage of every Borel subset of Y is a Borel subset of X. It is easy to see ([14, Proposition 2.6.2]) that it is enough to check this condition for any family $\mathcal{A} \subseteq \mathcal{B}(Y)$ that generates $\mathcal{B}(Y)$, e.g. just for open sets. Since we identify the function f with its graph $\{(x, f(x)) : x \in X\} \subseteq X \times Y$, the following result justifies why it is more common to call a Borel measurable function just "a Borel function".

Lemma 3.2 *A function $f : X \to Y$ between Polish spaces is Borel measurable if and only if it is a Borel subset of $X \times Y$.*

Proof See e.g. [14, Proposition 8.3.4]. □

Although there are many non-equivalent choices of a Polish topology τ on an infinite countable set X, each of them has 2^X as its Borel σ-algebra. Indeed, every subset Y of X, as the countable union $Y = \cup_{y \in Y} \{y\}$ of closed sets, is Borel. A remarkable generalisation of this trivial observation is the following.

Theorem 3.3 (Borel Isomorphism Theorem) *Every two Polish spaces X, Y of the same cardinality are Borel isomorphic, that is, there is a bijection $f : X \to Y$ such that the functions f and f^{-1} are Borel. (In fact, by Theorem 3.8, it is enough to require that just one of the bijections f or f^{-1} is Borel.)*

Proof See e.g. [14, Theorem 8.3.6] or [79, Theorem 13.10]. □

One crucial property of the Borel σ-algebra is that it can be generated by a countable family of sets.

Lemma 3.4 *For every Polish space X, there is a countable family $\mathcal{J} = \{J_n : n \in \omega\}$ with $\mathcal{B}(X) = \sigma_X(\mathcal{J})$.*

Proof Fix a countable dense subset $Y \subseteq X$ and a compatible metric d on X. Let $\mathcal{J}$ consist of all open balls in the metric d around points of Y with rational radii. Of course, $\mathcal{J} \subseteq \mathcal{B}$. Every open set is a union of elements of $\mathcal{J}$ and this union is countable as $\mathcal{J}$ is countable. Thus $\sigma_X(\mathcal{J})$ contains all open sets and, therefore, has to be equal to $\mathcal{B}$. □

Remark 3.5 We can additionally require in Lemma 3.4 that $\mathcal{J}$ is an algebra on X. Indeed, given any generating countable family $\mathcal{J} \subseteq \mathcal{B}$, we can enlarge it by adding all Boolean combinations of its elements. This does not affect the equality $\sigma_X(\mathcal{J}) = \mathcal{B}$ (as all added sets are in $\mathcal{B}$) and keeps the family countable.

For example, for $X := \mathbb{R}$, $d(x,y) := |x-y|$, and $Y := \mathbb{Q}$, our proof of Lemma 3.4 gives the family of open intervals with rational endpoints. If we want to have a generating algebra for $\mathcal{B}(\mathbb{R})$ as in Remark 3.5, it may be more convenient to use half-open intervals in the first step, namely to take

$$\mathcal{J} := \{[a,b) : a, b \in \mathbb{Q}\} \cup \{[a, \infty) : a \in \mathbb{Q}\} \cup \{(-\infty, b) : b \in \mathbb{Q}\}.$$

Then the algebra generated by $\mathcal{J}$ will precisely consists of finite unions of disjoint intervals in $\mathcal{J}$.

Note that, for every distinct $x, y \in X$, there is an open (and thus Borel) set U that *separates* x from y, that is, U contains x but not y (e.g. an open ball around x of radius $d(x,y)/2$ for some compatible metric d). Thus Borel sets separate points. In fact, we have the following stronger property.

Lemma 3.6 *Let X be a Polish space and let $\mathcal{J}$ be an algebra on X that generates $\mathcal{B}(X)$. Then for every pair of disjoint finite sets $A, B \subseteq X$ there is $J \in \mathcal{J}$ with $A \subseteq J$ and $B \cap J = \emptyset$.*

Proof For distinct $a, b \in X$, there is some $J_{a,b} \in \mathcal{J}$ that contains exactly one of a and b (as otherwise $\mathcal{B}(X) = \sigma_X(\mathcal{J})$ cannot separate a from b) and, by passing to the complement if necessary, we can assume that $J_{a,b}$ separates a from b. For $a \in A$, the set $J_a := \cap_{b \in B} J_{a,b}$ belongs to $\mathcal{J}$, contains a and is disjoint from B. Thus $J := \cup_{a \in A} J_a$ satisfies the claim. □

The *lexicographic order* $\leqslant_{\text{Lex}}$ on 2^ω is defined so that $(x_i)_{i \in \omega}$ comes before $(y_i)_{i \in \omega}$ if $x_i < y_i$ where $i \in \omega$ is the smallest index with $x_i \neq y_i$.

Lemma 3.7 *For every Polish space X there is a Borel injective map $I : X \to 2^\omega$. Furthermore, given any such map I, if we define $\preccurlyeq$ to consist of those pairs $(x, y) \in X^2$ with $I(x) \leqslant_{\text{Lex}} I(y)$, then we obtain a total order on X which is Borel (as a subset of X^2).*

Proof By Lemma 3.4, fix a countable family $\mathcal{J}$ that generates $\mathcal{B}$ and define

$$I(x) := (\mathbb{1}_{J_0}(x), \mathbb{1}_{J_1}(x), \ldots), \quad x \in X, \qquad (3.2)$$

where the *indicator function* $\mathbb{1}_J$ of $J \subseteq X$ assumes value 1 on J and 0 on $X \setminus J$. In other words, $I(x)$ records which sets J_i contain x. The Borel σ-algebra on the product space 2^ω is generated by the sets of the form $Y_{i,\sigma} := \{x \in 2^\omega : x_i = \sigma\}$ for $i \in \omega$ and $\sigma \in 2$ (as these form a pre-base for the product topology on 2^ω). Since $I^{-1}(Y_{i,1}) = J_i$ and $I^{-1}(Y_{i,0}) = X \setminus J_i$ are in $\mathcal{B}$, the map I is Borel. Also, I is injective by Lemma 3.6.

By the injectivity of I, we have that $\preccurlyeq$ is a total order on X. Also, the lexicographic order $\leqslant_{\text{Lex}}$ is a closed (and thus Borel) subset of $(2^\omega)^2$. Indeed, if $x \not\leqslant_{\text{Lex}} y$ then, with $i \in \omega$ being the maximum index such that $(x_0, \ldots, x_{i-i}) = (y_0, \ldots, y_{i-1})$, every pair (x', y') that coincides with (x, y) on the first $i+1$ indices (an open set of pairs) is not in $\leqslant_{\text{Lex}}$. Thus $\preccurlyeq$, as the preimage under the Borel map $I \times I : X \times X \to 2^\omega \times 2^\omega$ that sends (x, y) to $(I(x), I(y))$, is a Borel subset of X^2. (Our claim that the map $I \times I$ is Borel can be easily derived from (3.1).) □

If X is a Borel subset of $[0, 1)$ then another, more natural, choice of I in Lemma 3.7 is to map $x \in [0, 1)$ to the digits of its binary expansion where we do not allow infinite sequences of trailing 1's. Then $\preccurlyeq$ becomes just the standard order on $[0, 1)$.

While continuous images of Borel sets are not in general Borel, this is true all for countable-to-one Borel maps for which, moreover, a Borel right inverse exists (with the latter property called *uniformization* in descriptive set theory).

Theorem 3.8 (Lusin-Novikov Uniformization Theorem) *Let X, Y be Polish spaces and let $f : X \to Y$ be a Borel map. Let $A \subseteq X$ be a Borel set such that every $y \in Y$ has countably many preimages in A under f. Then $f(A)$ is a Borel subset of Y. Moreover, there are countably many Borel maps $g_n : f(A) \to A$, $n \in \omega$, each being a right inverse to f (i.e. the composition $f \circ g_n$ is the identity function on $f(A)$), such that for every $x \in A$ there is $n \in \omega$ with $g_n(f(x)) = x$.*

Proof This theorem, in the case when $X = Z \times Y$ for some Polish Z and $f = \pi_1$ is the projection on Y, can be found in e.g. [34, Theorem 18.10] or [79, Theorem

13.6]. To derive the version stated here, let $A' := (A \times Y) \cap f$. (Recall that we view the function $f : X \to Y$ as a subset $f \subseteq X \times Y$.) Then $A' \subseteq X \times Y$ is Borel by Lemma 3.2, $f(A) = \pi_1(A')$ is Borel by the above product version while the sequence of Borel right inverses $g'_n : f(A) \to A'$ for π_1 gives the required functions $g_n := \pi_0 \circ g'_n$. □

One can view the second part of Theorem 3.8 as a "Borel version" of the Axiom of Choice. If, say, $A = X$ and f is surjective, then finding one right inverse g_0 of f amounts to picking exactly one element from each of the sets $X_y := f^{-1}(y)$ indexed by $y \in Y$. Thus Theorem 3.8 gives that, if all dependences are "Borel" and each set X_y is countable, then a Borel choice is possible. This result was generalised to the case where each X_y is required to be only σ-compact (a countable union of compact sets), see e.g. [34, Theorem 35.46.ii].

3.4 Standard Borel Spaces

A *standard Borel space* is a pair $(X, \mathcal{A})$ where X is a set and there is a choice of a Polish topology τ on X such that $\mathcal{A}$ is equal to $\sigma_X(\tau)$, the Borel σ-algebra of (X, τ). For a detailed treatment of standard Borel spaces, see e.g. [34, Chapter 12] or [75, Chapter 3].

We will denote the Borel σ-algebra on a standard Borel space X by $\mathcal{B}(X)$ or $\mathcal{B}$, also abbreviating $(X, \mathcal{B})$ to X. It is customary not to fix a Polish topology τ on X (which, strictly speaking, requires checking that various operations defined on standard Borel spaces do not depend on the choice of topologies).

By the Borel Isomorphism Theorem (Theorem 3.3) combined with Theorem 3.1, every standard Borel space either consists of all subsets of a countable set X or admits a Borel isomorphism into e.g. the interval $[0, 1]$ of reals. Since this survey concentrates on the Borel structure, we could have, in principle, restricted ourselves to just these special Polish spaces. However, many constructions and proofs are much more natural and intuitive when written in terms of general Polish spaces.

A useful property that is often used without special mention is that every Borel subset induces a standard Borel space. This follows from the following result.

Lemma 3.9 *Let (X, τ) be a Polish space. Then for every $Y \in \mathcal{B}(X, \tau)$ there is a Polish topology $\tau' \supseteq \tau$ on X such that Y is a closed set with respect to τ' and $\mathcal{B}(X, \tau) = \mathcal{B}(X, \tau')$.*

Proof See e.g. [34, Theorem 13.1] or [79, Theorem 11.16]. □

Remark 3.10 In fact, any two nested Polish topologies $\tau \subseteq \tau'$ on a common set X generate the same Borel σ-algebra (so this conclusion could be omitted from the statement of Lemma 3.9). Indeed, the identity map $(X, \tau') \to (X, \tau)$ is continuous and thus Borel. By Theorem 3.8, its unique right inverse, which is the identity map $(X, \tau) \to (X, \tau')$, is Borel. Thus $\mathcal{B}(X, \tau') = \mathcal{B}(X, \tau)$.

Corollary 3.11 *If X is a standard Borel space and $Y \in \mathcal{B}(X)$, then $(Y, \mathcal{B}(X) \upharpoonright Y)$ is a standard Borel space, where for $\mathcal{A} \subseteq 2^X$ we denote $\mathcal{A} \upharpoonright Y := \{A \cap Y : A \in \mathcal{A}\}$.*

Proof Fix a Polish topology τ on X that generates $\mathcal{B}(X)$, that is, $\sigma_X(\tau) = \mathcal{B}(X)$. By Lemma 3.9, there is a Polish topology $\tau' \supseteq \tau$ for which Y is closed. Let $\tau'' := \{U \cap Y : U \in \tau'\}$. As it is easy to see, (Y, τ'') is a Polish space and $\sigma_Y(\tau'') = \sigma_X(\tau') \restriction Y$. By the second conclusion of Lemma 3.9 (or by Remark 3.10), we have that $\sigma_X(\tau') = \sigma_X(\tau)$, finishing the proof. □

4 Borel Graphs: Definition and Some Examples

For a short discussion of bounded-degree Borel graphs, see Lovász [52, Section 18.1].

A *Borel graph* is a triple $\mathcal{G} = (V, E, \mathcal{B})$ such that (V, E) is a graph (that is, $E \subseteq V^2$ is a symmetric and anti-reflexive relation), $(V, \mathcal{B})$ is a standard Borel space, and E is a Borel subset of $V \times V$. (As it follows from (3.1), the Borel σ-algebra on $V \times V$ depends only on $\mathcal{B}(V)$ but not on the choice of a compatible Polish topology τ on V.)

All our graph theoretic notation will also apply to Borel graphs and, when the underlying Borel graph $\mathcal{G}$ is clear, we usually remove any reference to $\mathcal{G}$ from notation (writing $N(x)$ instead of $N_\mathcal{G}(x)$, etc). Note that we use the calligraphic letter $\mathcal{G}$ to emphasise that it is a Borel graph (whereas G is used for general graphs).

Remark 4.1 If one prefers, then one can work with the edge set as a subset of $\binom{V}{2}$, the set of all unordered pairs of distinct elements of V. The standard σ-algebra on $\binom{V}{2}$ (which also makes it a standard Borel space) is obtained by taking all those sets $A \subseteq \binom{V}{2}$ for which $\pi^{-1}(A)$ is Borel subset of V^2, where $\pi : V^2 \to \binom{V}{2}$ is the natural projection mapping (x, y) to $\{x, y\}$. In terms of Polish topologies, if we fix a topology τ on V, then we consider the *factor topology* τ' on $\binom{V}{2}$ with respect to π (that is, the largest topology that makes π continuous) and take the Borel σ-algebra of τ'.

Here are some examples of Borel graphs.

Example 4.2 For every graph (V, E) whose vertex set is countable, the triple $(V, E, 2^V)$ is a Borel graph. Indeed, $(V, 2^V)$ is a standard Borel space (take, for example, the discrete topology on V). Then $\mathcal{B}(V \times V)$ contains all singleton sets (as closed sets) and, being closed under countable unions, it contains all subsets of $V \times V$, in particular, the edge set E. ∎

Example 4.3 Let $(\Gamma; S)$ be a *marked group*, that is Γ is a group generated by a finite set $S \subseteq \Gamma$ which is *symmetric*, that is, $S = S^{-1}$ where $S^{-1} := \{\gamma^{-1} : \gamma \in S\}$. (We do not assume that S is minimal in any sense.) Let $a : \Gamma \curvearrowright X$ be a (left) action of Γ on a Polish space X which is *Borel*, meaning for the countable group Γ that for every $\gamma \in \Gamma$ the bijection $a(\gamma, \cdot) : X \to X$, which maps $x \in X$ to $\gamma.x$, is Borel. Let the *Schreier graph* $\mathcal{S}(a; S)$ have X as the vertex set and

$$\{(x, \gamma.x) : x \in X, \ \gamma \in S\} \setminus \text{Diag}_X \qquad (4.1)$$

as the edge set. (Note that, regardless of the choice of a Polish topology on X, the diagonal $\text{Diag}_X = \{(x, x) : x \in X\}$ is a closed and thus Borel subset of X^2.) Thus $\mathcal{S}(a; S)$ is a Borel graph by Lemma 3.2. ∎

For example, the irrational rotation graph $\mathcal{R}_\alpha$ from Example 1.1 is a Borel graph, e.g. as the Schreier graph of the Borel action of the marked group $(\mathbb{Z},\{-1,1\})$ on $[0,1)$ given by $n.x := x + n\alpha \pmod 1$ for $n \in \mathbb{Z}$ and $x \in [0,1)$.

By Corollary 3.11, if $\mathcal{G}$ is a Borel graph and Y is a Borel subset of V, then $\mathcal{G} \upharpoonright Y$ is again a Borel graph. Here is one case that often arises in the context of Example 4.3 (and that we will need later in Section 6). The *free part* of a group action $a : \Gamma \curvearrowright X$ is

$$\text{Free}(a) := \{x \in X : \forall \gamma \in \Gamma \setminus \{e\} \ \gamma.x \neq x\}. \tag{4.2}$$

As it is trivial to see, $x \in X$ belongs to the free part if and only if the map $\Gamma \to X$ that sends $\gamma \in \Gamma$ to $\gamma.x$ is injective.

Lemma 4.4 *The free part of a Borel action $a : \Gamma \curvearrowright X$ of a countable group Γ is Borel.*

Proof Using the definition in (4.2), we see that

$$X \setminus \text{Free}(a) = \bigcup_{\gamma \in \Gamma \setminus \{e\}} \{x \in X : \gamma.x = x\} = \bigcup_{\gamma \in \Gamma \setminus \{e\}} \pi_0(\text{Diag}_X \cap \{(x, \gamma.x) : x \in X\})$$

is a Borel set by Lemma 3.2 and the Lusin-Novikov Uniformization Theorem (Theorem 3.8). $\square$

5 Basic Properties of Locally Finite Borel Graphs

Recall that, unless stated otherwise, we consider *locally finite* graphs only, that is, those graphs in which every vertex has finitely many neighbours. Note that we do not require that all degrees are uniformly bounded by some constant. This is already a very rich and important class in descriptive combinatorics, and is a very natural one from the point of view of finite combinatorics.

Lemma 5.1 *Let $(V, \mathcal{B})$ be a standard Borel space and let $G = (V, E)$ be a locally finite graph. Then the following are equivalent.*
(i) The set $E \subseteq V^2$ is Borel (i.e. $(V, E, \mathcal{B})$ is a Borel graph).
(ii) For every Borel set $Y \subseteq V$, its neighbourhood $N_G(Y)$ is Borel.
(iii) For every Borel set $Y \subseteq V$, the 1-ball $N_G^{\leqslant 1}(Y)$ around Y (i.e. the set of vertices at distance at most 1 from Y) is Borel.

Proof Let us show that (i) implies (ii). Take any Borel $Y \subseteq V$. Note that $N(Y)$ is the projection of $Z := (Y \times V) \cap E$ on the second coordinate. The projection is a continuous map and, as a map from Z to V, has countable (in fact, finite) preimages. Thus $N(Y) = \pi_1(Z)$ is Borel by Theorem 3.8.

The implication (ii) $\Rightarrow$ (iii) trivially follows from $N^{\leqslant 1}(Y) = N(Y) \cup Y$ (and the σ-algebra $\mathcal{B}$ being closed under finite unions).

Let us show that (iii) implies (i). Remark 3.5 gives us a countable algebra $\mathcal{J}$ on V that generates $\mathcal{B}$. For $J \in \mathcal{J}$, let A_J be the union of $J \times (V \setminus N^{\leqslant 1}(J))$ and its "transpose" $(V \setminus N^{\leqslant 1}(J)) \times J$. By (3.1), each set A_J is Borel. Recall that the

diagonal $\text{Diag}_V = \{(x,x) : x \in V\}$ is a closed and thus Borel subset of V^2. It is enough to prove that
$$E = V^2 \setminus (\text{Diag}_V \cup (\cup_{J \in \mathcal{J}} A_J)), \tag{5.1}$$
because this writes E as the complement of a countable union of Borel sets.

By definition, each A_J is disjoint from E and thus the forward inclusion in (5.1) is obvious. Conversely, take any $(x,y) \in V^2 \setminus E$. Suppose that $x \neq y$ as otherwise $(x,y) \in \text{Diag}_V$ and we are done. By Lemma 3.6 there is a set $J \in \mathcal{J}$ which contains x but is disjoint from the finite set $N^{\leqslant 1}(y)$. Then $y \notin N^{\leqslant 1}(J)$ and the pair (x,y) belongs to A_J, giving the required. □

Corollary 5.2 *Let $r \in \omega$. If $\mathcal{G} = (V, E, \mathcal{B})$ is a locally finite Borel graph, then so is its r-th power graph $\mathcal{G}^r$ (as a graph on the standard Borel space V).*

Proof Trivially (or by a very special case of the König Infinity Lemma, see e.g. [19, Lemma 8.1.2]), the graph $\mathcal{G}^r$ is locally finite.

In order to check that $\mathcal{G}^r$ is a Borel graph, we verify Condition (iii) of Lemma 5.1. By the definition of $\mathcal{G}^r$, we have for every $A \subseteq V$ that $N_{\mathcal{G}^r}^{\leqslant 1}(A) = N_{\mathcal{G}}^{\leqslant r}(A)$. Also, we can construct $N_{\mathcal{G}}^{\leqslant r}(A)$ from A by iteratively applying r times the 1-ball operation in $\mathcal{G}$. By Lemma 5.1(iii), this operation preserves Borel sets. Thus $N_{\mathcal{G}^r}^{\leqslant 1}(A)$ is Borel for every Borel $A \subseteq V$. We conclude that Condition (iii) of Lemma 5.1 is satisfied for $\mathcal{G}^r$ and thus this graph is Borel. □

Here is an important consequence to Corollary 5.2. Recall that the connectivity relation $\mathcal{E}_G$ of a graph G consists of all pairs of vertices that lie in the same connectivity component of G.

Corollary 5.3 *If $\mathcal{G} = (V, E, \mathcal{B})$ is a locally finite Borel graph, then $\mathcal{E}_\mathcal{G}$ is a Borel subset of V^2.*

Proof We have that $\mathcal{E}_\mathcal{G}$ is the union of the diagonal Diag_V and the edge sets of $\mathcal{G}^r$ over $r \geqslant 1$. As each $\mathcal{G}^r$ is a Borel graph by Corollary 5.2, the set $\mathcal{E}_\mathcal{G} \subseteq V^2$ is Borel. □

Remark 5.4 Addressing a question of the author, Chan [12] showed, under the set-theoretic assumption that the constructable continuum is the same as the continuum, that Lemma 5.1 fails for general locally countable graphs: namely Property (ii) (and thus also Property (iii)) does not imply Property (i). An example is given by the countable equivalence relation studied in [11, Section 9]. (Thus, combinatorially, the graph is a union of countable cliques.) However, it remains unclear if the above additional set-theoretic assumption is needed.

5.1 Borel Colourings

Let $\mathcal{G} = (V, E, \mathcal{B})$ be a Borel graph (which we assume to be locally finite).

A colouring $c : V \to X$ is *proper* if no two adjacent vertices get the same colour. Since we will consider only the case of countable colour sets X (when 2^X is the only σ-algebra making it a standard Borel space), we define $c : V \to X$ to be *Borel* if the preimage under c of every element of X is Borel. When we view c as a vertex colouring, this amounts to saying that each of (countably many) colour classes belongs to $\mathcal{B}(V)$.

Lemma 5.5 *Every locally finite Borel graph $\mathcal{G} = (V, E, \mathcal{B})$ admits a Borel proper colouring $c : V \to \omega$.*

Proof Fix a countable algebra $\mathcal{J} = \{J_0, J_1, \dots\}$ generating $\mathcal{B}(V)$, which exists by Remark 3.5. Define

$$A := \{(x, k) \in V \times \omega : x \in J_k \ \wedge \ N(x) \cap J_k = \emptyset\}.$$

Thus $(x, k) \in A$ if the $(k+1)$-st set J_k of $\mathcal{J}$ separates x from all its neighbours. For every $x \in V$, there is at least one $k \in \omega$ with $(x, k) \in A$ by Lemma 3.6, and we define $c(x)$ to be the smallest such $k \in \omega$.

Clearly, if two distinct vertices x and y get the same colour k then they are both in J_k and cannot be adjacent as their neighbourhoods are disjoint from J_k. Thus $c : V \to \omega$ is a proper vertex colouring.

It remains to argue that the map $c : V \to \omega$ is Borel. For $k \in \omega$, define

$$B_k := \{x \in V : c(x) \geqslant k\}.$$

Since $c^{-1}(k) = B_k \setminus B_{k+1}$, it is enough so show that each B_k is Borel. The complement of B_k is exactly the image of $A \cap (V \times k)$ under the projection π_0. Thus, by Theorem 3.8, it suffices to show that $A \subseteq V \times \omega$ is Borel. As it is easy to see, $A = \cup_{k \in \omega}((J_k \setminus N(J_k)) \times \{k\})$. Each set in this countable union is Borel by Lemma 5.1(ii). So A is Borel, finishing the proof. □

The main idea of the proof of the following result is very useful in Borel combinatorics: we first show that we can cover V by countably many sets that are sufficiently "sparse" (namely, independent in this proof) and then apply some parallel algorithm (namely, greedy colouring) where we take these sets one by one and process all vertices of the taken set in one go.

Theorem 5.6 (Kechris et al. [38]) *Every locally finite Borel graph $\mathcal{G}$ has a maximal independent set A which is Borel.*

Proof Let $c : V \to \omega$ be the proper Borel colouring returned by Lemma 5.5. We apply the greedy algorithm where we process colours $i \in \omega$ one by one and, for each i, add to A in parallel all vertices of colour i that have no neighbours in the current set A.

Formally, let $A_0 := \emptyset$ and, inductively for $i \in \omega$, define

$$A_{i+1} := A_i \cup (c^{-1}(i) \setminus N(A_i)).$$

Finally, define $A := \cup_{i \in \omega} A_i$.

As c is a proper colouring, each set A_i is independent by induction on i. Thus A, as the union of nested independent sets, is independent. Also, the set A is maximal independent. Indeed, if $x \notin A$ then, with $i := c(x)$, the reason for not adding x into $A_{i+1} \subseteq A$ was that x has a neighbour in A_i and thus a neighbour in $A \supseteq A_i$.

An easy induction on i shows by Lemma 5.1(ii) that each A_i is Borel. Thus $A = \cup_{i \in \omega} A_i$ is also Borel. □

Recall that a set A of vertices in a graph G is called *r-sparse* if the distance in G between every two distinct elements of A is larger than r.

Corollary 5.7 *Every locally finite Borel graph $\mathcal{G}$ has a maximal r-sparse set A which is Borel.*

Proof A set $A \subseteq V$ is (maximal) r-sparse in $\mathcal{G}$ if and only if it is (maximal) independent in $\mathcal{G}^r$. Thus the required Borel set A exists by Theorem 5.6 applied to $\mathcal{G}^r$, which is a locally finite Borel graph by Corollary 5.2. □

The *Borel chromatic number* $\chi_B(\mathcal{G})$ of an arbitrary Borel graph $\mathcal{G}$ is defined as the smallest cardinality of a standard Borel space Y for which there is a Borel proper colouring $c: V \to Y$. Trivially, $\chi_B(\mathcal{G})$ is at least the usual *chromatic number* $\chi(\mathcal{G})$, which is the smallest cardinality of a set Y with $\mathcal{G}$ admitting a proper vertex colouring $V \to Y$ (which need not be constructive in any way); for more on $\chi(G)$ for infinite graphs see e.g. the survey by Komjáth [41]. Since we restrict ourselves to locally finite graphs here, we have by Lemma 5.5 that both $\chi_B(\mathcal{G})$ and $\chi(\mathcal{G})$ are in $\omega \cup \{\omega\}$.

Theorem 5.8 (Kechris et al. [38]) *Every Borel graph $\mathcal{G} = (V, E, \mathcal{B})$ with finite maximum degree $d := \Delta(\mathcal{G})$ satisfies $\chi_B(G) \leqslant d + 1$.*

Proof One way to prove this result with what we already have is to iteratively keep removing Borel maximal independent sets from $\mathcal{G}$ that exist by Theorem 5.6. (Here we use Corollary 3.11 to show that each new graph is Borel; alternatively, we could have removed only edges touching the current independent set while keeping V unchanged.) Then the degree of each remaining vertex strictly decreases during each removal. Thus, after d removals, every remaining vertex is isolated and, after $d+1$ removals, the vertex set becomes empty.

Alternatively, we can take a countable partition $V = \cup_{i \in \omega} V_i$ into Borel independent sets given by Lemma 5.5 and, iteratively for $i \in \omega$, colour all vertices of the independent set V_i in parallel, using the smallest available colour on each. Clearly, we use at most $d + 1$ colours while an easy inductive argument on i combined with Lemma 5.1(ii) shows that the obtained colouring is Borel on each set V_i. □

Here is a useful consequence of the above results. (Recall that we identify a non-negative integer k with the set $\{0, \ldots, k-1\}$.)

Corollary 5.9 *For every Borel graph $\mathcal{G} = (V, E, \mathcal{B})$ of finite maximum degree d and every integer $r \geqslant 1$ there is a Borel colouring $c: V \to k$ with $k := 1 + d\sum_{i=0}^{r-1}(d-1)^i$ such that every colour class is r-sparse.*

Proof Apply Theorem 5.8 to the r-th power $\mathcal{G}^r$, which is a Borel graph by Corollary 5.2 and, trivially, has maximum degree at most $k-1$. □

The set of edges $E \subseteq V^2$ is a Borel set, so E is itself a standard Borel space by Corollary 3.11. In particular, it makes sense to talk about Borel edge k-colourings, meaning symmetric Borel functions $E \to k$. (Alternatively, one could have defined a Borel edge k-colouring as a symmetric Borel function $c : V^2 \to \{-1\} \cup k$ with $c(x,y) \geqslant 0$ if and only if $(x,y) \in E$, thus eliminating the need to refer to Corollary 3.11 here.)

The above results on independent sets and vertex colouring extend to matchings and edge colourings as follows.

Lemma 5.10 *The edge set of every locally finite Borel graph $\mathcal{G} = (V, E, \mathcal{B})$ can be partitioned into countably many Borel matchings.*

Proof Corollary 5.2 and Lemma 5.5 give a proper Borel vertex colouring $c : V \to \omega$ of $\mathcal{G}^2$, the square of $\mathcal{G}$. Thus each colour class $V_i := c^{-1}(i)$, $i \in \omega$, is 2-sparse in $\mathcal{G}$. It follows that, for each pair $i < j$ in ω, the set $M_{ij} := E \cap ((V_i \times V_j) \cup (V_j \times V_i))$ is a matching. Moreover, since each V_i is independent in $\mathcal{G}$, the Borel matchings M_{ij} over all $i < j$ in ω partition E, as required. □

Theorem 5.11 *Every locally finite Borel graph $\mathcal{G} = (V, E, \mathcal{B})$ has a maximal matching $M \subseteq E$ which is a Borel subset of V^2.*

Proof Let M_i', $i \in \omega$, be the matchings returned by Lemma 5.10. We construct M greedily, by taking for each $i \in \omega$ all edges in M_i' that are vertex disjoint from the current matching.

Formally, let $M_0 := \emptyset$ and, inductively for $i \in \omega$, define

$$M_{i+1} := M_i \cup \big(M_i' \setminus \big((\pi_0(M_i) \times V) \cup (V \times \pi_0(M_i))\big)\big).$$

As in Theorem 5.6, the set $M := \cup_{i \in \omega} M_i$ is a maximal matching in $\mathcal{G}$. Also, each M_{i+1} (and thus the final matching M) is Borel, which can argued by induction on i (using Theorem 3.8 to show that $\pi_0(M_i)$ is Borel). □

The *Borel chromatic index* $\chi_B'(\mathcal{G})$ is the smallest $k \in \omega \cup \{\omega\}$ such that there exists a Borel map $c : E \to k$ with no two intersecting edges having the same colour (equivalently, with each colour class being a matching). Similarly to how Theorem 5.8 was derived from Theorem 5.6, the following result can be derived from Theorem 5.11 by removing one by one maximal Borel matchings and observing that for every remaining edge the number of other edges that intersect it strictly decreases with each removal step.

Theorem 5.12 (Kechris et al. [38]) *Every Borel graph $\mathcal{G}$ with finite maximum degree $d := \Delta(\mathcal{G})$ satisfies $\chi_B'(G) \leqslant 2d - 1$.* □

Remark 5.13 Lemma 5.10 and Theorems 5.11–5.12 can be also deduced by applying the corresponding results on independent sets to the *line graph* $L(\mathcal{G})$ whose vertex set consists of unordered pairs $\{x, y\}$ with $(x, y) \in E$, where two distinct pairs are adjacent if they intersect. Let us just outline a proof that $L(\mathcal{G})$ is a Borel graph. Recall the definition of the Borel σ-algebra on $\binom{V}{2}$ from Remark 4.1. From this definition, it follows that the vertex set of the line graph $L(\mathcal{G})$ is a Borel subset of $\binom{V}{2}$ and thus is itself a standard Borel space by Corollary 3.11. To show that the line graph is Borel, we check Property (iii) of Lemma 5.1. By lifting all to V^2, it is enough to check a version of this property for every symmetric Borel set $A \subseteq E$.

Now, $Y := \pi_0(A) = \pi_1(A)$, the set of vertices covered by the edges in A, is Borel by Theorem 3.8 as the sizes of preimages under $\pi_0 : A \to Y$ are finite. The 1-ball of A in the line graph corresponds to the set of edges of $\mathcal{G}$ that intersect Y. The latter set is equal to $E \cap ((Y \times V) \cup (V \times Y))$ and is thus Borel. Finally (assuming we have verified all steps above), we can conclude by Lemma 5.1 that $L(\mathcal{G})$ is Borel.

Remark 5.14 Observe that there are locally countable Borel graphs that do not admit a Borel proper vertex colouring with countably many colours (see e.g. [36, Examples 3.13–16]); such graphs were completely characterised by Kechris et al. [38, Theorem 6.3] as containing a certain obstacle. On the other hand, Kechris et al. [38, Proposition 4.10] observed that, by the Feldman–Moore Theorem (Theorem 7.1 here) the statement of Lemma 5.10 (and thus of Theorem 5.11) remains true also in the locally countable case.

5.2 Local Rules

We will show in this section that all "locally defined" vertex labellings are Borel as functions. As a warm up, consider the degree function $\deg : V \to \omega$, which sends a vertex $x \in V$ to its degree $\deg(x) = |N(x)|$.

Lemma 5.15 *For every locally finite Borel graph* $\mathcal{G} = (V, E, \mathcal{B})$, *the degree function* $\deg_\mathcal{G} : V \to \omega$ *is Borel.*

Proof It is enough to show that for every $k \in \omega$ the set $D_k := \{x \in V : \deg(x) \geqslant k\}$ is Borel, because the set of vertices of degree exactly k is $D_k \setminus D_{k+1}$.

The most direct proof is probably to use a countable generating algebra $\mathcal{J} = \{J_i : i \in \omega\}$ from Remark 3.5. Note that a vertex $x \in V$ has degree at least k if and only if there are k pairwise disjoint sets in $\mathcal{J}$ with x having at least one neighbour in each of them. Indeed, if $y_0, \ldots, y_{k-1} \in N(x)$ are pairwise distinct then by Lemma 3.6 there are $A_0, \ldots, A_{k-1} \in \mathcal{J}$ with $A_i \cap \{y_0, \ldots, y_{k-1}\} = \{y_i\}$ for each $i \in k$ and if we let $B_0 := A_0$, $B_1 := A_1 \setminus A_0$, $B_2 := A_2 \setminus (A_0 \cup A_1)$, and so on, then $B_0, \ldots, B_{k-1} \in \mathcal{J}$ have the required properties. Thus, we can write D_k as the countable union of the intersections $\cap_{m \in k} N(J_{i_m})$ over all k-tuples $i_0, \ldots, i_{k-1}$ such that $J_{i_0}, \ldots, J_{i_{k-1}}$ are pairwise disjoint. By Lemma 5.1(ii), all neighbourhoods $N(J)$ for $J \in \mathcal{J}$ and thus the set D_k are Borel.

Alternatively, fix any Borel proper edge colouring $c : E \to \omega$ which exists by Theorem 5.12. Trivially, $\deg(x) \geqslant k$ if and only if there are at least k distinct colours under c at the vertex x. The set of vertices that belong to an edge of colour i is $\pi_0(c^{-1}(i))$, which is Borel by Theorem 3.8. Thus

$$D_k = \bigcup_{\substack{i_0, \ldots, i_{k-1} \in \omega \\ i_0 < \cdots < i_{k-1}}} \bigcap_{m \in k} \pi_0(c^{-1}(i_m))$$

is a Borel set. □

In order to make the forthcoming general statement (Lemma 5.17) stronger and better suitable for applications, we consider a version where we may have some additional structure on graphs. Namely, a *labelling* of a graph G is any function

ℓ from V to some countable set; then we say that a pair (G,ℓ) is a *labelled graph*. Vertex labellings allow us to encode many other types of structures on G such as, for example, edge colourings (see Remark 5.18 for a reduction).

Let ℓ be a labelling of a graph $G = (V, E)$. For $r \in \omega$, let F_r be the function on V which sends a vertex $x \in V$ to the isomorphism type of

$$(G, c, x) \restriction N^{\leqslant r}(x) := (G \restriction N^{\leqslant r}(x),\, c \restriction N^{\leqslant r}(x),\, x),$$

the labelled graph induced by the r-ball $N^{\leqslant r}(x)$ in G rooted at x, where isomorphisms have to preserve also the root and the vertex labelling. Since we consider only locally finite graphs, F_r assumes countably many possible values and thus is an example of a labelling. By a *local rule of radius r* (or an *r-local rule*) on (G, ℓ) we mean a function R on V whose value at any $x \in V$ depends only on $\mathsf{F}_r(x)$. In other words, r-local rules are exactly those functions that factor through F_r, that is, are representable as a composition $f \circ \mathsf{F}_r$ for some function f. A function on V is a *local rule* if it is an r-local rule for some $r \in \omega$. Unless stated otherwise, we assume that local rules and labellings are functions from V to ω, that is, their values are non-negative integers.

For example, the degree function deg or the number of triangles that contain a vertex are local rules of radius 1 (that do not depend on the labelling). An example of a 1-local rule that uses the labelling ℓ is, say, $x \mapsto |\ell(N(x))|$, the number of distinct ℓ-labels on the neighbours of x.

Let $\mathcal{G}$ be a graph with a labelling $\ell : V \to \omega$ such that, for every $x \in V$, its neighbours get pairwise distinct colours. Fix some special element not in V, denoting it by $\bot$. For a non-empty sequence $S = (s_0, \ldots, s_j) \in \omega^{j+1}$ of labels, let us define a function $f_S : V \to V \cup \{\bot\}$ as follows. Take any $x \in V$. If there is a *walk* in $\mathcal{G}$ of (edge) length j (i.e. a sequence $(x_0, \ldots, x_j)$ with $(x_i, x_{i+1}) \in E$ for each $i \in j$) that starts with x (i.e. $x_0 = x$) and is *S-labelled* (i.e. $\ell(x_i) = s_i$ for each $i \in j+1$) then let $f_S(x) := x_j$ be the final endpoint of this walk; otherwise let $f_S(x) := \bot$. By our assumption on ℓ, there can be at most one such walk, so $f_S(x)$ is well-defined. (Equivalently, we could have worked with partially defined functions, instead of using the special symbol $\bot$.) For convenience, if $S = ()$ is the empty sequence, then we define $f_{()}$ to be the identity function on V.

Lemma 5.16 *If $\mathcal{G}$ is a locally finite Borel graph with a Borel labelling $\ell : V \to \omega$ that is injective on $N(x)$ for every $x \in V$, then for every $j \in \omega$ and every sequence $S = (s_0, \ldots, s_j)$ in ω^{j+1} the function $f_S : V \to V \cup \{\bot\}$ is Borel.*

Proof Regardless of how we extend a Polish topology from V to $V \cup \{\bot\}$, a subset A of $V \cup \{\bot\}$ is Borel if and only if $A \cap V$ is Borel. Thus it is enough to check that the preimage $f_S^{-1}(V)$ is Borel and the restriction of f_S to $f_S^{-1}(V)$ is a Borel function.

We use induction on $j \in \omega$. If $j = 0$, then $f_{(s_0)}$ is the identity function on the Borel set $\ell^{-1}(s_0)$ and assumes value $\bot$ otherwise; so $f_{(s_0)}$ is indeed Borel. Suppose that $j \geqslant 1$. Let $f := f_{(s_0, \ldots, s_j)}$ and $g := f_{(s_0, \ldots, s_{j-1})}$.

Observe that, for every $x \in V$, there is an $(s_0, \ldots, s_j)$-labelled walk starting at x if and only if there is an $(s_0, \ldots, s_{j-1})$-labelled walk starting at x and its endpoint $g(x)$ has a neighbour labelled s_j. That is,

$$f^{-1}(V) = g^{-1}(V) \cap g^{-1}(N(\ell^{-1}(s_j))),$$

and this set is Borel by induction and Lemma 5.1(ii). Let $Y := g(g^{-1}(V))$ consist of the endpoints of all $(s_0, \ldots, s_{j-1})$-labelled walks in $\mathcal{G}$. This set is Borel by Theorem 3.8 as the bijective image under the Borel map g of the Borel set $g^{-1}(V)$. Let Y' consist of those vertices in Y that have a neighbour labelled s_j. Again by Theorem 3.8, Y' is Borel as the bijective image of

$$Y'' := \{(y,z) \in E : y \in Y \wedge \ell(z) = s_j\} = (Y \times \ell^{-1}(s_j)) \cap E$$

under the projection π_0 on the first coordinate. Moreover, the map $h : Y' \to Y''$ which is the (unique) right inverse of π_0 is Borel by the second part of Theorem 3.8. The composition $\pi_1 \circ h$ sends an element of Y' to its unique neighbour labelled s_j. Thus the function f is Borel since, on the Borel set $f^{-1}(V)$, it is the composition $\pi_1 \circ h \circ g$ of three Borel functions. □

Lemma 5.17 *Let $\mathcal{G} = (V, E, \mathcal{B})$ be a locally finite Borel graph with a Borel labelling $\ell : V \to \omega$. Then every local rule $\mathsf{R} : V \to \omega$ on $(\mathcal{G}, \ell)$ is a Borel function.*

Proof Let the local rule R have radius r. We can additionally assume that ℓ is a $2r$-sparse colouring of $\mathcal{G}$. Indeed, take any Borel proper vertex colouring $c : V \to \omega$ of $\mathcal{G}^{2r}$ (which exists by Corollary 5.2 and Lemma 5.5), replace ℓ by the labelling $(\ell, c) : V \to \omega^2$, which maps a vertex $x \in V$ to $(\ell(x), c(x))$, and update the local rule R to ignore the c-component of the labelling.

To prove the lemma, it is enough to show that the function F_r on V is Borel because each preimage under R is a countable union of some preimages under F_r. Note that F_0 is Borel since the (countable) vertex partition defined by F_0 is the same as the partition defined by the Borel function ℓ. So assume that $r \geqslant 1$.

Fix any particular feasible F_r-value $\mathcal{F}$ (a rooted labelled graph with each vertex at distance at most r from the root). By relabelling vertices, assume that the vertex set of $\mathcal{F}$ is k with 0 being the root. Let $\ell' : k \to \omega$ be the vertex labelling of $\mathcal{F}$. We assume that the function $\ell' : k \to \omega$ is injective as otherwise $\mathsf{F}_r^{-1}(\mathcal{F})$ is empty and thus trivially Borel.

For each $i \in k$, let $(s_0, \ldots, s_j)$ be the sequence of labels on some fixed shortest path P_i from the root 0 to i in $\mathcal{F}$ and let $f_i := f_{(s_0, \ldots, s_j)}$ be the function defined before Lemma 5.16 with respect to the labelled graph $(\mathcal{G}, \ell)$.

Observe that, for $x \in V$, the r-ball $\mathsf{F}_r(x)$ is isomorphic to $\mathcal{F}$ if and only if all the following statements hold:
(a) For every $i \in k$, we have $f_i(x) \neq \bot$ (that is, $\mathcal{G}$ has a walk starting from x labelled the same way as P_i is).
(b) For every distinct $i, j \in k$, the pair (i, j) is an edge in $\mathcal{F}$ if and only if $(f_i(x), f_j(x))$ is an edge of $\mathcal{G}$.
(c) For every $i \in k$ with distance at most $r - 1$ from the root in $\mathcal{F}$ and every $y \in N_\mathcal{G}(f_i(x))$ there is $j \in k$ with $y = f_j(x)$ and $j \in N_\mathcal{F}(i)$.

Indeed, the map $h : k \to V$ that sends i to $f_i(x)$ preserves the labels by Property (a) since we considered labelled walks when defining each f_i. Furthermore, h is injective; in fact, even the composition $\ell \circ h$ is injective since the endpoints of the paths P_i have distinct ℓ'-labels as distinct vertices of $\mathcal{F}$. Now, Property (b) states that h is a graph isomorphism from $\mathcal{F}$ to the subgraph induced by

$\{f_0(x), \ldots, f_{k-1}(x)\}$ in $\mathcal{G}$. Finally, Property (c) states that the breadth-first search of depth r from x in $\mathcal{G}$ does not return any vertices not accounted by $\mathcal{F}$.

Let X consist of those $x \in V$ that satisfy Property (a). By Lemma 5.16, the set $X = \cap_{i \in k} f_i^{-1}(V)$ is Borel.

Note that, for any distinct $i, j \in k$, the set

$$\begin{aligned} Y_{i,j} &:= \{x \in X : (f_i(x), f_j(x)) \in E\} \\ &= \pi_0\big(\{(x, x_i, x_j) \in X^3 : x_i = f_i(x)\} \\ &\quad \cap \ \{(x, x_i, x_j) \in X^3 : x_j = f_j(x)\} \cap (X \times E)\big) \end{aligned}$$

is Borel by Lemma 3.2 (applied to the functions f_i and f_j) and Theorem 3.8 (applied to the projection π_0, whose preimages are finite here as the graph $\mathcal{G}$ is locally finite). Thus the set Y of the elements in X that satisfy Property (b) is Borel since it is the intersection of the sets $Y_{i,j}$ over all edges $\{i, j\}$ of $\mathcal{F}$ and the sets $X \setminus Y_{i,j}$ over non-edges $\{i, j\}$ of $\mathcal{F}$.

Finally, the set $\mathsf{F}_r^{-1}(\mathcal{F})$ that we are interested in is Borel by Lemma 5.16 as the countable intersection with Y of $f_S^{-1}(\bot)$ over all sequences S of labels of length at most r that do not occur on walks in $\mathcal{F}$ that start with the root. □

Remark 5.18 Note that many other types of structures on $\mathcal{G}$, such as a Borel edge labelling $\ell' : E \to \omega$, can be encoded via some vertex labelling ℓ to be used as the input to the local rule R in Lemma 5.17. Namely, fix a Borel 2-sparse colouring $c : V \to \omega$ of $\mathcal{G}$ and let the label $\ell(x)$ of a vertex x be defined as the finite list $(c(x), \ell'(x, y_0), \ldots, \ell'(x, y_{d-1}))$ where $y_0, \ldots, y_{d-1}$ are all neighbours of x listed increasingly with respect to their c-colours. In fact, the same reduction also works for labellings $\ell' : E \to \omega$ that need not be *symmetric* (meaning that $\ell'(x, y) = \ell'(y, x)$ for all $(x, y) \in E$). More generally, any countably valued function defined on subsets of uniformly bounded diameter in $\mathcal{G}$ can be locally encoded by a vertex labelling.

In some cases, a single local rule does not work but a good assignment can be found by designing a sequence of local rules of growing radii that eventually stabilise at every vertex. The author is not aware of any commonly used name for functions arising this way so, for the purposes of this paper, we make up the following name (inspired by the term *finitary factor* from probability). For a labelled graph (G, ℓ), let us call a function $\mathsf{R} : V \to \omega$ *finitary* (with respect to (G, ℓ)) if there are local rules R_i, $i \in \omega$, on (G, ℓ) such that the sequence of functions $\mathsf{R}_i : V \to \omega$ eventually stabilises to R everywhere, that is, for every $x \in V$ there is $n \in \omega$ such that for every $i \geqslant n$ we have $\mathsf{R}(x) = \mathsf{R}_i(x)$. Note that we do not require that R_i "knows" if its value at a vertex x is the eventual value or not.

Example 5.19 Let $c : V \to \omega$ be a 2-sparse colouring of a graph G. For $x \in V$ let $\mathsf{R}(x) \in \omega$ be 0 if the component of x has no perfect matching and be the largest $i \in \{1, \ldots, \deg(x)\}$ such that the component of x has a perfect matching that matches x with the i-th element of $N(x)$ (where we order each neighbourhood by looking at the values of c, which are pairwise distinct by the 2-sparseness of c). Then $\mathsf{R} : V \to \omega$ is finitary on (G, c) as the following local rules R_r, $r \in \omega$, demonstrate. Namely, $\mathsf{R}_r(x)$ is the largest $i \in \{1, \ldots, \deg(x)\}$ such that there is a matching M in $G \restriction N^{\leqslant r}(x)$ that covers every vertex in $N^{\leqslant r-1}(x)$ and matches x to the i-th element of $N(x)$; if no such M exists then we let $\mathsf{R}_r(x) := 0$. Trivially, for every $x \in V$, if we increase r then $\mathsf{R}_r(x)$ cannot increase. Thus the values $\mathsf{R}_r(x)$, $r \in \omega$, eventually stabilise and, moreover, this final value can be easily shown to be exactly $\mathsf{R}(x)$. On the other hand, the function R is not r-local for any $r \in \omega$: e.g. a vertex at distance at least r from both endpoints of a finite path cannot decide by looking at distance at most r if a perfect matching exists or not.

Corollary 5.20 *For every locally finite Borel graph $\mathcal{G} = (V, E, \mathcal{B})$ with a Borel labelling $\ell : V \to \omega$, every function $\mathsf{R} : V \to \omega$ which is finitary with respect to $(\mathcal{G}, \ell)$ is Borel.*

Proof Fix local rules $\mathsf{R}_i : V \to \omega$, $i \in \omega$, that witness that R is finitary. By Lemma 5.17, each R_i is a Borel function. The function $\mathsf{R} : V \to \omega$ is Borel since a pointwise limit of Borel functions is Borel ([14, Proposition 2.1.5]). Alternatively, the last step follows from observing that, for every possible value $j \in \omega$, its preimage $\mathsf{R}^{-1}(j) = \cup_{i \in \omega} \cap_{m \geqslant i} \mathsf{R}_m^{-1}(j)$ is the lim inf (and also, in fact, lim sup) of the preimages $\mathsf{R}_i^{-1}(j)$, $i \in \omega$. □

5.3 Graphs with a Borel Transversal

In this section we present one application of Lemma 5.17 (namely, Theorem 5.23 below) which, informally speaking, says that if we can pick exactly one vertex inside each graph component in a Borel way then any satisfiable locally checkable labelling problem has a Borel satisfying assignment.

We define a *locally checkable labelling problem* (or an *LCL* for short) on labelled graphs to be a $\{0,1\}$-valued local rule C that takes as input graphs with ω^2-valued labellings. A labelling $a : V \to \omega$ *satisfies* (or *solves*) the LCL C on a labelled graph (G, ℓ) if C returns value 1 on every vertex when applied to the labelled graph $(G, (\ell, a))$. (Recall that the labelling $(\ell, a) : V \to \omega^2$ labels a vertex x with $(\ell(x), a(x))$.) We will identify the function $\mathsf{C}(G, (\ell, a)) : V \to \{0, 1\}$ with the set of $x \in V$ for which it assumes value 1 and, if the ambient labelled graph is understood, abbreviate this set to $\mathsf{C}(a)$. Thus a satisfies C if and only if $\mathsf{C}(a) = V$. We call the requirement that a vertex x belongs to $\mathsf{C}(\cdot)$ the *constraint at x*. Labellings $a : V \to \omega$ on which we check the validity of an LCL will usually be called *assignments*.

Let us give a few examples of LCLs. First, checking that $a : V \to \omega$ is a proper colouring can be done by the 1-local rule which returns 1 if and only if the colour of the root is different from the colour of any of its neighbours (and that the colour of the root is in the set $n = \{0, \ldots, n-1\}$ if we additionally want to require that a uses at most n colours). Note that the LCL ignores the labelling ℓ in this example. Second, suppose that a labelling a encodes some edge colouring $c : E \to \omega$ as in Remark 5.18. Then a 2-local rule can check if c is proper: each vertex x checks that all colours on ordered pairs $(x, y) \in E$ are pairwise distinct and that $c(x, y) = c(y, x)$ for every $y \in N(x)$; for this the knowledge of $N^{\leqslant 2}(x)$ is enough. Another example is checking that $a : V \to \{0, 1\}$ is the indicator function of a maximal independent set: this can be done by the obvious 1-local rule.

Let $\mathcal{G}$ be a Borel graph. A *transversal* for $\mathcal{G}$ is a subset $X \subseteq V$ such that X has exactly one vertex from every connectivity component of $\mathcal{G}$ (i.e. $|X \cap [x]| = 1$ for every $x \in V$). In the language of Borel equivalence relations, the existence of a Borel transversal can be shown (see e.g. [37, Proposition 6.4] or [79, Proposition 20.5]) to be equivalent to the statement that the connectivity relation $\mathcal{E}$ of $\mathcal{G}$ (which is Borel by Corollary 5.3) is *smooth*, meaning that there is a countable family of Borel sets $Y_n \subseteq V$, $n \in \omega$, such that, for all $x, y \in V$, we have $(x, y) \in \mathcal{E}$ if and only if every Y_n contains either both or none of x and y.

Here is an example of a large class of Borel graphs that have a Borel transversal.

Lemma 5.21 *Every (locally finite) Borel graph $\mathcal{G}$ with all components finite admits a Borel transversal.*

Proof Fix any Borel total order $\preccurlyeq$ on the vertex set V which exists by Lemma 3.7 and define a transversal X by picking the $\preccurlyeq$-smallest element in each component. Note that X is the countable union of the sets X_r, $r \in \omega$, where X_r consists of the $\preccurlyeq$-smallest vertices inside connectivity components of diameter at most r. Each X_r is Borel by Lemma 5.17 as its indicator function can be computed by an $(r+1)$-local rule. Thus the constructed transversal X is also Borel. □

Having a Borel transversal is a very strong property from the point of view of Borel combinatorics as the forthcoming results state. The first one is a useful auxiliary lemma stating, informally speaking, that if we can pick exactly one vertex in each component in a Borel way then we can enumerate each component in a Borel way.

Lemma 5.22 *Let $\mathcal{G} = (V, E, \mathcal{B})$ be a locally finite Borel graph, admitting a Borel transversal $X \subseteq V$. Then there are Borel functions $g_i : X \to V$, $i \in \omega$, such that g_0 is the identity function on X and, for every $x \in X$, the sequence $(g_i(x))_{i \in m}$ bijectively enumerates $[x]_\mathcal{G}$ and satisfies $\mathrm{dist}_\mathcal{G}(x, g_i(x)) \leqslant \mathrm{dist}_\mathcal{G}(x, g_j(x))$ for all $i < j$ in $m := |[x]_\mathcal{G}|$.*

Proof Fix a 2-sparse Borel colouring $c : V \to \omega$ which exists by Corollary 5.9.

The main idea of the proof is very simple: for each selected vertex $x \in X$ we order the vertices in the connectivity component of x first by the distance to x and then by the c-labellings of the shortest paths from x to them, and let $g_i(x)$ be the $(i + 1)$-st vertex in this order on $[x]$.

Formally, let $\preccurlyeq$ be the total ordering of $\omega^{<\omega}$, the set of all finite sequences of non-negative integers, first by the length and then lexicographically. Note that $\preccurlyeq$ is a *well-order* on $\omega^{<\omega}$ (that is, every non-empty subset of $\omega^{<\omega}$ has the $\preccurlyeq$-smallest element). For $S \in \omega^{<\omega}$, let $f_S : V \to V \cup \{\bot\}$ be the function defined before Lemma 5.16 (which sends $x \in V$ to the other endpoint of the S-coloured walk starting at x, if it exists). By Lemma 5.16, each function f_S is Borel.

We define functions g_i inductively, with $g_0 := f_{()} \restriction X$ being the identity function on X. Suppose that $i \geqslant 1$. For $x \in V$, let $g_i(x)$ be equal to $f_S(x)$ where $S \in \omega^{<\omega}$ is the $\preccurlyeq$-smallest sequence with $f_S(x) \in [x] \setminus \{g_0(x), \ldots, g_{i-1}(x)\}$ and let $g_i(x) := x$ if no such S exists (i.e. if we have already exhausted the whole component of x).

Let us argue by induction on $i \in \omega$ that the function g_i is Borel. As g_0 is clearly Borel, take any $i \geqslant 1$. For $S \in \omega^{<\omega}$, let $X_S := f_S^{-1}(V \setminus \cup_{j \in i} g_j(X)) \cap X$. For any given $S \in \omega^{<\omega}$, the set of $x \in X$ for which we use f_S when defining $g_i(x)$ is exactly $X_S \setminus (\cup_{R \prec S} X_R)$, since we have to exclude the already labelled vertices

$g_0(x), \ldots, g_{i-1}(x)$ and then any $R \prec S$ as it takes precedence over S. Each set X_S for $S \in \omega^{<\omega}$ is Borel by Theorem 3.8 and Lemma 5.16. It follows that the function g_i is Borel. The other claimed properties of the constructed functions g_i are obvious from the definition. □

Theorem 5.23 *Let $\mathcal{G} = (V, E, \mathcal{B})$ be a locally finite Borel graph, admitting a Borel transversal $X \subseteq V$. Let $\ell : V \to \omega$ be a Borel labelling, $n \in \omega$ and C be an LCL. If there is an assignment $V \to n$ that solves C on $(\mathcal{G}, \ell)$, then there is a Borel assignment $V \to n$ that solves C on $(\mathcal{G}, \ell)$.*

Proof As in the proof of Lemma 5.17, by replacing ℓ with (ℓ, d) for some 2-sparse Borel colouring d (and letting the updated LCL C ignore the d-component), we can additionally assume that ℓ is a 2-sparse colouring of $\mathcal{G}$. Let $g_i : X \to V$ for $i \in \omega$ be the Borel functions returned by Lemma 5.22.

Since we need to use these functions g_i as inputs to in our local rules, we encode them by a vertex labelling I as follows. Namely, we define $I : V \to \omega$ to map $y \in V$ to the smallest $i \in \omega$ with $y \in g_i(X)$. Thus I bijectively enumerates each component of $\mathcal{G}$ by an initial interval of ω. This function is Borel by Theorem 3.8 since $I^{-1}(0) = X$ and

$$I^{-1}(i) = g_i(X) \setminus \{x \in X : |[x]| \leqslant i\}, \quad \text{for each } i \geqslant 1.$$

(Note that, for every $i \in \omega$, the set of $x \in X$ whose connectivity component has at most i vertices is Borel by Lemma 5.17 since its indicator function can be computed by an i-local rule on $(\mathcal{G}, \mathbb{1}_X)$.)

Let t be the radius of the local rule C.

For every $r \in \omega$, we define an r-local rule $\mathsf{A}_r : V \to \{-1\} \cup n$ which on $(\mathcal{G}, (\ell, I))$ works as follows. Given $y \in V$, explore $N^{\leqslant r}(y)$, the r-ball around y. If it contains no vertex of X, then let $\mathsf{A}_r(y) := -1$ (which could be interpreted that the vertex y does not yet make any guess of its value in the set n). Otherwise, let x be the (unique) vertex from X that we have encountered. Let

$$k = k(r, y) := r - \mathrm{dist}(x, y) \quad \text{and} \quad Y = Y(r, y) := N^{\leqslant k}(x).$$

If $y \notin Y$, then we define $\mathsf{A}_r(y) := -1$. Suppose that $y \in Y$. Note that $Y = \{y_0, \ldots, y_{m-1}\}$, where $m := |Y|$ and y_i for $i \in m$ is defined as the unique vertex of Y labelled i by I (which is, of course, $g_i(x)$). Since $Y \subseteq N^{\leqslant r}(y)$, the local rule $\mathsf{A}_r(y)$ can identify all these vertices. Among all assignments $A : Y \to n$, pick one, denoting it $A = A(r, y)$, that satisfies C on every vertex of

$$Y' = Y'(r, y) := N^{\leqslant k-t}(x)$$

and, if there is more than one choice then choose the one for which the sequence $(A(y_0), \ldots, A(y_{m-1}))$ is lexicographically smallest. (Note that there is always at least one choice of A by our assumption that a global solution exists.) Finally, let $\mathsf{A}_r(y) := A(y)$, be the value of A on the vertex $y \in Y$. Note that in order to compute A (given y_i's), we need to know only the labelled graph induced by $N^{\leqslant t}(Y') \subseteq Y \subseteq N^{\leqslant r}(y)$. Thus A_r is indeed a local rule of radius r.

Informally speaking, these rules A_r are constructed so that each vertex x from the transversal X takes growing balls around itself and properly labels each with the lexicographically smallest assignment that looks satisfiable, given the current local information. Every other vertex y in the component of x has to follow these choices once y discovers enough information to compute x's choice for the value at y.

Let us show that the values of A_r, $r \in \omega$, eventually stabilise to some element of n on each vertex $y \in V$, by using induction on $I(y)$. Take any $x \in X$. Define $y_i := g_i(x)$ for $i \in \omega$. Note that y_0 is the special vertex $x \in X$. Take any $i \in \omega$. First, observe that, trivially, $\mathsf{A}_r(y_i) \neq -1$ for all $r \geqslant 2\operatorname{dist}(y_0, y_i) + t$. By induction pick $r_0 \in \omega$ such that A_r stabilises from r_0 on each of $y_0, \ldots, y_{i-1}$, that is, for all $r \geqslant r_0$ the restrictions of A_r to $\{y_0, \ldots, y_{i-1}\}$ are equal to each other. Consider the values $\mathsf{A}_r(y_i)$ for $r \geqslant r_0'$ where

$$r_0' := r_0 + d \text{ and } d := \max\{\operatorname{dist}(y_i, y_j) : j \in i\}.$$

Using the notation from the definition of A_r, it holds that $k(r, y_i) = r - \operatorname{dist}(y_0, y_i)$ is at least as large as $k(r_0, y_j) = r_0 - \operatorname{dist}(y_0, y_j)$ for each $j \in i$. Thus, for every $j \in i$, we have that $Y'(r, y_i) \supseteq Y'(r_0, y_j)$, as these sets are just balls around the special vertex $x = y_0$ of radii $k(r, y_i) - t$ and $k(r_0, y_j) - t$ respectively. Thus, when we compute $A = A(r, y_i)$, the constraints that this assignment has to satisfy include all constraints for $A(r_0, y_j)$. In the other direction, it holds for all $j \in i$ again by our choice of d that $Y'(r + d, y_j) \supseteq Y'(r, y_i)$, that is, $A(r+d, y_j)$ has to satisfy all constraints that are imposed on $A(r, y_i)$. Since we always go for the lexicographically minimal assignment, we conclude that $A(r, y_i)$ coincides with A_{r_0} on $\{y_0, \ldots, y_{i-1}\}$. Thus, for all $r \geqslant r_0'$, when we compute $A(r, y_i)$, we can equivalently view its values on $\{y_0, \ldots, y_{i-1}\}$ as fixed and, given this, we minimise the value at y_i. This value cannot decrease when we increase r (because any increase of the radius r just adds some extra constraints on $A(r, y_i)$). So the values at each $y \in V$ eventually stabilise as they come from the finite set n, as desired.

We define the final assignment a as the one to which the local rules A_r, $r \in \omega$, stabilise. It is finitary and thus Borel by Corollary 5.20.

It remains to check that the constructed assignment $a : V \to n$ solves the LCL C. Take any $y \in V$. Its t-ball $Z := N^{\leqslant t}(y)$ is finite and thus there is $r_0 \in \omega$ such that $a \upharpoonright Z = \mathsf{A}_r \upharpoonright Z$ for each $r \geqslant r_0$. Let x be the unique element of $[y] \cap X$. Take any

$$r \geqslant \max(r_0, 2\operatorname{dist}(x, y)) + t.$$

When we compute $\mathsf{A}_r(y)$, we have that $k(r, y) = r - \operatorname{dist}(x, y)$ is at least $\operatorname{dist}(x, y) + t$, so $y \in Y'(r, y)$, that is, the constraint at y is one of the constraints that the partial assignment $A(r, y)$ has to satisfy. For each element $z \in N^{\leqslant t}(y)$, we have $k(r, y) \geqslant k(r_0, z)$ and thus $Y(r, y) \supseteq Y(r_0, z)$. Recall that the final assignment a coincides with $A(r, y)$ on $N^{\leqslant t}(y) = Z$ by the choice of r_0. Since the LCL C of radius t is satisfied by $A(r, y)$ at $y \in Y'(r, y)$, the assignment a also satisfies the constraint at y. As $y \in V$ was arbitrary, the constructed assignment a solves the problem C. □

Remark 5.24 Bernshteyn [8] gave an example showing that Theorem 5.23 is false when $n = \omega$, that is, when we consider assignments $V \to \omega$ that can assume can assume infinitely many values. (Also, Bernshteyn [8] presents an alternative proof of Theorem 5.23, via the Uniformization Theorem for compact preimages.)

5.4 Borel Assignments without Small Augmenting Sets

Suppose that we look for Borel assignments $a : V \to \omega$ on a labelled graph (G, ℓ) that solve an LCL C where we want many vertices to satisfy another LCL P. As before, we denote the set of vertices where a satisfies P by $\mathsf{P}(G, (\ell, a))$ (or by $\mathsf{P}(a)$ if the context is clear), calling it the *progress set*. For an assignment a that solves the LCL C, an *r-augmenting set* is a set $R \subseteq V$ such that R is *connected* (meaning that $G \upharpoonright R$ is connected), $|R| \leqslant r$, and there is an assignment $b : R \to \omega$ such that $a \rightsquigarrow b$ solves C and satisfies $\mathsf{P}(G, (\ell, a)) \subsetneq \mathsf{P}(G, (\ell, a \rightsquigarrow b))$. Here,

$$a \rightsquigarrow b := b \cup (a \upharpoonright (\pi_0(a) \setminus \pi_0(b))) \tag{5.2}$$

denotes the function obtained from putting the functions a and b together, with b taking preference on their common domain $\pi_0(a) \cap \pi_0(b)$. We call the new assignment $a \rightsquigarrow b$ an *r-augmentation* of a. Thus, an r-augmentation strictly increases the progress set by changing the current assignment on a connected set of at most r vertices without violating any constraint of C.

As an example, suppose that C states that a encodes a set M of edges which is a matching (that is, $\Delta(M) \leqslant 1$) and $\mathsf{P}(x) = 1$ means that a vertex $x \in M$ is matched by M. Similarly to Remark 5.18, one can encode a matching M by letting $a(x) := \deg(x)$ if x is unmatched and otherwise letting $a(x) \in \deg(x)$ specify the unique M-match of x by its position in $N(x)$ with respect to the ordering of $N(x)$ coming from a fixed 2-sparse Borel colouring. Then the LCL C can be realised by a 2-local rule (checking that the value at x is consistent with the value at each $y \in N(x)$) while P is the 1-local rule that outputs 0 at x if and only if $a(x) = \deg(x)$. Note that if we change the encoding so that $a(x) = 0$ means that x is unmatched while $a(x) \in \{1, \ldots, \deg(x)\}$ encodes the M-match of x otherwise, then the progress function P, which verifies that $a(x) = 1$, becomes 0-local. One special example of an augmentation here is to replace $M \subseteq E$ by the symmetric difference $M \triangle P$, where $P \subseteq E$ is a (usual) augmenting path (that is, a path in G whose endpoints are unmatched and whose edges alternate between $E \setminus M$ and M). Under either of the above encodings, the set $V(P)$ is augmenting here.

The following standard result states that, informally speaking, we can eliminate all r-augmentations in a Borel way, additionally including a Borel "certificate" that we changed at most r assignment values per every vertex added to the progress set. Elek and Lippner [21] presented a version of it for matchings (when C checks that the current assignment encodes a matching M and augmenting sets are limited to augmenting paths). Their proof extends with obvious modifications to the general case and is presented here.

Theorem 5.25 *Let C and P be LCLs for labelled graphs. Let $\mathcal{G}$ be a locally finite Borel graph with a Borel labelling $\ell : V \to \omega$ and a Borel assignment $a_0 : V \to \omega$ that solves C on $(\mathcal{G}, \ell)$. Then for every $r \geqslant 1$ there is a Borel assignment $a : V \to \omega$ such that a solves C on $(\mathcal{G}, \ell)$ and admits no r-augmentation with respect to P. Additionally, there are r Borel maps*

$$f_j : \mathsf{P}(\mathcal{G}, (\ell, a)) \setminus \mathsf{P}(\mathcal{G}, (\ell, a_0)) \to V, \quad j \in r,$$

such that $f_j \subseteq \mathcal{E}_\mathcal{G}$ for each $j \in r$ and every vertex $x \in V$ with $a(x) \neq a_0(x)$ is in the image of at least one f_j.

Proof Let $t \in \omega$ be such that both C and P can be computed by a t-local rule.

As in the proof of Lemma 5.17, we can assume that $\ell : V \to \omega$ is an $(r + 2t)$-sparse colouring. Fix a sequence $(X_i)_{i \in \omega}$ where each X_i is a non-empty subset of ω of size at most r such that each such set appears as X_i for infinitely many values of the index i. Given a_0, we inductively define Borel assignments $a_1, a_2, \ldots : V \to \omega$, each solving the LCL C. (Also, we will define some auxiliary functions $f_{i,j}$ that will be used to construct the final functions f_j, $j < r$.)

Informally speaking, each new a_{i+1} is obtained from a_i by doing simultaneously all r-augmentations that can be supported on a connected set whose ℓ-labels are exactly X_i. Every two such sets are far away from each other by the sparseness of ℓ, so all these augmentations can be done in parallel without conflicting with each other.

Suppose that $i \geqslant 0$ and we have already defined a_i, a Borel assignment $V \to \omega$ that solves C. Let $\mathcal{X}_i$ be the family of subsets $S \subseteq V$ such that $\mathcal{G} \upharpoonright S$ is connected, $|S| \leqslant r$, and $X_i = \ell(S)$, that is, the set of ℓ-values seen on S is precisely X_i. Note that, since ℓ is $(r + 2t)$-sparse, ℓ is injective on each $S \in \mathcal{X}_i$ and the distance in $\mathcal{G}$ between any two sets from $\mathcal{X}_i$ is larger than $2t$; in particular, these sets are pairwise disjoint. Let $\mathcal{Y}_i$ consist of those S in $\mathcal{X}_i$ which are augmenting for a_i. Do the following for every $S \in \mathcal{Y}_i$. First, take the lexicographically smallest function $b_S : S \to \omega$ such that the assignment $a_i \rightsquigarrow b_S$ (which is obtained from a_i by letting the values of b_S supersede it on $\pi_0(b_S) = S$) satisfies C at every vertex and has a strictly larger progress set than a_i has, that is,

$$\mathsf{C}(a_i \rightsquigarrow b_S) = V \quad \text{and} \quad \mathsf{P}(a_i \rightsquigarrow b_S) \supsetneq \mathsf{P}(a_i). \tag{5.3}$$

Let $P_S := \mathsf{P}(a_i \rightsquigarrow b_S) \setminus \mathsf{P}(a_i) \neq \emptyset$. This set measures the progress made by the augmentation on S. Note that P_S is finite as a subset of $N^{\leqslant t}(S)$. Let $(f_{S,0}, \ldots, f_{S,r-1}) \in (S^{P_S})^r$ be the lexicographically smallest sequence with each $f_{S,i}$ being a function from P_S to S such that their combined images cover S, that is, $\cup_{i \in r} \pi_1(f_{S,i}) = S$. (When defining the lexicographical order on $(S^{P_S})^r$, we can use, for definiteness, the total ordering of $S \cup P_S \subseteq N^{\leqslant t}(S)$ given by the values of the $(r+2t)$-sparse colouring ℓ which is injective on this set.) Note that r functions are enough as $|S| \leqslant r$ while the "worst" case is when $P_S \neq \emptyset$ is a singleton. Having processed each $S \in \mathcal{Y}_i$ as above, we define

$$a_{i+1} := a_i \rightsquigarrow \cup_{S \in \mathcal{Y}_i} b_S.$$

In other words, a_{i+1} is obtained from a_i by replacing it by b_S on each $S \in \mathcal{Y}_i$. (Recall that these sets are pairwise disjoint.) Likewise, let

$$f_{i,j} := \cup_{S \in \mathcal{Y}_i} f_{S,j}, \quad j \in r. \tag{5.4}$$

Let us check, via induction on $i \in \omega$, the claimed properties of each constructed assignment a_{i+1}, namely, that a_{i+1} is Borel and solves C. Note that a_{i+1} is defined by a local rule on $(\mathcal{G}, (\ell, a_i))$ of radius $r + 2t$: by looking at $N^{\leqslant r}(x)$ we can check if $x \in S$ for some $S \in \mathcal{X}_i$ and, if such a set S exists, then $N^{\leqslant 2t}(S) \subseteq N^{\leqslant r+2t}(x)$ determines whether $S \in \mathcal{Y}_i$ and the value of a_{i+1} on x. (Note that we may need to look at distance as large as $2t$ from S because the new values of a_{i+1} on S can affect the values of C and P on $N^{\leqslant t}(S)$ which in turn can depend on the values of a_i on $N^{\leqslant 2t}(S)$.) Thus by induction and Lemma 5.17, the assignment $a_{i+1} : V \to \omega$

is Borel. Let us show that a_{i+1} satisfies C. Take any $x \in V$. If $a_{i+1} \upharpoonright N^{\leqslant t}(x) = a_i \upharpoonright N^{\leqslant t}(x)$, that is, a_{i+1} and a_i coincide on every vertex at distance at most t from x, then C returns the same value on x for a_{i+1} as for a_i, which is 1 by induction. Otherwise there is $y \in N^{\leqslant t}(x)$ which changes its value when we pass from a_i to a_{i+1}. Let S be the unique element of $\mathcal{Y}_i$ that contains y. One of the requirements when we defined $b_S : S \to \omega$ was that $a' := a_i \leadsto b_S$ satisfies C. In particular, a' satisfies C at the vertex x. Now a_{i+1} and a' are the same at $N^{\leqslant t}(x)$ because every element of $\mathcal{Y}_i \setminus \{S\}$ is at distance more than $2t$ from S and, by $S \cap N^{\leqslant t}(x) \neq \emptyset$, at distance more than t from x. Thus a_{i+1} satisfies C at x since a' does. As $x \in V$ was arbitrary, a_{i+1} solves C.

Note that, since t is an upper bound also on the radius of P, the same argument as above when applied to every element of $N^{\leqslant t}(S)$ shows by $\mathsf{P}(a_i \leadsto b_S) \supsetneq \mathsf{P}(a_i)$ that

$$\mathsf{P}(a_{i+1}) \cap N^{\leqslant t}(S) = \mathsf{P}(a_i \leadsto b_S) \cap N^{\leqslant t}(S) \supsetneq \mathsf{P}(a_i) \cap N^{\leqslant t}(S), \quad \text{for every } S \in \mathcal{Y}_i. \tag{5.5}$$

Let us show that, for every $x \in V$, the values $a_i(x)$, $i \in \omega$, stabilise eventually. Suppose that $a_{i+1}(x) \neq a_i(x)$ for some $i \in \omega$. Then $x \in S$ for some augmenting set $S \in \mathcal{Y}_i$. When we change a_i to b_S on S, the progress set strictly increases, so take any $y \in \mathsf{P}(a_i \leadsto b_S) \setminus \mathsf{P}(a_i)$. Of course, $y \in N^{\leqslant t}(S)$. By (5.5), we have $y \in \mathsf{P}(a_{i+1}) \setminus \mathsf{P}(a_i)$. This means that, every time the assignment at x changes, some extra vertex from $N^{\leqslant r+t}(x)$ is added to the progress set. This can happen only finitely many times by the local finiteness of $\mathcal{G}$ (and since no vertex is ever removed from the progress set by (5.5)). Thus the constructed assignments a_i stabilise at every vertex, as claimed.

Define $a : V \to \omega$ by letting $a(x)$ be the eventual value of $a_i(x)$ as $i \to \infty$.

This assignment $a : V \to \omega$ is finitary and thus Borel by Corollary 5.20. Also, it solves the LCL C (because, for every $x \in V$, all values of a on $N^{\leqslant t}(x)$ are the same as the values of the C-satisfying assignment a_i for sufficiently large i).

Let us show that no connected set $S \subseteq V$ of size at most r can be augmenting for a. This property depends on the values of a on the finite set $N^{\leqslant 2t}(S)$. Again, there is $i_0 \in \omega$ such that a and a_i coincide on this set for all $i \geqslant i_0$. Since the set $\ell(S)$ appears in the sequence $(X_i)_{i \in \omega}$ infinitely often, there is $j \geqslant i_0$ with $X_j = \ell(S)$. Thus S is in $\mathcal{X}_j$ but not in $\mathcal{Y}_j$ (otherwise at least one value on S changes when we define a_{j+1}). Thus S is not augmenting for a_j and, consequently, is not augmenting for a. We conclude that the constructed Borel assignment a admits no R-augmentation.

Finally, let us define $f_j := \cup_{i \in \omega} f_{i,j}$ for $j \in r$, where each $f_{i,j}$ was defined in (5.4). Note that when we define $f_{i,j}$ on some P_S then, by (5.5), every vertex of P_S moves to the progress set at Stage i and stays there at all later stages. Thus each f_j, as a subset of $V \times V$, is a function.

Note that each function $f_{i,j}$ moves any vertex in its domain by bounded distance in $\mathcal{G}$ (namely, at most $r+t$). So it can be encoded by a vertex labelling $\ell_{i,j}$ on V where for every $x \in V$ we specify if $f_{i,j}$ is defined on x and, if yes, the sequence of the ℓ-labels on the shortest (and then ℓ-lexicographically smallest) path from x to $f_{i,j}(x)$. Each $\ell_{i,j}$ can clearly be computed by a local rule so it is Borel by Lemma 5.17. Lemma 5.16 implies that each function $f_{i,j}$ is Borel. Thus each function f_j for $j \in r$ is Borel. Also, every vertex $x \in V$ where a differs from a_0 belongs to some $S \in \mathcal{Y}_i$

for some $i \in \omega$ and, by construction, x is covered by the image of $f_{i,j}$ for some $j \in r$. Thus the images of $f_0, \ldots, f_{r-1}$ cover all vertices where a differs from a_0, while the domain of each f_j is $\mathsf{P}(a) \setminus \mathsf{P}(a_0)$ by construction.

This finishes the proof of Theorem 5.25. □

6 Negative Results via Borel Determinacy

The greedy bound $\chi(G) \leqslant d+1$, where $d := \Delta(G)$ is the maximal degree of G, is not in general best possible and can be improved for many finite graphs. For example, Brooks' theorem [10] states that, for a connected graph G, we have $\chi(G) \leqslant d$ unless G is a clique or an odd cycle. (See also Molloy and Reed [65] for a far-reaching generalisation of this result.)

In contrast to these results, Marks [58] showed rather surprisingly that, for every $d \geqslant 3$, the greedy upper bound $d+1$ on the Borel chromatic number is best possible, even if we consider acyclic graphs only. This was previously known for $d = 2$: the irrational rotation graph $\mathcal{R}_\alpha$ of Example 1.1, is a 2-regular acyclic Borel graph whose Borel chromatic number is 3.

We present a slightly stronger version which follows directly from Marks' proof.

Theorem 6.1 (Marks [58]) *For every $d \geqslant 3$ there is a Borel acyclic d-regular graph $\mathcal{G} = (V, E, \mathcal{B})$ with a Borel proper edge colouring $\ell : E \to d$ such that for every Borel map $c : V \to d$ there is an edge $(x, y) \in E$ with $c(x) = c(y) = \ell(x, y)$. (In particular, $\chi_B(\mathcal{G}) \geqslant d+1$.)*

Proof We follow the presentation from Marks [57], generally adding more details. While the proof can be concisely written (the whole note [57] is only 1-page long), it is quite intricate.

Let
$$\Gamma := \langle \gamma_0, \ldots, \gamma_{d-1} \mid \gamma_0^2 = \cdots = \gamma_{d-1}^2 = e \rangle$$
be the group freely generated by d involutions $\gamma_0, \ldots, \gamma_{d-1}$, that is, $\Gamma = \mathbb{Z}_2 * \cdots * \mathbb{Z}_2$ is the free product of d copies of $\mathbb{Z}_2$, the cyclic group of order 2.

Let (G, λ) be the *(right) edge coloured Cayley graph* of $(\Gamma; \gamma_0, \ldots, \gamma_{d-1})$ whose vertex set is Γ and whose edges are given by right multiplication by the involutions $\gamma_0, \ldots, \gamma_{d-1}$, that is, for each $\beta \in \Gamma$ and $i \in d$ we connect β and $\beta\gamma_i$ by an edge, which gets colour i under λ. (Note that the colour of the edge $\{\beta, \beta\gamma_i\}$ does not depend on the choice of an endpoint since $\gamma_i^2 = e$.) The graph (G, λ) is isomorphic to the infinite edge d-coloured d-regular tree.

The group Γ naturally acts on itself. We consider the left action $a : \Gamma \curvearrowright \Gamma$ where the action $a(\gamma, \cdot)$ of $\gamma \in \Gamma$ is just the left multiplication by γ which maps $\beta \in \Gamma$ to $\gamma\beta \in \Gamma$. Note that we take the left multiplication for the action but the right multiplication when defining the Cayley graph G. This ensures that the automorphisms of the graph G that preserve the edge colouring λ are precisely the left multiplications by the elements of Γ:

$$\mathrm{Aut}(G, \lambda) = \{a(\gamma, \cdot) : \gamma \in \Gamma\}. \tag{6.1}$$

We view the elements of ω^Γ as functions $\Gamma \to \omega$ and call them *labellings*. (The proof, as it is written, also works if we replace ω^Γ by A^Γ for some finite set A of size

$(d-1)^2+1$.) The standard Borel structure on ω^Γ comes from the product topology where we view ω^Γ as the product of countably many copies of the discrete space ω.

The group Γ naturally acts on this space via the *(left) shift action* $s : \Gamma \curvearrowright \omega^\Gamma$ defined as follows. For $\gamma \in \Gamma$ and $x \in \omega^\Gamma$, the *(left) shift* $\gamma.x \in \omega^\Gamma$ of x is defined by
$$(\gamma.x)(\beta) := x(\gamma^{-1}\beta), \quad \beta \in \Gamma.$$
In other words, we just pre-compose the labelling $x : \Gamma \to \omega$ with the map $a(\gamma^{-1}, \cdot)$, the left multiplication by γ^{-1}. (We take the inverse of γ to get a left action, namely so that the identity $(\gamma\beta).x = \gamma.(\beta.x)$ always holds.) For each $\gamma \in \Gamma$, the γ-shift map $s(\gamma, \cdot) : \omega^\Gamma \to \omega^\Gamma$ is Borel; in fact, it is a homeomorphism of the product space ω^Γ as it just permutes the factors. Thus the action is Borel.

Let $\mathcal{S}$ be the *shift graph* on ω^Γ where a vertex $x \in \omega^\Gamma$ is adjacent to every element in $\{\gamma_0.x, \ldots, \gamma_{d-1}.x\} \setminus \{x\}$. In other words, $\mathcal{S}$ is the Schreier graph $\mathcal{S}(s; \{\gamma_0, \ldots, \gamma_{d-1}\})$ as defined in Example 4.3. The Borel graph $\mathcal{S}$ comes with the Borel edge colouring $E(\mathcal{S}) \to 2^d$ where the colour of an edge $(x, y) \in E(\mathcal{S})$ is the (non-empty) set of $i \in d$ such that $\gamma_i.x = y$.

The graph $\mathcal{S}$ has cycles and is not d-regular. (For example, the constant-0 labelling of Γ is an isolated vertex of $\mathcal{S}$.) Let X consist of those labellings $x \in \omega^\Gamma$ that give a proper vertex colouring of the Cayley graph G, that is,
$$\begin{aligned} X &:= \{x \in \omega^\Gamma : \forall \beta \in \Gamma \; \forall i \in d \; x(\beta) \neq x(\beta\gamma_i)\} \\ &= \bigcap_{\beta \in \Gamma} \bigcap_{i \in d} \bigcup_{\substack{k,m \in \omega \\ k \neq m}} \{x \in \omega^\Gamma : x(\beta) = k \wedge x(\beta\gamma_i) = m\}. \end{aligned}$$

The second formula for X makes it clear that this set is a Borel subset of ω^Γ. Also, X is an invariant set under the shift action s; this follows from (6.1) since a proper colouring remains proper when pre-composed with an automorphism of the graph.

The graph $\mathcal{S} \upharpoonright X$ is neither acyclic nor d-regular. For example, there are exactly two proper 2-colourings $\Gamma \to \{0, 1\}$ of the bipartite graph G and they form an isolated edge in $\mathcal{S} \upharpoonright X$. So we need to do another (final) trimming.

Let $Y := X \cap \text{Free}(s)$ be the intersection of X with the free part of the shift action s. In other words, Y consists of those proper vertex colourings of G that have no symmetries under the automorphisms of the edge-coloured graph (G, λ). The set Y is invariant under the action s, since X and the free part $\text{Free}(s)$ are. In particular, there are no edges connecting $X \setminus Y$ to Y and the induced subgraph $\mathcal{G} := \mathcal{S} \upharpoonright Y$ is d-regular and acyclic. The graph $\mathcal{G}$ comes with the Borel edge d-colouring ℓ, where, for each $x \in Y$ and $i \in d$, we colour the edge $\{x, \gamma_i.x\}$ by i. As we argued before, the set X is Borel. Also, the free part of the Borel action s is Borel by Lemma 4.4. Thus Y and the graph $\mathcal{G}$ are Borel. The following two claims clearly imply that the edge coloured graph $(\mathcal{G}, \ell)$ satisfies the theorem. (Note that, for all $x \in X$ and $i \in d$, we have $\gamma_i.x \neq x$ because $(\gamma_i.x)(e) = x(\gamma_i^{-1})$ is different from $x(e)$ as $x \in X$ is a proper colouring of G and assigns distinct colours to the adjacent vertices e and $\gamma_i^{-1} = \gamma_i$.)

Claim 6.2 *If $c : X \to d$ is a Borel map then there is $x \in X$ and $i \in d$ with $c(x) = c(\gamma_i.x) = i$.*

Claim 6.3 *There is a proper Borel d-colouring c of the graph $\mathcal{S} \upharpoonright (X \setminus Y)$.*

Proof of Claim 6.2. Given $c : X \to d$, we define for every $i \in d$ and $j \in \omega$ the game $G_{i,j}$ where two players, I and II, take turns to construct $x : \Gamma \to \omega$ which is a proper vertex colouring of G. Initially, we start with the partial colouring x which assigns value j to the identity (that is $x(e) = j$) and is undefined on $\Gamma \setminus \{e\}$. There are countably many rounds as follows. For convenience, let us identify each element $\beta \in \Gamma$ with the unique reduced word in $\gamma_0, \ldots, \gamma_{d-1}$ representing β. In *Round r* for $r = 1, 2, \ldots$, first Player I chooses the values of x at all reduced words of length r that begin with γ_i, and then Player II chooses the values of x at all other reduced words of length r (that is, those that begin with γ_m for some $m \in d \setminus \{i\}$). The restriction that applies to both players is that the current partial vertex colouring x of G is proper at every stage. Since there are infinitely many available colours, there is always a non-empty set of responses for a player. So the game continues for ω rounds. Since a value of x, once assigned, is never changed later, a run of the game gives a fully defined map $x : \Gamma \to \omega$ which, by the imposed restriction, is in fact an element of X. Player I wins the game if and only if $c(x) \neq i$, where $c : X \to d$ is the given Borel map.

In general, by using the Axiom of Choice one can design games of the above type (when two players construct an infinite sequence in countably many rounds) where none of the players has a winning strategy, that is, for every strategy of one player, there is a strategy of the other player that beats it. The groundbreaking result of Martin [61] (with a simplified proof presented in [62]) states that, for games with countably many choices in each round, if the set of winning sequences is Borel then the game is *determined*, i.e. one of the players has a winning strategy. Here, for the game $G_{i,j}$, the set of winning labellings for Player I is exactly $c^{-1}(i)$. Thus each game $G_{i,j}$ is determined.

Let us show that for every $j \in \omega$ there is $i \in d$ such that Player II has winning strategy in $G_{i,j}$. Suppose on the contrary that this is false. This implies by the Borel determinacy that, for every $i \in d$, Player I has a winning strategy for $G_{i,j}$; let us call this strategy S_i. Now, we let these d strategies play each against the others as follows. We start with $x(e) := j$. In *Stage r*, for $r = 1, 2, \ldots$, we use the round-r responses of the strategies $S_0, \ldots, S_{d-1}$ in parallel to define x on all reduced words of length r. Induction on r shows that, before this stage, we have a proper partial vertex colouring x defined on $N_G^{\leq r-1}(e)$ with $x(e) = j$. For every $i \in d$, this is a legal position of $G_{i,j}$ when Player I is about to define x on length-r reduced words beginning with γ_i. So we use the values specified by S_i on these words. Also, the new values assigned by two different strategies are never adjacent; in fact, they are $2r$ apart as the unique shortest path between them in the tree G has to go via the identity e. Thus, after Stage r, the new partial labelling x is a proper vertex colouring of $N_G^{\leq r}(e)$ and we can proceed to Stage $r+1$. The final labelling $x : \Gamma \to \omega$ is clearly a proper colouring of G. Since each S_i is a winning strategy, we have that $c(x) \neq i$. But this is impossible as c has to assign some colour to $x \in X$.

By the previous paragraph and the Pigeonhole Principle, there are $i \in d$ and distinct $j_0, j_1 \in \omega$ such that Player II has a winning strategy in G_{i,j_0} and G_{i,j_1}. Let these strategies be T_0 and T_1 respectively. Now, we let T_0 play against $\gamma_i.T_1$, the "γ_i-shifted" version of T_1, to construct a labelling $x : \Gamma \to \omega$ as follows. Initially, we let $x(e) := j_0$ and $x(\gamma_i) := j_1$, viewing it as *Stage* 0. Iteratively for each $r = 1, 2, \ldots$, *Stage r* uses the round-r responses of T_0 and $\gamma_i.T$ in parallel to colour all reduced words of length r that do not start with γ_i and, respectively, all $\beta \in \Gamma$ such that the reduced word of $\gamma_i \beta$ has length r and does not start with γ_i. The last set is represented precisely by reduced words of length $r+1$ that start with γ_i. Thus, an induction on $r \geqslant 1$ shows that the set D_r of vertices on which the partial colouring x is defined just before Stage r is represented by all reduced words of length $r-1$ and those of length r that being with γ_i. (This is true in the base case $r = 1$ since the initial colouring is defined on $D_1 = \{e, \gamma_i\}$.) This is exactly the information that the strategy T_0 needs to know in Round r. Note that $\gamma_i.D_r := \{\gamma_i.\beta : \beta \in D_r\}$ is the same set D_r so the round-r response of $\gamma_i.T_1$ can also be computed. Thus we can play T_0 and $\gamma_i.T_1$ in every stage, without any conflicts between the assigned values. After ω stages, $\gamma_i.T_1$ colours all reduced words beginning with γ_i and T_0 colours the rest of Γ and we get an everywhere defined function $x : \Gamma \to \omega$ which is also a proper colouring of the Cayley graph G. (Note that the only possibly conflicting edge $\{e, \gamma_i\}$ of G gets distinct colours j_0 and j_1 before Stage 1.)

Since x can be represented as a run of the game G_{i,j_0} where Player II applies the winning strategy T_0, it holds that $c(x) = i$. Also, $\gamma_i.x$ is a run of the game G_{i,j_1}, where the winning strategy T_1 is applied. Thus $c(\gamma_i.x) = i$. We see that this labelling x satisfies Claim 6.2. ∎

Next, we prove the remaining Claim 6.3 with its proof being fairly routine to experts.

Proof of Claim 6.3. Let a *generalised cycle* be a finite sequence

$$(x_0, \ldots, x_{m-1}; \gamma_{i_0}, \ldots, \gamma_{i_{m-1}}) \in (\omega^\Gamma)^m \times \Gamma^m$$

such that $m \geqslant 1$, $x_0, \ldots, x_{m-1}$ are pairwise distinct, $x_{j+1} = \gamma_{i_j}.x_j$ for every $j \in m$ where we denote $x_m := x_0$, and if $m = 2$ then $i_0 \neq i_1$. If $m \geqslant 3$ then this gives a usual cycle of length m in the graph $\mathcal{S}$ (with a direction and a starting vertex specified). The cases $m = 1$ and $m = 2$ correspond to "imaginary cycles": if we re-define the Schreier graph $\mathcal{S}$ as the natural d-regular multigraph with loops, then these would correspond to loops and pairs of multiple edges respectively.

Note that generalised cycles are minimal witnesses to non-freeness of the shift action $s : \Gamma \curvearrowright \omega^\Gamma$ in the following sense. Suppose that $y \in \omega^\Gamma$ is not in the free part. Then there are $x_0 \in [y]$ and non-identity $\gamma \in \Gamma$ with $\gamma.x_0 = x_0$. Writing $\gamma = \gamma_{i_{m-1}} \ldots \gamma_{i_0}$ as the reduced word in $\{\gamma_0, \ldots, \gamma_{d-1}\}$ and inductively on $j \in m-1$, letting $x_{j+1} := \gamma_{i_j}.x_j$, we get all properties of a generalised cycle except vertices can repeat here. Now, if there are repetitions among $x_0, \ldots, x_m$ then take two repeating vertices whose indices are closest and restrict to the subsequence between them.

Let us briefly argue that we can choose a Borel set $\mathcal{C}$ of vertex-disjoint generalised cycles in

$$\mathcal{S}' := \mathcal{S} \upharpoonright (X \setminus Y)$$

such that every component of $\mathcal{S}'$ contains at least one. (This claim also directly follows from [37, Lemma 7.3].) First, notice that the function S which corresponds to each $x \in X \setminus Y$ the shortest length of a generalised cycle in $[x]_{\mathcal{S}'} = [x]_\mathcal{S}$ is Borel by Corollary 5.20 as a finitary function. Namely, it is the pointwise limit of r-local rules $\mathsf{S}_r : X \setminus Y \to \omega \cup \{\omega\}$, where for $x \in \omega^\Gamma$ we define $\mathsf{S}_r(x)$ to be the minimum length of a generalised cycle inside $N^{\leqslant r}(x)$ (which is not required to pass through

x) and to be ω if none exists. Note that the function S is invariant (that is, assumes the same value for all vertices in a graph component) even though S_r need not be. Now, for each $m \geqslant 1$, define the graph $\mathcal{G}_m$ whose vertices are all generalised cycles of length m in the components where the function S assumes value m, where two vertices of $\mathcal{G}_m$ are adjacent if the corresponding cycles have at least one common vertex. This graph, whose vertex set is a subset of the Polish space $(\omega^\Gamma)^m \times \Gamma^m$, can be routinely shown to be Borel with the tools that we have already presented. Also, the maximum degree of $\mathcal{G}_m$ can be bounded by $m\,d^m$. Thus, by Theorem 5.6, we can choose a Borel maximal independent set I_m in $\mathcal{G}_m$. The union $\mathcal{C} := \cup_{m \geqslant 1} I_m$ satisfies the claim by the maximality of I_m. (In fact, more strongly, we picked a maximal subset of vertex-disjoint shortest generalised cycles inside each component of $\mathcal{S}'$.)

For every chosen generalised cycle $(x_0, \ldots, x_{m-1}; \gamma_{i_0}, \ldots, \gamma_{i_{m-1}}) \in \mathcal{C}$ and for each $j \in m$, define $c(x_j) := i_j$. This partially defined vertex colouring of $\mathcal{S}'$ has the property that, for every edge $\{x, \gamma_i.x\}$ of $\mathcal{S}'$, if $c(x) = i$ then $c(\gamma_i.x) \neq i$. (Note that $\mathcal{S}'$ does not have any loops by definition, although we used "imaginary loops" when defining generalised cycles of length 1.)

Now, for every uncoloured vertex x of $\mathcal{S}'$ take a shortest path P from x to a generalised cycle in $\mathcal{C}$ and, if there is more than one choice of P then choose one where the sequence of edge colours on P is lexicographically smallest. Define $c(x) := i$, where i is the smallest element of $\{0, \ldots, d-1\}$ such that $(x, \gamma_i.x)$ is the first edge of P. Note that $c(\gamma_i.x)$ is the colour of either the second edge on P (if P has at least two edges) or the edge coming out of $\gamma_i.x$ in the (unique) generalised cycle in $\mathcal{C}$ containing $\gamma_i.x$. In either case, $c(\gamma_i.x)$ cannot be i as otherwise P is not a shortest path.

This colouring c is finitary on the labelled graph $(\mathcal{S}', \ell')$, where ℓ' is a (Borel) vertex labelling encoding both the edge labelling of $\mathcal{S}'$ as well as the initial partial colouring of all vertices on the generalised cycles from $\mathcal{C}$. Indeed, c can be built as the nested union of partial local colourings C_r, $r \in \omega$, where each vertex x computes its (final) colour if $N^{\leqslant r}(x)$ contains at least one vertex covered by $\mathcal{C}$ and declares $\mathsf{C}_r(x)$ undefined otherwise. By Corollary 5.20, c is Borel. Thus c is the required colouring of $\mathcal{S}'$. Claim 6.3 is proved. ∎

This finishes the proof of Theorem 6.1. □

Among many further results, Marks [58, Theorem 1.4] proved that the greedy upper bound for edge colouring of Theorem 5.12 is best possible even for acyclic Borel graphs that additionally admit a Borel bipartition.

Theorem 6.4 (Marks [58]) *For every $d \geqslant 3$, there is a Borel acyclic d-regular graph $\mathcal{G}$ such that $\chi_B(\mathcal{G}) = 2$ and $\chi'_B(\mathcal{G}) = 2d - 1$.* □

Remark 6.5 In fact, the bipartite graph in Theorem 6.4 constructed by Marks also does not admit a Borel perfect matching. Previously, such a graph for $d = 2$ was constructed by Laczkovich [43]. Of course, it does not admit a Borel edge 2-colouring and satisfies Theorem 6.4 for $d = 2$.

Thus the Borel chromatic number of a Borel graph cannot be bounded by some function of its chromatic number. However, some bounds can be shown under certain additional assumptions. For example, Weilacher [81] showed by building upon some earlier results of Miller [64] that if each component of a Borel graph $\mathcal{G}$ is 2-ended then $\chi_B(\mathcal{G}) \leqslant 2\chi(\mathcal{G}) - 1$ (and that, under these assumptions, this bound is in fact best possible). The same bound $\chi_B(\mathcal{G}) \leqslant 2\chi(\mathcal{G}) - 1$ was established by Conley, Jackson, Marks, Seward and Tucker-Drob [15] under the assumption that the Borel asymptotic dimension of $\mathcal{G}$ is finite. The proofs from both papers can be presented so that we first find a Borel partition $V = A \cup B$ with the induced graphs $\mathcal{G} \restriction N^{\leqslant 1}(A)$ and $\mathcal{G} \restriction B$ having finite components only. Then, by Lemma 5.21 and Theorem 5.23, there are proper Borel colourings $a : N^{\leqslant 1}(A) \to k$ and $b : B \to \{k, \ldots, 2k-1\}$ of these graphs, where $k := \chi(\mathcal{G})$. We use a on A and b on B, except we can save one colour by recolouring the independent set $b^{-1}(2k-1)$ where we use the colouring a on its vertices with at least one neighbour in A and assign colour 0 to the rest.

7 Borel Equivalence Relations

An equivalence relation $\mathcal{E}$ on a standard Borel space X is called *Borel* if it is Borel as a subset of X^2. When we have some notion of isomorphism on a set of structures, it often leads to a Borel equivalence relation. For example, if $X = R^{n \times n}$ encodes $n \times n$ matrices then the similarity relation can be shown to be Borel (by combining Proposition 20.3 and Example 20.6.(b) from [79]).

If we view Borel maps as "computable" then many "computational" questions translate to descriptive set theory problems. For example, the existence of a Borel *selector* for $\mathcal{E}$ (a map $s : X \to X$ such that $s \subseteq \mathcal{E}$ and $s(X)$ is a transversal of $\mathcal{E}$) can be interpreted as being able to "compute" one canonical representative from each equivalence class. (To connect this to Section 5.3, observe that, by e.g. [79, Proposition 20.3], a Borel equivalence relation admits a Borel selector if and only if it admits a Borel transversal.) The Jordan canonical form is an example of a Borel selector for the above matrix similarity relation (see [79, Example 20.6.(b)]). Also, if we have a Borel *reduction* from $(X, \mathcal{E})$ to another Borel equivalence relation $(X', \mathcal{E}')$, that is, a Borel map $r : X \to X'$ such that, for $x, y \in X$, we have $(x, y) \in \mathcal{E}$ if and only if $(r(x), r(y)) \in \mathcal{E}'$, then we could say that the relation $\mathcal{E}$ as not "harder to compute modulo r" than $\mathcal{E}'$. A lot of effort in this area went into understanding the hierarchy of possible Borel equivalence relations under the Borel (and some other kinds of) reducibility, see e.g. the survey by Hjorth [33] that concentrates on this aspect.

A promising field for applying combinatorial methods is the theory of *countable* Borel equivalence relations (*CBERs* for short), where each equivalence class is countable. A more general result of Miller [64, Theorem C] implies that every CBER is the connectivity relation of some locally finite Borel graph. So, various questions of descriptive set theory can be approached from the graph theory point of view.

One example of an important and actively studied property is as follows. A CBER $(X, \mathcal{E})$ is called *hyperfinite* if there are Borel equivalence relations $F_m \subseteq X \times X$, $m \in \omega$, such that $F_0 \subseteq F_1 \subseteq F_2 \subseteq \ldots$, $\cup_{m \in \omega} F_m = \mathcal{E}$ and each F_m has finite equivalence classes (or, equivalently, all of size at most m, see [37, Remark 6.10]). Slaman and Steel [74] and Weiss [82] showed that the hyperfiniteness of $\mathcal{E}$ is equivalent to being generated by some Borel action of $\mathbb{Z}$. The latter means that there is a Borel bijection $\phi : X \to X$ such that for every $x \in X$ the equivalence class $[x]_\mathcal{E}$ of

x is exactly $\{\phi^n(x) : n \in \mathbb{Z}\}$. Such a function ϕ is easy to find for finite equivalence classes (which generate a smooth equivalence relation by Lemma 5.21). Thus the main point here is that every infinite class can be bijectively exhausted from any of its elements by applying Borel functions "next" (namely, ϕ) and "previous" (namely, the inverse ϕ^{-1}). Although this reformulation of hyperfiniteness looks somewhat similar to smoothness, these two properties behave quite differently. For example, Conley, Jackson, Marks, Seward and Tucker-Drob [16] showed that there is Borel graph $\mathcal{G}$ satisfying Theorem 6.1 (resp. Theorem 6.4) such that its connectivity relation $\mathcal{E}_\mathcal{G}$ is hyperfinite. See e.g. [37, Section 6] for a detailed discussion of hyperfiniteness including the proof of the Slaman–Steel–Weiss Theorem.

Although each CBER as a graph is just a union of countable cliques, the following theorem of Feldman and Moore [23] (see e.g. [37, Theorem 1.3]) gives a very useful symmetry breaking tool (in particularly, allowing us to identify vertices from the local point of view of any vertex x by the edge colourings of shortest paths from x and then apply results like an edge coloured version of Lemma 5.16).

Theorem 7.1 (Feldman and Moore [23]) *For every countable Borel equivalence relation $\mathcal{E} \subseteq X^2$ on a standard Borel space X there is a Borel map $c : \mathcal{E} \to \omega$ such that every colour class is a matching.* □

Given the edge colouring $c : \mathcal{E} \to \omega$ returned by the Feldman–Moore Theorem, we can encode each matching $c^{-1}(i)$ by an involution $\phi_i : X \to X$ that swaps every pair $x, y \in X$ with $c(x, y) = i$ and fixes every remaining element of X. Thus if Γ is the group generated by the bijections $\phi_i : X \to X$ for $i \in \omega$, then the equivalence classes of $\mathcal{E}$ are exactly the orbits of the action of Γ on X. Thus every CBER comes from a Borel action of a countable group, giving another very fruitful connection.

The books by Gao [27] and Kechris and Miller [37] provide an introduction to Borel equivalence relations (from the point of view of group actions). See also the recent survey of results on CBERs by Kechris [35].

8 Baire Measurable Combinatorics

Recall that a subset A of a Polish space X has *the property of Baire* if it is the symmetric difference of a Borel set and a set which is *meager* (that is, a countable union of nowhere dense sets). Such a set A is also often called *Baire measurable*. An equivalent (and, from some points of view, more natural) definition is that A is the symmetric difference of an open set and a meager set. Note that this property is not determined by the Borel σ-algebra $\mathcal{B}(X)$ alone (that is, it depends in general on the topology on X). For an introduction to these concepts from the descriptive set theory point of view, we refer to [34] or [79].

Meager (resp. Baire measurable) sets can be considered as topologically "negligible" (resp. "nice"). One can show by using the Axiom of Choice that there are subsets $\mathbb{R}$ without the property of Baire, see e.g. [34, Example 8.24] or [69, Chapter 5]. So various questions to find satisfying assignments $a : V \to \omega$ in a Borel graph $\mathcal{G}$ with each preimage in $\mathcal{T}$ are meaningful and interesting. A typical strategy is to construct a Borel assignment apart of a meager set Y of vertices. Ideally, the remaining set Y is *invariant* (i.e. $Y = [Y]$); then $a \restriction Y$ can be defined independently of the rest of the vertex set, e.g. by using the Axiom of Choice. The reader should be aware that, in general locally finite Borel graphs, the neighbourhood of a meager set (resp. a set with the property of Baire) need not be meager (resp. have the property of Baire); however, these properties do hold in many natural situations (e.g. when $\mathcal{G}$ is the Schreier graph of a marked group acting by homeomorphisms). So, usually, one works with (partially defined) Borel assignments.

One method of how to deal with this technical issue is the following useful lemma of Marks and Unger [59, Lemma 3.1]: for every locally finite Borel graph $\mathcal{G}$ and any function $f : \omega \to \omega$ there are Borel sets A_n, $n \in \omega$, such that each A_n is $f(n)$-sparse and the complement of their union, $V \setminus \cup_{n\in\omega} A_n$, is a meager and invariant set. It is used by Marks and Unger [59, Theorem 1.3] to prove the following "topological" version of Hall's marriage theorem. Let us say that a bipartite graph G with a bipartition $V = B_0 \cup B_1$ satisfies $\text{Hall}_{\varepsilon,n}$ if for every finite set X in a one part we have $|N(X)| \geqslant |X|$ (which is the usual Hall's marriage condition) and, additionally, if $|X| \geqslant n$ and X is connected in $\mathcal{G}^2$ then $|N(X)| \geqslant (1+\varepsilon)|X|$.

Theorem 8.1 (Marks and Unger [59]) *If $\varepsilon > 0$, $n \in \omega$, V is a Polish space, and $\mathcal{G} = (V, E, \mathcal{B})$ is a locally finite Borel graph with a Borel bipartition $V = B_0 \cup B_1$ satisfying $\text{Hall}_{\varepsilon,n}$ then $\mathcal{G}$ has a Borel matching such that the set of unmatched vertices is meager and invariant.*

In brief, the proof of Theorem 8.1 proceeds as follows. Given ε and n, choose a fast growing sequence $f(0) \ll f(1) \ll f(2) \ll \ldots$ and let A_i, $i \in \omega$, be the sets returned by [59, Lemma 3.1] for this function f. Starting with the empty matching $M_0 := \emptyset$ and $\mathcal{G}_0 := \mathcal{G}$, we have countably many stages indexed by $i \in \omega$. At Stage i, every vertex x of $\mathcal{G}_i$ in the $f(i)$-sparse set A_i picks a neighbour $y_x \in N(x)$ such that $\mathcal{G}_i$ has a perfect matching containing the edge $\{x, y_x\}$, say we take the largest such y_x with respect to some fixed Borel total order on V. Define $M_{i+1} := \{\{x, y_x\} : x \in A_i \cap V(\mathcal{G}_i)\}$ and let $\mathcal{G}_{i+1}$ be obtained from $\mathcal{G}_i$ by removing all vertices matched by M_{i+1}. A combinatorial argument shows by induction on $i \in \omega$ that $\mathcal{G}_i$ satisfies $\text{Hall}_{\varepsilon_i, f(i)}$, where $\varepsilon_i := \varepsilon - \sum_{j \in i} 8/f(j) > 0$. In particular, each graph $\mathcal{G}_i$ satisfies the usual Hall's marriage condition. Thus, by Rado's theorem (Theorem 2.1), y_x exists for every x (and, by induction, we can carry out each stage). Moreover, the function $x \mapsto y_x$ of Stage i is Borel by induction on i as it can be computed by a finitary rule on $(\mathcal{G}_i, M_i)$, as it was demonstrated in Example 5.19. (In fact, we do not need to refer to Rado's theorem at all: when defining y_x, we can instead require that the graph obtained from $\mathcal{G}_i$ by removing the adjacent vertices x and y_x satisfies Hall's marriage condition.) Finally, $M := \cup_{i \in \omega} M_i$ has all the required properties.

Note that the matching M returned by Theorem 8.1 can be extended to cover all vertices, using Rado's theorem (Theorem 2.1); this application of the Axiom of Choice is restricted to a meager set. Thus if we encode the final perfect matching via a vertex labelling $\ell : V \to \omega$ as in Remark 5.18 then each preimage of ℓ has the property of Baire.

A notable general result that is very useful in Baire measurable combinatorics is that every CBER $\mathcal{E}$ on a Polish space X can be made hyperfinite by removing an $\mathcal{E}$-invariant meager Borel subset of X; for more details see e.g. [37, Theorem 12.1]

9 μ-Measurable Combinatorics

Suppose that we have a Borel locally finite graph $\mathcal{G} = (V, E, \mathcal{B})$ and a measure μ on $(V, \mathcal{B})$. Let $\mathcal{B}_\mu$ be the μ-completion of $\mathcal{B}$ which is the smallest σ-algebra on V containing Borel sets and all μ-null sets (i.e. arbitrary subsets of Borel sets of μ-measure 0). Recall from the Introduction that the sets in $\mathcal{B}_\mu$ are called μ-measurable or just measurable.

Of course, the presence of a measure μ makes the set of questions that can be asked and the tools that can be applied much richer. For example, the problems that we considered in this paper also make sense in the measurable setting, where we look for assignments a that are measurable as functions from $(V, \mathcal{B}_\mu)$ to ω, meaning here that $a^{-1}(i) \in \mathcal{B}_\mu$ for each $i \in \omega$.

Another studied possibility is to consider the so-called approximate versions: for example, the approximate μ-measurable chromatic number is the smallest k such that for every $\varepsilon > 0$ there is a Borel set of vertices $A \subseteq V$ of measure at most ε such that the graph induced by $V \setminus A$ can be coloured with at most k colours in a Borel way. For example, the approximate measurable chromatic number of the irrational rotation graph $\mathcal{R}_\alpha$ from Example 1.1 is 2 because the Lebesgue measure of one colour class, namely $X_2 = [0, c)$, in the constructed Borel 3-colouring of $\mathcal{R}_\alpha$ can be chosen to be arbitrarily small. The survey by Kechris and Marks [36] gives an overview of such results as well.

For the reader who, inspired by this paper, would like to read more on the topic, let us point some technical subtleties that are sometimes not mentioned explicitly in the literature. The measure μ is (almost) always assumed to be σ-finite, meaning that V can be covered by countably many sets of finite μ-measure. This implies many important properties such as the regularity of μ ([14, Proposition 8.1.2]), being able to talk about the product of μ with other measures without the complications of going through the so-called complete locally determined products (see [24, Chapter 25]), etc. Also, there is a simple trick (see e.g. [31, Proposition 3.2]) that allows us to construct another measure ν on $(V, \mathcal{B})$ with $\mathcal{B}_\nu \subseteq \mathcal{B}_\mu$ such that ν is quasi-invariant (meaning that the saturation $[N]$ of any ν-null set $N \subseteq V$ is a ν-null set). Thus it is enough to find a measurable satisfying assignment when the measure is quasi-invariant. The advantage of the quasi-invariance of ν is that the neighbourhood and the saturation of any ν-null (resp. ν-measurable) set is also ν-null (resp. ν-measurable). The reader should also be aware that the measurability of an edge labelling $c: E \to \omega$ is understood as the measurability of the vertex labelling $\ell: V \to \omega$ that encodes c under some fixed local rule as in Remark 5.18. This is equivalent to requiring that, there is a Borel μ-null set $N \subseteq V$ such that $c \upharpoonright E \cap (V \setminus N)^2$ is Borel. Note that it is not a good idea to define the measurability of $c: E \to \omega$ with respect the product measure $\mu \times \mu$ on $V^2 \supseteq E$: if μ is atomless then $(\mu \times \mu)(E) = 0$ by Tonelli's theorem and every subset of E is $(\mu \times \mu)$-measurable.

A particularly important case is when μ is a probability measure (that is, $\mu(V) = 1$) which is moreover invariant, meaning that every Borel map $f: V \to V$ with $f \subseteq \mathcal{E}_\mathcal{G}$ preserves the measure μ. (The reader should be able to show that it is enough to check the above property only for involutions f with $f \subseteq E$, that is, functions that come from matchings in $\mathcal{G}$.) In this case, the quadruple $(V, E, \mathcal{B}, \mu)$ is now often called a graphing. (The reader should also be aware of another different usage of this term, where a *graphing* of an equivalence relation $\mathcal{E}$ means a Borel graph $\mathcal{G}$ whose connectivity relation $\mathcal{E}_\mathcal{G}$ coincides with $\mathcal{E}$.)

The invariance of μ is a measure analogue of the obvious fact from finite combinatorics that any bijection preserves the sizes of finite sets. It has many equivalent reformulations such as, for example, the Mass Transport Principle (see e.g. [52, Section 18.4.1]).

For an invariant probability measure μ on the vertex set V, one can define a new measure η on edges where the η-measure of a Borel subset $A \subseteq E$ is defined to be $\int_V |\{y \in V : (x,y) \in A\}| \, \mathrm{d}\mu(x)$, the average A-degree. Then, in fact, the invariance of μ is equivalent to η being *symmetric* (meaning that $\eta(A)$ is always the same as the η-measure of the "transpose" $\{(x,y) : (y,x) \in A\}$ of A), see e.g. [52, Section 18.2]. Interestingly, the η-measurability of an edge labelling $c : E \to \omega$ now coincides with the μ-measurability of the vertex labelling $\ell : V \to \omega$ that encodes c under the Borel reduction of Remark 5.18.

Graphings can serve as the local limits of bounded degree graphs, roughly speaking as follows. An *r-sample* from a graphing is obtained by sampling a random vertex $x \in V$ under the probability measure μ and outputting $\mathsf{F}_r(x)$, the isomorphism type of the rooted graph induced by the r-ball around x. Also, each finite graph (V, E) can be viewed as a graphing $(V, E, 2^V, \nu)$ where ν is the uniform measure on the finite set V. Then the *local* convergence can be described by a metric where two graphings are "close" if the distributions of their samples are "close" to each other, see [52, Section 18] for an introduction to graphings as limit objects.

Also, the Schreier graph of any Borel probability measure-preserving action of a marked group is a graphing. It contains a lot of information about the action and thus is an important object of study in *measured group theory*. Even for such a simple group as the integers $\mathbb{Z}$, its measure-preserving actions (which are specified by giving just one measure-preserving transformation) form a very beautiful and deep subject that is the main focus of the classical ergodic theory. It is hard to pick a good starting introductory point for this vast area. The reader is welcome to consult various surveys and textbooks, e.g. [1, 25, 26, 39, 51, 71], and pick one (or its part) that looks most interesting.

A special but important case of a graphing is the Schreier graph $\mathcal{S}$ of the shift action of a marked group $(\Gamma; S)$ on the product measure space X^Γ where X is a standard probability space, e.g. $\{0,1\}$ or $[0,1]$ with the uniform measure. Measurable labellings of $\mathcal{S}$ are exactly the so-called *factors of IID* labellings with vertex seeds from X of the Cayley graph of (Γ, S). This connection gives a way of applying methods of descriptive combinatorics for constructing various invariant processes on vertex-transitive countable graphs. This is a very active area of discrete probability (see e.g. the book by Lyons and Peres [56]) where even the case of trees has many tantalising unsolved questions (see e.g. Lyons [54]).

As a showcase of how the results that we have proved can be used in the measurable setting, let us present a very brief outline of the following "measurable" version of Hall's marriage theorem by Lyons and Nazarov [55, Remark 2.6] (whose detailed proof can be found in [29, Theorem 3.3]).

Theorem 9.1 (Lyons and Nazarov [55]) *Let $\varepsilon > 0$ and let $\mathcal{G} = (V, E, \mathcal{B}, \mu)$ be a bipartite graphing with a Borel bipartition $V = B_0 \cup B_1$ such that $\mu(B_0) = \mu(B_1) =$*

$1/2$ and for every measurable X inside a part it holds that

$$\mu(N(X)) \geqslant \min((1+\varepsilon)\mu(X), 1/4+\varepsilon).$$

Then $\mathcal{G}$ has a Borel matching that covers all vertices except a null set.

In order to prove Theorem 9.1, we start with the empty matching M_0 and, iteratively for each $i \in \omega$, augment M_i to M_{i+1} in a Borel way using augmenting paths of (edge) length at most $2i+1$, until none remains. This can done by the result of Elek and Lippner [21]. (Alternatively, we can use Theorem 5.25 here, applying all possible $(2i+2)$-augmentations to M_i, with the rest of the proof being the same.) It is a nice combinatorial exercise to show that for every $\varepsilon > 0$ there is $c > 0$ such that if a finite bipartite graph G with both parts of the same size n is an ε-*expander* (i.e. the neighbourhood of any set X in a part has size at least $\min((1+\varepsilon)|X|, (1/2+\varepsilon)n)$), then any matching without augmenting paths of length at most $2i+1$ covers all except at most $(1-c)^i n$ vertices of G. The proof of this statement from [55] extends from finite graphs to the measurable setting since all inequalities used by it come from double counting and, by the invariance of μ, also apply when the relative sizes of sets are formally replaced by their measures. Thus the measure of the set X_i of vertices unmatched by the Borel matching M_i in $\mathcal{G}$ is at most $(1-c)^i$. When we do augmentations to construct M_{i+1}, we change the current matching on at most $2i+2$ vertices per one new matched vertex by the second part of Theorem 5.25. Again, by the invariance of μ, the measure of vertices where M_i and M_{i+1} differ is at most $(2i+2)(1-c)^i$. Define the final Borel matching

$$M := \liminf\nolimits_i M_i = \cup_{i \in \omega} \cap_{j \geqslant i} M_j$$

to be the pointwise limit of the matchings M_i where they stabilise. Let X be the (Borel) set of vertices unmatched by M. For every $i \in \omega$ the following clearly holds: if a vertex is not matched by M then it is not matched by M_i or it witnesses at least one change (in fact, infinitely many changes) after Stage i. Thus the Union Bound gives that $\mu(X) \leqslant (1-c)^i + \sum_{j \geqslant i}(2j+2)(1-c)^j$. Since this inequality is true for every i and its right-hand side can be made arbitrarily small by taking i sufficiently large, we conclude that X has measure 0, as desired. (The reader may have recognised the last two steps as a veiled application of the Borel–Cantelli Lemma.)

10 Borel Colourings from LOCAL Algorithms

Under rather general settings, local rules are equivalent to the so-called deterministic LOCAL algorithms that were introduced by Linial [49, 50]. Their various variants have been actively studied in theoretical computer science; for an introduction, we refer the reader to the book by Barenboim and Elkin [5]. Here we briefly discuss this connection and present a result of Bernshteyn [7] that efficient LOCAL algorithms can be used to find satisfying assignments that are Borel.

Suppose that we search for an assignment $a : V \to \omega$ that solves a given LCL C on a graph G (for example, we would like a to be a proper vertex colouring with k colours) where, just for the clarity of presentation, we assume that C is defined on unlabelled graphs. A *deterministic* LOCAL *algorithm with r rounds* is defined as follows. Each vertex x of G is a processor with unlimited computational power. There are r synchronous rounds. In each round, every vertex can exchange any amount of information with each of its neighbours. After r rounds, every vertex x has to output its own value $a(x)$, with all vertices using the same algorithm for the communications during the rounds and the local computations.

Clearly, the final value $a(x)$ is some function of the r-ball of x so the produced assignment is given by some r-local rule. Conversely, for every r-local rule A, a possible strategy is that each vertex x collects all current information from all its neighbours in each round and computes $A(x)$ at the end of r rounds, by knowing everything about its whole r-ball $N^{\leqslant r}(x)$. Given this equivalence, we will use mostly the language of local rules.

Such a rule need not exist if there are symmetries. For example, if the input graph is vertex-transitive then all vertices produce the same answer so there is no chance to find, for example, a proper vertex colouring. Let us assume here that each vertex x is given the order n of the graph and its unique identifier $\ell(x) \in n$. Thus we evaluate the rule A, that may depend on n, on the labelled graph (G, ℓ).

The corresponding algorithmic question is, for a given family of graphs $\mathcal{H}$ (which we assume to be closed under adding isolated vertices) and an LCL C, to estimate $\mathsf{Det}_{\mathsf{C},\mathcal{H}}(n)$, the smallest r for which there is an r-local rule A such that, for every graph $G \in \mathcal{H}$ with n vertices and every bijection $\ell : V \to n$, the assignment $A(G, \ell)$ solves C on G. If there is some input (G, ℓ) as above which admits no C-satisfying assignment, then we define $\mathsf{Det}_{\mathsf{C},\mathcal{H}}(n) := \omega$. Note that the value of $\mathsf{Det}_{\mathsf{C},\mathcal{H}}(n)$ will not change if we modify the above definition by allowing to take any graph $G \in \mathcal{H}$ with at most n vertices and any injection $\ell : V \to n$ (because we can always add isolated vertices to G and extend ℓ to a bijection).

Note that if $\mathsf{Det}_{\mathsf{C},\mathcal{H}}(n)$ is finite then it is at most $n - 1$. Indeed, by being able to see at distance up to $n - 1$ (and knowing n), each vertex x knows its injectively labelled component $(G, \ell) \restriction [x]$; thus a possible $(n - 1)$-local rule A is that $x \in V$ computes the lexicographically smallest (under the ordering of $[x]$ given by the values of ℓ) C-satisfying assignment $a : [x] \to \omega$ and outputs $a(x)$ as its value.

The following result is a special case of [7, Theorem 2.10] whose proof, nonetheless, contains the main idea.

Theorem 10.1 (Bernshteyn [7]) *Let $\mathcal{H}$ be a family of graphs with degrees bounded by d which is closed under adding isolated vertices. Let C be an LCL on unlabelled graphs such that $\mathsf{Det}_{\mathsf{C},\mathcal{H}}(n) = o(\log n)$ as $n \to \infty$. Then every Borel graph $\mathcal{G} = (V, E, \mathcal{B})$ such that $\mathcal{G} \restriction N^{\leqslant r}(x) \in \mathcal{H}$ for every $x \in V$ and $r \in \omega$, admits a Borel assignment $a : V \to \omega$ that solves C on $\mathcal{G}$.*

Proof Let t be the radius of C. Fix some sufficiently large n, namely we require that $1 + d \sum_{i=0}^{s-1}(d-1)^i \leqslant n$, where $s := 2(\mathsf{Det}_{\mathsf{C},\mathcal{H}}(n) + t)$. This is possible since $s = o(\log n)$ by our assumptions. Take any local rule A of radius $r := \mathsf{Det}_{\mathsf{C},\mathcal{H}}(n)$ that works for all graphs from $\mathcal{H}$ on at most n vertices. Fix a Borel $2(r+t)$-sparse colouring $c : V \to n$ of $\mathcal{G}$ which exists by Corollary 5.9 (and our choice of n).

Apply the rule A to $(\mathcal{G}, c)$, viewing $c : V \to n$ as the identifier function, to obtain a labelling $a : V \to \omega$. (Thus, informally speaking, we run the algorithm pretending that the graph $\mathcal{G}$ has n vertices.) Note that a is well-defined for every vertex $x \in V$

since, by our assumption on $\mathcal{G}$, the subgraph induced by the r-ball $N^{\leqslant r}(x)$ belongs to $\mathcal{H}$ and is injectively labelled by the $2r$-sparse n-colouring c. By Lemma 5.17, the function a is Borel, so it is remains to check that it solves the LCL C. Take any $x \in V$. Consider $H := (\mathcal{G}, c) \upharpoonright N_\mathcal{G}^{\leqslant r+t}(x)$, the labelled subgraph induced in $\mathcal{G}$ by the $(r+t)$-ball around x. If we apply the rule A to H then we obtain the same assignment a on $N_H^{\leqslant t}(x) = N_\mathcal{G}^{\leqslant t}(x)$, because for every vertex in this set its c-labelled r-balls in H and $\mathcal{G}$ are the same. The graph H of diameter at most $2(r+t)$ is injectively labelled by the $2(r+t)$-sparse colouring c and, in particular, has at most n vertices. By the correctness of A, the assignment a satisfies the C-constraint at x on H (and also on $\mathcal{G}$). Thus $a : V \to \omega$ is the required Borel assignment. □

Let us point some algorithmic results when $\mathcal{H}$ consists of graphs with maximum degree bounded by a fixed integer d while n tends to ∞. The deterministic LOCAL complexities of a proper vertex $(d+1)$-colouring, a proper edge $(2d-1)$-colouring, a maximal independent set and a maximal matching are all $O(\log^* n)$, where $\log^* n$ is the *iterated logarithm* of n, the number of times needed to apply the logarithm function to n to get a value at most 1. (For references and the best known bounds as functions of (n, d), we refer the reader to [13, Table 1.1].) These are exactly the problems for which we showed the existence of a Borel solution in Theorems 5.6–5.12. In fact, Theorems 5.6–5.12 for bounded degree graphs are, by Theorem 10.1, direct consequences of the above mentioned results on the existence of efficient local algorithms. While there seems to be a large margin (between the running time of $O(\log^* n)$ for known algorithms and the $o(\log n)$-assumption of Theorem 10.1), in fact, Chang, Kopelowitz and Pettie [13] showed that, for LCLs on bounded-degree graphs, if there is a LOCAL algorithm with $o(\log n)$ rounds that solves the problem then there is one with $O(\log^* n)$ rounds. (In fact, the proof of the last result is similar to the proof of Theorem 10.1: fix a large constant r, generate a proper colouring a of the r-th power of the input order-n graph using $O(\log^* n)$ rounds and then "simulate" the $o(\log n)$ algorithm using the values of a in lieu of the vertex identifiers.)

On the other hand, deterministic algorithms using only $o(\log n)$ rounds seem to be rather weak, e.g. for colouring problems when the number of colours is even slightly below the trivial greedy bound. One (out of many) results demonstrating this is by Chang et al. [13, Theorem 4.5] who showed that vertex d-colouring of trees (of maximum degree at most d) requires $\Omega(\log n)$ rounds. The last result should be compared with Theorem 6.1 here.

Let us also briefly discuss another general (and much more difficult) transference result of Bernshteyn [7] that randomised LOCAL algorithms that require $o(\log n)$ rounds on n-vertex graphs give measurable assignments. In an *r-round randomised* LOCAL *algorithm* A, each vertex x generates at the beginning its own random seed $s(x)$, a uniform element of m, independently of all other choices. Vertices can share any currently known seeds during each of the r communication rounds. Thus each obtained value A(x) can also depend on the generated seeds inside $N^{\leqslant r}(x)$. Equivalently, once the function $s : V \to m$ has been generated, the resulting assignment A is computed by some local rule on (G, s). Given an LCL C, we say that the algorithm *solves* C on an n-vertex graph $G = (V, E)$ if, for every vertex x of G the probability that C(x) fails is at most $1/n$. Let $\mathsf{Rand}_{\mathsf{C},\mathcal{H}}(n)$ be the smallest r such that some

r-round randomised LOCAL algorithm (for some $m = m(\mathsf{C}, \mathcal{H}, n)$) solves C for every n-vertex graph in $\mathcal{H}$. Note that we do not need any identifier function here, since $s(x)$ uniquely identifies a vertex x with probability $(1-1/m)^{n-1}$ which can be made arbitrarily close to 1 by choosing m sufficiently large. Under these conventions, Bernshteyn [7, Theorem 2.14] proved, roughly speaking, that if $\mathsf{Rand}_{\mathsf{C},\mathcal{H}}(n) = o(\log n)$, then the corresponding LCL on bounded-degree Borel graphs admits a satisfying assignment which is μ-measurable for any probability measure μ on $(V, \mathcal{B})$ (resp. Baire measurable for any given Polish topology τ on V with $\sigma_V(\tau) = \mathcal{B}(V)$). This result has already found a large number of applications, see [7, Section 3].

11 Borel Results that Use Measures or Baire Category

Of course, every result that an LCL admits no μ-measurable (resp. no Baire measurable) solution on some Borel graph $\mathcal{G}$ automatically implies that no Borel solution exists for $\mathcal{G}$ either. There is a whole spectrum of techniques for proving results of this type (such as ergodicity that was briefly discussed in Section 1 in the context of Example 1.1). We refer the reader to the survey by Kechris and Marks [36] that contains many further examples of this kind.

Rather surprisingly, measures have turned to be useful in proving also the existence of full Borel colourings for some problems. Let us briefly discuss a few such (very recent) results.

One is a result of Bernshteyn and Conley [9] who proved a very strong Borel version of the theorem of Hajnal and Szemerédi [32]. Recall that the original Hajnal-Szemerédi Theorem states that if G is a finite graph of maximum degree d and $k \geqslant d+1$ then G has an *equitable colouring* (that is, a proper colouring $c : V \to k$ such that every two colour classes of c differ in size at most by 1). For a Borel graph $\mathcal{G}$, let us call a Borel k-colouring, given by a partition $V = V_0 \cup \cdots \cup V_{k-1}$, *equitable* if for every $i, j \in k$ there is a Borel bijection $g : V_i \to V_j$ with $g \subseteq \mathcal{E}$. Of course, an obvious obstacle here is the existence of a finite component whose size is not divisible by k. The main result of Bernshteyn and Conley [9, Theorem 1.5] is that this is the only obstacle. Very briefly, the proof in [9] runs a Borel version of the algorithm of Kierstead, Kostochka, Mydlarz and Szemerédi [40] that finds equitable colourings in finite graphs, where a current proper k-colouring gets "improved" from the point of view of equitability via certain recolouring moves. This iterative procedure gives only a partial colouring as it is unclear how to colour the set X of components that contain vertices that change their colour infinitely often. However, Bernshteyn and Conley [9] proved via the Borel–Cantelli Lemma that if μ is an arbitrary probability measure on $(V, \mathcal{B})$ which is invariant, then $\mu(X) = 0$. Note that the conclusion holds even though the definition of X does not depend on μ. This means that $\mathcal{G} \upharpoonright X$ does not admit any invariant probability measure. This is, by a result of Nadkarni [68], equivalent to X being *compressible*, meaning that there is a Borel set $A \subseteq V$ intersecting every component of $\mathcal{G} \upharpoonright X$ and a Borel bijection $f : X \to X \setminus A$ with $f \subseteq \mathcal{E}_\mathcal{G}$. (Note that the converse direction in Nadkarni's result is easy: if such a function f exists and μ is an invariant probability measure on $(X, \mathcal{B})$, then the Borel sets $A, f(A), f(f(A)), \ldots$ are disjoint and have the same μ-measure, which is thus 0; however, then the saturation X of the μ-null set A cannot have positive μ-measure, a contradiction.) Then a separate argument shows that the remaining compressible graph $\mathcal{G} \upharpoonright X$ admits a full Borel equitable colouring.

Another result that we would like to discuss comes from a recent paper of Conley and Tamuz [17]. Call a colouring $c : V \to 2$ *unfriendly* if every $x \in V$ has at least as many neighbours of the other colour as of its own, that is,

$$|\{y \in N(x) : c(y) \neq c(x)\}| \geqslant |\{y \in N(x) : c(y) = c(x)\}|.$$

Such a colouring trivially exists for every finite graph: take, for example, one that maximises the number of non-monochromatic edges (i.e. a max-cut colouring). Also, the Axiom of Choice (or the Compactness Principle) shows that every locally finite graph admits an unfriendly colouring. Interestingly, Shelah and Milner [73] showed that there are graphs with uncountable vertex degrees that have no unfriendly colouring; the case of locally countable graphs is open.

Conley and Tamuz [17] showed that every Borel bounded-degree graph admits a Borel unfriendly colouring provided the graph has *subexponential growth* meaning that

$$\forall \varepsilon > 0 \; \exists n_0 \in \omega \; \forall n \geqslant n_0 \; \forall x \in V \quad |N^{\leqslant n}(x)| \leqslant (1+\varepsilon)^n, \qquad (11.1)$$

or, informally, that n-balls have size at most $(1 + o(1))^n$ uniformly in n. Like in [9], they use a family of augmenting moves so that, for every invariant probability measure μ on $(V, \mathcal{B})$, the set X where the constructed colourings do not stabilise has μ-measure 0. Here, each move is very simple: if more than half of neighbours of a vertex x have the same colour as x, then we change the colour of x. We do these moves in stages so that the moves made in one stage do not interfere with each other (similarly as in the proof of Theorem 5.25). Since the graph has subexponential growth, for every $x \in V$ there is a probability measure μ_x which is very close to being invariant and satisfies $\mu_x(\{x\}) > 0$. (For example, take the discrete measure that is supported on $[x]$ and puts weight $(1-\varepsilon)^i \mu_x(\{x\})$ on each vertex at distance i from x for all $i \in \omega$.) Since there happens to be some leeway when applying the Borel–Cantelli Lemma in the invariant case, it can also be applied to μ_x provided $\varepsilon > 0$ is sufficiently small. As $\{x\}$ has positive measure in μ_x, it cannot belong to X. Thus $X = \emptyset$, as desired.

This idea was used also by Thornton [77] to prove that every Borel graph of maximum degree d and of subexponential growth admits a Borel orientation of edges so that each out-degree is at most $d/2 + 1$.

It is not clear if compressibility helps for the last two problems. In particular, it remains an open problem if, for example, every 3-regular Borel graph admits an unfriendly Borel 2-colouring.

12 Graphs of Subexponential Growth

Recall that the notion of subexponential growth was defined in (11.1). Unfriendly 2-colouring is one example of a problem which admits a Borel solution for every graph of subexponential growth but this becomes an open problem or a false statement when the growth assumption is removed. Let us just point to some general results which show that Borel graphs of subexponential growth are indeed more tractable.

Csóka, Grabowski, Máthé, Pikhurko and Tyros [18] showed that such graphs admit a Borel satisfying assignment for every LCL where the existence of global solution can be established by the symmetric Lovász Local Lemma. The Local Lemma, introduced in a paper of Erdős and Lovász [22], is a very powerful tool for proving the existence of a satisfying assignment. A special case of it is as follows. Suppose that we have a collection of bad events $\{B : B \in V\}$, each having probability at most p and being a function of some finite set supp(B) of binary random variables. Define the *dependency graph* D on V, where two events are connected if they share at least one variable. The Local Lemma gives that if

$$p < \frac{(\Delta - 1)^{\Delta - 1}}{\Delta^{\Delta}}, \tag{12.1}$$

where Δ is the maximum degree of D, then there is an assignment of variables such that no bad event occurs. (Remarkably, Shearer [72] showed that the bound in (12.1) is, in fact, best possible.)

The Borel version of this result from [18] is quite technical to state; informally speaking it states that if bad events and variables are indexed by elements of some standard Borel space V so that the corresponding dependency graph $\mathcal{D}$ on V is Borel and has subexponential growth, and (12.1) holds, then there is a Borel assignment of variables such that no bad event occurs. For other "definable" versions of the Local Lemma, see Bernshteyn [6, 7] and Kun [42].

An efficient randomised algorithm that finds an assignment in finite graphs whose existence is guaranteed by the Local Lemma was found in a breakthrough work of Moser and Tardos [66, 67]. Actually, one example of a good algorithm is very simple: start with any initial assignment and, as long as there is an occurrence of some bad event B, pick one such B arbitrarily and re-sample all variables in supp(B). This can be adopted to the Borel setting using the ideas of the proof of Theorem 5.25 as follows. Fix some sufficiently large r_0. Take an r_0-sparse Borel colouring c of the dependency graph $\mathcal{D}$. For each colour i, generate a uniform random sequence of binary bits $b_i := (b_{i,j})_{j \in \omega}$. (Note that we have to generate only countably many bits.) Let the initial assignment of variables be $x \mapsto b_{c(x),0}$. At each iteration, find via Theorem 5.6 a Borel set I of currently occurring bad events in which every two have disjoint sets of variables and, moreover, I is a maximal set with this property. (Note that the subexponential growth assumption implies that $\mathcal{D}$ has finite maximum degree.) Reassign the value of every x in the (disjoint) union $\cup_{B \in I}$supp(B) to the next bit of the sequence $b_{c(x)}$ that has not been used by the variable x yet. The problem with this naive adaptation is that variables which are far away in $\mathcal{D}$ can depend on each other and the estimates of Moser and Tardos do not apply because long chains of interdependent bad events may have now very different probabilities.

The key idea of [18] is that, because of the subexponential growth assumption, if some variable $y \in V$ is resampled many times at some finite stage of this procedure then there are another variable x and an integer $r \leqslant r_0/2$ such that, among all re-sampled bad events B, the number of *internal* ones (those with supp(B) $\subseteq N^{\leqslant r}(x)$) is very small compared with the number of *boundary* ones (those with supp(B) intersecting both $N^{\leqslant r}(x)$ and its complement). The internal resamples obey the Moser–Tardos estimates because, by $r \leqslant r_0/2$, they never use the same random bit twice. On the other hand, the number k of boundary resamples is so small that we can afford to take the Union Bound over all possible ways of how at most k uncontrollable boundary events can pop up around x during the run of the algorithm.

This means that, there is an assignment of binary bits $b_{i,j}$ and some $n \in \omega$ such that **every** variable is resampled at most n times. It follows by the finiteness of $\Delta(\mathcal{D})$ that this procedure stabilises for every variable. The final colouring, as a finitary function, is Borel by Corollary 5.20.

The following general application of the Borel Local Lemma from [18] was observed by Bernshteyn [7, Theorem 2.15]: if $\mathsf{Rand}_{\mathcal{H},\mathsf{C}}(n) = O(\log n)$ and $\mathcal{G}$ is a Borel graph of subexponential growth with every ball belonging to the graph family $\mathcal{H}$ then there is a Borel assignment $a : V \to \omega$ solving the LCL C. (Recall that $\mathsf{Rand}_{\mathcal{H},\mathsf{C}}(n)$, as defined in Section 10, is the smallest number of rounds in a randomised LOCAL algorithm that fails any one C-constraint with probability at most $1/n$ for every n-vertex graph in $\mathcal{G}$.) The proof idea is as follows. Fix large n and a suitable randomised algorithm that uses $r := \mathsf{Rand}_{\mathcal{H},\mathsf{C}}(n)$ rounds for some m. This gives an r-local rule A that can be evaluated on V once we have some seed function $s : V \to m$. We view the values of s as variables. For every $x \in V$, let the "bad" event B_x state that the assignment returned by A fails the C-constraint at x. If the radius of the local rule C is t, then B_x depends only on the values of s in $N^{\leqslant r+t}(x)$. In particular, B_x and B_y can share a variable only if the distance between x and y is at most $2(r+t)$. Thus the maximum degree of the corresponding dependency graph $\mathcal{D}$ is at most the maximum size of a ball in $\mathcal{G}$ of radius $2(r+t)$. This is at most $o(n)$ since $r = O(\log n)$, t is a constant and $\mathcal{G}$ has subexponential growth. Also, the probability of each B_x (for a random uniform function $s : N^{\leqslant r+t}(x) \to m$) is at most $1/n$. Thus the Borel Local Lemma from [18] (which also works if the bits $b_{i,j}$ are m-ary instead of binary) gives that there is a Borel assignment $s : V \to m$ with no bad event B_x occurring. Then the evaluation of A on $(\mathcal{G}, s)$ satisfies the LCL C and is a Borel function by Lemma 5.17.

Thus any LCL that can be solved by a randomised LOCAL algorithm of radius $O(\log n)$ on graphs of order $n \to \infty$ admit Borel solutions on any Borel graph of subexponential growth. For some examples of such problems that are interesting from the point of view of Borel combinatorics, see Section 3 in [7]. One is the result of Molloy and Reed [65] who proved that, for $d \geqslant d_0$ with k_d being the maximum integer with $(k+1)(k+2) \leqslant d$, if each 1-ball in a graph G of maximum degree d can be properly coloured with $c \geqslant d - k_d$ colours, then in fact the whole graph G can be properly coloured with c colours. For large d, this is a far-reaching generalisation of Brooks' theorem [10] which, for $d \geqslant 3$, corresponds to the case $c = d$ of the above implication. Bamas and Esperet [2, 3] proved that, for a fixed large d, a c-colouring of an n-vertex graph G whose existence is guaranteed by the above result of Molloy and Reed, can in fact be found by a randomised LOCAL algorithm using $o(\log n)$ rounds. Putting all together, we conclude that every Borel graph $\mathcal{G}$ of maximum degree $d \geqslant d_0$ and subexponential growth has Borel chromatic number at most $\max(d - k_d, \chi(\mathcal{G}))$. Note that the last statement fails if we remove the growth assumption, even when we look at d-colourings only: indeed, the graph $\mathcal{G}$ given by Theorem 6.1 has Borel chromatic number $d + 1$ while the whole graph $\mathcal{G}$, as thus each of its 1-balls, is bipartite (as $\mathcal{G}$ has no cycles at all).

13 Applications to Equidecomposability

In order to demonstrate how some of the above results are applied, let us very briefly discuss the question of equidecomposability, where the methods of descriptive combinatorics have been recently applied with great success.

Two subsets A and B of $\mathbb{R}^n$ are called *equidecomposable* if it is possible to find a partition $A = A_0 \cup \cdots \cup A_{m-1}$ and isometries $\gamma_0, \ldots, \gamma_{m-1}$ of $\mathbb{R}^n$ so that $\gamma_0.A_0, \ldots, \gamma_{m-1}.A_{m-1}$ partition the other set B (or, in other words, we can split A into finitely many pieces and rearrange them using isometries to form a partition of B). The most famous result about equidecomposable sets is probably the *Banach-Tarski Paradox* [4]: in $\mathbb{R}^3$, the unit ball and two disjoint copies of the unit ball are equidecomposable.

Equidecompositions are often constructed to show that certain kinds of isometry-invariant means do not exist. For example, the Banach-Tarski Paradox implies that every finitely additive isometry-invariant mean defined on all bounded subsets of $\mathbb{R}^3$ must be identically 0. We refer the reader to the monograph by Tomkowicz and Wagon [78] on the subject.

The connection to descriptive combinatorics comes from a well-known observation that if one fixes the set of isometries $S = \{\gamma_0, \ldots, \gamma_{m-1}\}$ to be used, then an equidecomposition between A and B is equivalent to a perfect matching in the bipartite graph

$$G := \big(A \sqcup B, \{ (a,b) \in A \times B : \exists \gamma \in S \; \gamma.a = b \} \big). \tag{13.1}$$

A fairly direct application of Theorem 8.1 of Marks and Unger [59] gives a new proof of the important result of Dougherty and Foreman [20], whose original proof was very complicated, that doubling a ball in the Banach-Tarski Paradox can be done with pieces that have the property of Baire.

Of course, doubling a ball is impossible with Lebesgue measurable pieces because equidecompositions have to preserve the Lebesgue measure (as it is invariant under isometries). Interestingly, the obvious necessary condition for $A \subseteq \mathbb{R}^n$, $n \geqslant 3$, to be equidecomposable with Lebesgue measurable pieces to, say, the cube $[0,1]^n$ (namely, A is Lebesgue measurable of measure 1, and finitely many congruents of each of A and $[0,1]^n$ are enough to cover the other set) was shown by Grabowski, Máthé and Pikhurko [29] to be sufficient. The proof carefully chooses isometries $\gamma_0, \ldots, \gamma_{m-1}$, applies Theorem 9.1 of Lyons and Nazarov [55] to the graph G defined in (13.1) (after removing a null set from A and B to make these sets and the graph G Borel) and fixes the remaining null sets of unmatched vertices using the Axiom of Choice. The hardest part here was to show the existence of suitable isometries such that the bipartite graph G is a "measure expander", although the current version of [29, Section 6.5] points out a few different proofs of this step, all relying on some version of the spectral gap property.

The Borel version of the above result, say, if every two bounded Borel subsets of $\mathbb{R}^n$, $n \geqslant 3$, with non-empty interior and the same Lebesgue measure are equidecomposable with Borel pieces, remains open. (The examples by Laczkovich [47, 48] show that this statement is false for $n \leqslant 2$.) However, if the sets have "small" boundary then the following very strong results can be proved in every dimension;

in particular, they all apply to the famous *Circle Squaring Problem* of Tarski [76]. (We refer the reader to corresponding papers for all missing definitions.)

Theorem 13.1 *Let $n \geqslant 1$ and $A, B \subseteq \mathbb{R}^n$ be bounded sets with non-empty interior such that $\mu(A) = \mu(B)$ (where μ denotes the Lebesgue measure on $\mathbb{R}^n$) and $\dim_{\mathcal{M}}(\partial A), \dim_{\mathcal{M}}(\partial B) < n$ (i.e. their topological boundaries have upper Minkowski dimension less than n).*

1. (Laczkovich [46, 45]): *The sets A and B are equidecomposable using translations.*

2. (Grabowski, Máthé and Pikhurko [30]): *The sets A and B are equidecomposable using translations with pieces that are both Lebesgue and Baire measurable.*

3. (Marks and Unger [60]): *If, additionally, the sets A and B are Borel then they are equidecomposable using translations with Borel pieces.*

Very briefly, some of the key steps in the above results are as follows. All papers assume $A, B \subseteq [0,1)^n$ and work modulo 1 (i.e. inside the torus $\mathbb{T}^n := \mathbb{R}^n/\mathbb{Z}^n$). Given A and B, we first choose a large integer d and then some vectors $\boldsymbol{x}_0, \ldots, \boldsymbol{x}_{d-1} \in \mathbb{T}^n$ (a random choice will work almost surely). In particular, we assume that $\boldsymbol{x}_i$'s are linearly independent over the rationals. Let $\mathcal{G}$ be the Schreier graph of the natural action of the additive marked group $\Gamma \cong \mathbb{Z}^d$ generated by $S := \{\sum_{i \in d} \varepsilon_i \boldsymbol{x}_i : \varepsilon_i \in \{-1, 0, 1\}\}$ on the torus $\mathbb{T}^n$. In other words, $\mathcal{G}$ has $[0,1)^n$ for the vertex set with distinct $\boldsymbol{x}, \boldsymbol{y} \in [0,1)^n$ being adjacent if their difference modulo 1 belongs to S. Thus each component of $\mathcal{G}$ is a copy of the (3^d-1)-regular graph on $\mathbb{Z}^d$. We fix large N and look for a bijection $\phi : A \to B$ such that for every $a \in A$ the distance in the graph $\mathcal{G}$ between a and $\phi(a)$ is at most N. If we succeed, then we have equidecomposed the sets A and B using at most $(2N+1)^d$ parts.

The deep papers of Laczkovich [44, 46, 45] show that the assumption $\dim_{\mathcal{M}}(\partial A) < n$ translates into the set A being really well distributed inside each component of $\mathcal{G}$ and this property in turn shows that if N is large then the required bijection ϕ exists by Rado's theorem (Theorem 2.1). This crucially uses the Axiom of Choice (since Rado's theorem does).

The equidecompositions built in [30] come from some careful augmenting local algorithms of growing radii (tailored specifically to $\mathbb{Z}^d$-actions) and showing that the set where they do not stabilise is null (by the Borel–Cantelli Lemma) and meager (by adopting the proof of Theorem 8.1).

Marks and Unger [60] approached this problem in a novel way via real-valued flows in the graph $\mathcal{G}$. The advantage of working with flows (versus matchings) is that, for example, any convex combination of feasible flows is again a feasible flow. First, Marks and Unger showed that there is a real-valued uniformly bounded Borel flow f from A to B (which can be viewed as a fractional version of the required bijection ϕ). Secondly, they proved that one can *round* f to a flow h (which has the same properties as f and, additionally, assumes only integer values). Finally, they showed that the flow h can be converted into the desired Borel bijection $\phi : A \to B$ via a local rule. In this approach the second step is the most difficult one. This step relies on the unpublished result of Gao, Jackson, Krohne, and Seward announced in [28] (for a proof see [60, Theorem 5.5]) that $\mathcal{G}$ (or, more generally, the Schreier graph coming from any free Borel action of $\mathbb{Z}^d$) admits a Borel family $\mathcal{C}$ of finite connected vertex sets that cover all vertices and whose boundaries in $\mathcal{G}$ are sufficiently far from each other. (In other words, the family $\mathcal{C}$ provides a certificate of hyperfiniteness of $\mathcal{E}_\mathcal{G}$ that has an extra boundary separation property.) One can arrange $\mathcal{C}$ to arrive in ω-many stages so that, for any newly arrived set $C \in \mathcal{C}$, every previous set is either deep inside C or far away from C. The rounding algorithm from the point of view of any vertex x is to wait until some $C \in \mathcal{C}$ containing x arrives and then round all f-values inside C in agreement with the other vertices of C, making sure not to override any rounding made inside any earlier $C' \in \mathcal{C}$ with $C' \subseteq C$. Thus the final integer-valued flow h can be computed by a finitary rule on $\mathcal{G}$ (depending of the real-valued Borel flow f) and is Borel by a version of Corollary 5.20.

More recently, Máthé, Noel and Pikhurko [63] strengthened the results from [60] by proving that, additionally, we can require that the pieces themselves have boundary of upper Minkowski dimension less than n (and, in particular, are Jordan measurable). Also, it is shown in [63] that if the sets A and B in Theorem 13.1 are, say, open then each piece can additionally be a Boolean combination of F_σ-sets (i.e. countable unions of closed sets). These improvements started with the new result that, in order to find an integer-valued flow h inside some $C \in \mathcal{C}$ that will be compatible with all future rounding steps, we need to know only some local information, namely, the ball around C of sufficiently large but finite radius. Thus the new rounding algorithm does not need to know the flow f (which may depend on the whole component of $\mathcal{G}$).

14 Concluding Remarks

Due to the limitation on space, we have just very briefly touched on some very exciting topics, with each of Sections 7–9 deserving a separate introductory paper (perhaps even a few, as is the case of μ-measurable combinatorics). Also, there are some other topics that we have not even mentioned (the Borel hierarchy, analytic graphs and equivalence relations; treeability; the cost of a measure-preserving group action; combinatorial cost; classical/sofic entropy and other invariants; the local-global convergence of bounded degree graphs; "continuous combinatorics" on zero-dimensional Polish spaces, etc). Also, we have presented hardly any open questions; we refer the reader to e.g. the survey by Kechris and Marks [36] that contains quite a few of them.

Nonetheless, the author hopes that this paper will be helpful in introducing more researchers to this dynamic (in both meanings of the word) and exciting area.

Acknowledgements

The author is grateful to Jan Grebík, Václav Rozhoň and the anonymous referee for the very useful comments and/or discussions.

References

[1] M. Abért, D. Gaboriau, and A. Thom, *Measured group theory*, Oberwolfach Reports **13** (2017), no. 3, 2347–2397.

[2] É. Bamas and L. Esperet, *Distributed coloring of graphs with an optimal number of colors*, E-print arxiv:1809.08140, 2018.

[3] _____, *Distributed coloring of graphs with an optimal number of colors*, 36th International Symposium on Theoretical Aspects of Computer Science, LIPIcs. Leibniz Int. Proc. Inform., vol. 126, Schloss Dagstuhl. Leibniz-Zent. Inform., Wadern, 2019, pp. Art. No. 10, 15pp.

[4] S. Banach and A. Tarski, *Sur la décomposition des ensembles de points en parties respectivement congruentes.*, Fund. Math. **6** (1924), 244–277.

[5] L. Barenboim and M. Elkin, *Distributed graph coloring*, Synthesis Lectures on Distributed Computing Theory, vol. 11, Morgan & Claypool Publishers, Williston, VT, 2013.

[6] A. Bernshteyn, *Measurable versions of the Lovász Local Lemma and measurable graph colorings*, Adv. Math. **353** (2019), 153–223.

[7] _____, *Distributed algorithms, the Lovász Local Lemma, and descriptive combinatorics*, E-print arxiv:2004.04905v3, 2020.

[8] _____, *Local colouring problems on smooth graphs*, E-print arxiv:2012.11031, 2020.

[9] A. Bernshteyn and C. T. Conley, *Equitable colorings of Borel graphs*, E-print arxiv:1908.10475, 2019.

[10] R. L. Brooks, *On colouring the nodes of a network*, Proc. Cambridge Philos. Soc. **37** (1941), 194–197.

[11] W. Chan, *Equivalence relations which are Borel somewhere*, J. Symb. Logic **82** (2017), 893–930.

[12] _____, *Constructibility level equivalence relation*, Manuscript, 2pp, available at https://williamchan-math.github.io, 2020.

[13] Y.-J. Chang, T. Kopelowitz, and S. Pettie, *An exponential separation between randomized and deterministic complexity in the local model*, SIAM J. Computing **48** (2019), 122–143.

[14] D. L. Cohn, *Measure theory*, second ed., Birkhäuser Advanced Texts: Basel Textbooks, Birkhäuser/Springer, New York, 2013.

[15] C. Conley, S. Jackson, A. S. Marks, B. Seward, and R. Tucker-Drob, *Borel asymptotic dimension and hyperfinite equivalence relations*, E-print arxiv:2009.06721, 2020.

[16] C. Conley, S. Jackson, A. S. Marks, B. Seward, and R. Tucker-Drob, *Hyperfiniteness and Borel combinatorics*, J. Europ. Math. Soc **22** (2020), 877–892.

[17] C. T. Conley and O. Tamuz, *Unfriendly colorings of graphs with finite average degree*, Proc. London Math. Soc. **121** (2020), 828–832.

[18] E. Csóka, L. Grabowski, A. Máthé, O. Pikhurko, and K.Tyros, *Borel version of the local lemma*, E-print arxiv:1605.04877, 2016.

[19] R. Diestel, *Graph theory*, 5th ed., Springer, Berlin, 2017.

[20] R. Dougherty and M. Foreman, *Banach-Tarski decompositions using sets with the property of Baire*, J. Amer. Math. Soc. **7** (1994), 75–124.

[21] G. Elek and G. Lippner, *Borel oracles. An analytical approach to constant-time algorithms*, Proc. Amer. Math. Soc. **138** (2010), 2939–2947.

[22] P. Erdős and L. Lovász, *Problems and results on 3-chromatic hypergraphs and some related questions*, Infinite and Finite Sets (A. Hajnal et al, ed.), Colloq. Math. Soc. János. Bolyai, vol. 11, North-Holland, Amsterdam, 1975, pp. 609–627.

[23] J. Feldman and C. C. Moore, *Ergodic equivalence relations, cohomology, and von Neumann algebras. I*, Trans. Amer. Math. Soc. **234** (1977), 289–324.

[24] D. H. Fremlin, *Measure theory. Vol. 2*, Torres Fremlin, Colchester, 2003, Broad foundations, Corrected second printing of the 2001 original.

[25] A. Furman, *A survey of measured group theory*, Geometry, rigidity, and group actions, Chicago Lectures in Math., Univ. Chicago Press, Chicago, IL, 2011, pp. 296–374.

[26] D. Gaboriau, *Orbit equivalence and measured group theory*, Proceedings of the International Congress of Mathematicians. Volume III, Hindustan Book Agency, New Delhi, 2010, pp. 1501–1527.

[27] S. Gao, *Invariant descriptive set theory*, Pure and Applied Mathematics (Boca Raton), vol. 293, CRC Press, Boca Raton, FL, 2009.

[28] S. Gao, S. Jackson, E. Krohne, and B. Seward, *Forcing constructions and countable borel equivalence relations*, E-print arxiv:1503.07822, 2015.

[29] Ł. Grabowski, A. Máthé, and O. Pikhurko, *Measurable equidecompositions for group actions with an expansion property*, accepted by J. Eur. Math. Soc., E-print arxiv:1601.02958, 2016.

[30] _____, *Measurable circle squaring*, Annals of Math. (2) **185** (2017), 671–710.

[31] J. Grebík and O. Pikhurko, *Measurable versions of Vizing's theorem*, Adv. Math. **374** (2020), 107378, 40.

[32] A. Hajnal and E. Szemerédi, *Proof of a conjecture of P. Erdős*, Combinatorial theory and its applications, II (Proc. Colloq., Balatonfüred, 1969), North-Holland, Amsterdam, 1970, pp. 601–623.

[33] G. Hjorth, *Borel equivalence relations*, Handbook of set theory. Vols. 1, 2, 3, Springer, Dordrecht, 2010, pp. 297–332.

[34] A. S. Kechris, *Classical descriptive set theory*, Graduate Texts in Mathematics, vol. 156, Springer-Verlag, New York, 1995.

[35] _____, *The theory of countable Borel equivalence relations*, Manuscript, 156pp, available at http://www.math.caltech.edu/~kechris/papers/lectures on CBER03.pdf, 2019.

[36] A. S. Kechris and A. S. Marks, *Descriptive graph combinatorics*, Manuscript, 112pp, available at http://www.math.caltech.edu/~kechris/papers/combinatorics20.pdf, 2019.

[37] A. S. Kechris and B. D. Miller, *Topics in orbit equivalence*, Lecture Notes in Mathematics, vol. 1852, Springer, 2004.

[38] A. S. Kechris, S. Solecki, and S. Todorcevic, *Borel chromatic numbers*, Adv. Math. **141** (1999), 1–44.

[39] D. Kerr and H. Li, *Ergodic theory*, Springer Monographs in Mathematics, Springer, Cham, 2016, Independence and dichotomies.

[40] H. A. Kierstead, A. V. Kostochka, M. Mydlarz, and E. Szemerédi, *A fast algorithm for equitable coloring*, Combinatorica **30** (2010), 217–224.

[41] P. Komjáth, *The chromatic number of infinite graphs—a survey*, Discrete Math. **311** (2011), 1448–1450.

[42] G. Kun, *Expanders have a spanning Lipschitz subgraph with large girth*, E-print arxiv:1303.4982, 2013.

[43] M. Laczkovich, *Closed sets without measurable matching*, Proc. Amer. Math. Soc. **103** (1988), 894–896.

[44] _____, *Equidecomposability and discrepancy; a solution of Tarski's circle-squaring problem*, J. Reine Angew. Math. **404** (1990), 77–117.

[45] _____, *Decomposition of sets with small boundary*, J. Lond. Math. Soc. **46** (1992), 58–64.

[46] _____, *Uniformly spread discrete sets in* $\mathbf{R}^d$, J. Lond. Math. Soc. **46** (1992), 39–57.

[47] _____, *Decomposition of sets of small or large boundary*, Mathematika **40** (1993), 290–304.

[48] _____, *Equidecomposability of Jordan domains under groups of isometries*, Fund. Math. **177** (2003), 151–173.

[49] N. Linial, *Distributive graph algorithms global solutions from local data*, 28th Annual Symposium on Foundations of Computer Science, IEEE, 1987, pp. 331–335.

[50] _____, *Locality in distributed graph algorithms*, SIAM J. Computing **21** (1992), 193–201.

[51] C. Löh, *Ergodic theoretic methods in group homology: A minicourse on L^2-Betti numbers in group theory*, Springer, 2020.

[52] L. Lovász, *Large networks and graph limits*, Colloquium Publications, Amer. Math. Soc., 2012.

[53] A. Lubotzky, *Discrete groups, expanding graphs and invariant measures*, Modern Birkhäuser Classics, Birkhäuser Verlag, Basel, 2010, With an appendix by Jonathan D. Rogawski, Reprint of the 1994 edition.

[54] R. Lyons, *Factors of IID on trees*, Combin. Probab. Computing **26** (2017), 285–300.

[55] R. Lyons and F. Nazarov, *Perfect matchings as IID factors on non-amenable groups*, European J. Combin. **32** (2011), 1115–1125.

[56] R. Lyons and Y. Peres, *Probability on trees and networks*, Cambridge Series in Statistical and Probabilistic Mathematics, vol. 42, Cambridge University Press, New York, 2016.

[57] A. S. Marks, *A short proof that an acyclic n-regular Borel graph may have Borel chromatic number $n + 1$*, Unpublished manuscript, 1 page, available at http://math.ucla.edu/ ~marks/notes/003-short_coloring.pdf, 2013.

[58] A. S. Marks, *A determinacy approach to Borel combinatorics*, J. Amer. Math. Soc. **29** (2016), 579–600.

[59] A. S. Marks and S. Unger, *Baire measurable paradoxical decompositions via matchings*, Adv. Math. **289** (2016), 397–410.

[60] _____, *Borel circle squaring*, Annals of Math. (2) **186** (2017), 581–605.

[61] D. A. Martin, *Borel determinacy*, Annals of Math. (2) **102** (1975), 363–371.

[62] _____, *A purely inductive proof of Borel determinacy*, Recursion theory (Ithaca, N.Y., 1982), Proc. Sympos. Pure Math., vol. 42, Amer. Math. Soc., Providence, RI, 1985, pp. 303–308.

[63] A. Máthé, J. Noel, and O. Pikhurko, *Circle squarings with pieces of small boundary and low Borel complexity*, In preparation, 2021.

[64] B. D. Miller, *Ends of graphed equivalence relations. I*, Israel J. Math. **169** (2009), 375–392.

[65] M. Molloy and B. Reed, *Colouring graphs when the number of colours is almost the maximum degree*, J. Combin. Theory (B) (2014), 134–195.

[66] R. A. Moser, *A constructive proof of the Lovász local lemma*, STOC'09—Proceedings of the 2009 ACM International Symposium on Theory of Computing, ACM, New York, 2009, pp. 343–350.

[67] R. A. Moser and G. Tardos, *A constructive proof of the general Lovász local lemma*, J. ACM **57** (2010), Art. 11, 15pp.

[68] M. G. Nadkarni, *On the existence of a finite invariant measure*, Proc. Indian Acad. Sci. Math. Sci. **100** (1990), 203–220.

[69] J. C. Oxtoby, *Measure and category*, second ed., Graduate Texts in Mathematics, vol. 2, Springer-Verlag, New York, 1980.

[70] R. Rado, *A theorem on independence relations*, Quart. J. Math. Oxford **13** (1942), 83–89.

[71] Y. Shalom, *Measurable group theory*, European Congress of Mathematics, Eur. Math. Soc., Zürich, 2005, pp. 391–423.

[72] J. B. Shearer, *On a problem of Spencer*, Combinatorica **5** (1985), 241–245.

[73] S. Shelah and E. C. Milner, *Graphs with no unfriendly partitions*, A tribute to Paul Erdős, Cambridge Univ. Press, Cambridge, 1990, pp. 373–384.

[74] T. A. Slaman and J. R. Steel, *Definable functions on degrees*, Cabal Seminar 81–85, Lecture Notes in Math., vol. 1333, Springer, Berlin, 1988, pp. 37–55.

[75] S. M. Srivastava, *A course on Borel sets*, Graduate Texts in Mathematics, vol. 180, Springer-Verlag, New York, 1998.

[76] A. Tarski, *Problème 38*, Fund. Math. **7** (1925), 381.

[77] R. Thornton, *Orienting Borel graphs*, E-print arxiv:2001.01319, 2020.

[78] G. Tomkowicz and S. Wagon, *The Banach-Tarski paradox*, 2d ed., Cambridge University Press, 2016.

[79] A. Tserunyan, *Introduction to descriptive set theory*, Available at https://faculty.math.illinois.edu/~anush/lecture_notes.html, Version from 12 November 2019, 2019.

[80] M. Viana and K. Oliveira, *Foundations of ergodic theory*, Cambridge Studies in Advanced Mathematics, vol. 151, Cambridge Univ. Press, 2016.

[81] F. Weilacher, *Descriptive chromatic numbers of locally finite and everywhere two ended graphs*, E-print arxiv:2004.02316, 2020.

[82] B. Weiss, *Measurable dynamics*, Conference in modern analysis and probability (New Haven, Conn., 1982), Contemp. Math., vol. 26, Amer. Math. Soc., Providence, RI, 1984, pp. 395–421.

Mathematics Institute and DIMAP
University of Warwick
Coventry CV4 7AL, UK

Codes and designs in Johnson graphs with high symmetry

Cheryl E. Praeger

Abstract

The Johnson graph $J(v,k)$ has, as vertices, all k-subsets of a v-set $\mathcal{V}$, with two k-subsets adjacent if and only if they share $k-1$ common elements of $\mathcal{V}$. Subsets of vertices of $J(v,k)$ can be interpreted as the blocks of an incidence structure, or as the codewords of a code, and automorphisms of $J(v,k)$ leaving the subset invariant are then automorphisms of the corresponding incidence structure or code. This approach leads to interesting new designs and codes. For example, numerous actions of the Mathieu sporadic simple groups give rise to examples of Delandtsheer designs (which are both flag-transitive and anti-flag transitive), and codes with large minimum distance (and hence strong error-correcting properties). The paper surveys recent progress, explores links between designs and codes in Johnson graphs which have a high degree of symmetry, and discusses several open questions.

Key-words: designs, codes in graphs, Johnson graph, 2-transitive permutation group, neighbour-transitive, Delandtsheer design, flag-transitive, antiflag-transitive.

Mathematics Subject Classification (2010): 05C25, 20B25, 94B60.

1 Introduction

The Johnson graphs are ubiquitous in mathematics, perhaps because of their many useful properties. They are distance transitive, and indeed geodesic-transitive, they underpin the Johnson association schemes – and 'everyone's favourite graph', the Petersen Graph, occurs as the complement of one of them[1]. Also, the class of Johnson graphs played a key role in Babai's recent breakthrough [2] to a quasipolynomial bound on the complexity of graph isomorphism testing[2].

Our focus in this paper is studying Johnson graphs as 'carrier spaces' of error-correcting codes and combinatorial designs, and in particular using group theory to find surprisingly rich families of examples of codes and designs with 'high symmetry'.

The Johnson graph $J(\mathcal{V}, k)$ (or $J(v,k)$) is based on a set $\mathcal{V}$ of v elements, called *points*. Its vertices are the k-subsets of $\mathcal{V}$, and distinct k-subsets are adjacent precisely when their intersection has size $k-1$. For distinct k-subsets γ, γ', their distance $d(\gamma, \gamma')$ (length of shortest path from γ to γ') in $J(\mathcal{V}, k)$ is therefore $k - |\gamma \cap \gamma'|$. The *complementing map* τ which maps each k-subset of $\mathcal{V}$ to its complement induces a graph isomorphism $\tau : J(\mathcal{V}, k) \to J(\mathcal{V}, v-k)$. For this reason we may, and sometimes we do, replace $J(\mathcal{V}, k)$ by $J(\mathcal{V}, v-k)$ in our analysis and thereby assume that $k \leqslant v/2$. The symmetric group $\text{Sym}(\mathcal{V})$ acts as automorphisms on $J(v,k)$, and if $k \neq v/2$ then $\text{Sym}(\mathcal{V})$ is the full automorphism group, while if $k = v/2$ then the

---
[1] The Petersen graph is the complement of the Johnson graph $J(5,2)$; a whole book has been written about it [17].
[2] Babai, whose work relies on both group-theoretic and combinatorial techniques, found that "in a well-defined sense, the Johnson graphs are the only obstructions to effective canonical partitioning". See also Helfgott's lecture [15].

automorphism group is $\mathrm{Sym}(\mathcal{V}) \times \langle \tau \rangle$, (see, for example, [28] for $k \neq v/2$ and [12] for $k = v/2$). We will work with symmetry provided by $\mathrm{Sym}(\mathcal{V})$.

The link between $J(\mathcal{V}, k)$ and combinatorial designs is fairly clear: points of the design are elements of $\mathcal{V}$, and if each block of the design is incident with exactly k points, then the blocks are vertices of $J(\mathcal{V}, k)$. Similarly, a code in $J(\mathcal{V}, k)$ is a subset of vertices. We next make some further brief comments on each of these links in turn.

1.1 Codes in Johnson graphs

In 1973, Philippe Delsarte [7] introduced the notion of a code in a distance-regular graph: a vertex subset $\mathcal{C}$ is considered to be a code, its elements are the codewords, and distance between codewords is the natural distance in the graph. In particular the *minimum distance* $\delta(\mathcal{C})$ of $\mathcal{C}$ is the minimum length of a path between distinct codewords, and the *automorphism group* $\mathrm{Aut}(\mathcal{C})$ is the setwise stabiliser of $\mathcal{C}$ in the automorphism group of the graph.

Delsarte defined a special type of code, now called a completely-regular code, 'which enjoys combinatorial (and often algebraic) symmetry akin to that observed for perfect codes' (see [24, page 1], and also Section 2), and he posed explicitly the question of existence of completely-regular codes in Johnson graphs. Such codes in Johnson graphs were studied by Meyerowitz [25, 26] and Martin [22, 23], but disappointingly, not many were found with good error-correcting properties (large distance between distinct codewords). In joint work with Liebler [21], the stringent regularity conditions imposed for complete regularity, were replaced by a 'local transitivity' property: a *neighbour-transitive code* in $J(\mathcal{V}, k)$ was defined as a vertex-subset $\mathcal{C}$ such that $\mathrm{Aut}(\mathcal{C})$ (*which we will take here as the setwise stabiliser of $\mathcal{C}$ in $\mathrm{Sym}(\mathcal{V})$, even if $k = v/2$*) is transitive both on $\mathcal{C}$ and on the set $\mathcal{C}_1$ of 'code-neighbours' (the non-codewords which are adjacent in $J(\mathcal{V}, k)$ to some codeword). If $\delta(\mathcal{C}) \geqslant 3$, for a neighbour-transitive code $\mathcal{C}$, it turns out (see [21, Theorem 1.2]) that $\mathrm{Aut}(\mathcal{C})$ has an even stronger property, namely, $\mathrm{Aut}(\mathcal{C})$ is transitive on the set of triples

$$(u, u', \gamma) \text{ where } \gamma \in \mathcal{C},\ u \in \gamma,\ \text{and}\ u' \in \mathcal{V} \setminus \gamma. \tag{1.1}$$

A code $\mathcal{C}$ with this property is called *strongly-incidence-transitive*, and we will say that $\mathcal{C}$ *has the SIT-property* and is an *SIT-code*. We will describe surprisingly rich classes of SIT-codes in Johnson graphs arising from both combinatorial and geometric constructions, and we will mention some open problems.

We observe that a code consisting of a single codeword γ trivially has the SIT-property, since its automorphism group is $\mathrm{Sym}(\gamma) \times \mathrm{Sym}(\mathcal{V} \setminus \gamma)$. To avoid such trivial cases, we will always therefore assume that *SIT-codes $\mathcal{C}$ have size at least* 2. We call such codes *nontrivial*. 'Nontriviality' still allows the following examples:

- $k = v/2$ and $\mathcal{C}$ consists of a single k-subset γ together with its complement. Here the automorphism group is $\mathrm{Sym}(\gamma) \wr \mathrm{Sym}(2)$, it has the SIT-property and is transitive on $\mathcal{V}$.

- $\mathcal{C} = \binom{\mathcal{V}}{k}$ is the *complete code* containing all k-subsets of $\mathcal{V}$. Here the automorphism group $\mathrm{Sym}(\mathcal{V})$ has the SIT-property and of course is transitive on $\mathcal{V}$.

Let $\mathcal{C}$ be a nontrivial SIT-code in $J(v,k)$. If $k = v/2$ and $\mathcal{C}$ consists of a single k-subset and its complement, then as we noted above its automorphism group is transitive on $\mathcal{V}$. In all other cases there exist distinct codewords $\gamma, \gamma' \in \mathcal{C}$ with $\gamma' \neq \mathcal{V} \setminus \gamma$. Then for $G = \text{Aut}(\mathcal{C})$, the stabiliser G_γ has two orbits in $\mathcal{V}$, namely γ and $\mathcal{V} \setminus \gamma$, and similarly $G_{\gamma'}$ is transitive on γ' and $\mathcal{V} \setminus \gamma'$. Also, at least one of γ', $\mathcal{V} \setminus \gamma'$ meets both γ and $\mathcal{V} \setminus \gamma$, and hence $\langle G_\gamma, G_{\gamma'} \rangle$ is transitive on $\mathcal{V}$. It follows from this discussion that every nontrivial SIT-code has automorphism group transitive on $\mathcal{V}$. We record this fact, and also a property on the minimum distance proved in [21].

Lemma 1.1 *[21, Lemma 2.1]* *Let $\mathcal{C}$ be an SIT-code in $J(\mathcal{V}, k)$. If $|\mathcal{C}| \geq 2$, then $\text{Aut}(\mathcal{C})$ is transitive on $\mathcal{V}$. If $\mathcal{C}$ is not the complete code $\binom{\mathcal{V}}{k}$, then the minimum distance $\delta(\mathcal{C}) \geq 2$.*

1.2 Designs in Johnson graphs: Delandtsheer designs

As mentioned above a vertex-subset $\mathcal{C}$ of $J(\mathcal{V}, k)$ can be interpreted as the block-set of a design. We will usually assume some additional regularity properties on $\mathcal{C}$. For $1 \leq t < k < v$, a $t-(v, k, \lambda)$ *design* $\mathcal{D} = (\mathcal{V}, \mathcal{C})$ consists of a point-set $\mathcal{V}$ of size v, and a subset $\mathcal{C}$ of k-subsets of $\mathcal{V}$ (called blocks) with the property that each t-subset of $\mathcal{V}$ is contained in exactly λ blocks. A point-block pair (u, γ) is called a *flag*, or an *antiflag* according as $u \in \gamma$ or $u \notin \gamma$, respectively. If $t = 2$ and $\lambda = 1$, then $\mathcal{D}$ is called a *linear space*. The automorphism group $\text{Aut}(\mathcal{D})$ is the setwise stabiliser of $\mathcal{C}$ in $\text{Sym}(\mathcal{V})$ – hence the same group occurs whether we regard $\mathcal{C}$ as a code or the block set of a design.

In 1984, Delandtsheer [5] classified all antiflag-transitive linear spaces $\mathcal{D} = (\mathcal{V}, \mathcal{C})$, proving that the linear spaces with this property are the projective and affine spaces, the Hermitian unitals, and two exceptional examples. In her paper she noticed that all, apart from the two exceptional examples, possessed a stronger property, namely the block set $\mathcal{C}$ has the SIT-property when viewed as a code in $J(\mathcal{V}, k)$. Requiring $\mathcal{C}$ to have the SIT-property implies both flag-transitivity and antiflag-transitivity of $\mathcal{D}$, and it is even more restrictive, as illustrated by the two exceptions in [5]. Here is a simpler example: the unique $1-(4, 2, 2)$ design, which is the edge set of a 4-cycle in the complete graph $K_4 = J(4, 2)$, is flag-transitive and antiflag-transitive, but does not have the SIT property. It is only very recently that we noticed that the SIT-property for codes coincides with this property observed by Delandtsheer for antiflag-transitive linear spaces. We want to consider designs, not necessarily linear spaces, with the SIT-property. Trivially, if $\mathcal{C}$ contains only one block then, as we remarked above, $\mathcal{C}$ has the SIT-property. We avoid this case and consider only designs with more than one block. We therefore (appropriately) refer to any design $\mathcal{D} = (\mathcal{V}, \mathcal{C})$ in $J(\mathcal{V}, k)$ as a *Delandtsheer design* if $\mathcal{C}$ has the SIT-property and $|\mathcal{C}| \geq 2$.

Thus each new Delandtsheer design corresponds to a new SIT code, and conversely.

1.3 Imprimitive examples and a dichotomy

For each nontrivial strongly incidence-transitive (SIT) code in $J(\mathcal{V},k)$, the automorphism group is transitive on $\mathcal{V}$ and apart from the complete code, the minimum distance is at least 2, see Lemma 1.1. In Section 2 we show how to produce an infinite family of SIT-codes via a *blow-up* procedure: Construction 2.1 provides a general method for building SIT-codes and Delandtsheer designs $\mathcal{C}$ from smaller codes/designs $\mathcal{C}_0$ possessing the SIT-property. As long as the input code $\mathcal{C}_0$ has size at least two, the output 'blow-up' code $\mathcal{C}$ admits as automorphism group a full wreath product in its transitive imprimitive action. Moreover $\mathcal{C}$ is the block set of a Delandtsheer 1-design, but never of a 2-design, see Lemmas 2.2 and 2.3.

Importantly, Proposition 2.4 shows that the imprimitive SIT-codes are precisely those arising from Construction 2.1, and by Theorem 2.5, each imprimitive SIT-code can be obtained by applying Construction 2.1 to a primitive SIT-code.

Despite this structural link between the imprimitive and primitive strongly incidence-transitive codes, the behaviour of these two sub-families of examples is vastly different, especially from the viewpoint of designs. To begin with, the primitive examples have very restricted automorphism groups.

Theorem 1.2 *[21, Theorem 1.2] Let $\mathcal{C}$ be a strongly incidence-transitive code in $J(\mathcal{V},k)$ such that $|\mathcal{C}| \geqslant 2$, and suppose that $\mathrm{Aut}(\mathcal{C})$ is primitive on $\mathcal{V}$. Then $\mathrm{Aut}(\mathcal{C})$ is 2-transitive on $\mathcal{V}$.*

A subgroup $G \leqslant \mathrm{Sym}(\mathcal{V})$ is called *2-transitive* if it is transitive on the ordered pairs of distinct points from $\mathcal{V}$. We note that Theorem 1.2 is valid even if the condition $2 \leqslant k \leqslant v-2$ in [21, Theorem 1.2] is removed, since $k=1$ or $k=v-1$ implies that $\mathcal{C} = \binom{\mathcal{V}}{k}$ is a complete code with 2-transitive automorphism group $\mathrm{Aut}(\mathcal{C}) = \mathrm{Sym}(\mathcal{V})$.

If a design $(\mathcal{V},\mathcal{C})$ in $J(\mathcal{V},k)$ has automorphism group 2-transitive on $\mathcal{V}$ then there is a constant λ such that each point-pair is contained in exactly λ blocks, that is, $(\mathcal{V},\mathcal{C})$ is a $2-(v,k,\lambda)$ design. Thus Theorem 1.2 leads to a striking dichotomy between the imprimitive and primitive Delandtsheer designs:

Theorem 1.3 *Let $\mathcal{D} = (\mathcal{V},\mathcal{C})$ be a Delandtsheer design in $J(\mathcal{V},k)$ with automorphism group G. If G is primitive on $\mathcal{V}$ then $\mathcal{D}$ is a $2-(v,k,\lambda)$ design, for some λ. On the other hand if G is not primitive on $\mathcal{V}$, then $\mathcal{D}$ is a $1-(v,k,\lambda)$ design, for some λ, but is never a 2-design.*

1.4 Primitive examples: towards a classification

Thus understanding the primitive SIT-codes and Delandtsheer designs, is of central importance. Numerous actions of the sporadic Mathieu groups [27] have been shown to give rise to examples of Delandtsheer designs which, when viewed as SIT-codes in Johnson graphs, have large minimum distances and hence strong error-correcting properties. Other natural geometrical examples of Delandtsheer designs come from subspaces and classical unitals [21] (not all of them linear spaces), binary quadratic forms [18], and more exotic geometrical examples linked to cones, cylinders and maximal arcs have been constructed by Durante [11]. Yet more examples

	Group	Degree
Mathieu	M_n	$n \in \{11, 12, 22, 23, 24\}$, or
	M_{11}	12
alternating	A_7	15
projective	$PSL(2, 11)$	11
sporadic	HS	176, or
	Co_3	276

Table 1: 'Sporadic' 2-transitive permutation groups

projective	$PSL(n, q) \leqslant G \leqslant P\Gamma L(n, q)$ on $PG(n - 1, q)$
rank 1	the Suzuki, Ree and Unitary groups
affine	$G \leqslant A\Gamma L(\mathcal{V})$ acting on $\mathcal{V} = \mathbb{F}_q^n$
symplectic	$Sp(2n, 2)$ on $Q^\varepsilon(2)$, $\varepsilon = \pm$

Table 2: Infinite families of 2-transitive permutation groups

arise from symmetric 2-designs [18]. We discuss some of these examples and give references for locating other known examples.

The fact that all finite 2-transitive permutation groups are known explicitly, as a consequence of the finite simple group classification (see for example [10, Chapter 7.7]), suggests that an obvious strategy for finding all SIT-codes and Delandtsheer designs is to analyse all these groups, or families of groups, one by one. Significant progress has been made, which we will outline. We also point out some major open cases.

We may subdivide the finite 2-transitive permutation groups according to whether or not they lie in an infinite family of 2-transitive groups. Those which do not lie in an infinite family we call *sporadic*; these are listed in Table 1. Note that the Mathieu group M_{11} occurs in both line 1 and line 2 of Table 1, corresponding to its two 2-transitive actions of degrees 11 and 12 (with point stabilisers M_{10} and $PSL(2, 11)$), respectively. The infinite families of finite 2-transitive groups G, not containing the alternating group, are listed in Table 2.

When determining examples,

we will from now on assume that $3 \leqslant k \leqslant v/2$.

We may do this because, if $k = 1, 2, v - 2$ or $v - 1$, then the 2-transitive group G is transitive on k-subsets of $\mathcal{V}$, and hence the only G-SIT-code in $J(v, k)$ is the complete code consisting of all k-subsets of $\mathcal{V}$. Also if $v/2 < k \leqslant v - 3$ and $\mathcal{C}$ is an SIT-code in $J(v, k)$, then $\{\mathcal{V} \setminus \gamma \mid \gamma \in \mathcal{C}\}$ is an SIT-code in $J(v, v - k)$ with the same automorphism group as $\mathcal{C}$, and we have $3 \leqslant v - k < v/2$. In these cases for each G-SIT-code $\mathcal{C}$ in $J(v, k)$, the pair $(\mathcal{V}, \mathcal{C})$ is a Delandtsheer $2 - (v, k, \lambda)$ design, for some λ. To determine the value of λ we first note that the number of blocks is $b = |\mathcal{C}| = |G : G_\gamma|$ for $\gamma \in \mathcal{C}$. Then, counting the number of triples (u, u', γ) with γ a block and u, u' distinct points of γ, gives

$$\lambda = \frac{bk(k-1)}{v(v-1)}. \tag{1.2}$$

Analysis of the sporadic cases was completed in [27], yielding 27 strongly-incidence-transitive (code, group) pairs in $J(v,k)$ with $3 \leqslant k \leqslant v/2$, which we list in Table 3 in Section 3. Table 3 also gives details of the corresponding Delandtsheer designs, many but not all of which are well-known. They include the Steiner systems from the Mathieu groups and the 11-point biplane.

This leaves the infinite families of 2-transitive groups G to be considered. We discuss the 'projective' and 'rank 1' lines of Table 2 in Section 4. These cases are completely resolved using work in [11, 21]. The examples are well known, and comprise the classical unital, designs of projective subspaces of given dimension, designs of Baer sublines of a projective line, and one exceptional example, see Table 4.

The 'affine' line of Table 2 deserves special mention because the classification here is still open. As in the projective linear case, we obtain examples by taking the set of affine subspaces of given dimension. Also, in [11, Section 3.2], Durante constructed additional infinite families of examples geometrically in affine spaces over fields of order $q \in \{4, 16\}$: from cylinders in $\mathrm{AG}(n,4)$ with base the hyperoval in $\mathrm{PG}(2,4)$ or its complement, and from unions of two or four parallel hyperplanes, for $q = 4$ or $q = 16$, respectively. Moreover, Durante [11, Theorem 27] showed that when $q \in \{4, 16\}$ these are the only examples with $3 \leqslant k \leqslant v/2$. By the results in [21, Section 6] and [11] it was believed that the classification of the affine SIT-codes was complete. However Mark Ioppolo discovered an error in the proof of [21, Proposition 6.6] in the case $q = 2$. He showed [18, Lemma A.1] that the analysis in [11, 21] completed the classification for affine spaces over all fields of size at least 3, and produced [18, Example 7.11, Theorem 7.13], two additional infinite families of affine SIT-codes and Delandtsheer designs with $q = 2$. They are related to the symplectic symmetric 2-designs arising in Kantor's classification of 2-transitive symmetric designs [20]. Ioppolo [18, Theorem 7.9] also showed how to use the blowup Construction 2.1 to produce larger examples admitting an affine group (not 2-transitive). The known affine examples are summarised in Table 5 in Section 5, and in particular we discuss there what is required to resolve the following problem.

Problem 1 *Complete the classification of SIT-codes and Delandtsheer designs admitting a 2-transitive affine group over a field of order 2.*

The last infinite family of 2-transitive groups, in the 'symplectic' line of Table 2, corresponds to the Jordan–Steiner actions of the symplectic groups $G = \mathrm{Sp}(2n, 2)$ on nondegenerate quadratic forms. Investigating this was the major topic of Ioppolo's thesis [18], and we discuss his findings in Section 6. He identifies two distinct infinite families of examples arising from subspace actions of the symplectic groups [18, Chapter 4], and shows that, for any further examples of G-SIT codes, the stabiliser of a codeword G_γ is an almost simple group acting absolutely irreducibly on the underlying space $\mathbb{F}_2^{2n}$, see [18, Theorem 8.2] or [3, Theorem 1.4]. In the case where G_γ is almost simple, there is at least one further example: one in $J(136, 10)$ with automorphism group S_{10} constructed in [18, Section 5.4]. On the other hand G_γ is not a sporadic group or an exceptional group of Lie type, [18, Theorem 8.5]. It would be very nice to see a full classification.

Problem 2 *Complete the classification of SIT-codes and Delandtsheer designs admitting a 2-transitive symplectic group in one of its Jordan–Steiner actions.*

In the final Section 7 we discuss several related families of codes and designs. These include pairwise transitive designs and binary linear codes.

Remark I began the study of SIT-codes in 2005 in collaboration with R. A. (Bob) Liebler. Sadly, Bob died in July 2009 while hiking in California, some years before our joint paper [21] was completed. When proof-reading the paper [21], I failed to identify several misprints introduced during the typesetting process: namely, around 5% of expressions of the form $\binom{\mathcal{V}}{k}$ had been changed to $\frac{\mathcal{V}}{k}$. This affects the following statements in [21]: Lemma 3.2 parts (a) and (c), Proposition 6.1 (ii), and Example 8.1. There are also a few instances of these misprints in the proofs in [21].

2 Completely regular codes and a blow-up Construction

A code $\mathcal{C}$ in $J(\mathcal{V}, k)$ determines a distance partition $\{\mathcal{C}_0, \ldots, \mathcal{C}_{r-1}\}$ of the vertex set of $J(\mathcal{V}, k)$, where $\mathcal{C}_0 = \mathcal{C}$ is the code itself, $\mathcal{C}_1$ is the set of code-neighbours, and in general $\mathcal{C}_i$ is the set of k-subsets which are at distance i from at least one codeword in $\mathcal{C}$, and at distance at least i from every codeword. For the last non-empty set $\mathcal{C}_{r-1}$, the parameter r is called the *covering radius* of $\mathcal{C}$. The code is *completely-regular* if, for any $i, j \in \{0, \ldots, r-1\}$ and $\gamma \in \mathcal{C}_i$, the number of vertices of $\mathcal{C}_j$ which are adjacent to γ in $J(\mathcal{V}, k)$ is independent of the choice of γ in $\mathcal{C}_i$, and depends only on i and j. Further, $\mathcal{C}$ is called *completely-transitive* if Aut($\mathcal{C}$) (which fixes each $\mathcal{C}_i$ setwise) is transitive on each $\mathcal{C}_i$.

For our blow-up construction we begin with a code $\mathcal{C}_0$ in $J(\mathcal{U}, k_0)$ and essentially replace each element of $\mathcal{U}$ with a set of constant size $a > 1$ to obtain a code in $J(\mathcal{V}, ak_0)$ with base set $\mathcal{V}$ of size $a|\mathcal{U}|$. These blow-up codes are a generalisation of the *groupwise complete designs* introduced by Martin in [22] which correspond to the special case where $\mathcal{C}_0$ is the complete code $\binom{\mathcal{U}}{k_0}$. Martin [22, Theorem 2.1] determined all groupwise complete designs that are completely-regular codes, and as noted in [21, Remark 4.5], most of the codes we introduce below are *not* completely regular. We note that complete codes provide somewhat trivial examples of strongly incidence-transitive codes and Delandtsheer designs.

Construction 2.1 *Let $\mathcal{U} = \{U_1 | U_2 | \ldots | U_{v_0}\}$ be a partition of the v-set $\mathcal{V}$ with v_0 parts of size a, where $v = av_0, a > 1, v_0 \geqslant 2$, and let $k = ak_0$ where $1 \leqslant k_0 \leqslant v_0 - 1$. For a code $\mathcal{C}_0$ in $J(\mathcal{U}, k_0)$ with $|\mathcal{C}_0| \geqslant 2$, the blow-up code $\mathcal{C}(a, \mathcal{C}_0)$ in $J(\mathcal{V}, k)$ is the set of all k-subsets of $\mathcal{V}$ of the form $\cup_{U \in \gamma_0} U$, for some $\gamma_0 \in \mathcal{C}_0$.*

This construction, in the special case where $\mathcal{C}_0$ is the complete code $\binom{\mathcal{U}}{k_0}$, was introduced by Bill Martin [22] in 1994. He called such codes $\mathcal{C}(a, \mathcal{C}_0)$ *groupwise complete designs*, and in particular he determined the groupwise complete designs which are completely-regular codes in $J(\mathcal{V}, k)$. The general Construction 2.1 above is [21, Example 4.4] except that we allow v_0 to take any value at least 2. In [21, Example 4.4], $v_0 \geqslant 4$ is assumed, but we note that [21, Lemma 4.6] is valid for all $v_0 \geqslant 2$. In particular Construction 2.1 allows the possibility $k_0 = 1$, in which case the smaller Johnson graph $J(\mathcal{U}, 1)$ is the complete graph with vertex set $\mathcal{U}$. In all cases the construction preserves the property of strong incidence-transitivity. It is inherently an 'imprimitive construction' in the sense that the natural automorphism group preserving the code leaves the nontrivial partition $\mathcal{U}$ invariant. Recall that we define $\mathrm{Aut}(\mathcal{C}_0)$ as the stabiliser of $\mathcal{C}_0$ in $\mathrm{Sym}(\mathcal{U})$.

Lemma 2.2 [21, Lemma 4.6] Let $\mathcal{C} = \mathcal{C}(a, \mathcal{C}_0)$ as in Construction 2.1, and let $A = \mathrm{Aut}(\mathcal{C}_0)$.

(a) Then $\delta(\mathcal{C}) = a\,\delta(\mathcal{C}_0)$ and $\mathrm{Aut}(\mathcal{C})$ contains $\mathrm{Sym}(a) \wr A$ in its imprimitive action on $\mathcal{V}$;

(b) if $a \geqslant 3$, or if $a = 2 \leqslant \delta(\mathcal{C}_0)$, then $\mathcal{C}$ is $(\mathrm{Sym}(a) \wr A)$-strongly incidence-transitive if and only if $\mathcal{C}_0$ is A-strongly incidence-transitive;

(c) in particular if $\mathcal{C}_0$ is strongly incidence-transitive, then $\mathcal{C}$ is also strongly incidence-transitive.

Thus Construction 2.1 produces many strongly incidence-transitive codes, and hence also many Delandtsheer designs. It turns out that essentially all are 1-designs and none are 2-designs, even if the input $(\mathcal{U}, \mathcal{C}_0)$ is a 2-design.

Lemma 2.3 Let $\mathcal{C} = \mathcal{C}(a, \mathcal{C}_0)$ be a code in $J(\mathcal{V}, k)$ as in Construction 2.1, for some strongly incidence-transitive code $\mathcal{C}_0$ in $J(\mathcal{U}, k_0)$, where $|\mathcal{C}_0| \geqslant 2$, $v = |\mathcal{V}| = a|\mathcal{U}| = av_0$, and $k = ak_0$. Then $(\mathcal{U}, \mathcal{C}_0)$ is a Delandtsheer $1 - (v_0, k_0, r)$ design, for some r, and $(\mathcal{V}, \mathcal{C})$ is a Delandtsheer $1 - (v, k, r)$ design but is not a 2-design.

Proof The fact that $\mathcal{C}$ is strongly incidence-transitive follows from Lemma 2.2(c). We note that $|\mathcal{C}| = |\mathcal{C}_0| \geqslant 2$, from Construction 2.1, and hence, by definition, both $(\mathcal{U}, \mathcal{C}_0)$ and $(\mathcal{V}, \mathcal{C})$ are Delandtsheer designs. Also, by Lemma 1.1, $A = \mathrm{Aut}(\mathcal{C}_0)$ is transitive on $\mathcal{U}$ and $\mathrm{Sym}(a) \wr A$ is transitive on $\mathcal{V}$, and hence both are 1-designs, say, $(\mathcal{U}, \mathcal{C}_0)$ is a $1 - (v_0, k_0, r)$ design. Then it is easily seen from the construction that $(\mathcal{V}, \mathcal{C})$ is a $1 - (v, k, r)$ design with the same parameter r. It remains to show that $(\mathcal{V}, \mathcal{C})$ is not a 2-design.

Consider distinct points u, u' in the same class U_i of $\mathcal{U}$. Then as U_i lies in exactly r blocks of $\mathcal{C}_0$, the pair $\{u, u'\}$ lies in exactly r blocks of $\mathcal{C}$. Suppose that $(\mathcal{V}, \mathcal{C})$ is a $2 - (v, k, \lambda)$ design. Let u, u' lie in distinct classes U, U' of $\mathcal{U}$. Then by the definition of $\mathcal{C}$, the set of blocks of $\mathcal{C}$ containing $\{u, u'\}$ is in one-to-one correspondence with the set of blocks of $\mathcal{C}_0$ containing $\{U, U'\}$. Since this set has size $\lambda > 0$, independent of the choices of u, u' it follows that $(\mathcal{U}, \mathcal{C}_0)$ is also a $2 - (v_0, k_0, \lambda)$ design (with the same λ). In particular $k_0 > 1$, since otherwise there would be no blocks of $\mathcal{C}_0$ containing $\{U, U'\}$. Moreover λ must be equal to the parameter r (the number of blocks of $\mathcal{C}$ containing a point pair within a single class). Counting triples (U, U', γ_0), with distinct $U, U' \in \mathcal{U}$ and $\gamma_0 \in \mathcal{C}_0$ such that $U, U' \in \gamma_0$, we have $v_0(v_0 - 1)\lambda = |\mathcal{C}_0|\,k_0(k_0 - 1) = v_0 r(k_0 - 1)$, and since $\lambda = r$, this implies that $v_0 = k_0$. This means, however, that $|\mathcal{C}_0| = 1$, which is a contradiction. □

It turns out that the family of examples from Construction 2.1 essentially exhausts all possibilities for G-SIT-codes, and Delandtsheer designs, in $J(\mathcal{V}, k)$ for which the automorphism group G is not primitive on the base set $\mathcal{V}$. This assertion

Codes and designs   329

is proved in [19] and draws together a number of results from [21] concerning the
larger class of neighbour-transitive codes. It is instructive to repeat the argument
here.

Proposition 2.4 [19, Theorem 2] *Suppose that $\mathcal{C} \subseteq \binom{\mathcal{V}}{k}$ is a code in $J(\mathcal{V}, k)$, where $2 \leqslant k \leqslant v - 2$ and $|\mathcal{C}| \geqslant 2$, and suppose that $G \leqslant \mathrm{Aut}(\mathcal{C})$ is not primitive on $\mathcal{V}$. Then the following are equivalent:*

(a) $\mathcal{C}$ is a G-SIT code;

(b) G is transitive on $\mathcal{V}$ and, for some divisor a of $\gcd(v, k)$ with $a \geqslant 2$, G leaves invariant a partition $\mathcal{U}$ of $\mathcal{V}$ with $|\mathcal{U}| = v/a$ parts of size a, and $\mathcal{C} = \mathcal{C}(a, \mathcal{C}_0)$ as in Construction 2.1, for some G_0-SIT-code $\mathcal{C}_0$ in $J(\mathcal{U}, k_0)$, where $k = ak_0$ and G_0 is the group induced by G on $\mathcal{U}$.

Proof Let $\mathcal{C}, G$ be as in the statement. Then $|\mathcal{C}| \geqslant 2$. Also, since G is not primitive on $\mathcal{V}$, it follows in particular that $G \neq \mathrm{Sym}(\mathcal{V})$ or $\mathrm{Alt}(\mathcal{V})$. Suppose first that $\mathcal{C}$ is G-strongly incidence-transitive. If $\mathcal{C}$ were the complete code $\binom{\mathcal{V}}{k}$, then G would be transitive on the k-subsets of $\mathcal{V}$. By [10, Theorem 9.4B], such a group G is either (i) 2-transitive on $\mathcal{V}$, or (ii) $k = 2$, $v \equiv 3 \pmod 4$, a prime power, and a point stabiliser in G has orbits in $\mathcal{V}$ of lengths $1, (v-1)/2, (v-1)/2$. Now each 2-transitive permutation group is primitive, and also each group in case (ii) is primitive on $\mathcal{V}$. This contradicts our assumption that G is imprimitive on $\mathcal{V}$. Therefore $\mathcal{C}$ is not the complete code and so G is transitive on $\mathcal{V}$ and $\delta(\mathcal{C}) \geqslant 2$, by Lemma 1.1. Hence by assumption, G is imprimitive on $\mathcal{V}$, and so by [21, Proposition 4.7], $\mathcal{C}$ is as in Example 4.1 or Example 4.4 of [21] relative to some G-invariant nontrivial partition $\mathcal{U}$ of $\mathcal{V}$ with $|\mathcal{U}| = v/a$ parts of size a. If $\mathcal{C}$ comes from [21, Example 4.4], then since this set of examples arises from our Construction 2.1 we have $\mathcal{C} = \mathcal{C}(a, \mathcal{C}_0)$ for some code $\mathcal{C}_0$ in $J(\mathcal{U}, k/a)$. Now G leaves $\mathcal{U}$ invariant and the group G_0 induced by G on $\mathcal{U}$ must preserve $\mathcal{C}_0$, so $G \leqslant \mathrm{Sym}(a) \wr G_0$ and $G_0 \leqslant \mathrm{Aut}(\mathcal{C}_0)$. Moreover, since $\mathcal{C} = \mathcal{C}(a, \mathcal{C}_0)$ is a G-SIT-code, the small code $\mathcal{C}_0$ must also be a G_0-SIT-code. Thus part (b) holds. Suppose now that $\mathcal{C}$ comes from [21, Example 4.1]. Then since $\delta(\mathcal{C}) \geqslant 2$, it follows from [21, Lemma 4.3] that $\mathcal{C}$ is as in 'Line 1 of Table 3 for [21, Example 4.1] with $k = a$'. This means that $\mathcal{U}$ has parts of size $a = k$ and $\mathcal{C}$ is equal to the set of parts of $\mathcal{U}$. Thus $\mathcal{C}$ is a blow-up code $\mathcal{C}(a, \mathcal{C}_0)$ as in Construction 2.1, where $k_0 = 1$ and the small code $\mathcal{C}_0$ consists of all the singletons from $\mathcal{U}$. Since $\mathcal{C}$ is a G-SIT-code, the stabiliser G_γ of any part γ of $\mathcal{U}$ is transitive on $\mathcal{V} \setminus \gamma$ and hence the group G_0 induced by G on $\mathcal{U}$ is 2-transitive, and $\mathcal{C}_0$ is a G_0-SIT-code. Thus again part (b) holds.

Conversely, assume that part (b) holds. Then $\mathcal{C}$ is strongly incidence transitive by Lemma 2.2. □

We deduce that an imprimitive strongly incidence-transitive code can be obtained directly by applying Construction 2.1 to a primitive SIT-code.

Theorem 2.5 *Suppose that $\mathcal{C} \subseteq \binom{\mathcal{V}}{k}$ is a G-SIT-code in $J(\mathcal{V}, k)$, where $|\mathcal{V}| = v$, $2 \leqslant k \leqslant v - 2$, $|\mathcal{C}| \geqslant 2$, and $G \leqslant \mathrm{Aut}(\mathcal{C})$, such that G is imprimitive on $\mathcal{V}$.*

(a) Then $v = av_0$ and $k = ak_0$ for some $a \geqslant 2$, there exists a strongly incidence-transitive code $\mathcal{C}_0$ in $J(\mathcal{U}, k_0)$, where $|\mathcal{U}| = v_0$, $1 \leqslant k_0 \leqslant v_0 - 1$, such that $\mathcal{C} = \mathcal{C}(a, \mathcal{C}_0)$ and $\mathrm{Aut}(\mathcal{C}_0)$ is primitive on $\mathcal{U}$.

(b) Moreover, $\mathrm{Aut}(\mathcal{C}_0)$ is 2-transitive on $\mathcal{U}$, $\mathrm{Aut}(\mathcal{C}) = \mathrm{Sym}(a) \wr \mathrm{Aut}(\mathcal{C}_0)$, and either

  (i) $k_0 = 1$, $\mathcal{C}_0 = \binom{\mathcal{U}}{1}$ is the complete code, $\mathcal{C} = \mathcal{U}$, and $(\mathcal{V}, \mathcal{C})$ is a Delandtsheer $1 - (v, k, 1)$-design; or

  (ii) $k_0 \geqslant 2$, $(\mathcal{U}, \mathcal{C}_0)$ is a Delandtsheer $2 - (v_0, k_0, \lambda_0)$ design, for some λ_0, and $(\mathcal{V}, \mathcal{C})$ is a Delandtsheer $1 - (v, k, r)$ design where $r = (v_0 - 1)\lambda_0/(k_0 - 1)$.

Proof (a) By Proposition 2.4, $\mathcal{C} = \mathcal{C}(a, \mathcal{C}_0)$ for some strongly incidence-transitive code $\mathcal{C}_0$ in $J(\mathcal{U}, k_0)$ where $v = av_0, k = ak_0$. Suppose that $\mathrm{Aut}(\mathcal{C}_0)$ is imprimitive on $\mathcal{U}$. Then a further application of Proposition 2.4 shows that $\mathcal{C}_0 = \mathcal{C}(a', \mathcal{C}'_0)$ for some strongly incidence-transitive code $\mathcal{C}'_0$ in $J(\mathcal{U}', k'_0)$ where $v_0 = a'|\mathcal{U}'|, k_0 = a'k'_0$. It follows from Construction 2.1 that $\mathcal{C} = \mathcal{C}(aa', \mathcal{C}'_0)$. Thus we see recursively that, if a is maximal such that $\mathcal{C} = \mathcal{C}(a, \mathcal{C}_0)$ for some strongly incidence-transitive code $\mathcal{C}_0$ in $J(\mathcal{U}, k_0)$ with $v = av_0$ and $k = ak_0$, then $\mathrm{Aut}(\mathcal{C}_0)$ is primitive on $\mathcal{U}$.

(b) Suppose that $\mathcal{C} = \mathcal{C}(a, \mathcal{C}_0)$ with $\mathcal{C}_0$ as in part (a). Since $\mathcal{C}_0$ is strongly incidence-transitive and $|\mathcal{C}_0| = |\mathcal{C}| \geqslant 2$ with $A = \mathrm{Aut}(\mathcal{C}_0)$ primitive on $\mathcal{V}$, the group A is 2-transitive on $\mathcal{V}$ by Theorem 1.2. If $k_0 \geqslant 2$ this implies that $(\mathcal{U}, \mathcal{C}_0)$ is a Delandtsheer $2 - (v_0, k_0, \lambda_0)$ design, for some λ_0. This means that each $U_i \in \mathcal{U}$ lies in $r = |\mathcal{C}_0|k_0/v_0 = (v_0 - 1)\lambda_0/(k_0 - 1)$ blocks in $\mathcal{C}_0$, and hence each point of $\mathcal{V}$ lies in r blocks of $\mathcal{C}$. On the other hand if $k_0 = 1$ then $\mathcal{C}_0 = \binom{\mathcal{U}}{1}$ is the complete code with $v_0 = |\mathcal{U}|$ blocks, and $\mathcal{C} = \mathcal{U}$ with each point lying in a unique block.

Finally we show that $\mathrm{Aut}(\mathcal{C})$ is equal to $X := \mathrm{Sym}(a) \wr A$ where $A = \mathrm{Aut}(\mathcal{C}_0)$. Now $\mathrm{Aut}(\mathcal{C})$ contains X by Lemma 2.2. Note that X contains a transposition so its only primitive overgroup is $\mathrm{Sym}(\mathcal{V})$, which does not leave $\mathcal{C}$ invariant. Therefore $\mathrm{Aut}(\mathcal{C})$ is imprimitive. If $\mathcal{U}'$ is a nontrivial $\mathrm{Aut}(\mathcal{C})$-invariant partition of $\mathcal{V}$, then considering the action of X we see that each part of $\mathcal{U}$ (on which X induces the full group $\mathrm{Sym}(a)$) must be contained in a part of $\mathcal{U}'$, that is, $\mathcal{U}$ is a refinement of $\mathcal{U}'$. Then the set of parts of $\mathcal{U}$ contained in a fixed part of $\mathcal{U}'$ forms a block of imprimitivity for the induced action of X on $\mathcal{U}$, and since X induces the primitive group A on $\mathcal{U}$, it follows that this block of imprimitivity has size 1 and $\mathcal{U}' = \mathcal{U}$. Thus $\mathrm{Aut}(\mathcal{C})$ leaves $\mathcal{U}$ invariant. This implies that $\mathrm{Aut}(\mathcal{C})$ induces a subgroup of $\mathrm{Sym}(\mathcal{U})$ leaving $\mathcal{C}_0$ invariant. It follows that $\mathrm{Aut}(\mathcal{C}) = X$. □

3 Sporadic SIT-codes and Delandtsheer 2-designs

Although the codes arising from Construction 2.1 can have arbitrarily large minimum distance (since $\delta(\mathcal{C}(a, \mathcal{C}_0)) = a\delta(\mathcal{C}_0) \geqslant a$) by Lemma 2.2, the codes are not very large since $|\mathcal{C}(a, \mathcal{C}_0)| = |\mathcal{C}_0|$ remains fixed as a grows. Thus the interesting strongly incidence-transitive codes in $J(\mathcal{V}, k)$ have automorphism groups primitive on $\mathcal{V}$, as we discussed in Subsection 1.4. In this section we examine the SIT-codes and Delandtsheer 2-designs admitting a sporadic 2-transitive group. It turns out that each of the sporadic 2-transitive groups, as listed in Table 1, acts on at least one SIT-code and Delandtsheer design. The approach to this classification in [27] was computational, using the computer system GAP [13]. Note that, if $\mathcal{C}$ is a G-SIT-code and

Codes and designs   331

$\gamma \in \mathcal{C}$, then G_γ has two orbits in $\mathcal{V}$, namely γ and its complement. For each of the sporadic 2-transitive groups G in Table 1,

- we started with a list $\mathcal{L}$ of representatives from the conjugacy classes of maximal subgroups of G (apart from the stabilisers of points of $\mathcal{V}$). For each subgroup $H \in \mathcal{L}$,

- we computed the orbits of H on $\mathcal{V}$, and discarded any subgroups with more than two orbits on $\mathcal{V}$;

- if H had two orbits γ and $\mathcal{V} \setminus \gamma$, with say $2 \leqslant |\gamma| \leqslant v/2$, then we checked whether or not H is transitive on $\gamma \times (\mathcal{V} \setminus \gamma)$, and if not then H was discarded;

- if H was transitive on $\gamma \times (\mathcal{V} \setminus \gamma)$, then we enumerated the corresponding code $\mathcal{C} := \{\gamma^g \mid g \in G\}$ in $J(v, |\gamma|)$, determined the minimal distance $\delta(\mathcal{C})$, and checked if H was the full stabiliser of $\mathcal{C}$ in G (since later in this process the subgroup H might not be maximal in G); in this case we would have found an interesting SIT-code listed in Table 3;

- finally, if H was transitive on $\mathcal{V}$, then we appended to $\mathcal{L}$ a representative of each of the conjugacy classes of maximal subgroups of H, and continued with these steps on the next listed subgroup in $\mathcal{L}$.

Since the groups in Table 1 were explicitly available in GAP as permutation groups of reasonable degree v, the initial list $\mathcal{L}$ could either be constructed by explicit computation, or by inspecting the Atlas of Finite Group Representations [31]. The procedure terminates because the groups are finite, and the fact that it yields all examples is justified by [27, Lemma 1]. To aid with an understanding of the beautiful geometric and group-theoretic structures underpinning the examples in Table 3, mathematical arguments were given in [27, Section 2] in most cases (with occasionally reference to computations to finish off the case).

Some of the Delandtsheer designs corresponding to these SIT-codes in Table 3 are very familiar, such as the Steiner systems in lines 4, 8, 12, 15, 24, and 26 associated with the Mathieu groups. Others might not be so easily identifiable: the design in line 21 for the Conway group Co_3 is the $2 - (276, 100, 1458)$ design found by Haemers et al [14] in 1993. This and other 2-designs for Co_3 were constructed and the parameters found using the DESIGN package for GAP [29].

4   Classical SIT-codes and Delandtsheer designs

In this section we describe the SIT-codes and Delandtsheer designs admitting either a 'projective' or a 'rank 1' group from Table 2. There are three infinite families of examples and one exceptional case. We list them in Table 4 and describe how they were classified. In the table, $\binom{m}{r}_q$ denotes the number of r-dimensional subspaces of $\mathbb{F}_q^m$.

Line	G	v	k	$\delta(\mathcal{C})$	$|\mathcal{C}|$	Delandtsheer design	G_γ
1	$L_2(11)$	11	5	3	11	2-$(11,5,2)$ biplane	A_5
2	A_7	15	7	4	15	planes of $PG(3,2)$	$L_2(7)$
3	M_{11}	12	6	3	22	totals	A_6
4	M_{22}	22	6	4	77	3-$(22,6,1)$ design	$2^4 : A_6$
5			7	4	176	heptads	A_7
6			8	4	330	octads	$2^3 : L_3(2)$
7			10	4	616	decads	$M_{10} \cong A_6 \cdot 2_3$
8	$M_{22}.2$	22	6	4	77	3-$(22,6,1)$ design	$2^4 : S_6$
9			7	3	352	heptads	A_7
10			8	4	330	octads	$2 \times 2^3 : L_3(2)$
11			10	4	616	decads	$A_6 \cdot (2^2)$
12	M_{23}	23	7	4	253	4-$(23,7,1)$ design	$2^4 : A_7$
13			8	4	506	octads	A_8
14			11	4	1288	endecads	M_{11}
15	M_{24}	24	8	4	759	5-$(24,8,1)$ design	$2^4 : A_8$
16			12	4	2576	duum	M_{12}
17	HS	176	50	36	176	2-$(176,50,14)$	$U_3(5) : 2$
18			56	32	1100	2-$(176,56,110)$	$L_3(4).2$
19	Co_3	276	6	3	708400	2-$(276,6,280)$	$3^{1+4}_+ : 4S_6$
20			36	24	170775	2-$(276,36,2835)$	$2 \cdot Sp_6(2)$
21			100	50	11178	2-$(276,100,1458)$	HS
22			126	36	655776	2-$(276,126,136080)$	$U_3(5) : S_3$
23	A_7	15	3	2	35	lines of $PG(3,2)$	$(A_3 \times A_4).2$
24	M_{11}	11	5	2	66	4-$(11,5,1)$ design	S_5
25	M_{11}	12	6	2	110	halves of quadrisect.	$3^2 : Q_8$
26	M_{12}	12	6	2	132	5-$(12,6,1)$ design	$A_6.2$
27	M_{24}	24	12	2	35420	5-$(24,12,660)$	$2^6 : 3.(S_3 \times S_3)$

Table 3: Sporadic 2-transitive SIT-codes $\mathcal{C}$ with group G and $3 \leqslant k \leqslant v/2$.

Line	G	v	k	$\delta(\mathcal{C})$	λ	Blocks of $\mathcal{D}(\mathcal{C})$
1	$P\Gamma U(3,q)$	q^3+1	$q+1$	q	1	Classical unital
2	$P\Gamma U(3,3)$	28	12	6	11	'Bases'
3	$P\Gamma L(2,q)$	$q+1$	q_0+1	q_0-1	$\frac{q_0-1}{2}$	Baer sublines, $q = q_0^2$
4	$P\Gamma L(n,q)$	$\frac{q^n-1}{q-1}$	$\frac{q^s-1}{q-1}$	q^{s-1}	$\binom{n-2}{s-2}_q$	Subspaces, $2 \leqslant s < n$

Table 4: Classical 2-transitive SIT-codes $\mathcal{C}$ and Delandtsheer $2-(v,k,\lambda)$ designs $\mathcal{D}(\mathcal{C})$ with group G and $3 \leqslant k \leqslant v/2$

Constructions for the SIT-codes $\mathcal{C}$ in Table 4 giving G, v, k and $\delta(\mathcal{C})$ are given in [21, Examples 7.1, 7.3, 8.1, and Theorem 1.3(b)]. It follows from Theorem 1.2 that in each case $\mathcal{D}(\mathcal{C}) = (\mathcal{V}, \mathcal{C})$ is a Delandtsheer $2 - (v, k, \lambda)$ design, and the value of λ is determined using (1.2). More details on the classical unital may be found in [4, 30]. The fact that there are no other examples from groups in the rank 1 and projective lines of Table 2 needs some comment.

This classification is given for the 2-transitive Suzuki, Ree and unitary groups in [21, Theorem 1.3], showing that only the unitary groups yield any examples, and that these examples are the ones in Lines 1 and 2 of Table 4. The case of 2-transitive subgroups of $P\Gamma L(2, q)$ acting on the projective line $PG(1, q)$ was dealt with in [21, Proposition 7.2], which showed that examples arise only if $q = q_0^2$ and the blocks are the Baer sublines as in Line 3 of Table 4.

This leaves the case where $PSL(n, q) \leqslant G \leqslant P\Gamma L(n, q)$ with $n \geqslant 3$, acting on the point set $\mathcal{V}$ of the projective space $PG(n - 1, q)$. Here the restrictions required for a subset $\gamma \subseteq \mathcal{V}$ to be a codeword of a G-SIT-code, or equivalently, a block of a Delandtsheer design admitting G, may be interpreted as a condition on the lines of the projective geometry $PG(n - 1, q)$. For a pair of points $u \in \gamma$ and $u' \in \mathcal{V} \setminus \gamma$, there is a unique line ℓ of $PG(n - 1, q)$ containing u and u', so transitivity of G on the triples in (1.1) implies that G is transitive on the projective lines which meet both γ and its complement. Hence such 'shared lines' meet γ in a constant number of points, say x, where $0 < x < |\ell| = q + 1$. Thus each projective line ℓ meets γ in 0, or x or $q + 1$ points. A subset γ of $\mathcal{V}$ with these properties is called a *subset of class* $[0, x, q + 1]_1$. If $x = 1$ then each line containing two distinct points of γ must be contained in γ, and so γ is a projective subspace of $PG(n - 1, q)$ – and the set of subspaces of a given dimension is an example, as in Line 4 of Table 4. Similarly if $x = q$ then each line containing two distinct points of $\mathcal{V} \setminus \gamma$ lies in $\mathcal{V} \setminus \gamma$, and so γ is a subspace complement: however in this case $k = |\gamma| > v/2$ so we do not list these examples separately.

In the case where $1 < x < q$ we restrict analysis to the case where $2 \leqslant x \leqslant (q + 1)/2$, since $J(v, k) \cong J(v, v - k)$ and if γ is a subset of class $[0, x, q]_1$, then $\mathcal{V} \setminus \gamma$ is a subset of class $[0, q + 1 - x, q]_1$. Here, by [21, Proposition 7.4], $x \in \{2, \sqrt{q} + 1\}$. Using this information it was shown by Durante [11, Theorem 15] that there are no further examples of SIT-codes admitting projective groups. In fact Durante [11, Theorem 14] described all subsets γ of $PG(n - 1, q)$ of class $[0, x, q + 1]_1$, showing that either (i) γ or its complement is a subspace, or (ii) $n = 3$, q is even, and γ is a proper x-maximal arc or the complement of a proper $(q + 1 - x)$-maximal arc. Applying a result of Delandtsheer and Doyen [6, Theorem], among the examples in case (ii), the only ones for which G_γ is transitive on 'shared lines' are the hyperoval in $PG(2, 4)$, for which $(k, x) = (6, 2)$, and the dual of a regular hyperoval, for which $(k, x) = (q(q - 1)/2, q/2)$. Since $x \in \{2, \sqrt{q}\}$, this leaves only $q = 4$ with γ a hyperoval in $PG(2, 4)$. However, it was noted in [21, Remark 7.5] that, although the stabiliser in $PGL(3, 4)$ of a hyperoval γ is transitive on 'shared lines', the more restrictive SIT-property fails and this is not an additional example.

5 Affine SIT-codes and Delandtsheer designs

Line	q	n	k	$\delta(\mathcal{C})$	λ	Blocks of $\mathcal{D}(\mathcal{C})$
1	any	$n \geqslant 2$	q^s	$q^{s-1}(q-1)$	$\binom{n-1}{s-1}_q$	Subspaces, $1 \leqslant s < n$
2	16	1	4	3	1	Baer sublines
3	16	$n \geqslant 2$	4.16^{n-1}	?	?	Four parallel hyperplanes
4	4	$n \geqslant 2$	2.4^{n-1}	?	?	Two parallel hyperplanes
5	4	$n=2$	6	3	6	Hyperovals in $AG(2,4)$
6	4	$n \geqslant 3$	?	?	?	Cylinders in $AG(n,4)$ with base as in Line 5
7	2	$n \geqslant 4$ n even	$2^{n-1}+$ $\varepsilon 2^{n/2-1}$	?	?	Blocks of symplectic design $\mathcal{S}^\varepsilon$, where $\varepsilon = \pm$
8	2	$n \geqslant 4$ n even	$2^{n-1}+$ $\varepsilon 2^{n/2-1}$	?	?	Union of all symplectic designs on $\mathcal{V}$ of type ε

Table 5: Affine 2-transitive SIT-codes $\mathcal{C}$ and Delandtsheer $2-(q^n,k,\lambda)$ designs $\mathcal{D}(\mathcal{C})$ with 2-transitive group $G \leqslant A\Gamma L(n,q)$, $q \geqslant 4$, and $k \leqslant q^n/2$.

As we discussed in Subsection 1.4, there are many interesting examples of affine SIT-codes and Delandtsheer designs. Let $G \leqslant A\Gamma L(n,q)$ be 2-transitive on the point-set $\mathcal{V}$ of $AG(n,q)$, and suppose that $\mathcal{C}$ it a G-SIT-code in $J(q^n,k)$ with $3 \leqslant k \leqslant q^n/2$. Combining [21, Propositions 6.1 and 6.6] and [18, Lemma A.1], either $\mathcal{C}$ is the set of all affine subspaces of fixed dimension, as in Line 1 of Table 5, or the field size $q \in \{2,4,16\}$. If $n=1$ then the only non-subspace example is the set of Baer sublines of $AG(1,16)$, by [21, Proposition 6.1], as in Line 2 of Table 5. Suppose then that $n \geqslant 2$, and that γ is not a subspace, so $q \in \{2,4,16\}$.

For $\gamma \in \mathcal{C}$ and points $u \in \gamma$ and $u' \in \mathcal{V} \setminus \gamma$, there is a unique affine line ℓ of $AG(n,q)$ containing u and u', and transitivity of G on the triples in (1.1) implies that G is transitive on the affine lines which meet both γ and its complement. Hence such 'shared lines' meet γ in a constant number x of points, where $0 < x < |\ell| = q$, and therefore each line meets γ in 0, or x or q points, that is to say, γ is a *subset of class* $[0,x,q+1]_1$ in $AG(n,q)$. For $q \geqslant 4$, Durante [11] determined all subsets of class $[0,x,q+1]_1$: in Theorem 22 of [11] if $n \geqslant 3, q > 4$, Theorem 24 if $n \geqslant 3, q = 4$, and Proposition 18 if $n = 2$. He then examined all these examples and determined which of them have the SIT-property in [11, Theorem 27]. The examples give all possibilities for $q = 4, 16$, and are as in Lines 3–6 of Table 5. Unfortunately he does not give the parameters $\delta(\mathcal{C})$ and λ for $\mathcal{D}(\mathcal{C})$, but we have these for Line 5 deduced from [21, Example 6.7] (with the help of John Bamberg).

Problem 3 *Determine the minimum distance $\delta(\mathcal{C})$ and the parameter λ for the affine SIT-codes and Delandtsheer designs in Lines $3,4,6,7,8$ of Table 5.*

There remains the exceptional case of $q=2$, studied by Ioppolo in [18, 19]. The 2-transitive affine group $G \leqslant AGL(n,2)$ contains the translation group $T \cong C_2^n$ as its unique minimal normal subgroup, and Ioppolo showed that the structure of a G-SIT-code $\mathcal{C}$ in $J(2^n,k)$ depends on the intersections $M(\gamma) = G_\gamma \cap T$, for $\gamma \in \mathcal{C}$. Note that $M(\gamma) \neq T$ since $\gamma \neq \mathcal{V}$. Since G is transitive on $\mathcal{C}$, these subgroups $M(\gamma)$ form a conjugacy class $\mathcal{M}$ of G, and the code $\mathcal{C}$ is partitioned into pairwise disjoint subcodes

$\mathcal{C}(M) = \{\gamma \in \mathcal{C} | M(\gamma) = M\}$, called *components*, for $M \in \mathcal{M}$. Each component $\mathcal{C}(M)$ is itself an $N_G(M)$-SIT-code, by [18, L:emma 7.2]. If $2^a = |M| > 1$, then the group $N_G(M)$ acts imprimitively on $\mathcal{V}$ preserving the partition $\mathcal{U}$ into M-orbits, and hence, by Proposition 2.4, $\mathcal{C}(M) = \mathcal{C}(2^a, \mathcal{C}_0)$ as in Construction 2.1, with $\mathcal{C}_0$ a G_0-SIT-code in $J(2^{n-a}, k_0)$, where $k = 2^a k_0$ and G_0 is the group induced by $N_G(M)$ on the set $\mathcal{U}$ of M-orbits in $\mathcal{V}$. Since $M \neq T$, the group G_0 is an affine group on $\mathcal{U}$, and by the definition of $\mathcal{C}(M)$, each codeword $\gamma_0 \in \mathcal{C}_0$ intersects the translation group for $\mathcal{C}_0$ trivially: such a code $\mathcal{C}_0$ is said to be *translation-free*. Since G permutes the components $\mathcal{C}(M)$ transitively, it follows that each affine SIT-code $\mathcal{C}$ in $J(2^n, k)$ corresponds, up to isomorphism, to a unique affine translation-free SIT-code $\mathcal{C}_0$ in $J(2^{n-a}, k/2^a)$, for some $a \geqslant 0$.

The affine translation-free SIT-codes are therefore central to understanding the affine SIT-codes over $\mathbb{F}_2$, and they have been completely classified by Ioppolo [18, Chapter 7] (see also [19]). The examples are as in Lines 7–8 of Table 5. For the examples in Line 7, which are constructed in [18, Example 7.11], the Delandtsheer design $\mathcal{D}(\mathcal{C})$ is the symplectic symmetric design $\mathcal{S}^\varepsilon(\mathcal{V})$, where n is even and $\varepsilon \in \{\pm\}$. The automorphism group of $\mathcal{S}^\varepsilon(\mathcal{V})$, and of the corresponding SIT-code, is the symplectic affine group $A\mathrm{Sp}(n,2) = C_2^n \rtimes \mathrm{Sp}(n,2)$ (or $C_2^4 \rtimes S_6$ if $n=4$). It is shown in [18, Theorem 7.13] that every affine translation-free SIT-code in $J(2^n, k)$ has n even and $k = 2^{n-1} + \varepsilon 2^{n/2-1}$, for some $\varepsilon \in \{+, -\}$, and is a disjoint union of several copies of $\mathcal{S}^\varepsilon(\mathcal{V})$ permuted transitively by some overgroup of $A\mathrm{Sp}(n,2)$ in $\mathrm{AGL}(n,2)$. Since $\mathrm{Sp}(n,2)$ is a maximal subgroup of $\mathrm{GL}(n,2)$ (or S_6 maximal in $\mathrm{GL}(4,2) \cong A_8$ if $n = 4$), by [1], it follows that the only other examples are the union of all images of a code in Line 7 of Table 5 under elements of $\mathrm{AGL}(n,2)$, or equivalently, under elements of $\mathrm{GL}(n,2)$ (since the translation subgroup leaves the codes $\mathcal{S}^\varepsilon(\mathcal{V})$ invariant). These are the examples in Line 8 of Table 5. Note that, for each ε, there is one copy of $\mathcal{S}^\varepsilon(\mathcal{V})$ for each nondegenerate alternating form on $\mathcal{V}$.

It is not clear whether it is possible to use a translation-free code $\mathcal{C}_0$ to build a component code $\mathcal{C}(M) = \mathcal{C}(2^a, \mathcal{C}_0)$, with $M \neq 1$, in such a way that $\mathrm{Aut}(\mathcal{C}(M)) \cap \mathrm{AGL}(n,2)$ acts strongly incidence transitively, noting that $\mathrm{Aut}(\mathcal{C}(M)) = \mathrm{Sym}(2^a) \wr \mathrm{Aut}(\mathcal{C}_0)$. *The task remaining to solve Problem 1 is essentially to resolve this issue.* The reason is that each example $\mathcal{C}(M) = \mathcal{C}(2^a, \mathcal{C}_0)$ of an H-SIT code in $J(2^{a+n}, k)$, with $a > 0$, $\mathcal{C}_0$ as in Table 5 line 7 or 8, and $C_2^{a+n} \trianglelefteq H \leqslant \mathrm{AGL}(a+n,2)$, can be used to construct an affine G-SIT-code $\mathcal{C}$ for any 2-transitive group G satisfying $H < G \leqslant \mathrm{AGL}(a+n,2)$ by taking $\mathcal{C} = \cup_{g \in G} \mathcal{C}(M)^g$. And each such code would arise in this way.

6 Symplectic SIT-codes and Delandtsheer designs

The last infinite family of 2-transitive permutation groups to consider are the symplectic groups $G = \mathrm{Sp}(2n, 2)$, with $n \geqslant 2$, (line 'symplectic' of Table 2). The group G is the isometry group of a nondegenerate alternating form B on $V = \mathbb{F}_2^{2n}$, and G acts on the associated set $\mathcal{Q}$ of all nondegenerate quadratic forms $\phi : V \to \mathbb{F}_2$ which polarise to B, that is to say:

$$B(x,y) = \phi(x+y) - \phi(x) - \phi(y), \text{ for all } x, y \in V.$$

For $g \in G$ and $\phi \in \mathcal{Q}$, we write ϕ^g for the function $\phi^g(x) = \phi(xg^{-1})$, for $x \in V$, and note that $\phi^g \in \mathcal{Q}$. We let $\text{sing}(\phi) = \{x \in V \mid \phi(x) = 0\}$, the set of ϕ-*singular vectors*. The group G has exactly two orbits in $\mathcal{Q}$, which we denote $\mathcal{Q}^+$ and $\mathcal{Q}^-$: a form $\phi \in \mathcal{Q}^+$ if the maximum dimension of a subspace of V contained in $\text{sing}(\phi)$ is n, while for $\phi \in \mathcal{Q}^-$ this dimension is $n-1$. We say that ϕ has type $+$ or $-$, respectively. For $\varepsilon \in \{+, -\}$, this G-action on $\mathcal{Q}^\varepsilon$ is 2-transitive, and $|\mathcal{Q}^\varepsilon| = 2^{n-1}(2^n + \epsilon.1)$. These actions are known as the *Jordan-Steiner actions* of G. We sometimes write ε instead of $\varepsilon.1$.

For a given $\varepsilon \in \{+, -\}$, Ioppolo constructed two families of G-SIT codes in $J(\mathcal{Q}^\varepsilon, k)$, for certain k, as follows. The first family is based on nondegenerate subspaces of V of fixed dimension.

Construction 6.1 *[3, Construction 1.2]* Let $\varepsilon' \in \{+, -\}$, $d \in \mathbb{Z}$ with $1 \leqslant d < n$, and $k = 2^n(2^d + \varepsilon')(2^{n-d} + \varepsilon\varepsilon')$. Define the code

$$\Gamma(n, d, \varepsilon, \varepsilon') = \{\gamma(U) \mid U < V, U \text{ nondegenerate}, \dim(U) = 2d\}$$

in $J(\mathcal{Q}^\varepsilon, k)$, where

$$\gamma(U) = \{\phi \in \mathcal{Q}^\varepsilon \mid \varphi|_U \text{ has type } \varepsilon' \text{ and } \varphi|_{U^\perp} \text{ has type } \varepsilon\varepsilon'\}.$$

The second family uses totally isotropic subspaces of fixed dimension.

Construction 6.2 *[3, Construction 1.3]* Let $c, d \in \mathbb{Z}$ with $c = 0$ or 1 and $1 \leqslant d \leqslant n$ such that $(d, \varepsilon) \neq (n, -)$, and let $k = 2^{n-1}(2^{n-d} + \varepsilon)$. Define the code

$$\Gamma(n, d, \varepsilon, c) = \{\gamma(U) \mid U < V, U \text{ totally isotropic}, \dim(U) = d\}$$

in $J(\mathcal{Q}^\varepsilon, k)$, where

$$\gamma(U) = \{\phi \in \mathcal{Q}^\varepsilon \mid \dim(\text{sing}(\phi) \cap U) = d - c\}.$$

Ioppolo showed that these are the only examples which are 'geometrically based' in the following sense. The SIT-property requires a codeword stabiliser G_γ to have two orbits in $\mathcal{Q}^\varepsilon$, namely γ and its complement, and analysis of the various cases considers the possible maximal subgroups M of G containing G_γ. The family of maximal subgroups of $G = \text{Sp}(2n, 2)$ have been classified into various families by Aschbacher [1]. One of these families consists of the subspace stabilisers appearing in Constructions 6.1 and 6.2. The union of several of the families of maximal subgroups, including these subspace stabilisers, is called the family of *geometric subgroups*, and Aschbacher proved that each maximal subgroup M of G that is not geometric, in this sense, is an almost simple group, that is, $T \trianglelefteq M \leqslant \text{Aut}(T)$ where T is a nonabelian simple group, called the socle of M, and T is absolutely irreducible on $\mathbb{F}_2^{2n}$. The major result of Ioppolo's thesis ([18, Theorem 4.3], or see [3, Theorem 1.4]) is that the only SIT-codes for which a codeword stabiliser G_γ is contained in a maximal geometric subgroup are those from Constructions 6.1 and 6.2. It is not known (to us) whether the corresponding Delandtsheer designs are previously known 2-designs.

Questions 6.3 *Are the Delandtsheer designs corresponding to the SIT-codes in Constructions 6.1 and 6.2 previously known 2-designs?*

Ioppolo's work shows that for any other examples the codeword stabiliser itself must be almost simple. There is at least one additional example with $3 \leqslant k \leqslant v/2$ where G_γ is a maximal almost simple group, namely an $\mathrm{Sp}(8,2)$-SIT-code in $J(136, 10)$ based on $\mathcal{V} = \mathcal{Q}^+$, with $G_\gamma = \mathrm{Sym}(10)$. Here the underlying space $\mathbb{F}_2^8$ is the deleted permutation module for the natural action of $\mathrm{Sym}(10)$. This example is the only one where $\mathbb{F}_2^{2n}$ is the deleted permutation module for $\mathrm{Sym}(2n+2)$, [18, Theorem 8.3], and it is the only one for dimensions $2n \leqslant 12$, [18, Section 6.2]. Moreover, there are no examples if T is a sporadic simple group or an exceptional group of Lie type [18, Theorem 6.25 and Appendix C.2], and if T is a classical simple group then existence of a corresponding SIT-code places very strong restrictions on T [18, Chapter 6]. Nevertheless, there is much work to be done yet to deal completely with Problem 2.

7 Related codes and designs

In this final section we discuss several families of designs and codes with links to Delandtsheer designs and SIT-codes.

7.1 Transitivity properties on flags and antiflags of designs

Flag-transitive 2-designs have been studied extensively, particularly following the seminal result of D. G. Higman and J. E. McLaughlin [16] that a flag-transitive $2 - (v, k, 1)$ design (linear space) is point-primitive. It is however Delandtsheer's work [5] on antiflag-transitive linear spaces which is most relevant to the theme of this exposition, especially her observation that most examples in her classification possessed the stronger SIT-property: hence our definition of a Delandtsheer design $(\mathcal{V}, \mathcal{C})$ with block size k as one in which the block set $\mathcal{C}$ is an SIT-code in $J(\mathcal{V}, k)$. Proposition 2.4 together with Theorem 1.2 essentially reduce the problem of classifying the Delandtsheer designs to the case of (point) 2-transitive Delandtsheer 2-designs. And the discussion in Sections 3, 4, 5 and 6 outlines the current status of this classification problem.

It would be interesting to know if progress could be made in understanding the intermediate family where the stringent symmetry conditions for Delandtsheer designs are relaxed somewhat, but not as far as simply antiflag-transitivity.

Problem 4 *Investigate designs that are both flag-transitive and antiflag-transitive.*

Motivated by a problem in graph theory, investigations have begun [8, 9] of the family of pairwise transitive designs which are in particular both flag-transitive and antiflag-transitive. This family of designs is defined by its symmetry: a design $\mathcal{D} = (\mathcal{V}, \mathcal{B})$ is said to be *pairwise transitive* if its automorphism group is transitive on the following six (possibly empty) sets of ordered pairs from $\mathcal{V} \cup \mathcal{B}$: collinear point-pairs (that is, distinct points contained in some block of $\mathcal{B}$), non-collinear point-pairs, incident point-block pairs, non-incident point-block pairs, intersecting block-pairs, and non-intersecting block-pairs. To compare pairwise transitive designs with Delandtsheer designs we make the following observations.

(a) The families of pairwise transitive designs and Delandtsheer designs have *significant overlap*, for example, both contain the designs of points and hyperplanes of projective and affine geometries (see Tables 4 and 5, and [9, Tables 1 and 2]), as well as the symplectic symmetric designs $\mathcal{S}^\varepsilon(\mathcal{V})$ described in Section 5 and [9, Table 1], and other 2-transitive symmetric designs (Lines 1 and 17 of Table 3 and [9, Table 1]).

(b) *Neither family is contained in the other.* For example most Delandtsheer designs arising from small dimensional subspaces of projective and affine geometries are not pairwise transitive, since there is more than one kind of nontrivial block-intersection and hence the group is not transitive on pairs of intersecting blocks. On the other hand the exceptional design consisting of the 21 points of PG(2, 4) and one of the three PSL(3, 4)-orbits of 56 hyperovals as blocks, forms a pairwise transitive design [9, Theorem 4.6] which, as discussed in Section 4, is not a Delandtsheer design (see [21, Remark 7.5]).

(c) The two exceptional antiflag-transitive linear spaces in Delandtsheer's classification [5], which are not Delandtsheer designs, are both 2-transitive nondesarguesian affine planes, namely the nearfield plane of order 9 and Hering's plane of order 27. Neither of these is pairwise transitive, by [9, Theorem 1.1]. Thus there are examples of antiflag-transitive 2-designs which are *neither Delandtsheer designs not pairwise transitive designs.*

(d) By Lemma 2.3 each code $\mathcal{C}(a, \mathcal{C}_0)$, arising from Construction 2.1 with $\mathcal{C}_0$ an SIT-code in $J(\mathcal{U}, k_0)$, is a Delandtsheer 1-*design, but not a* 2-*design*. However the group $G = \text{Sym}(a) \wr \text{Aut}(\mathcal{C}_0)$ is not transitive on pairs of collinear points unless $k_0 = 1$ in which case $\mathcal{C}$ is the set of parts of the partition $\mathcal{U}$. Thus these designs are not G-pairwise transitive if $k_0 > 1$, but are indeed pairwise transitive if $k_0 = 1$, see [8, Example 2.1].

(e) Let $\mathcal{V} = \mathbb{F}_q^n$ and $0 \neq u \in \mathcal{V}$, and consider the subset $\mathcal{C}$ of $J(\mathcal{V}, q)$ consisting of all hyperplanes of AG(n, q) which do not contain a line in the direction $\langle u \rangle$. Then for $G = [q^n].U < \text{AGL}(n, 1)$, where U is the stabiliser in GL(n, q) of $\langle u \rangle$, the design $(\mathcal{V}, \mathcal{C})$ is a G-pairwise transitive $1 - (q^n, q^{n-1}, q^{n-1})$ design, by [8, Lemma 3.4], but it is not a Delandtsheer design (the SIT-property is easily seen to fail for γ an $(n-1)$-subspace complementing $\langle u \rangle$).

The pairwise transitive 2-designs have been classified by Devillers and the author in [9, Theorem 1.1]. Since all pairs of distinct points are collinear, the automorphism group G is 2-transitive on points, and since G is transitive on intersecting block-pairs and on non-intersecting block-pairs, there are at most two possible values for the intersection size of a block-pair. Thus a pairwise transitive 2-design is either symmetric (if all block-pairs intersect nontrivially) or quasisymmetric if there exist non-intersecting block-pairs, [9, Lemma 2.4]. The 2-transitive symmetric designs were classified by Kantor [20], while the quasisymmetric pairwise transitive 2-designs were classified in [9].

The examples given in (d) and (e) show that Delandtsheer 1-designs and pairwise transitive 1-designs exist, and that neither family contains the other; while the examples in (c) illustrate that these two families do not cover all the designs addressed in Problem 4. Nevertheless, a better understanding of both families would be helpful.

Problem 5 *[8, Problem 1.11]* Classify pairwise transitive 1-designs.

In particular, a classification of the sub-family of pairwise transitive 1-designs for which the block set can be partitioned into parallel classes would have significant graph theoretic application. Such a design would be an affine design (see [8, Section 2.2]). A bipartite graph Γ with ordered bipartition $(B|B')$ corresponds to a design $\mathcal{D}$ with point set B, block set B' and incidence given by adjacency; also Γ is said to be *locally (G, s)-distance transitive* with automorphism group G if, for each vertex x, the stabiliser G_x is transitive on the set of vertices at distance i from x, for $i = 1, \ldots, s$. If the design $\mathcal{D}$ is affine, then the graph Γ is locally $(G, 4)$-distance transitive if and only if $\mathcal{D}$ is G-pairwise transitive, [8, Proposition 2.7(ii)].

7.2 SIT-codes and codes in binary Hamming graphs

Some of the examples of sporadic SIT-codes and Delandtsheer designs in Table 3 have surprisingly large minimum distance, and this prompted us to think in [27] about links between codes in Johnson graphs and codes in binary Hamming graphs - the traditional block codes. The *binary Hamming graph* $H(v, 2)$ has as vertices the ordered v-tuples with entries from $\{0, 1\}$, and edges those pairs of v-tuples which agree in all but one entry. If we write $\mathcal{V} = \{1, 2, \ldots, v\}$, then each vertex γ of the Johnson graph $J(v, k)$ can be identified with the binary v-tuple $h(\gamma)$ with i-entry 1 if and only if $i \in \gamma$, (see [27, Section 1.3]). This allows $J(v, k)$ to be identified with the set of all weight k vertices of $H(v, 2)$, and each code $\mathcal{C}$ in $J(v, k)$ to be identified with a constant weight code $H(\mathcal{C})$ in $H(v, 2)$. Two vertices at distance d in $J(v, k)$ correspond to two vertices in $H(v, 2)$ at distance $2d$ so the minimum distance of $H(\mathcal{C})$ is equal to $2\delta(\mathcal{C})$. Moreover $\mathrm{Aut}(\mathcal{C})$, in its action on entries, acts as automorphisms of the code $H(\mathcal{C})$ in $H(v, 2)$, so for an SIT code $\mathcal{C}$ in $J(v, k)$, $\mathrm{Aut}(\mathcal{C})$ is transitive on the codewords of $H(\mathcal{C})$. The neighbours of $H(\mathcal{C})$ in $H(v, 2)$ have weights $k \pm 1$, yielding two $\mathrm{Aut}(\mathcal{C})$-orbits on code neighbours for an SIT code $\mathcal{C}$.

Lemma 7.1 *Let $\mathcal{C}$ be an SIT-code in $J(v, k)$. Then $\mathrm{Aut}(\mathcal{C})$ has two orbits on the code-neighbours of $H(\mathcal{C})$, namely those of weight $k - 1$ and those of weight $k + 1$.*

Proof Extend the map h to act on each subset $\nu \subseteq \mathcal{V}$, so that $h(\nu)$ is the v-tuple with i-entry 1 if and only if $i \in \nu$. Since $\mathcal{C}$ is an SIT-code, $\mathrm{Aut}(\mathcal{C})$ is transitive on the set of triples (u, u', γ) such that $\gamma \in \mathcal{C}$, $u \in \gamma$ and $u' \in \mathcal{V} \setminus \gamma$. In particular $\mathrm{Aut}(\mathcal{C})$ is transitive on pairs (u, γ) with $\gamma \in \mathcal{C}$, $u \in \gamma$ (the flags of the corresponding Delandtsheer design $\mathcal{D}(\mathcal{C})$). Since each code-neighbour of weight $k - 1$ is of the form $h(\gamma \setminus \{u\})$ for some flag (u, γ) of $\mathcal{D}(\mathcal{C})$, it follows that $\mathrm{Aut}(\mathcal{C})$ is transitive on the set of all code-neighbours of weight $k - 1$. An analogous proof, using anti-flags of $\mathcal{D}(\mathcal{C})$, shows that $\mathrm{Aut}(\mathcal{C})$ is transitive on the set of all code-neighbours of weight $k + 1$.
□

In fact, the conclusion of Lemma 7.1 holds if the assumption on $\mathcal{C}$ is weakened to simply requiring $\mathrm{Aut}(\mathcal{C})$ to be transitive on flags and on antiflags, as in Problem 4.

Acknowledgements

The author is grateful to Mark Ioppolo, John Bamberg, and an anonymous referee for their careful reading of the manuscript. The paper forms part of Australian Research Council grant DP200100080.

References

[1] M. Aschbacher, On the maximal subgroups of the finite classical groups, *Invent. Math.* **76** (1984), 469–514.

[2] L. Babai, Graph isomorphism in quasipolynomial time, in *STOC'16 – Proceedings of the 48th Annual ACM SIGACT Symposium on Theory of Computing* ACM, New York (2016), pp. 684–697.

[3] J. Bamberg, A. C. Devillers, M. Ioppolo and C. E. Praeger, Codes and designs in Johnson graphs from symplectic actions of quadratic forms, preprint, 2021.

[4] S. Barwick and G. Ebert, *Unitals in projective planes*, Springer Monographs in Mathematics, Springer, New York (2008).

[5] A. Delandtsheer, Finite antiflag-transitive linear spaces, *Mitt. Math. Sem. Giessen* **164** (1984), 65–75.

[6] A. Delandtsheer and J. Doyen, A classification of line-transitive maximal (v,k)-arcs in finite projective spaces, *Arch. Math.* **55** (1990), 187–192.

[7] P. Delsarte, An algebraic approach to the association schemes of coding theory, *Philips Res. Rep. Suppl.* **10** (1973), vi+97 pp.

[8] A. Devillers, M. Giudici, C.H. Li, and C.E. Praeger, Locally s-distance transitive graphs and pairwise transitive designs, *J. Combin. Theory Ser. A* **120** (2013), 1855–1870.

[9] A. Devillers and C. E. Praeger, Pairwise transitive 2-designs, *J. Combin. Theory Ser. A* **132** (2015), 246–270.

[10] J. D. Dixon and B. Mortimer, *Permutation groups*, Springer-Verlag, New York (1996).

[11] N. Durante, On sets with few intersection numbers in finite projective and affine spaces, *Electron. J. Combin.* **21, no. 4** (2014), Paper 4.13, 18 pp.

[12] A. Ganesan, On the automorphism group of the Johnson graph, *Ars Comb.* **136** (2018), 391–396.

[13] The GAP Group, GAP – Groups, Algorithms and Programming, Version 4.4, 2004. http://www.gap-system.org.

[14] W. H. Haemers, C. Parker, V. Pless, and V. Tonchev, A design and a code invariant under the simple group Co_3, *J. Comb. Theory Ser. A* **62 (2)** (1993), 225–233.

[15] H. A. Helfgott, J. Bajpai and D. Dona, Graph isomorphisms in quasi-polynomial time. (Translation from the French original of an expository paper associated with Helfgott's Bourbaki seminar (Jan 14, 2017), with additions: solutions, further problems. Main text (42 pages) by Helfgott and Appendix B (24 pages) by Bajpai and Dona, arXiv 1710.04574.

[16] D. G. Higman and J. E. McLaughlin, Geometric ABA-groups, *Illinois J. Math.* **5** (1961), 382–397.

[17] D. A. Holton and J. Sheehan, *The Petersen graph*, Australian Mathematical Society Lecture Series, No. 7, Cambridge University Press, Cambridge (1993).

[18] M. Ioppolo, Codes in Johnson graphs associated with quadratic forms over $\mathbb{F}_2$, PhD Thesis, University of Western Australia, Perth, Australia, 2020.

[19] M. Ioppolo and C. E. Praeger, Translation-free Delandtsheer designs in affine spaces over the field of order two, preprint, 2021.

[20] W. M. Kantor, Classification of 2-transitive symmetric designs, *Graphs and Combinatorics* **1** (1985), 165–166.

[21] R. A. Liebler and C. E. Praeger, Neighbour-transitive codes in Johnson graphs, *Des. Codes Cryptogr.* **73** (2014), 1–25.

[22] W. J. Martin, Completely regular designs of strength one, *J. Algebraic Combin.* **3** (1994), 177–185.

[23] W. J. Martin, Completely regular designs, *J. Combin. Des.* **6** (1998), 261–273.

[24] W. J. Martin, Completely regular codes: a viewpoint and some problems, in *Proceedings* Com^{2}MaC *Workshop on Distance-Regular Graphs and Finite Geometry (Busan, Korea, July 24 - 26, 2004)* (2004), pp. 43–56.

[25] A. Meyerowitz, Cycle-balanced partitions in distance-regular graphs, *J. Combin. Inform. System Sci.* **17** (1992), 39–42.

[26] A. Meyerowitz, Cycle-balance conditions for distance-regular graphs, in *Proceedings of the 2000* Com^{2}MaC *Conference on Association Schemes, Codes and Designs (Pohang) Discrete Math.* **264** (2003), 149–165.

[27] M. Neunhöffer and C. E. Praeger, Sporadic neighbour-transitive codes in Johnson graphs, *Designs, Codes and Crypt.* **72** (2014), 141–152.

[28] M. Ramras and E. Donovan, The automorphism group of a Johnson graph, *SIAM J. Disc. Math.* **25, No. 1** (2011), 267–270.

[29] L. H. Soicher, The DESIGN package for GAP, Version 1.7, 2019, https://gap-packages.github.io/design, 2003-2019. Accessed 23 June 2020.

[30] D. E. Taylor, Unitary block designs, *J. Combinatorial Theory Ser. A* **16** (1974), 51–56.

[31] R. Wilson, P. Walsh, J. Tripp, I. Suleiman, R. Parker, S. Norton, S. Nickerson, S. Linton, J. Bray, and R. Abbott, ATLAS of Group Representations, http://brauer.maths.qmul.ac.uk/Atlas/v3/, 2013. Accessed 23 June, 2020.

Centre for the Mathematics of Symmetry and Computation
The University of Western Australia
35 Stirling Highway, Perth, WA 6009, Australia
cheryl.praeger[at]uwa.edu.au

Maximal subgroups of finite simple groups: classifications and applications

Colva M. Roney-Dougal

Abstract

This paper surveys what is currently known about the maximal subgroups of the finite simple groups. After briefly introducing the groups themselves, if their maximal subgroups are completely determined then we present this classification. For the remaining finite simple groups our current knowledge is only partial: we describe the state of play, as well as giving some results that apply more generally. We also direct the reader towards computational resources for the construction of maximal subgroups.

After this, we present three sample applications, selected because they combine group theoretical and combinatorial arguments, and because they use either or both of the detailed classifications and the looser statements that can be made about all maximal subgroups. In particular, we discuss results relating to generation, and the generating graph; results concerning bases; and some applications to computational complexity, in particular to graph colouring and other problems with no known polynomial-time solution.

1 Introduction

The Classification of Finite Simple Groups was perhaps the most impressive achievement in 20th century mathematics. A simple statement is as follows.

Theorem 1.1 (Classification of Finite Simple Groups) *Let G be a finite simple group. Then G is one of the following:*

(a) *A cyclic group, of prime order.*

(b) *An alternating group A_n, of degree $n \geq 5$.*

(c) *A group of Lie type, either classical or exceptional.*

(d) *One of 26 sporadic groups.*

After naming the simple groups, two natural questions arise.

1. What is the subgroup structure of these groups?

2. How are these groups represented, by means of permutations or matrices? Phrased differently, what objects have these groups as groups of automorphisms?

These two questions are intimately linked, since any transitive permutation action of a group G is equivalent to the action of G on the set of right cosets of a point stabiliser in G.

A useful first step towards understanding all of the subgroups of a group is to determine its *maximal* subgroups. This paper will describe both what is known about the maximal subgroups of the finite simple groups, and give three examples of areas towards the intersection of group theory and combinatorics where this information has been put to fascinating use.

One might hope that information about the maximal subgroups of the finite simple groups would be sufficient to determine the maximal subgroups of any finite group, but thus turns out not to be the case. However, it is *almost* the case. Each group G acts on itself by conjugation, and this yields a homomorphism ρ from G to the group $\mathrm{Aut}(G)$ of automorphisms of G, where each $g \in G$ is mapped to the automorphism $x \mapsto g^{-1}xg$. The image of ρ is the group $\mathrm{Inn}(G)$ of *inner automorphisms* of G, which can easily be shown to be a normal subgroup of $\mathrm{Aut}(G)$. The kernel of ρ is the centre $Z(G)$ of G, so the first isomorphism theorem shows that $\mathrm{Inn}(G)$ is isomorphic to $G/Z(G)$. It follows that if T is a nonabelian simple group then $T \cong \mathrm{Inn}(T) \trianglelefteq \mathrm{Aut}(T)$, and it is easy to check that T is the unique minimal normal subgroup of $\mathrm{Aut}(T)$.

Definition 1.2 A group G such that $T \trianglelefteq G \leq \mathrm{Aut}(T)$, for some nonabelian simple group T, is an *almost simple group*. The group T is the socle of G, denoted $\mathrm{Soc}(G)$.

If G is almost simple with socle T, and M is a maximal subgroup of G, then $M \cap T \trianglelefteq M$, and so M is a subgroup of $\mathrm{N}_G(M \cap T)$, the normaliser of $M \cap T$ in G. In most cases, $M \cap T$ is a proper subgroup of T, and so is not normal in G, which implies that M is equal to $\mathrm{N}_G(M \cap T)$ (since M is maximal). Often, $M \cap T$ is maximal in T, so many of the maximal subgroups of G are of the form $\mathrm{N}_G(H)$, where H is a maximal subgroup of T. Given such a maximal subgroup H of T, it is relatively straightforward to determine whether $\mathrm{N}_G(H)$ is maximal in G.

However, it is also possible that $M \cap T$ is not maximal in T. For an example of this behaviour, consider the normaliser M of a 7-cycle in the symmetric group S_7, which is a semidirect product $7 \rtimes 6$ of a cyclic group of order 7 by a cyclic group of order 6. It is not too hard to check that M is a maximal subgroup of S_7. However, $M \cap A_7$ is *not* maximal in A_7, being contained in the representation of $\mathrm{PSL}_3(2)$ as a group of automorphisms of the Fano plane, $\mathrm{PG}_2(2)$.

It turns out that this is the only complication: work of Kovács [57], and of Aschbacher and Scott [4] reduces the problem of finding the maximal subgroups of an arbitrary group to knowing the maximal subgroups of the almost simple extensions of its composition factors, together with solving some cohomological problems, which can be done computationally [25, 34]. Thus we shall now describe the maximal subgroups of the almost simple groups.

2 The maximal subgroups

The maximal subgroups of the cyclic groups of prime order are trivial. We therefore start this section by describing the maximal subgroups of the alternating groups, then the classical groups of Lie type, the exceptional groups of Lie type, and finally the sporadic groups.

2.1 Alternating groups

In this section we discuss the classification of the maximal subgroups of the alternating groups A_n and the symmetric groups S_n, for $n \geq 5$. If $n \neq 6$ then $\mathrm{Aut}(A_n) = S_n$, whilst the isomorphism $A_6 \cong \mathrm{PSL}_2(9)$ means that some additional automorphisms arise naturally: see Subsection 2.2.

The first family of maximal subgroups is easy, both to describe and to classify. A permutation group $G \leq S_n$ is *transitive* if for all $\alpha, \beta \in \{1, \ldots, n\}$ there exists $g \in G$ such that $\alpha^g = \beta$, and *intransitive* otherwise. For $1 \leq k \leq n$, the group A_n is transitive on the k-subsets of $\{1, \ldots, n\}$. Furthermore, stabilising a k-subset is equivalent to stabilising its complement, an $(n-k)$-subset. Hence, for each k in $\{1, \ldots, \lfloor n/2 \rfloor\}$, and for $X \in \{A_n, S_n\}$, we find a unique conjugacy class of subgroups of type $(S_k \times S_{n-k}) \cap X$. It is a relatively easy exercise to show that these subgroups are maximal if and only if $k \neq n-k$. Thus, it remains to classify the transitive maximal subgroups of A_n and S_n.

The next family of maximal subgroups is almost as easy. A subset Δ of $\{1, \ldots, n\}$ is a *block* for a permutation group $G \leq S_n$ if for all $g \in G$ either $\Delta^g \cap \Delta = \emptyset$, or $\Delta^g = \Delta$. That is, each element of G either permutes the elements of Δ amongst themselves, or maps all of them outside Δ. If G is a transitive subgroup of S_n, and there exists a block for G of size greater than one and less than n, then G is *imprimitive*; all other transitive groups are *primitive*.

If Δ is a block for G, then it is easy to check that each translate Δ^g is also a block, so $\ell := |\Delta|$ is a divisor of n. It is not too hard to show that S_n is transitive on all partitions of $\{1, \ldots, n\}$ into parts of size ℓ, so for each proper nontrivial divisor ℓ of n the stabilisers in S_n of such partitions form a single conjugacy class of imprimitive subgroups. One may also with not too much effort show that these imprimitive groups are always maximal in S_n. The group A_n also has a unique conjugacy class of such stabilisers for each value of ℓ, and it turns out that there is a unique non-maximal example, namely the imprimitive groups stabilising four blocks of size two in A_8.

Thus, the main work is to classify the primitive maximal subgroups of A_n and S_n. The first step is the further case distinction given by the O'Nan–Scott Theorem. There are many different versions of this theorem, some giving much more detail than others, but since we are only interested in *maximal* subgroups the following will suffice.

Theorem 2.1 *Let G be a primitive maximal subgroup of $X = S_n$ or $X = A_n$, with $n \geq 5$. Then G is one of the following.*

(a) $G = \mathrm{AGL}_k(p) \cap X$, with $n = p^k$ and p prime: the affine case.

(b) $G = (T^k.(\mathrm{Out}(T) \times S_k)) \cap X$, with T a nonabelian simple group, $k \geq 2$ and $n = |T|^{k-1}$: the diagonal case.

(c) $G = (S_m \wr S_k) \cap X$, with $n = m^k$, $m \geq 5$ and $k \geq 2$: the product action case.

(d) $T \trianglelefteq G \leq \mathrm{Aut}(T)$, with T a nonabelian simple group: the almost simple case.

We shall not give details of the structure of the groups in each of these cases, see textbooks such as [20], [32] or [97] for much more information.

Two questions then arise. Firstly, which of the groups listed above are, in fact, maximal? Secondly, whilst it is clear for which n the groups of affine, diagonal and product action type arise, the description of the almost simple case gives no indication as to the value of n: what *are* the almost simple primitive groups?

The first of these questions is completely answered by Liebeck, Praeger and Saxl in [61]. They show that the group $G = \mathrm{AGL}_k(p)$ is maximal in S_{p^k} whenever G contains an odd permutation (namely, when p is odd), and that $G \cap \mathrm{A}_{p^k}$ is maximal except when $p^k \in \{7, 11, 17, 23\}$. (As an aside, we note that if $n \le 4$ then $n = p^k$ for a prime p, and S_n is *equal* to the group $\mathrm{AGL}_k(p)$). Furthermore, the groups of diagonal and product action type are maximal whenever they occur. Thus the most detailed work in [61] is to consider the almost simple case, and here the result consists of several pages of tables listing containments of almost simple primitive groups.

Despite the monumental achievement of [61], this still leaves our second question: which almost simple groups actually occur? To answer this, one uses the straightforward result that a transitive permutation group G is primitive if and only if the point stabiliser G_α is maximal in G. Thus, to classify the almost simple primitive groups, we need to classify the maximal subgroups of the almost simple groups! Fortunately, the problem is less circular than it seems, as if G is a maximal subgroup of S_n then certainly $|G| < n!$, so we can bootstrap our solutions.

The classification of the primitive permutation groups of low degree is one of the oldest problems in group theory. We shall only briefly summarise the parts of the history that are most relevant to our current question, namely determining the almost simple maximal subgroups of A_n and S_n: see Short's book [89] for a considerably more detailed exposition. We should also mention that determining the full list of groups of affine type when $n = p^k$ is a very different type of problem to that of determining the other types of maximal subgroup, as this is where by far the greatest number of examples arise.

The earliest significant progress was made by Jordan, who in 1871 counted the primitive permutation groups of degree n for $n \le 17$, and stated that a transitive group of degree 19 is A_{19}, S_{19}, or a group of affine type [47]. By 1912, the classification up to degree 20 had been completed by Martin [79] and Bennet [8]. After a long pause, the birth of practical symbolic computation in the 1960s gave rise to new approaches, and by 1970 Sims in [90] had redetermined the primitive groups of degree up to 20. Sims also classified the primitive groups of degree up to 50: this list was never published, but was widely circulated in manuscript form, and the resulting groups formed one of the earliest databases in computational group theory, eventually becoming part of GAP [36] and MAGMA [11].

The next dramatic leap forward came as a result of the announcement of the Classification of Finite Simple Groups (CFSG), after which Dixon and Mortimer used the O'Nan–Scott Theorem to classify the primitive groups with insoluble socles of degree less than 1000 [31] (these are the non-affine groups). For these degrees, MAGMA can compute the maximal subgroups of A_n and S_n, using the database of primitive groups and the results in [61].

More recently, the author classified all primitive groups of degree up to 2500 in [85], and this was extended to degree 4095 by Coutts, Quick and the author in

[28]. Both of these classifications are available in both MAGMA and GAP, as part of the primitive groups database. Thus, for any given degree $n \leq 4095$, the interested reader can use the database of primitive groups, together with [61], to determine the maximal subgroups of A_n and S_n. In as yet unpublished work, Ben Stratford has extended the classification to degree $2^{13} - 1$, and he is working on a classification of the non-affine primitive groups of degree up to one million.

Thus for all degrees which are likely to be encountered by the average user, the groups are available. In the rest of this subsection we shall therefore concentrate on results which hold for *all* n, and on methods which enable one to prove things about all permutation groups without using such tremendously detailed classifications.

We shall summarize the two types of result we shall present for general n as: the primitive subgroups of A_n and S_n, other than A_n and S_n themselves, are *small* and there are *relatively few* of them. These two facts are often sufficient for many applications. There is a further family of beautiful results that we would love to have discussed, which can be summarised as saying that the majority of the elements of S_n belong to few transitive, and even fewer primitive, subgroups of S_n (permutations belonging to no proper primitive subgroup of S_n other than A_n are called *Jordan elements*). A good starting point for the interested reader is [22].

The question of bounding the order of a proper primitive subgroup of A_n or S_n has a lengthy history. Perhaps the oldest result still in regular use is that of Bochert from 1889 [10], which states that if a primitive group G has index greater than 2 in S_n, then this index is at least $\lfloor (n + 1)/2 \rfloor!$. This bound was dramatically improved in 1980, when Praeger and Saxl proved in [82] that if $G \leq S_n$ is primitive and does not contain A_n, then $|G| \leq 4^n$.

One especially useful theorem in this field, due to its nice mix of concision and strength, is due to Maróti [78]. In Case (a), below, notice that $r = 1$ if and only if G is almost simple, so that the almost simple examples are precisely the actions of A_m and S_m on the cosets of intransitive maximal subgroups.

Theorem 2.2 ([78]) *Let G be a primitive permutation group of degree n. Then one of the following holds.*

(a) *G is an almost simple or product action subgroup of $S_m \wr S_r$, where the action of S_m is on k-subsets of $\{1, \ldots, m\}$, so that G has degree $\binom{m}{k}^r$.*

(b) *G is one of the Mathieu groups M_{11}, M_{12}, M_{23} or M_{24}, with their 4-transitive actions.*

(c) *$|G| \leq n \cdot \prod_{i=0}^{\lfloor \log n \rfloor - 1}(n - 2^i) < n^{1 + \lfloor \log n \rfloor}$.*

(In the above theorem statement, and throughout the paper, all logarithms are to the base 2).

A second extremely useful result, which concerns itself only with almost simple primitive groups, and hence gives a tighter bound, is the following, due to Liebeck in 1984.

Theorem 2.3 ([58]) *Let G be an almost simple primitive subgroup of S_n, and let $T = \mathrm{Soc}(G)$. Then one of the following holds.*

(a) $T = A_m$, acting on k-subsets of $\{1, \ldots, m\}$, or on partitions of $\{1, \ldots, m\}$ into subsets of size $\ell > 1$, where ℓ properly divides m.

(b) T is a simple classical group acting on an orbit of subspaces of the natural module, or (if $T = \mathrm{PSL}_d(q)$) on pairs of subspaces of complementary dimensions.

(c) $|G| < n^9$.

A small number of remarks are in order. Firstly, the groups in Case (a) include the almost simple groups in Case (a) of Maróti's theorem, together with the actions on cosets of imprimitive maximal subgroups. The groups in Case (b) will be examined in more detail in the next subsection, and in particular the natural module for a classical group will be defined. Using results that we shall see in Section 4, Liebeck observed (see [20, p116]) that one can replace the 9 in Case (c) with a number just slightly greater than 6, with the precise bound coming from the Mathieu group M_{24} in its natural action on 24 points.

We move on to the idea that there are "few" primitive maximal subgroups, and mention just two asympotic results. The first is due to Liebeck, Martin and Shalev.

Theorem 2.4 ([60]) *The symmetric group S_n has $n^{o(1)}$ conjugacy classes of primitive maximal subgroups.*

Here $o(1)$ denotes a number that tends to 0 as n tends to infinity. The author has been unable to find a linear or sublinear bound on the number of conjugacy classes of primitive maximal subgroups that features explicit constants; it would be nice to have one.

Finally, Cameron, Neumann and Teague proved that for almost all n the only primitive groups of degree n are S_n and A_n.

Theorem 2.5 ([24]) *Let $e(x)$ denote the number of integers $n \leq x$ such that there exists a primitive group of degree n, other than A_n or S_n. Then $e(x) \sim 2x/\log x$.*

2.2 Classical groups

We shall give a rather superficial introduction to the classical groups, as this article will require few of their properties. See any of [12], [56], [95] or [97] for significantly more detailed information.

A *classical form* on a finite vector space $V = \mathbb{F}^d$ is a nondegenerate form $\beta : V \times V \to \mathbb{F}$ or $Q : V \to F$ of one of the following types.

1. A *unitary form*: β is linear in the first variable, additive in the second, the order of $\mathbb{F}$ is a square, and $\beta(u, v) = \beta(v, u)^\sigma$ for all $\lambda \in \mathbb{F}$ and $u, v \in V$, where σ is the (unique) field automorphism of $\mathbb{F}$ of order two.

2. A *symplectic form*: β is bilinear, $\beta(u, v) = -\beta(v, u)$ for all $u, v \in V$, and $\beta(v, v) = 0$ for all $v \in V$.

3. A *quadratic form*: $Q(\lambda v) = \lambda^2 Q(v)$ for all $\lambda \in \mathbb{F}$ and $v \in V$, and the associated form β defined by $\beta(u, v) = Q(u) + Q(v) - Q(u + v)$ is bilinear.

A form β is *nondegenerate* if $\beta(x,v) = 0$ for all $v \in V$ implies that $x = 0$, whilst a quadratic form is nondegenerate if its associated form β is nondegenerate. To enable us to speak uniformly about all of the families of classical groups, we often also allow the zero map $\beta : V \times V \to \{0\}$ to count as a classical form.

An *isometry* of a classical form β or Q is an element $g \in \mathrm{GL}(V)$ such that $\beta(ug, vg) = \beta(u, v)$ or $Q(ug) = Q(u)$, for all $u, v \in V$. It is clear that the set of all such isometries forms a subgroup of $\mathrm{GL}(V)$; notice that if the form β is identically zero then this subgroup is $\mathrm{GL}(V)$ itself.

The term *classical group* is rarely used with a precise definition. It includes the groups of all isometries of each classical form (including the zero form), together with various quotients and extensions by automorphisms of these groups: see [12] or [56] for much more discussion of the various chains of subgroups and quotients that arise. The *natural module* for a classical group is the vector space V, equipped with the corresponding form.

It turns out that the groups of isometries of distinct forms on V of the same type are conjugate in $\mathrm{GL}(V)$, and in particular are isomorphic. This enables the definition of the isometry groups $\mathrm{GL}(V)$ of the zero form, $\mathrm{Sp}(V)$ of a symplectic form, $\mathrm{GU}(V)$ of a unitary form and $\mathrm{GO}^\varepsilon(V)$, where $\varepsilon = \{+, -, \circ\}$, of a quadratic form. Here $\varepsilon = \circ$ if and only if the dimension is odd, whilst for even dimension the sign of ε depends on the dimension of a maximal subspace on which the form is identically zero. We warn the reader that notation for the orthogonal groups is unfortunately inconsistent in the literature, and we follow [12].

These isometry groups are not in general simple, but their derived group $\mathrm{SL}(V)$, $\mathrm{Sp}(V)$, $\mathrm{SU}(V)$ or $\Omega^\varepsilon(V)$ is usually quasisimple, which means in particular that the quotient of their derived group by its centre is usually simple. There are some exceptions to these assertions for small dimensions and fields: see [26, Chapter 2] or [56, Prop 2.9.2] for the full list.

The simple classical groups therefore act faithfully on projective space, equipped with the corresponding classical form. Since the maximal subgroups of a quasisimple group are in natural bijection with the maximal subgroups of its simple quotient, many authors work with the linear groups, rather than the projective quotients.

The outer automorphism groups of the classical groups are all small and soluble, and are generally formed as a semidirect product of up to three groups, most of which are cyclic. See the ATLAS [26] for a brief introduction, or [12] or [56] for a more extensive discussion, including how to pick canonical representatives of the automorphisms.

We shall not give an extensive historical survey of the classification of maximal subgroups of classical groups in this article, as there are already several excellent sources for this history. Firstly, we would like to mention Oliver King's 2005 survey [50], which appeared in the same series of conference proceedings as the present article. Secondly, there is a useful survey [55] by Kleidman and Liebeck from 1988, which also discusses the alternating and exceptional groups.

The key result when analysing the maximal subgroups, or indeed the full subgroup lattice, of the finite classical groups is Aschbacher's Theorem [1]. We shall not give a precise statement of this theorem, as it requires too much technical setup, but in spirit it is similar to the O'Nan–Scott Theorem. It divides the set of *all* subgroups of a finite classical group (other than certain 4-dimensional symplectic groups and 8-dimensional orthogonal groups with $\varepsilon = +$) into nine classes, known as *Aschbacher classes*. Eight of these classes, denoted $\mathcal{C}_1$ to $\mathcal{C}_8$, contain the groups stabilising some natural geometry on the space (V, β) or (V, Q), and so these classes are called *geometric*. The ninth class consists of groups which are almost simple, modulo scalars, and is generally denoted $\mathcal{S}$ or $\mathcal{C}_9$.

The maximal subgroups of the almost simple 4-dimensional symplectic groups are described by Aschbacher in [1], and the maximal subgroups of the almost simple 8-dimensional orthogonal groups with $\varepsilon = +$ are fully classified by Kleidman in [52].

For the geometric subgroups, Aschbacher's Theorem yields restrictions on the dimension d, the field $\mathbb{F}$, and the type of classical form for which the groups arise. For Class $\mathcal{S}$ there is no such information. The Aschbacher classes are not pairwise disjoint, and it is not the case that being the maximal member of one of these classes guarantees maximality in the corresponding classical group.

We are therefore left with the same two questions as we asked for the alternating groups, but with Class $\mathcal{S}$ in place of the almost simple primitive groups.

1. When is a maximal member of one of these classes in fact a maximal subgroup?

2. Which groups actually appear in Class $\mathcal{S}$, for a given classical group?

Let us start by concentrating on the dimensions for which the answers to both questions are known. A classification of the maximal subgroups of the simple classical groups in dimension at most 12 was given in the PhD thesis of Peter Kleidman [51], without full proofs. This was a remarkable achievement, and the list was widely circulated and used by a great many authors. Kleidman intended to publish a subsequent book, which would include complete proofs, and also cover the almost simple groups, but unfortunately this was not published. However, the task was eventually carried out by Bray, Holt and the author in [12].

In dimension at least 13, the geometric maximal subgroups of the almost simple groups are classified by Kleidman and Liebeck in [56], completely answering our first question for these classes. In [56], Kleidman and Liebeck also give detailed descriptions of the conjugacy class stabilisers in the outer automorphism group, and the structure of the maximal subgroup in the simple group (similar information also appears in [12]).

This leaves Class $\mathcal{S}$. For dimensions 13, 14 and 15 the maximal subgroups in Class $\mathcal{S}$ are classified by Schröder in [88]. In dimensions 16 and 17 the maximal subgroups of Class $\mathcal{S}$ are studied by Rogers in [84]: for non-orthogonal groups, and for orthogonal groups with $\varepsilon = -$, a complete classification is given, but some small questions remain about maximal subgroups of certain non-simple groups in the exceptional orthogonal cases. The results in [88] and [84] are currently being prepared for publication.

The MAGMA function `ClassicalMaximals` takes as input the type of a classical group, a dimension and a field size, and returns a list of conjugacy class representatives of the maximal subgroups, organised by their Aschbacher class. The geometric maximal subgroups are given by the explicit generating matrices constructed in [44] and [45], whilst many of the groups in Class $\mathcal{S}$ are taken from MAGMA's database of finite quasisimple matrix groups, based on work by Steel [91] and by Hiß and Malle

[43]. Up to dimension 17, the list returned by ClassicalMaximals is guaranteed to be complete, whilst beyond that it returns only the geometric maximal subgroups.

In the remainder of this section we shall discuss what is known, and what remains to be proved, in dimension at least 18. Firstly, papers by Hiß and Malle [43] and by Lübeck [66] provide lists containing all quasisimple groups in class $\mathcal{S}$, for dimension up to 250. Thus in principle the work carried out in [12], [88] and [84] could be continued to higher dimensions, although various issues become harder as the dimension increases.

The very most that one could hope for would be a theorem similar to that of [61]: a statement that said that a quasisimple group had maximal normaliser in some almost simple classical group *unless* it was contained in some known list of exceptions. Unfortunately, for the groups in Class $\mathcal{S}$, there are considerably more possible obstructions to maximality than there are for almost simple primitive permutation groups, and whilst progress is being made, there is long way to go. As an example, we would like to mention [41] and [42] for the analysis of when a member of Class $\mathcal{S}$ can be contained in a member of Class $\mathcal{C}_2$. We refer the reader to [76] for a vision of what might be achieved.

In the absence of such a theorem, there are results of a similar flavour to those discussed in Subsection 2.1: groups in Class $\mathcal{S}$ are *small* and there are *few of them*. The order of $\mathrm{GL}_d(q)$ is (approximately) q^{d^2}, and the other classical groups have similar orders, so the bound given in the following theorem, due to Liebeck, is dramatically smaller.

Theorem 2.6 ([59]) *Let G be a finite almost simple classical group of dimension d over $\mathbb{F}_q$. Let H be a maximal subgroup of G that does not contain $\mathrm{Soc}(G)$, and does not lie in one of the geometric Aschbacher classes. Then either H is isomorphic to A_m or S_m, with $m \in \{d+1, d+2\}$, or $|H| \leq q^{3d}$.*

The representations of A_m and S_m in Theorem 2.6 are not at all mysterious. The group S_m can be represented by *permutation matrices* in $\mathrm{GL}_m(q)$: matrices with a unique 1 in each row and column, and all other entries 0, that permute the ordered basis vectors in the same way as the original permutations act on $\{1, \ldots, m\}$. This action stabilises the subspace $U = \langle (1, 1, \ldots, 1) \rangle$, and also stabilises the subspace S of all vectors of weight 0, which is $(m-1)$-dimensional. If $p \nmid m$ then S gives the required action, but notice that if $p \mid m$ then $U \leq S$, so we find an action of S_m on the $(m-2)$-dimensional space S/U.

There are other versions of Theorem 2.6 which yield smaller bounds than q^{3d}, but at the cost of a larger list of exceptions.

We finish this subsection with a result due to Häsä, which shows that there are not so many maximal subgroups in a classical group.

Theorem 2.7 ([39]) *Let G be a finite almost simple classical group of dimension d over $\mathbb{F}_q$. Let $m(G)$ denote the number of conjugacy classes of maximal subgroups of G not containing $\mathrm{Soc}(G)$. Then $m(G) < 2d^{5.2} + d \log \log q$.*

2.3 Exceptional groups

The remaining finite simple groups of Lie type are known as the *exceptional groups*. Similarly to the classical groups, they are often most easily constructed via groups of matrices over finite fields, although sometimes constructions over objects other than fields, or as finite subgroups of algebraic groups, are more revealing. The classical groups are each parametrised by two integers, a dimension d and a field size q, but the exceptional groups come with a fixed Lie rank, and only the field size can vary. The Lie rank is used rather than the dimension, since the dimension of the smallest linear representation may depend on the characteristic of the field.

The exceptional groups fall into two main types: the *untwisted* groups are defined over every finite field, whilst the *twisted* groups require the field order to be certain powers, and sometimes also for the field to have a specified characteristic.

The smaller rank groups are subgroups of various classical groups in dimension at most 8, and their maximal subgroups are known. These are as follows.

- The *Suzuki groups* $^2B_2(q)$ (also known as $Sz(q)$) are defined only for q an odd power of 2, and are subgroups of $Sp_4(q)$. The maximal subgroups of the simple Suzuki groups were determined by Suzuki himself in [94]; the almost simple groups are dealt with in [12].

- The groups $G_2(q)$ are subgroups of $Sp_6(q)$ when q is even and of $\Omega_7(q)$ otherwise. The maximal subgroups of almost simple groups G with socle $G_2(q)$ were determined by Kleidman in [54] for q odd, by Cooperstein in [27] for q even and G simple, and by Aschbacher in [2] otherwise.

- The *small Ree groups* $^2G_2(q)$ (also known as $R(q)$), are defined only for q an odd proper power of 3, and are subgroups of $G_2(q)$. Their maximal subgroups were determined by Kleidman in [54].

- The groups $^3D_4(q)$ are defined for q a cube, and are subgroups of $\Omega_8^+(q)$. Their maximal subgroups were determined by Kleidman in [53].

All of these maximal subgroups are described in the tables in [12].

The *big Ree groups* $^2F_4(q)$ are defined only for q an odd power of 2. The maximal subgroups of the almost simple groups with socle $^2F_4(q)$ were determined by Malle in [77]. These groups are considerably larger than the previous examples, and first occur in dimension 26.

This leaves only four families of groups: $F_4(q)$, $^2E_6(q)$ with q a square, and $E_r(q)$ for $r \in \{6, 7, 8\}$. For these groups, full classifications of maximal subgroups are still not published, but in an exciting, very recent development, David Craven has just announced a complete classification.

The state of play until earlier this year was as follows. Thanks to a great many papers by a great many people, there are detailed descriptions of the types of maximal subgroups that arise: the survey article by Liebeck and Seitz [62] is a good starting point. The only unknown maximal subgroups had similar properties to Aschbacher Class $\mathcal{S}$ for the classical groups: almost simple, and very small. Thanks to Liebeck, Seitz and many others, it is known which simple groups could be the socles of such groups, but there were no useful bounds on the number of conjugacy classes, and little information about what almost simple extensions arise.

2.4 Sporadic groups

In contrast to the rest of this section, the problem of classifying the maximal subgroups of the almost simple sporadic groups is evidently a finite one, and the task is almost complete. An excellent survey of the maximal subgroups of the sporadic groups by Wilson [98] has recently appeared, and we direct the reader there for more details.

The ATLAS [26] describes the maximal subgroups of the almost simple sporadic groups, with the exception of seven socle types: the three Fischer groups Fi_{22}, Fi_{23} and Fi_{24}, the Janko group J_4, the Thompson group Th, the baby Monster $\mathbb{B}$ and the Monster $\mathbb{M}$.

Wilson's book [97] lists the maximal subgroups of the almost simple sporadic groups, although [98] points out two mistakes: one in the list of maximal subgroups of $\mathbb{B}$, and one in the list of maximal subgroups of Co_1. The lists re-appear, in somewhat more compressed form, in [98].

The only almost simple sporadic group whose maximal subgroups are not fully determined is $\mathbb{M}$. The published results on maximal subgroups of $\mathbb{M}$ include complete classifications, except for the possibility of maximal subgroups with socle $PSL_2(8)$, $PSL_2(13)$, $PSL_2(16)$ and $PSU_3(4)$. Of these, $PSL_2(8)$ and $PSL_2(16)$ were considered in unpublished work of P. E. Holmes, and papers by Wilson on $PSL_2(8)$ and $PSU_3(4)$ are in progress. The most recent case to appear was in [99], which proves that no maximal subgroup of the Monster has socle $PSU_3(8)$.

3 Generation and the generating graph

One of the surprising facts about the finite simple groups is that they can all be generated by only two elements. This is an easy exercise for the alternating groups, was proved by Steinberg [92] for the groups of Lie type, and was shown by Aschbacher and Guralnick in [3] for the sporadic groups. We sadly still have no proof that does not depend on the classification.

Furthermore, Dalla Volta and Lucchini showed in [29] that each almost simple group can be generated by at most three elements, and that an almost simple group G requires three generators if and only if $G/\operatorname{Soc}(G)$ requires three generators. Since these quotients are straightforward to describe, the determination of the *minimal size* of generating sets of the almost simple groups is complete.

The *nature* of the generating sets is still somewhat mysterious. For example, the following landmark result by Liebeck and Shalev in 1995 has created a world of related theorems.

Theorem 3.1 ([63]) *Let G be an almost simple group, and let $P(G)$ denote the probability that two randomly chosen elements of G generate a subgroup containing $\operatorname{Soc}(G)$. Then $P(G) \to 1$ as $|G| \to \infty$.*

Whilst the proof is necessarily complex, and builds on pre-classification work by Dixon [30], who proved the theorem for the case $G = A_n$, and on a paper [48] by

Kantor and Lubotzky, which dealt with the case of G a classical group, the basic idea is quite simple. It uses results in a similar vein to those presented in Section 2, although the results used were not exactly the ones we have presented, some of which are much more recent! We now briefly explain the idea.

Let G be nonabelian and simple, for ease of description. If two elements of G *fail* to generate G, then they must both lie in some maximal subgroup M of G. The probability of this happening, for a specific subgroup M, is $|G:M|^{-2}$. Since M is a maximal subgroup of the simple group G, it is equal to its own normaliser in G, so by the Orbit-Stabiliser Theorem there are $|G:M|$ conjugates of M in G. Hence the probability that two elements lie in a common conjugate of M can be bounded above by $|G:M|^{-1}$. Summing over all conjugacy classes of maximal subgroups of G gives an upper bound on $1 - P(G)$, and hence a lower bound on $P(G)$.

As described in Section 2, the maximal subgroups of the finite simple groups are either known, and so their orders and numbers of conjugacy classes can be bounded explicitly, or are known to be small, almost simple, and (excluding until recently the exceptional groups) relatively few. With some effort, and not a little inspiration in the case of the exceptional groups, it is possible to use this insight to bound $P(G)$.

Whilst Theorem 3.1 places no lower bound on the probability, this is given by the following, due to Menezes, Quick and the author.

Theorem 3.2 ([80]) *Let G be a finite simple group. Then*

$$P(G) \geq 53/90,$$

with equality if and only if G is A_6.

Hence, although we have no classification-free proof that simple groups are 2-generated, in fact they are *overwhelmingly* so – almost any pair of elements will do. However, it is not the case that these generating pairs are spread evenly around the group. Consider, for example, the alternating group A_n, and let x be a 3-cycle. If x is to generate A_n with some other element y, then y must contain a cycle of length at least $n-2$, since $\langle x, y \rangle$ must be transitive. A straightforward counting argument then shows that the probability that a randomly-chosen y generates with x tends rapidly to zero as n grows. Hence, although there are many generating pairs for A_n, very few of them feature x.

One lovely combinatorial object which is used to analyse the structure of the generating pairs is the *generating graph* $\Gamma(G)$, whose definition is due to Liebeck and Shalev in [64]. The graph $\Gamma(G)$ has as vertices the nonidentity elements of G, and two vertices form an edge if and only if they generate G. Many algebraic facts about G correspond to nice combinatorial properties of $\Gamma(G)$, and as we shall now see, the original definition was made for this reason.

The *clique number* of a graph Γ is the size of the largest complete induced subgraph. Notice that a clique in $\Gamma(G)$ is a subset S of G such that every pair of distinct elements of S generates G. For a group G, let $m(G)$ denote the minimal index of a proper subgroup of G. In [64], Liebeck and Shalev prove that there exists an absolute constant α such that if G is a finite simple group then

$$1 - \frac{\alpha}{m(G)} \leq P(G).$$

Bounds on $P(G)$ are statements about edge density in $\Gamma(G)$, so Turán's theorem immediately yields that there exists an absolute constant c such that if G is a nonabelian finite simple group then $\Gamma(G)$ contains a clique of size at least $c \cdot m(G)$.

This lower bound is far from tight, in general. For example, the minimal index of a proper subgroup of A_n is n, but Blackburn showed in [9] that if n is sufficiently large and congruent to 2 modulo 4 then the clique number of $\Gamma(A_n)$ is 2^{n-2}; he also proved a similar result for S_n. In Stringer's PhD thesis [93], it was shown that "sufficiently large" could be taken to be "at least 22". It would be fascinating to see whether more combinatorial arguments about the structure of graphs $\Gamma(G)$ could be used to prove stronger lower bounds on the clique number, for other simple groups G.

Further *upper* bounds on clique numbers have been proved by using the observation that the clique number of any graph is at most its chromatic number, and in turn the chromatic number of $\Gamma(G)$ is bounded above by $\sigma(G)$, the smallest number of proper subgroups of G whose union is G. This number $\sigma(G)$ is called the *covering number* of G. See [69] and [70] for much more information about these parameters.

A *coclique* in a graph Γ is a subset of the vertices such that no two vertices are adjacent (cocliques are sometimes known as sets of *independent* vertices). In the generating graph $\Gamma(G)$, this corresponds to a subset of G such that every pair of elements generates a proper subgroup of G. There are some obvious cocliques in any generating graph: firstly, the elements of any proper subgroup form a coclique, and (slightly less obviously) since any two involutions generate a dihedral group, if the group G has even order and is not dihedral then the set of all involutions forms a coclique in $\Gamma(G)$. However, it is far from clear whether these cocliques are *maximal*, in the sense that all other vertices of $\Gamma(G)$ are adjacent to at least one vertex in the coclique.

Saunders in [87] completely classfied the cocliques of $\Gamma(\mathrm{PSL}_2(p))$ of size larger than $129(p-1)/2$, where p is prime. It turns out that they are only the maximal subgroups and the set of all involutions. This was done by a detailed analysis of the full subgroup lattice of $\mathrm{PSL}_2(p)$. More recently, Kelsey and the author in [49] considered G equal to S_n or A_n, and showed that for $n > 6$ the maximal intransitive subgroups $G \cap (S_k \times S_{n-k})$ are maximal cocliques if and only if $G = A_n$ or $\gcd(n,k) = 1$. Analysis of the maximal imprimitive subgroups is work in progress, and in general the question of determining the maximal cocliques of $\Gamma(G)$ is wide open.

Much of the inspiration for work on the generating graph was prompted by questions of connectivity. What can be said about the isolated vertices of $\Gamma(G)$? What can be said about the connected components? If G has a normal subgroup N such that G/N is not cyclic, then no element of N can be in a 2-element generating set, and so each element of N is an isolated vertex of $\Gamma(G)$. In [13], Breuer, Guralnick and Kantor conjectured that the converse is also true, and in dramatic recent progress, this conjecture (and much more) has been proved by Burness, Guralnick and Harper. See their paper [16] for an excellent description of some of the history of the problem, and the key results which go into the proof.

Theorem 3.3 ([16]) *Let G be a finite group. Then the following are equivalent,* for $|G| \geq 3$.

(a) *Every proper quotient of G is cyclic.*

(b) $\Gamma(G)$ *contains no isolated vertices.*

(c) $\Gamma(G)$ *has diameter at most two.*

It is conjectured that there is another equivalent condition to the one above for $|G| \geq 4$: the graph $\Gamma(G)$ contains a Hamiltonian cycle. This was shown to be the case for sufficiently large simple groups, for sufficiently large symmetric groups, and for almost simple sporadic groups in [14], which is also where this conjecture appears. On a related note, [68] begins the study of when $\Gamma(G)$ is Eulerian, and shows that $\Gamma(A_n)$ and $\Gamma(S_n)$ are Eulerian if and only if n and $n-1$ are not equal to a prime congruent to 3 modulo 4.

The above theorem completely answers the question of connectivity when there are no isolated vertices in $\Gamma(G)$. As for the general case, Lucchini showed in [67] that if G is soluble, then the induced subgraph on the nonisolated vertices of $\Gamma(G)$ is connected with diameter at most three, and gives sufficient conditions for the diameter to be two.

There has also been some interesting work on connectivity in the *complement* of $\Gamma(G)$, where two elements are joined if and only if they do *not* generate G. In fact, for nonabelian groups, it makes sense to consider a further refinement. The *commuting graph* has vertices $G \setminus \{1\}$, and an edge between two vertices if and only if they commute; whilst the *non-commuting, non-generating graph* $\Xi(G)$ has vertices $G \setminus Z(G)$, and an edge between two vertices if and only if they do not commute and do not generate G. The union of these two graphs with $\Gamma(G)$ is the complete graph on $G \setminus \{1\}$. Since the identity is not a vertex of the commuting graph, it need not be connected, and Giudici and Parker showed in [37] that the diameter of the commuting graph, if connected, may be unbounded. Conversely, Cameron, Freedman and the author have recently shown in [21] that if G is nilpotent then $\Xi(G)$ either has no edges, or the induced subgraph on the nonisolated vertices of $\Xi(G)$ has diameter two or three. Furthermore [21] gives necessary and sufficient conditions for the diameter to be three. This is intriguingly similar to Lucchini's results for $\Gamma(G)$ when G is soluble, from the previous paragraph.

Many other graph-theoretic questions have been studied for $\Gamma(G)$, but we shall conclude this section by asking to what extent $\Gamma(G)$ determines G. If $\Gamma(G)$ has isolated vertices, then possibly not much can be said: for example, it is not too hard to see that the generating graphs of the dihedral and quaternion groups of order 8 are isomorphic, both having one isolated vertex, and the remaining six vertices of degree 4 (the complement of a perfect matching on these six vertices).

Lucchini, Maróti and the author showed in [71] that if H is a sufficiently large simple group, or a symmetric group, and $\Gamma(G) \cong \Gamma(H)$ for some group G, then $G \cong H$. They also show that if $\Gamma(G)$ is sufficiently large and contains no isolated vertices, then $\Gamma(G)$ determines whether or not G is solvable, and if so determines G up to isomorphism.

One way to approach the question of whether $\Gamma(G)$ determines G would be to understand the automorphism group of the graph $\Gamma(G)$. This was not studied much until recently, perhaps because examination of even fairly trivial cases shows that $\mathrm{Aut}(\Gamma(G))$ is exceptionally large: the group $\mathrm{Aut}(\Gamma(\mathrm{A}_5))$ has order $2^{31} \cdot 3^7 \cdot 5$. However, in [23] an explanation is found for this large order, and a description is given of $\mathrm{Aut}(\Gamma(G))$.

The key observation is that in $\Gamma(G)$ there are, in general, many sets of elements with the same neighbours: for example, if g has prime order p, then every nonidentity element of $\langle g \rangle$ generates G with exactly the same other elements as g, and so there is a set of $p-1$ elements with identical neighbours. For $x, y \in G$, we write $N_\Gamma(x)$ for the neighbours of x, and write $x \equiv_\Gamma y$ if $N_\Gamma(x) = N_\Gamma(y)$, i.e. if x and y generate G with the same elements of G. Vertices in the same $\equiv_\Gamma$-equivalence class may be independently permuted by the automorphism group of any graph Γ so $\mathrm{Aut}(\Gamma(G))$ contains a subgroup that is the direct product of various symmetric groups, one for each $\equiv_\Gamma$-equivalence class.

One may define a quotient graph, denoted $\overline{\Gamma}_w(G)$, which has one vertex for each $\equiv_\Gamma$-class $N_\Gamma(x)$, labelled by the size $|N_\Gamma(x)|$, with an edge between $N_\Gamma(x)$ and $N_\Gamma(y)$ if and only if x and y are adjacent in Γ. It is shown in [23] that the group $\mathrm{Aut}(\Gamma(G))$ is completely determined by the size of the equivalence classes and $\mathrm{Aut}(\overline{\Gamma}_w(G))$ (where we require the automorphisms to preserve the vertex labels). Using this, a complete description is given of $\mathrm{Aut}(\Gamma(G))$ in the case where G is soluble and $\Gamma(G)$ has no isolated vertices. It is asked whether if G is insoluble and $\Gamma(G)$ has no isolated vertices then $\mathrm{Aut}(G) = \mathrm{Aut}(\overline{\Gamma}_w(G))$, and for the same groups G it is conjectured that $x \equiv_\Gamma y$ if and only if x and y belong to exactly the same maximal subgroups of G. This latter conjecture has just been proved by Kelsey and the author in [49], for the case $G \in \{\mathrm{A}_p, \mathrm{S}_p\}$, with p a prime not equal to $(q^d - 1)/(q-1)$ for any prime power q.

If the reader is interested in further information about questions relating to generation, we highly recommend Burness's excellent survey article [15].

4 Base size

For our second application of the results in Section 2, we present a variety of recent work relating to the *base size* of a permutation group.

First, the definition. A *base* for a subgroup G of S_n is a sequence $\underline{\beta} = (\alpha_1, \ldots, \alpha_k)$ of points of $\Omega = \{1, \ldots, n\}$ such that the pointwise stabiliser in G of all of the points in $\underline{\beta}$ is trivial. That is,

$$\bigcap_{\alpha \in \underline{\beta}} G_\alpha = 1. \tag{4.1}$$

The size of the smallest possible base for G is called the *base size* of G, and is denoted $b(G)$.

For example, if $G \le \mathrm{S}_n$ is generated by a single n-cycle, then the stabiliser of any point is the identity, and so we can take $\underline{\beta} = (\alpha)$ for any $\alpha \in \{1, \ldots, n\}$, and the base size is 1. Conversely, if $G = \mathrm{S}_n$ then to reach the identity subgroup we must stabilise $n-1$ points, as otherwise there are still nontrivial permutations of the remaining points, so $b(G) = n - 1$. As an example where we do not have free choice for the base points, consider the dihedral group of order eight, acting as symmetries of the square. Here, once we have fixed one vertex, the opposite corner of the square is also fixed, but the remaining two vertices can still be interchanged. Hence a base of minimal size consists of two adjacent vertices.

The concept of a base is due to Sims [90], and bases are a key tool in computational group theory, due to the following elementary lemma.

Lemma 4.1 *Let G be a subgroup of S_n, let $\underline{\beta}$ be a base for G, and let $g, h \in G$. If $\alpha^g = \alpha^h$ for all $\alpha \in \underline{\beta}$, then $g = h$.*

That is, the sequence of base images $(\alpha_1^g, \ldots, \alpha_k^g)$ uniquely determines the group element g. This means that (with some additional data structures called strong generating sets which we shall not describe here), when working with a group on a computer it is possible to represent group elements by their sequences of base images, rather than as permutations.

The example of the cyclic group above demonstrates that this can be a much more compact representation of group elements than storing the full permutation. This begs the question: what bounds can be put on the base size, relative to n?

First, we make some combinatorial observations. Fix a base $\underline{\beta}$ for G of size k, say, and let G^i denote the stabiliser in G of the first $i-1$ points of $\underline{\beta}$, so that $G^1 = G$ and $G^{k+1} = 1$. Then $\underline{\beta}$ is called *irredundant* if the index of G^{i+1} in G^i is greater than one, for all i. Since this index is therefore at least two, and a base of minimal size is certainly irredundant, we deduce that if k is the size of an irredundant base then

$$2^{b(G)} \leq 2^k \leq |G|. \tag{4.2}$$

In another direction, since G^{i+1} is a point stabiliser in G^i, it cannot have index greater than n (by the Orbit-Stabiliser Theorem), and so

$$|G| \leq n^{b(G)}. \tag{4.3}$$

Notice that these inequalities immediately tell us that G is small if and only if $b(G)$ is small.

We shall concentrate on the base size of primitive groups, since the base sizes of imprimitive and intransitive groups are somewhat wild. Recalling Theorem 2.2, the following theorem of Liebeck should appear both beautiful and natural (however, note that it appeared nearly two decades before Maróti's result, we are being seriously ahistorical!).

Theorem 4.2 ([58]) *Let G be a finite primitive group of degree n. Then one of the following holds:*

(a) *G is as described in Case (a) of Theorem 2.2;*

(b) *$b(G) < 9 \log n$.*

In recent work [81], Mosciatiello and the author have improved the upper bound in Case (b) of Theorem 4.2 to say that if G is not the Mathieu group M_{24} in its natural action on 24 points then $b(G) \leq \lceil \log n \rceil + 1$. Furthermore, we show that that there are infinitely many groups G which attain the bound. That is, either G has a rather precisely described structure, or each element of G is uniquely determined by the image of at most $\lceil \log n \rceil + 1$ points!

However, with a few more definitions under our belt, it is possible to make considerably stronger statements about $b(G)$. A faithful primitive action of an almost simple group G with socle G_0 is *standard* if it is equivalent to one of the following:

(a) G_0 is A_m for some m and the action is on subsets or partitions of $\{1,\ldots,m\}$; or

(b) G_0 is classical with natural module V, and either

  (i) each maximal subgroup M of G_0 containing $G_\alpha \cap G_0$, where G_α is the point stabiliser, is the stabiliser in G_0 of a nondegenerate, totally singular or nonsingular subspace of V; or

  (ii) $G_0 = \mathrm{Sp}_{2d}(q)$, q is even and the point stabiliser is a maximal orthogonal subgroup of G.

Otherwise, G is *non-standard*.

In Case (a) above, the point stabiliser is a maximal intransitive or imprimitive subgroup; whilst in Case (b)(i) the intersection of the point stabiliser and the socle is contained in a member of Aschbacher's Class $\mathcal{C}_1$, and in Case (b)(ii) the point stabiliser is a member of Class $\mathcal{C}_8$. In fact, whilst orthogonal groups in odd dimension $2d+1$ are usually only defined over fields $\mathbb{F}_q$ of odd order, this is because if q is even then $\mathrm{GO}_{2d+1}(q)$ fixes a 1-dimensional subspace, and is isomorphic to the symplectic group $\mathrm{Sp}_{2d}(q)$. If one permits these orthogonal groups to exist, then Case (b)(ii) above disappears.

In 1993, Cameron and Kantor conjectured in [22] that there exists an absolute constant c such that every non-standard almost simple primitive group G satisfies $b(G) \leq c$. This conjecture was proved in 1999 by Liebeck and Shalev [65]. In a sequence of papers, culminating in 2009 paper by Burness, Liebeck and Shalev [18], and a 2011 paper by Burness, Guralnick and Saxl [17], it was shown that one may take the constant c to be 7. Furthermore, there is a unique example with $b(G) = 7$, namely the Mathieu group M_{24} in its natural action on 24 points (which confirms a further conjecture of Cameron in [20]). It is by combining this result with (4.3) that one can deduce the strengthening of Theorem 2.3 that we mentioned.

The next highly influential conjecture regarding base size is due to Laslo Pyber. From (4.3) we deduce that $b(G) \geq (\log|G|)/(\log n)$. In 1993, Pyber conjectured in [83] that there exists a constant c such that for all primitive groups G of degree n,

$$b(G) \leq c \frac{\log|G|}{\log n}.$$

Many authors worked on various cases of this conjecture, we mention just a couple of papers from which the interested reader can find others. The case of groups of diagonal type was settled by Fawcett in [35], and building on work of Benbenishty and the proof of the Cameron-Kantor conjecture, Burness and Seress in [19] completed the proof for all non-affine primitive groups. The conjecture was finally proved in 2018 by Duyan, Halasi and Maróti [33]: see their paper for the full list of previous work. Slightly more recently again, in [38] a version was given with all constants explicit: the base size $b(G)$ of a primitive group G of degree n satisfies

$$b(G) \leq 2\frac{\log |G|}{\log n} + 24,$$

and this bound is asymptotically best possible.

5 Graph isomorphism and related problems

At the beginning of Section 4, we mentioned that a key motivation for studying base size is computational complexity, so let us finish with some discussion of this. It should be intuitively clear that the number of points required to uniquely determine each group element will be a key factor in the running time of a great many algorithms, but rather than enumerating a long list of methods we focus just a few results, of a combinatorial nature.

A *decision problem* is any problem with a yes/no answer. Complexity class P consists of all decision problems that can be solved in time that is bounded above by a polynomial function of the input size. Complexity class NP consists of all decision problems for which the answer "yes" can be *checked* in time polynomial in the input size, given some kind of certificate that the answer is indeed yes. For example, a certificate that a graph is Hamiltonian could be a list of the edges in a Hamiltonian circuit: all we need to do is check that the edges do indeed form such a circuit. It is clear that P ⊆ NP, since if the answer can be computed in polynomial time then we can just take an empty certificate, and work from scratch. It is a famous open problem to determine whether P = NP.

One very well studied example is the *graph isomorphism problem*, which takes as input two graphs, and asks one to determine whether or not they are isomorphic. There is no known polynomial-time solution to this problem, so it is not known to lie in P. (For the purposes of polynomial-time statements, we can take the input size to be the number of vertices.) Conversely, it is clear that the graph isomorphism problem lies in NP, since an explicit bijection between the two graphs can easily be checked in polynomial time. Graph isomorphism is not known to be NP-complete (the NP-complete problems are the hardest problems in NP), and indeed if the graph isomorphism problem is not in P then it is thought to be one of the easier problems in NP \ P.

In 1982, Luks showed in [73] that the graph isomorphism problem is polynomial-time reducible to a somewhat more general problem, known as the *string isomorphism problem*, which is as follows. One is given two functions f and g from a finite set Ω to a set Σ, and a permutation group $G \leq \mathrm{Sym}(\Omega)$, and one is asked to determine whether there exists an element σ of G such that $f^\sigma = g$, where the action is $f^\sigma(\alpha) = f(\alpha^{\sigma^{-1}})$. If so, then we return all such σ. The action of σ on f may look a little odd, but if one represents f and g by $|\Omega|$-tuples of elements of Σ then σ is just permuting the coordinates.

Let's see how the graph isomorphism problem can be reduced to the string isomorphism problem. Suppose we have two graphs, Γ_1 and Γ_2, both with vertex set $\{1, \ldots, m\}$ (if the vertex sets have different sizes then the problem is easy!). We let Ω be the set of 2-subsets of $\{1, \ldots, m\}$, let $\Sigma = \{0, 1\}$, and let f and g be the characteristic functions of the edge sets of Γ_1 and Γ_2. That is, f and g return 1 if $\{i, j\}$ is an edge in their graph, and 0 otherwise. We let G be the subgroup of the symmetric group on Ω given by the natural action of G on 2-subsets of $\{1, \ldots, m\}$, so that G is the automorphism group of the complete graph K_m. Then each graph isomorphism from Γ_1 to Γ_2 is an element σ of G, and this element σ naturally induces a string isomorphism from f to g. Hence the graph isomorphism problem reduces in polynomial time to the string isomorphism problem.

Theorem 5.1 ([73]) *Fix an integer k, and let Ξ be the set of all graphs of degree at most k. Then the graph isomorphism problem for graphs in Ξ can be solved in polynomial time.*

Luks proved this theorem by showing that if a permutation group G has no composition factor of order greater than k, for some fixed k, then the string isomorphism problem is solvable in polynomial time.

We give a very brief sketch of some of the ideas in Luks' solution to the string isomorphism problem; a slightly more detailed exposition is given in [40]. First, one reduces to the case of G being transitive on Ω, since otherwise one can recursively solve the problem for the restriction of G to an orbit Δ of Ω, and then solve a slightly modified problem for the action of the pointwise stabiliser $G_{(\Delta)}$ on $\Omega \setminus \Delta$. Gluing these two solutions together is comparatively straightforward.

Next, one considers a minimal block system $\mathcal{B}$, and the (primitive) induced action $G^{\mathcal{B}}$. We shall use the following 1982 theorem of Babai, Cameron and Palfy, which we state in somewhat less generality than it was proved.

Theorem 5.2 ([7]) *Fix an integer m. Then there exists a constant $c(m)$ such that if $H \leq S_n$ is primitive and has no nonabelian composition factor of order greater than m, then $|H| \leq n^{c(m)}$.*

Let K be the kernel of the action of G on $\mathcal{B}$. Since $G^{\mathcal{B}}$ is primitive, it follows from Theorem 5.2 and our assumption on the composition factors of G that there are only polynomially many cosets of K in G. Furthermore, K is intransitive, and we have already seen how to solve the problem for intransitive groups. Carefully defining the strings to be examined, and then gluing these recursive solutions together, gives the result.

In a dramatic breakthrough, Babai in [5, 6] proved that the string isomorphism problem, and hence the graph isomorphism problem, can be solved in *quasipolynomial time*: time $2^{O((\log n)^c)}$, for some absolute constant c. Babai did so by replacing the polynomial bound in Theorem 5.2 by the bound $2^{\log n (\log n + 1)}$ in Theorem 2.2(c), and then performing an *extremely* careful analysis of the groups in Theorem 2.2(a). Helfgott then re-analysed Babai's approach, and was able to show that one can take $c = 3$, see [40].

Luks in [74] had studied other permutation problems believed to lie strictly between P and NP, and developed what is now called the *Luks hierarchy*. When analysing the complexity of permutation group algorithms, each subgroup of S_n is input via a set of generating permutations; any such generating set can be taken to have size at most $O(n^2)$, by Sims' work on bases and strong generating sets. Thus we measure the complexity as a function of n.

The easiest problem in the Luks hierarchy is the graph isomorphism problem. At least as difficult as the graph isomorphism problem, and all equivalent to each other, are three permutation group problems: the problem of finding the intersection of two subgroups of S_n; the problem of finding the setwise stabiliser of some subset Δ of $\{1, \ldots, n\}$ in a group $G \leq S_n$; and the problem of finding the centraliser in a group $G \leq S_n$ of an element $x \in S_n$. These three problems (and others which are equally difficult) form what is known as the *Luks class*. At least as hard as all of these is the *normaliser problem*: given two subgroups G and H of S_n, find $N_H(G)$. A special case of the normaliser problem, whose difficulty relative to the Luks class is unknown, is the problem of computing $N_{S_n}(G)$, that is the case $H = S_n$.

The setwise stabiliser problem is readily seen to be a special case of the string isomorphism problem: given $G \leq S_n$, and a subset Δ of $\{1, \ldots, n\}$, take $f = g$ to be the characteristic function of Δ (so $f(\alpha) = 1$ if and only if $\alpha \in \Delta$). Thus Babai's result shows that all of the problems in the Luks class can be solved in quasipolynomial time.

There are two obvious open problems left. One is to try to reduce Helfgott's constant down from $c = 3$ to $c = 1$ (which would be a polynomial-time solution) or, conversely and somewhat more dramatically, to show that no such reduction is possible (which would prove that $\mathsf{P} \neq \mathsf{NP}$). The other, which seems more tractable, is the normaliser problem.

For a long time, it was not even known whether the normaliser problem could be solved in time that was simply exponential in n, since a naive algorithm involves looking through up to all $2^{n \log n}$ elements of S_n for normalising elements. However, Wiebking in [96] has recently shown that the normaliser problem can be solved in time $2^{O(n)}$.

One way to look for more tractable special cases is to restrict the group H. In this vein, Luks and Miyazaki in [75] used somewhat similar ideas to those of [73] to show the following. Let Γ_d be the class of permutation groups for which every nonabelian composition factor is a subgroup of S_d (this is closely related to the class of groups considered in Theorem 5.2, but includes imprimitive and intransitive groups as well; in particular all soluble groups are in Γ_d). Then Luks and Miyazaki showed that for fixed d, if $H \in \Gamma_d$ then $N_H(G)$ can be computed in polynomial time.

The other way to look for more tractable special cases is to restrict the group G. Since Babai showed in particular that the intersection problem can be solved in quasipolynomial time, if all one seeks is a *quasipolynomial* solution to the normaliser problem, it suffices to study the special case $N_{S_n}(G)$. In very recent work the author and Siccha used this observation to prove the following.

Theorem 5.3 ([86]) *Let $G \leq S_n$ be primitive. Then one can compute $N_{S_n}(G)$, and hence $N_H(G)$, in time $2^{O((\log n)^3)}$.*

One key tool in the proof is the fact, due to Derek Holt and the author in [46], that a subnormal subgroup of a primitive group G of degree $n > 3$ can be generated by at most $\log n$ elements (one could instead use the asymptotic result due to Lucchini, Menegazzo and Morigi in [72] that such a G can be generated by $O(\log n/(\sqrt{\log \log n}))$ elements).

Special methods are developed to compute the normaliser of groups lying in Case (a) of Theorem 4.2. For groups in Case (b) of the theorem, one first observes that an element σ of S_n lies in $N_{S_n}(G)$ if and only if σ conjugates each of the $O(\log n)$ generators of G to elements of G. Next, notice that by Theorem 4.2, one can test whether an element g^σ lies in G by looking at $9 \log n$ base point images of g^σ, so one can test whether $\sigma \in N_{S_n}(G)$ without completely describing the permutation σ.

Acknowledgements

The author would like to thank Derek Holt and Veronica Kelsey for carefully reading a draft of this article, and making many helpful suggestions. The author would also like to thank the anonymous referee, whose thoughtful comments greatly improved the paper.

References

[1] M. Aschbacher, On the maximal subgroups of the finite classical groups, *Invent. Math.* **76** (1984), 469–514.

[2] M. Aschbacher, Chevalley groups of type G_2 as the group of a trilinear form, *J. Algebra* **109** (1987), 193–259.

[3] M. Aschbacher and R. Guralnick, Some applications of the first cohomology group, *J. Algebra* **90** (1984), 446–460.

[4] M. Aschbacher and L. Scott, Maximal subgroups of finite groups, *J. Algebra* **92** (1985), 44–80.

[5] L. Babai, Graph isomorphism in quasipolynomial time, arXiv e-print 1512.03547.

[6] L. Babai. Graph isomorphism in quasipolynomial time, In *Proceedings of STOC'16*, (ACM, New York, 2016), 684–697.

[7] L. Babai, P. J. Cameron and P. P. Pálfy, On the orders of primitive groups with restricted nonabelian composition factors, *J. Algebra* **79** (1982), 161–168.

[8] E. R. Bennett, Primitive groups with a determination of the primitive groups of degree 20, *Amer. J. Math.* **34** (1912), 1–20.

[9] S. R. Blackburn, Sets of permutations that generate the symmetric group pairwise, *J. Combin. Theory Ser. A* **113** (2006), 1572–1581.

[10] A. Bochert, Über die Transitivitätsgrenze der Substitutionengruppen, welche die Alternierende ihres Grades nicht einhalt, *Math. Ann.* **33** (1889), 572–583.

[11] W. Bosma, J. Cannon and C. Playoust, The Magma algebra system. I. The user language, *J. Symbolic Comput.* **24** (1997), 235–265.

[12] J. N. Bray, D. F. Holt and C. M. Roney-Dougal, *The maximal subgroups of the low-dimensional finite classical groups*, Lond. Math. Soc. Lecture Note Ser., 407, Cambridge Univ. Press, Cambridge (2013).

[13] T. Breuer, R. M. Guralnick and W. M. Kantor, Probabilistic generation of finite simple groups II, *J. Algebra* **320** (2008), 443–494.

[14] T. Breuer, R. M. Guralnick, A. Lucchini, A. Maróti and G. P. Nagy, Hamiltonian cycles in the generating graphs of finite groups, *Bull. Lond. Math. Soc.* **42** (2010), 621–633.

[15] T. C. Burness, Simple groups, generation and probabilistic methods, in *Groups St Andrews 2017 in Birmingham* (eds. C. M. Campbell, C. W. Parker, M. R. Quick, E. F. Robertson and C. M. Roney-Dougal), *London Math. Soc. Lecture Note Ser.*, 455, Cambridge Univ. Press, Cambridge (2019), pp. 200–229.

[16] T. C. Burness, R. M. Guralnick and S. Harper, The spread of a finite group, arXiv e-print. 2006.01421.

[17] T. C. Burness, R. M. Guralnick and J. Saxl, On base sizes for symmetric groups, *Bull. Lond. Math. Soc.* **43** (2011), 386–391.

[18] T. C. Burness, M. W. Liebeck and A. Shalev, Base sizes for simple groups and a conjecture of Cameron, *Proc. Lond. Math. Soc.* **98** (2009), 116–162.

[19] T. C. Burness and Á. Seress, On Pyber's base size conjecture, *Trans. Amer. Math. Soc.* **367** (2015), 5633–5651.

[20] P. J. Cameron, *Permutation groups*, Lond. Math. Soc. Student Texts, 45, Cambridge Univ. Press, Cambridge (1999).

[21] P. J. Cameron, S. D. Freedman and C. M. Roney-Dougal. The non-commuting, non-generating graph of a nilpotent group. *Electronic J. Combin.*, to appear.

[22] P. J. Cameron and W. M. Kantor, Random permutations: some group-theoretic aspects, *Combin. Probab. Comput*, **2** (1993), 257–262.

[23] P. J. Cameron, A. Lucchini and C. M. Roney-Dougal, Generating sets of finite groups, *Trans. Amer. Math. Soc.* **370** (2018), 6751–6770.

[24] P. J. Cameron, P. M. Neumann and D. N. Teague, On the degrees of primitive permutation groups, *Math. Z.* **180** (1982), 141–149.

[25] J. Cannon and D. F. Holt, Computing maximal subgroups of finite groups, *J. Symbolic Comput.* **37** (2004), 589–609.

[26] J. H. Conway, R. T. Curtis, S. P. Norton, R. A. Parker and R. A. Wilson, *An ATLAS of Finite Groups*, Clarendon Press, Oxford (1985; reprinted with corrections 2003).

[27] B. N. Cooperstein, Maximal subgroups of $G_2(2^n)$, *J. Algebra* **70** (1981), 23–36.

[28] H. J. Coutts, M. R. Quick and C. M. Roney-Dougal, The primitive groups of degree less than 4096, *Comm. Algebra* **39** (2011), 3526–3546.

[29] F. Dalla Volta and A. Lucchini, Generation of almost simple groups, *J. Algebra* **178** (1995), 194–223.

[30] J. D. Dixon, The probability of generating the symmetric group, *Math. Z.* **110** (1969), 199–205.

[31] J. D. Dixon and B. Mortimer, The primitive permutation groups of degree less than 1000, *Math. Proc. Cambridge Philos. Soc.* **103** (1988), 213–238.

[32] J. D. Dixon and B. Mortimer, *Permutation Groups*, Graduate Texts in Mathematics, 163, Springer-Verlag, New York (1996).

[33] H. Duyan, Z. Halasi and A. Maróti, A proof of Pyber's base size conjecture, *Adv. Math.* **331** (2018), 720–747.

[34] B. Eick and A. Hulpke, Computing the maximal subgroups of a permutation group I, in *Groups and computation, III (Columbus, OH, 1999)* de Gruyter, Berlin (2001), pp. 155–168.

[35] J. B. Fawcett, The base size of a primitive diagonal group, *J. Algebra* **375** (2013), 302–321.

[36] The GAP Group, *GAP – Groups, Algorithms, and Programming, Version 4.11.0*; 2020, https://www.gap-system.org.

[37] M. Giudici and C. Parker, There is no upper bound for the diameter of the commuting graph of a finite group, *J. Combin. Theory Ser. A* **120** (2013), 1600–1603.

[38] Z. Halasi, M. W. Liebeck and A. Maróti, Base sizes of primitive groups: bounds with explicit constants, *J. Algebra* **521** (2019), 16–43.

[39] J. Häsä, Growth of cross-characteristic representations of finite quasisimple groups of Lie type, *J. Algebra* **407** (2014), 275–306.

[40] H. A. Helfgott, Isomorphismes de graphes en temps quasi-polynomial (d'après Babai et Luks, Weisfeiler-Leman), *Astérisque*, **407** (2019), 135–182.

[41] G. Hiß, W. J. Husen and K. Magaard, Imprimitive irreducible modules for finite quasisimple groups, *Mem. Amer. Math. Soc.* **234** (2015).

[42] G. Hiß and K. Magaard, Imprimitive irreducible modules for finite quasisimple groups. II, *Trans. Amer. Math. Soc.* **371** (2019), 833–882.

[43] G. Hiß and G. Malle, Low-dimensional representations of quasi-simple groups, *LMS J. Comput. Math.* **4** (2001), 22–63. Corrigenda: *LMS J. Comput. Math.* **5** (2002), 95–126.

[44] D. F. Holt and C. M. Roney-Dougal, Constructing maximal subgroups of classical groups, *LMS J. Comput. Math.* **8** (2005), 46–79.

[45] D. F. Holt and C. M. Roney-Dougal, Constructing maximal subgroups of orthogonal groups, *LMS J. Comput. Math.* **13** (2010), 164–191.

[46] D. F. Holt and C. M. Roney-Dougal, Minimal and random generation of permutation and matrix groups, *J. Algebra* **387** (2013), 195–214.

[47] C. Jordan, *Traité des substitutions et des equations algébriques*, Gauthier-Villers, Paris (1871).

[48] W. M. Kantor and A. Lubotzky, The probability of generating a finite classical group, *Geom. Dedicata* **36** (1990), 67–87.

[49] V. Kelsey and C. M. Roney-Dougal, Maximal cocliques in the generating graphs of the alternating and symmetric groups, arXiv e-print 2007.12021.

[50] O. H. King, The subgroup structure of finite classical groups in terms of geometric configurations, in *Surveys in combinatorics, 2005* (ed. B. S. Webb), Lond. Math. Soc. Lecture Note Ser., 327, Cambridge Univ. Press, Cambridge (2005), pp. 29–56.

[51] P. B. Kleidman, The maximal subgroups of the low-dimensional classical groups, PhD Thesis, University of Cambridge, 1987.

[52] P. B. Kleidman, The maximal subgroups of the finite 8-dimensional orthogonal groups $P\Omega_8^+(q)$ and of their automorphism groups, *J. Algebra* **110** (1987), 173–242.

[53] P. B. Kleidman, The maximal subgroups of the Steinberg triality groups $^3D_4(q)$ and of their automorphism groups, *J. Algebra* **115** (1988), 182–199.

[54] P. B. Kleidman, The maximal subgroups of the Chevalley groups $G_2(q)$ with q odd, the Ree groups $^2G_2(q)$, and their automorphism groups, *J. Algebra* **117** (1988), 30–71.

[55] P. B. Kleidman and M. W. Liebeck, A survey of the maximal subgroups of the finite simple groups, *Geom. Dedicata* **25** (1988), 375–389.

[56] P. B. Kleidman and M. W. Liebeck, *The subgroup structure of the finite classical groups*, Lond. Math. Soc. Lecture Note Ser., 129, Cambridge Univ. Press, Cambridge (1990).

[57] L. G. Kovács, Maximal subgroups in composite finite groups, *J. Algebra* **99** (1986), 114–131.

[58] M. W. Liebeck, On minimal degrees and base sizes of primitive permutation groups, *Arch. Math. (Basel)* **43** (1984), 11–15.

[59] M. W. Liebeck, On the orders of maximal subgroups of the finite classical groups, *Proc. London Math. Soc. (3)* **50** (1985), 426–446.

[60] M. W. Liebeck, B. M. S. Martin and A. Shalev, On conjugacy classes of maximal subgroups of finite simple groups, and a related zeta function, *Duke Math. J.* **128** (2005), 541–557.

[61] M. W. Liebeck, C. E. Praeger and J. Saxl, A classification of the maximal subgroups of the finite alternating and symmetric groups, *J. Algebra* **111** (1987), 365–383.

[62] M. W. Liebeck and G. M. Seitz, A survey of maximal subgroups of exceptional groups of Lie type, in *Groups, combinatorics & geometry (Durham, 2001)*, World Sci. Publ., River Edge, NJ (2003), pp. 139–146.

[63] M. W. Liebeck and A. Shalev, The probability of generating a finite simple group, *Geom. Dedicata* **56** (1995), 103–113.

[64] M. W. Liebeck and A. Shalev, Simple groups, probabilistic methods, and a conjecture of Kantor and Lubotzky, *J. Algebra* **184** (1996), 31–57.

[65] M. W. Liebeck and A. Shalev, Simple groups, permutation groups, and probability, *J. Amer. Math. Soc.* **12** (1999), 497–520.

[66] F. Lübeck, Small degree representations of finite Chevalley groups in defining characteristic, *LMS J. Comput. Math.* **4** (2001), 135–169.

[67] A. Lucchini, The diameter of the generating graph of a finite soluble group, *J. Algebra* **492** (2017), 28–43.

[68] A. Lucchini and C. Marion. Alternating and symmetric groups with Eulerian generating graph, *Forum Math. Sigma* **5** (2017), 30pp.

[69] A. Lucchini and A. Maróti, Some results and questions related to the generating graph of a finite group, in *Ischia group theory 2008* World Sci. Publ., Hackensack, NJ (2009), pp. 183–208.

[70] A. Lucchini and A. Maróti, On the clique number of the generating graph of a finite group, *Proc. Amer. Math. Soc.* **137** (2009), 3207–3217.

[71] A. Lucchini, A. Maróti and C. M. Roney-Dougal, On the generating graph of a simple group, *J. Aust. Math. Soc.* **103** (2017), 91–103.

[72] A. Lucchini, F. Menegazzo and M. Morigi, Asymptotic results for primitive permutation groups and irreducible linear groups, *J. Algebra* **223** (2000), 154–170.

[73] E. M. Luks, Isomorphism of graphs of bounded valence can be tested in polynomial time, *J. Comput. System Sci.* **25** (1982), 42–65.

[74] E. M. Luks, Permutation groups and polynomial-time computation, in *Groups and Computation, 1991* (eds. L. Finkelstein and W. M. Kantor), *DIMACS Ser. Discrete Math. Theoret. Comput. Sci*, 11, Amer. Math. Soc., Providence, RI (1993), pp. 139–175.

[75] E. M. Luks and T. Miyazaki, Polynomial-time normalizers, *Discrete Math. Theor. Comput. Sci.* **13** (2011), 61–96.

[76] K. Magaard, Some remarks on maximal subgroups of finite classical groups, In *Finite simple groups: thirty years of the atlas and beyond*, Contemp. Math. **694** (Amer. Math. Soc., Providence, RI 2017), 123–137.

[77] G. Malle, The maximal subgroups of $^2F_4(q^2)$, *J. Algebra* **139** (1991), 52–69.

[78] A. Maróti, On the orders of primitive groups, *J. Algebra* **258** (2002), 631–640.

[79] E. N. Martin, On the imprimitive substitution groups of degree fifteen and the primitive substitution groups of degree eighteen, *Amer. J. Math.* **23** (1901), 259–286.

[80] N. E. Menezes, M. Quick and C. M. Roney-Dougal, The probability of generating a finite simple group, *Israel J. Math.* **198** (2013), 371–392.

[81] M. Mosciatello and C. M. Roney-Dougal, Base sizes of primitive permutation groups, *In preparation*.

[82] C. E. Praeger and J. Saxl, On the orders of primitive permutation groups, *Bull. London Math. Soc.* **12** (1980), 303–307.

[83] L. Pyber, Asymptotic results for permutation groups, in *Groups and Computation, 1991* (eds. L. Finkelstein and W. M. Kantor), *DIMACS Ser. Discrete Math. Theoret. Comput. Sci*, 11, Amer. Math. Soc., Providence, RI (1993), pp. 197–219.

[84] D. Rogers, Maximal subgroups of classical groups in dimensions 16 and 17, PhD Thesis, University of Warwick, 2017.

[85] C. M. Roney-Dougal, The primitive groups of degree less than 2500, *J. Algebra* **292** (2005), 154–183.

[86] C. M. Roney-Dougal and S. Siccha, Normalisers of primitive permutation groups in quasipolynomial time, *Bull. Lond. Math. Soc.* **52** (2020), 358–366.

[87] J. Saunders, Maximal cocliques in $PSL_2(q)$, *Comm. Algebra* **47** (2019), 3921–3931.

[88] A. K. Schröder, The maximal subgroups of the Classical Groups in Dimension 13, 14 and 15, PhD Thesis, University of St Andrews, 2015.

[89] M. W. Short, *The primitive soluble permutation groups of degree less than 256*, Lecture Notes in Mathematics, 1519, Springer-Verlag, Berlin–Heidelberg (1992).

[90] C. C. Sims, Computational methods for permutation groups, in *Computational Problems in Abstract Algebra* (ed. J. Leech), Pergamon, (1970), pp. 169–183.

[91] A. K. Steel, Construction of ordinary irreducible representations of finite groups, PhD Thesis, University of Sydney, 2012.

[92] R. Steinberg, Lectures on Chevalley Groups, *Yale University Mathematics Department*, 1968.

[93] L. Stringer, Pairwise generating sets for the symmetric and alternating groups, PhD Thesis, Royal Holloway University of London, 2008.

[94] M. Suzuki, On a class of doubly transitive groups, *Ann. of Math.* **75** (1962), 105–145.

[95] D. E. Taylor, *The geometry of the classical groups*, Sigma Series in Pure Mathematics, 9, Heldermann Verlag, Berlin (1992).

[96] D. Wiebking, Normalizers and permutational isomorphisms in simply-exponential time, arXiv e-print 904.10454.

[97] R. A. Wilson, *The Finite Simple Groups*, Graduate Texts in Mathematics, 251, Springer-Verlag London, Ltd., London (2009).

[98] R. A. Wilson, Maximal subgroups of sporadic groups, in *Finite simple groups: thirty years of the* ATLAS *and beyond Contemp. Math.*, 694, Amer. Math. Soc., Providence, RI (2017), pp. 57–72.

[99] R. A. Wilson, The uniqueness of $PSU_3(8)$ in the Monster, *Bull. Lond. Math. Soc.* **49** (2017), 877–880.

School of Mathematics & Statistics
University of St Andrews
Fife, KY16 9SS
United Kingdom
colva.roney-dougal@st-andrews.ac.uk